America's Military Biomedical Complex

America's Military Biomedical Complex

Law, Ethics, and the Drive for Scientific Innovation

EFTHIMIOS PARASIDIS

Chief Justice Thomas J. Moyer Professor for the Administration of Justice and Rule of Law, and Professor of Public Health, Ohio State University, Columbus, OH, USA

OXFORD
UNIVERSITY PRESS

OXFORD
UNIVERSITY PRESS

Oxford University Press is a department of the University of Oxford. It furthers the University's objective of excellence in research, scholarship, and education by publishing worldwide. Oxford is a registered trade mark of Oxford University Press in the UK and in certain other countries.

Published in the United States of America by Oxford University Press
198 Madison Avenue, New York, NY 10016, United States of America.

Library of Congress Cataloging-in-Publication Data
Names: Parasidis, Efthimios, 1975– author.
Title: America's military biomedical complex : law, ethics, and the drive
for scientific innovation / Efthimios Parasidis.
Description: New York : Oxford University Press, 2025. |
Includes bibliographical references and index.
Identifiers: LCCN 2024031958 (print) | LCCN 2024031959 (ebook) |
ISBN 9780199351459 (hardback) | ISBN 9780197751299 (epub) |
ISBN 9780199351466 (updf) | ISBN 9780199351473 (online)
Subjects: LCSH: Medical innovations—Law and legislation—United States. |
Medicine—Research—Law and legislation—United States. |
Medical technology—Law and legislation—United States. |
Medical ethics—United States. | Technological innovations—Law and
legislation—United States. | Military art and science—Moral and ethical
aspects—United States. | Science and state—United States. | Biomedical materials.
Classification: LCC KF3821 .P37 2025 (print) | LCC KF3821 (ebook) |
DDC 344.7303/21—dc23/eng/20240716
LC record available at https://lccn.loc.gov/2024031958
LC ebook record available at https://lccn.loc.gov/2024031959

DOI: 10.1093/9780199351473.001.0001

Printed by Integrated Books International, United States of America

Note to Readers
This publication is designed to provide accurate and authoritative information in regard to the subject matter covered. It is based upon sources believed to be accurate and reliable and is intended to be current as of the time it was written. It is sold with the understanding that the publisher is not engaged in rendering legal, accounting, or other professional services. If legal advice or other expert assistance is required, the services of a competent professional person should be sought. Also, to confirm that the information has not been affected or changed by recent developments, traditional legal research techniques should be used, including checking primary sources where appropriate.

(Based on the Declaration of Principles jointly adopted by a Committee of the American Bar Association and a Committee of Publishers and Associations.)

Contents

PART IV. 1991–2024: FROM PROTECTING TO ENHANCING THE FIGHTING FORCE

PART V. LOOKING AHEAD: *JUS IN PRAEPARATIONE BELLUM*

Preface

The scope of this book expanded substantially from when it was commissioned nearly a decade ago. Initially I planned to examine the development of laws and ethical codes in military science, from the World War II to the present day, and offer policy recommendations for balancing current projects with ethical issues such as informed consent, autonomy, and justice. As I marched through the archives and some obscure secondary sources, I realized that these topics had been discussed among academics, military leaders, and government officials dating back to the American Revolutionary War, and that modern debates would be enriched by considering this longer history. I then judged that military biomedical developments would be best situated if assessed in relation to laws and ethical codes governing civilian biomedical endeavors, and if placed within the broader political economy of their respective eras. I thought that a synthesis of this information into one book would be a valuable addition to the scholarly literature, and would make for good reading.

What resulted was an ambitious project, a book that traces the evolution of laws and ethical codes within the biomedical establishment, considered alongside developments in military science, military doctrine, and national security policy. This book does not pass retrospective judgment, but rather highlights moral quandaries faced by decision makers at key moments in military science. Coupled with this historical reflection, I introduce the concept of *jus in praeparatione bellum*—justice in war preparations—a doctrine of restraint and responsibility that expands upon just war theory and aims to elucidate just and unjust means of preparing for war. The book applies *jus in praeparatione bellum* to contemporary military science pursuits, and proposes new laws and policies to guide military science projects. I have sought to explain concepts clearly and have avoided legalese and technical jargon; readers may approach this book without any background or expertise in law, ethics, or military science.

As I was working on the manuscript, I was invited to join a research team as a bioethics consultant on a grant from the US Air Force to study whole exome sequencing, and was asked to join DARPA's Measuring Biological Aptitude program as an ethics team member. I also was invited to participate in several meetings that brought together academics, military officials, and government leaders to discuss ethical and legal issues in military science. This included meetings held at the Stockholm International Peace Research Institute, the Johns Hopkins University Applied Physics Laboratory, and the University of Pennsylvania Center for Ethics and the Rule of Law. I am grateful for each of these opportunities, which gave me a birds-eye view of cutting-edge military science discussions. This book does not contain any non-public information from these meetings and research teams.

Writing about military science is incredibly challenging. It is a secretive subject matter, and oftentimes project details are not publicly revealed until decades after their completion. Like any broad historical survey, this book relies on the work of others, though I have prioritized the use of primary sources. This book is not intended to be an encyclopedic account: that would take volumes. I have attempted to strike a balance between including robust factual details, maintaining a flow between examples and chapters, and offering the reader an appropriate level of information that can help elucidate key concepts. Several books have been written on scores of individual facts outlined in this book, but my primary goal has been to create one book that unearths broader themes that can get lost by focusing on a narrow issue or time period. This has not been an easy task: the first draft of this book contained twice as many words as the final manuscript. I apologize in advance if I have omitted important events or have made factual or interpretational mistakes along the way.

Efthimios Parasidis
Columbus, Ohio
June 2024

Acknowledgments

I am grateful to have had the wise guidance of several editors at Oxford University Press as this book has matured. Among this distinguished group, special thanks to Lindsay Glick and Brian Stone. Lindsay provided exceptional advice regarding the scope and structure of the manuscript, and shepherded me up to the production phase. She has been an unwavering advocate, and consistently encouraged me to dig deeper into the core issues of the book. I likewise am immensely grateful for Brian's editorial and managerial assistance, and his support for the book as the scope expanded. I thank Getsy Deva Kirubai for her work in moving the book through production, and Lori Jacobs for exceptional work at copy-editing the manuscript. During earlier stages of this project, I had the good fortune to work with Jaime Berezin, Alex Flach, Jennifer Gong, and David Lipp. I have been honored to work with each of you.

Generous funding for this book was provided by the Greenwall Foundation. During my first year working on the project, I received a three-year fellowship from the Foundation's Faculty Scholars in Bioethics program. I used the funding for partial course relief from my university teaching responsibilities, which afforded me dedicated time for thinking, writing, researching, and traveling to archives. More importantly, the twice annual scholar meetings—which include fellows and alums—have been a constant source of scholarly inspiration and support. Several Greenwall colleagues have read draft chapters and have provided sage advice, including Baruch Brody, Alta Charo, Glenn Cohen, Barbara Evans, Nic Evans, Gidon Felsen, Christine Grady, Jeremy Greene, Michelle Groman, Aaron Kesselheim, Craig Konnoth, Bernie Lo, Anna Mastroianni, Amy McGuire, Michelle Mello, Maria Merritt, Wangui Muigai, Kimani Paul-Emile, Ronit Stahl, Dan Sulmasy, Keith Wailoo, Miranda Waggoner, and Leslie Wolf. Special thanks to my Greenwall book writing club members—Jennifer Blumenthal-Barby, Mara Buchbinder, and Lori Freedman—we met regulatory to read and comment on draft chapters of our respective books.

I also thank the Robert B. Silvers Foundation. In 2019, I was named as one of the inaugural recipients of the Robert B. Silvers Grants for Works-in-Progress. I was honored to receive the award, which provided generous funding that I used for travel for archival research. The Silvers grant came at the perfect time, and helped propel me to complete the manuscript.

I likewise am grateful for my wonderful colleagues at Ohio State University, and for the opportunity to present aspects of this work at various conferences, including at Harvard University, the University of Oxford, Stanford University, the Johns Hopkins University Applied Physics Laboratory, the University of Connecticut, Case Western University, the University of Houston, the University of Kentucky, and Wake Forest University. Some of the analysis in this book builds off work that I published in academic journals. For helpful comments on ideas explored in the book, special thanks to Arthur Anderson, Micah Berman, Cinnamon Carlarne, Amy Cohen, Hank Greely, Michael Gross, Erin Hahn, Jon Harkness, Mohamed Helal, Ana Iltis, Eric Juengst,

Margot Kaminski, Nancy King, Max Mehlman, Alan Michaels, Jonathan Moreno, Magda Parasidis, Guy Rub, Peter Shane, Marc Spindelman, Chris Walker, and Patti Zettler. Of this group, Jonathan Moreno was the first to encourage me to write a book on this topic, and provided outstanding advice at various stages. Throughout my research, I spoke with several service members and veterans who preferred to remain anonymous; I am grateful for their willingness to share their experiences and insights. I also thank the librarians and archivists with whom I worked at the National Archives and the US Army War College, and Ohio State librarian Jamie Aschenbach for assistance in tracking down permissions and public domain information for images used in the book. Thanks to the Ohio State Center for Interdisciplinary Law and Policy Studies for funding to procure copyright licenses.

Many have observed that writing is a solitary endeavor. While true, I cherish the moments when I am sitting at my desk and engaged creatively with my work. It is one of my happy places. I am grateful for my parents, sister, and extended family for a lifetime of love. Above all, I am grateful for my wife and life partner, Magda, and our two remarkable children, Anaïs and Savva. Their radiant souls are an eternal source of inspiration, love, and joy.

Abbreviations

ABA	American Bar Association
ACLU	American Civil Liberties Union
ADHD	attention-deficit/hyperactivity disorder
AEC	Atomic Energy Commission
AMA	American Medical Association
APA	American Psychological Association
ARPA	Advanced Research Projects Agency
AUMF	Authorization for Use of Military Force
AVIP	anthrax vaccine immunization program
BARDA	Biomedical Advanced Research and Development Authority
BT	botulinum toxoid
BW	biological weapons; biological warfare
CBRN	chemical, biological, radiological, and nuclear
CDC	Centers for Disease Control and Prevention
CIA	Central Intelligence Agency
CMR	Committee on Medical Research
CW	chemical weapons; chemical warfare
DARPA	Defense Advanced Research Projects Agency
DHEW	Department of Health, Education, and Welfare
DoD	Department of Defense
DOJ	Department of Justice
EEG	electroencephalogram
EIT	enhanced interrogation technique
EPA	Environmental Protection Agency
EUA	Emergency Use Authorization
FBI	Federal Bureau of Investigation
FDA	Food and Drug Administration
FDCA	Federal Food, Drug, and Cosmetic Act
FISA	Foreign Intelligence Surveillance Act
FISC	Foreign Intelligence Surveillance Court
fMRI	functional magnetic resonance imaging
FOIA	Freedom of Information Act
FTCA	Federal Tort Claims Act
GAO	General Accounting Office; Government Accountability Office
GATT	General Agreement on Tariffs and Trade
Gulf War PAC	Presidential Advisory Committee on Gulf War Veterans' Illnesses
HHS	Department of Health and Human Services

ICRC	International Committee of the Red Cross
IED	improvised explosive device
IMTFE	International Military Tribunal for the Far East
IOM	Institute of Medicine
IRB	institutional review board
ISI	Inter-Services Intelligence
ISIS	Islamic State of Iraq and Syria
JSOC	Joint Special Operations Command
MASH	mobile army surgical hospital
MIT	Massachusetts Institute of Technology
MUST	medical unit, self-contained, transportable
NAFTA	North American Free Trade Agreement
NATO	North Atlantic Treaty Organization
NDRC	National Defense Research Committee
NIH	National Institutes of Health
NRC	National Research Council
NSA	National Security Agency
NSC	National Security Council
NSF	National Science Foundation
NYU	New York University
ODNI	Office of the Director of National Intelligence
OSRD	Office of Scientific Research and Development
OSS	Office of Strategic Services
OT&E	operational test and evaluation
PB	pyridostigmine bromide
PHS	Public Health Service
POW	prisoner of war
PREP Act	Public Readiness and Emergency Preparedness Act of 2005
PTSD	posttraumatic stress disorder
RDI	rendition, detention, and interrogation
SERE	survival, evasion, resistance, and escape
SIGAR	Special Inspector General for Afghanistan Reconstruction
SIPRI	Stockholm International Peace Research Institute
SSCI	Senate Select Committee on Intelligence
TBI	traumatic brain injury
tDCS	transcranial direct current stimulation
UCMJ	Uniform Code of Military Justice
US	United States
USAMRIID	US Army Medical Research Institute of Infectious Diseases
USDA	US Department of Agriculture
USSOCOM	US Special Operations Command
USSR	Union of Soviet Socialist Republics
USU	Uniformed Services University

VA	Veterans Administration; Department of Veterans Affairs
WHO	World Health Organization
WMA	World Medical Association
WTO	World Trade Organization
WWI	World War I
WWII	World War II

Introduction

This is a book about war and science. It provides an intimate look at how war brings together a nation's political, academic, and industrial elite: how each facet contributes to military goals, and how the collective situates the military at the forefront of science and technology. It demonstrates how the quest for scientific innovation often is intertwined with a quest for military superiority, and how war alters the moral landscape of how we think about science, ethics, law, and society. This book also unravels how these endeavors reflect fundamental elements of human nature—such as fear, ambition, greed, conquest, revenge, othering, and tribalism—and how these characteristics interplay with international relations, political economy, and a military ethos of duty, honor, and loyalty. Just as war is an engine of innovation that propels scientific advancements, war also transcends conventional boundaries, blurring the line between moral and immoral, ethical and unethical, legal and illegal.

This book questions the power of ethics and the force of law in the face of real or perceived threats to a nation's population, land, economy, or political interests. It details scores of examples where existential threats have driven researchers to engage in scientific pursuits deemed essential to national security despite knowing that the projects would harm or exploit certain groups. This includes, for example, locking service members into gas chambers and inundating them with chemical weapons after luring them to the experiments under false pretenses, conducting clandestine radiation experiments at civilian hospitals on patients who included pregnant women and children, administering LSD and other drugs to unsuspecting service members and civilians to see if the substances could be used to facilitate interrogations, and secretly exposing countless American communities to radiation and biological agents to further military science and assess the impact of these weapons on humans and the environment.

Some of these ventures resulted in scientific achievements that benefited military personnel, civilians, and society. Others floundered or were grounded in irrational theories. Although scientists, military leaders, and government officials generally believed that they were well-intentioned in their efforts—seeking to protect service members and the homeland, and further scientific or national security goals—in unpacking these events, it becomes clear that decision-making, particularly in times of crisis, rarely escapes the detrimental effects of cognitive biases and often maintains a broad interpretation of actions deemed necessary for the common good. Even well-meaning decision makers can go wildly astray.

Reflecting on the history of war and science, this book traverses this mercurial moral compass. It seeks to realign that compass to better harmonize contemporary scientific and national security concerns—such as the ethical duties of health professionals involved in interrogation, how to structure guidelines for approval and administration of biomedical countermeasures, and how to develop protocols

America's Military Biomedical Complex. Efthimios Parasidis, Oxford University Press.
 DOI: 10.1093/9780199351473.003.0001

governing research and operational integration of military human enhancements—with fundamental principles of justice, fairness, and human dignity. Applying these principles means treating individuals with respect and protecting their welfare, providing honest and transparent information regarding anticipated risks and benefits, balancing individual liberties with military goals, ensuring equitable distribution of benefits and burdens across subpopulations, affording adequate healthcare and other benefits for individuals harmed by research or military technologies, and instituting mechanisms that provide robust oversight of researchers, government actors, and military leaders.

Although national security comes at a price, the central theme of this book is that the laws and ethical codes that govern military science do not adequately balance military goals with fundamental human rights. More is needed to protect the health and welfare of military personnel and society, and more can be provided without compromising national security or the need for operational flexibility.

For military science, a key question is whether laws and ethical codes that apply in civilian science—such as those governing patient autonomy, medical product approval, and research protections—should be limited in military affairs. Historically, this question has consistently been answered on an ad hoc basis, often in the face of an imminent national security concern. As this book details, decision makers typically have erred on the side of loosening civilian guidelines for military matters—a form of military exceptionalism that permits actions on military issues that would be unethical or illegal if applied in civilian contexts.

Laws and ethical codes that govern military science reflect a society's norms, values, and aspirations. The common narrative is that law and ethics fail to keep pace with technological advancements. While this statement holds true in some instances, in others, laws and ethical codes have been created to facilitate scientific or military goals: that is, rather than being simply reactive, laws and ethical codes have been used proactively to promote the progress of science and national security.

At times, this has included structuring military-specific exemptions to rules governing civilian research, patient privacy, and the practice of medicine. In other instances, courts have refused to sanction untoward practices by upholding an expansive view of governmental immunities—such as sovereign immunity, the political question doctrine, and the state secrets privilege—doctrines that substantially preclude civilian court review of military affairs and leave injured individuals without legal recourse to address harms that they have suffered. The net result is a bifurcated legal and ethical framework whereby actions generally deemed inappropriate are permitted in the interests of national security.

At its core, the existing framework is comprised of an expansive post-9/11 national security state constructed upon a Cold War-era legal regime. In many cases, secrecy trumps accountability, safety measures are curtailed, and the potential for short-term gain prevails over long-term impact. In some instances laws have been crafted to permit hazardous experiments, conceal the studies from public review, and limit accountability for affiliated harms. Although such practices may reflect fundamental aspects that distinguish civilian matters from war and warfighting—for example, service members may be ordered to risk their lives to fulfill a military mission, and violent conduct that is customarily prohibited in society sometimes is sanctioned during war—they also permit the commodification and exploitation of certain lives for the benefit of others. The moral and ethical implications are significant, and the burdens of these pursuits often have fallen disproportionally on disenfranchised or marginalized groups.

History is rarely monocausal. In the scores of examples detailed in this book—which focuses on the development of military science in the United States, but also draws on case studies from throughout the world—I have been cautious in attributing cause and effect, and of presuming that innovations advance in one direction over time. Like scientific developments, ethical doctrines and legal codes have not progressed linearly. Rather, this book illustrates the oscillations of science, national security, law, and ethics. It underscores magnificent successes and regrettable errors. Although all historical accounts are revisionist to some extent, this book aims to evaluate actions by the societal, legal, and ethical norms of their time. I highlight contemporaries who were proponents of prevailing policies and others who were critics. For each, I have been mindful that individual decisions may have been driven by personal ambitions rather than political, scientific, or military aims. My purpose is not to condemn decisions or decision makers, but rather to provide a context to understand why certain paths were pursued and others were not.

For clarity, although I use the term "military science" frequently throughout the book, I intend for the reader to apply the term in the context of the health sciences. This includes developments in medicine, biology, chemistry, physics, public health, and related disciplines. I've excluded non-biomedical innovations regarding the development of warships, fighter planes, tanks, weapons, and related areas. As detailed in later chapters, this line becomes particularly amorphous concerning certain technologies, such as atomic and chemical weapons, or brain-computer interfaces that link medical devices with weapons systems or flight controls. In referring to members of the armed forces, for the sake of simplicity I often use the term "warfighter" or "service member," though I recognize that countless members of the armed forces do not fight in war and that many engage in humanitarian assistance. Sometimes, I use the term "soldier," "sailor," "airman," or "marine" to refer to an individual in a specific service (Army, Navy, Air Force, Marines).

This book expands upon existing scholarship to offer a comprehensive and nuanced analysis of the arc of ethical and legal guidelines governing the historical development of military science. Advancements from these fields have helped address the health and public health concerns of the armed forces, and have contributed to weapons research and development. Often, this work has provided benefits beyond the battlefield. Indeed, returning to the founding of America, the military has frequently been at the forefront of medicine and science. Whereas several advancements have been disruptive scientific innovations with transformational impact, others were lockstep improvements that developed slowly over time. The sum of these activities comprises what I call the military biomedical complex.

The five parts of this book are structured around key patterns in the progression of military science, and each part corresponds to a fundamental shift in the normative and operational foundations for how science has been integrated into military pursuits. In Part I (1775–1917), *Maintaining a Healthy Fighting Force*, I explain how, for nearly 140 years, the primary goal of military science was to promote the health of service members and veterans. This included the provision of medical care, preventive health measures, and medical research. In Part II (1917–1946), *Beyond Disease Prevention and Medical Care*, I detail how utilization of gas warfare during World War I (WWI) motivated a shift in military science, whereby physicians and researchers expanded their repertoire beyond healthcare to encompass weapons development, including chemical and biological weapons. This shift continued after WWI and accelerated during World War II (WWII). In Part III (1946–1991), *Proliferation of the Military Research Enterprise*, I explore the vast expansion of military science after WWII and throughout the Cold War. During this era, military science proliferated throughout civilian and military institutions with scant legal or ethical guidelines. This included an elaborate network of clandestine research involving the military, intelligence agencies, universities, and industry. Meaningful legal and ethical frameworks would not be established until the latter part of the period. In Part IV (1991–2024), *From Protecting to Enhancing the Fighting Force*, I explain how, soon after legal and ethical protections were codified, military exigencies contributed to a rollback of certain protections for military personnel. Also during this period, the military biomedical complex expanded to encompass research and development of human enhancements, including pharmaceuticals and brain-computer interfaces that augment human cognition and physiology. In Part V, *Looking Ahead: Jus in Praeparatione Bellum*, I offer a new normative framework for examining military science endeavors and detail several policy recommendations that aim to better harmonize national security with individual rights and societal goals.

While scores of books have explored some of the topics I have taken on in this book—such as the development of atomic and biological weapons, Cold War mind control programs, or biomedical human enhancements—to my knowledge, no book has attempted to bridge the broad historical development of military science with an analysis of how laws and ethical codes have co-evolved. I provide firsthand accounts of rationales underlying why decisions were made, as well as contemporary examples of proponents and critics of military science pursuits.

As detailed throughout the first four parts of the book, my assessment of the historical record involves unearthing decision-making patterns and situating the patterns within broader ethical, legal, and political debates of the time. Coupled with this descriptive and analytical approach, in Part V of the book, I advocate for new policies to (1) recalibrate the approval and administration guidelines for biomedical countermeasures, (2) strengthen the medical autonomy of service members, (3) expand military science ethics review, and (4) limit the government's ability to use immunity doctrines to evade responsibility for harms caused by military research or integration of biomedical technologies into military operations. I group these recommendations under a normative concept—*jus in praeparatione bellum*—justice in war preparations.

Most of the events discussed in this book are condemned to obscurity and rarely are taught in school or discussed in mainstream media outlets. Some examples may be shocking to those unfamiliar with the history of military science. My goal is to tell this longer history and to unpack the multifaceted aspects of the synthesis and evolution of science, law, ethics, and national security. Examining the long view is particularly helpful for contemplating how to restructure contemporary operational guidelines and military policy. Although nations are not trapped by their histories, and we may not find our future in our past, our past must be scrutinized in its historical context to help us understand and critique our present and consider the realm of human possibilities. The story begins with the birth of the American nation.

General George Washington and other leaders of the Continental Army recognized that good health was essential to the success of the revolutionaries. Within days of Washington taking command of the Army, the Continental Congress created a department to handle military medical affairs. The Army's medical leaders were accomplished civilian physicians, but they did not have experience in military medicine and fought incessantly over organizational and operational matters. In addition, widespread resource shortages severely hindered wartime health efforts. Disease was the leading cause of death in the war—responsible for nine in ten wartime casualties—and smallpox outbreaks devastated soldier ranks and brought the Army to the brink of a mutiny.

A key turning point during the war was Washington's order that mandated smallpox variolation—a procedure that attempts to build immunity by exposing a person to a mild form of the smallpox virus, typically by placing diseased tissue into an open wound of the person being variolated. The mandate was controversial. Individuals undergoing variolation are contagious and must be quarantined, and some die or suffer long-term disabilities such as blindness. Many physicians, soldiers, and civilians protested the mandate, claiming that the order violated a person's freedom, was based on uncertain clinical grounds, and was an affront to God's sole right to dictate the circumstances under which a person was to die. Others were eager to undergo the procedure in hopes of warding off the devastating effects of the disease. Ultimately, Washington persisted, the mandate was widely enforced, and variolation-induced smallpox immunity was a critical factor that helped propel the Continental Army to victory.

For nearly a century after the founding of the American Republic, military medicine floundered due to a lack of funding, a poorly organized system, and the inability of physicians to treat disease effectively. Medical interventions rarely were helpful and often exacerbated health conditions or caused adverse health consequences. Medical schools across the country maintained low educational standards, while apothecaries frequently sold quack medicines with wild curative claims. The inability of health professionals to cure disease or offer meaningful redress from health ailments led individuals to eschew medical care and ignore preventive health measures. Mistrust of health professionals was not merely a wartime concern.

Despite the lackluster state of medical affairs, the period between the late eighteenth and late nineteenth century included several medical achievements, chief among them the development of anesthesia, the recognition of the importance of sanitary and preventive health measures, and the understanding that medical treatments should be documented and assessed over time. After the Civil War, Union Army physicians published a multivolume treatise on battlefield medicine, widely viewed as America's first significant contribution to academic medicine.

During the late 1800s and early 1900s, medicine underwent a major transformation with the advent of the germ theory of disease and the discovery of causal agents for myriad health conditions. A scientific revolution followed. Medical researchers were heralded as cutting-edge professionals serving the public good, medical schools tightened their entry and educational requirements, and physicians were revered for their abilities to treat disease and prevent infections. Despite the achievements, military personnel and segments of the public were slow to embrace innovations from newfound medical knowledge due to long-standing skepticism toward health professionals. This resulted in countless preventable casualties. In addition to wartime health concerns, injured or sick veterans often languished without adequate benefits and medical care. Calls for reforms to military medicine and veterans' healthcare persisted for decades, yet the government consistently underfunded both.

Notwithstanding the challenges and limitations, the military was at the forefront of medical and scientific research by the early twentieth century. It established dozens of medical boards to study infectious and communicable diseases, and medical schools founded by the Army and Navy were of the first in the nation to expand their educational programs to include best practices for conducting medical research. At the same time, however, innovative military physicians often were shunned by their superiors, many of whom were set in their traditional methods and felt threatened by the young upstarts and their novel theories and practices. It would take years for society and the military to integrate medical and research developments into the military biomedical complex. Nevertheless, the United States had constructed a comprehensive military healthcare infrastructure by WWI, and the fighting force was healthier than ever.

The widespread use of chemical weapons during WWI altered the trajectory of military science and military medical affairs. This fundamental shift marks the beginning of Part II of the book. Whereas for over 140 years, American scientists and health professionals focused their efforts on maintaining a healthy fighting force, the advent of gas warfare created an urgency within military and civilian leadership to dedicate significant scientific efforts to develop gas warfare countermeasures and new chemical weapons for offensive purposes. For the first time in US history, physicians were intimately involved in the creation of weapons of mass destruction. This dynamic created clinical and ethical challenges for military physicians and researchers, and opened military personnel to new sources of service-related health risks: warfighters often were compelled or coerced into risky chemical warfare studies, and many suffered research-related harms.

Deployment of chemical weapons during WWI occurred despite the 1899 Hague Convention, whereby several nations pledged to abstain from deploying gas weapons. Germany, the first nation to institute a gas attack, defended its actions and characterized chemical weapons as an "extraordinarily mild method of war."[1] After the German offensive, military and scientific leaders in Britain, France, the United States, and elsewhere sought to develop methods to defend against gas warfare and launch gas attacks. Each nation drew upon its academic and industrial elite to help further chemical warfare efforts. Within a short period, several warring nations were deploying chemical weapons, each justifying the deployment of the chemicals under the principles of military necessity and national security.

At a broader level, gas warfare initiated a disruptive paradigm shift that expanded the scope of military medical affairs beyond disease prevention and medical care to encompass weapons development. In the decades following WWI—as nations around the globe continued with chemical weapons research and development—a series of international agreements attempted to instill a global norm against the further deployment of poison gases and other deleterious substances, including biological agents. Rather than curtailing development in these fields, the doctrines had the perverse effect of encouraging clandestine research and stockpiling. It marked the beginning of a perilous arms race.

Meanwhile, due largely to underfunding, the military struggled to accommodate the healthcare needs of service members and veterans. These struggles—particularly concerning WWI veterans—led to the creation of the Veterans Administration (VA), which was charged with providing healthcare and benefits for veterans. Contemporaneously, the military continued to play an integral role in developing medical and preventive health research.

The paradigm shift that began during WWI accelerated significantly during WWII. Universities dedicated cutting-edge laboratories to further war research, and faculty applied their expertise to military science. The line between military and civilian research largely disappeared. The drive to weaponize atomic energy, chemicals, and biological agents brought together America's entire scientific enterprise in a quest to create military advantages. Meanwhile, the ethical implications of research with human subjects—including military personnel, conscientious objectors, prisoners, and individuals housed in mental institutions—received little attention.

WWII also ushered in a new era for military science and a new model for government-sponsored research. A cadre of professors and researchers, led by Vannevar Bush, former dean of MIT's School of Engineering, encouraged President Franklin Delano Roosevelt to create a civilian-led agency that coordinated wartime research. The group contended that the military had ineptly synthesized American scientific know-how and that academic scientists were best suited to lead wartime research. Roosevelt adopted the proposal and afforded the agency near limitless resources.

Under the new model, the government entered into contracts directly with universities and covered research overhead costs. Previously, professors would obtain research grants, and the university was responsible for covering the costs of facilities and other research-related matters. Over time, the agency's mandate expanded beyond medical research to encompass weapons research and development. In light of the new funding model, universities encouraged their faculty to pursue government grants.

The government funded studies at more than one hundred institutions nationwide, the link between academia and the military grew stronger, and the line between civilian and military science was blurred. Under the premise of wartime necessity, nontherapeutic studies were conducted using orphans, prisoners, service members, and individuals housed in mental institutions. No federal statutes or regulations protected research participants, and departmental policies that implemented disclosure and consent requirements were rarely applied or enforced.

The breadth of wartime research was stunning. Prisoners were offered favorable treatment and early release for participating in medical research regarding antimalarial treatments and other therapies. Service members were ordered to stand in fields alongside monkeys and other animals as low-flying planes sprayed chemical weapons; the studies sought to test protective clothing and the lethality of the weapons. In other projects with similar aims, service members were ordered into gas chambers and inundated with mustard gas, experiments which were aptly termed "man-break" tests. Many of the men were recruited for the studies under false pretenses and told that they would be charged with espionage if they disclosed their participation in the research. The results of some of the studies were published in leading medical journals, and the research community did not raise any ethical concerns.

Participants in the experiments suffered research-related harms such as blindness, intense vomiting, internal and external bleeding, and damage to their lungs and respiratory system. Many endured long-term health problems such as cancer, asthma, and psychological disorders. Countless veterans maintained the secrecy of the studies for more than half a century due to fear of prosecution, and it would take decades for the government to publicly acknowledge the existence of the experiments.

During the Nuremberg trials that followed WWII, German physicians who engaged in gruesome concentration camp experiments pointed to the publicized American studies as part of their defense, arguing that wartime research with prisoners was justifiable and common practice. The Germans also highlighted the absence of laws or ethical codes that prohibited such studies. During the trials, American experts falsely testified that the United States had long maintained ethical codes for medical and scientific research; in fact, the American Medical Association and other groups had long protested against codifying ethical principles due to their fear of hindering physician discretion and the progress of science.

Even after the Nuremberg tribunal issued ten principles that are now commonly referred to as the Nuremberg Code, American doctors and researchers dismissed the principles as a code for barbarians and an unnecessary intrusion into physician and researcher independence. This moral stance would become embedded in the scientific and military establishments for decades, the ramifications of which are the focus of Part III of the book, which examines the proliferation of the military research enterprise.

Part III begins with the post-WWII scramble to exploit the scientific riches of Germany and Japan. Although postwar periods in the United States historically were characterized by a decrease in military size and spending, emerging geopolitical concerns pushed post-WWII America into an extraordinary state of military proliferation. The United States and other nations recruited German scientists, many of whom participated in concentration camp experiments. The US government also granted Japanese researchers immunity from war crimes prosecution in exchange for copious details on biological warfare, much of which was derived from horrific experiments on Chinese nationals and other war prisoners. During WWII, Japanese scientists injected humans with bacteria and viruses to study experimental vaccines, conducted field tests whereby humans tied to stakes would be exposed to biological agents, and placed humans in gas chambers and exposed them to toxic pathogens. In some instances, Japanese researchers dissected living humans to examine the impact of biological agents on the human body; children and infants were subjected to some of the studies.

In the case of Germany, during the Nuremberg trials, US lawyers led the prosecution against some Nazi scientists for war crimes and crimes against humanity, but behind the scenes, the US government created a comprehensive program to bring many

Germans to work for the US military. At the time, a few American leaders questioned the wisdom of employing Nazi scientists and Japanese biological warfare experts, and embedding them into the US research enterprise. Ultimately, national security and the desire to attain scientific superiority trumped any moral, ethical, or operational considerations.

As Cold War fears intensified in the decades following WWII, research flourished across the United States, largely unhindered by legal or regulatory guidelines and with little attention to moral or ethical concerns. This included medical research at hospitals and universities, industrial research for commercial applications, and military research to further weapons development. Although various policies afforded some protections for research participants, the policies only covered a small subset of research and, within that subset, were rarely enforced. Throughout the country, researchers routinely engaged in risky experiments on orphans, children with disabilities, prisoners, and individuals housed in mental institutions. Within these vulnerable groups, African Americans often were subjected to the most dangerous experiments. Although some contemporaries raised ethical objections, they were overshadowed by the prevailing view that prioritized the pursuit of science and viewed vulnerable groups, many of whom received care in state-funded institutions, as individuals who owed a debt to society.

In the context of military science, massive investments flowed into atomic, biological, and chemical weapons research. The breadth and depth of the work was extraordinary. The military administered field tests with atomic weapons to prepare service members for nuclear war, physicians conducted biomedical radiation research at hospitals across the country, and scientists examined the environmental impact of radiation by secretly inundating large swaths of America with intentional radiation releases. Other studies included weaponization of pathogens, field tests with biological weapons, development of mind-altering drugs for use in military and intelligence operations, and formulation of noxious chemicals that could devastate enemy forces, land, and food supplies.

Research participants included service members, veterans, prisoners, pregnant women, children, hospital patients, and others, though many were coerced or compelled into participation or were unsuspecting research subjects. The projects were developed to promote military preparedness and examine the impact of weapons of mass destruction on humans and the environment. The work also contributed to operational decisions to utilize certain weapons, such as Agent Orange during the Vietnam War, which later was found to have caused countless health ailments among warfighters and Vietnamese citizens. At times, military and intelligence agencies paused to consider legal and ethical considerations, though the concerns typically were sidestepped in a race to generate innovative solutions to Cold War problems. Secrecy classifications and legal doctrines shielded wrongful actions and created nearly insurmountable obstacles for the injured.

During the 1970s, revelations of government abuses in research and military endeavors unmasked several clandestine projects and set in motion events that reoriented the legal foundation for medical research and military science. In 1972, a journalist published details of the Tuskegee Syphilis Study and how the government engaged in four decades of experiments on poor African American men. Researchers withheld treatment from the men even though, for over twenty years, physicians knew that penicillin cured syphilis, a devastating disease that causes blindness, insanity, and even death. Throughout the Tuskegee study, the men were told they were being treated for "bad blood" and were never informed that they were participants in an experiment.

Following the exposé, new regulations governing research with human subjects were implemented, and Congress passed the National Research Act. Among its provisions, the act established the National Commission for the Protection of Human Subjects of Biomedical and Behavioral Research, a group charged with issuing recommendations on ethical principles to frame federal protections for research subjects.

In addition to revelations regarding the Tuskegee study, during the 1975 Church Committee hearings—which investigated abuses by American law enforcement and intelligence agencies—the Central Intelligence Agency (CIA) admitted that it had engaged in clandestine chemical and biological warfare experiments, and that the agency violated US policy by maintaining chemical and biological weapons for operational use. The hearings also revealed some details from the CIA's mind control program, where, for over two decades, the agency secretly drugged unsuspecting civilians and military personnel to study whether drugs could be used as truth serums or to facilitate interrogations.

In 1979, the National Commission issued its recommendations in a document now commonly known as the Belmont Report, which outlined ethical principles to guide legal protections in research. The core principles were justice, respect for persons, and beneficence. Following issuance of the report, the federal government instituted new guidelines governing research with human subjects that included informed consent, fair selection of research subjects, and an independent assessment of a research project's risks and benefits. The rules were codified and adopted by federal agencies in 1991.

Contemporaneously, as the Cold War ended and the twentieth century drew to a close, the military biomedical complex was evolving to combat the worldwide proliferation of weapons of mass destruction and new threats from terrorists and non-state actors. This point marks the transition to Part IV of the book. Just as new rules governing medical research were being adopted across the country, the military coordinated with federal regulators to create military-specific provisions that carved out exceptions to the guidelines. Specifically, after a series of closed-door meetings with the military, the Food and Drug Administration (FDA) altered its regulations to waive informed consent requirements in instances of military exigency. The new rule was crafted to assist in preparations for the Persian Gulf War, and the military applied the waivers to mandate biodefense countermeasures that were not FDA-approved as prophylaxis for chemical or biological warfare. Military personnel sued the FDA and the Department of Defense (DoD) in an attempt to invalidate the informed consent waivers, but a federal court ruled in favor of the government and dismissed the case.

Following the war, speculation arose that the countermeasures were a contributing factor to Gulf War Illness, a debilitating disease that impacted nearly 250,000 service members. After that, the FDA revoked the regulation that granted the military special exemptions, but only after Congress enacted a new law granting the US President the authority to waive informed consent rules in military missions. The new law firmly embedded a divergent policy whereby mandates for experimental medical products were permissible in the military even though they would be illegal in a civilian context.

Military-tailored exemptions expanded following the 9/11 attacks, a series of anthrax-laden letters mailed during the fall of 2001, and US involvement in post-9/11 wars in Afghanistan, Iraq, and the global war on terror. The laws provided the FDA with new regulatory mechanisms to authorize the use of countermeasures without the need to conduct a standard safety and efficacy review. This included the Animal Rule, which permits approval of certain medical products without conducting human clinical trials, and the Emergency Use Authorization (EUA) pathway, which maintains a lower bar to market for certain medical products during a declared public health or national security emergency. The laws also granted broad legal immunities for manufacturers of medical countermeasures approved via the accelerated pathways and earmarked significant funding for biosecurity, an allocation that has vastly expanded the military biomedical complex.

Also during this period, the military instituted anthrax and smallpox vaccination programs. Both were controversial. In the case of the anthrax vaccine, litigation that challenged the vaccine mandate spanned military and civilian courts and lasted more than a decade. A federal court paused the mandate because the FDA failed to follow its own guidelines on vaccine approval; thereafter, the FDA utilized its newly minted EUA pathway to allow the military to restart the program. Questions also were raised regarding the safety of the vaccines and whether the government intentionally overstated the risks of a biological attack to support unnecessary immunization programs and increase society's perception of national security threats to further the political agenda of President George W. Bush's administration.

By the beginning of the twenty-first century, America's military biomedical complex had become a massive enterprise that provided world-class medical care and conducted cutting-edge medical and scientific research. The US government allocates billions of dollars annually in research funding for military science, investments that intimately link the military with universities, hospitals, and industry. Although the military has been well positioned to support the medical and public health needs of twenty-first-century conflicts, the physical and psychological ravages of war have continued to plague military personnel during combat and upon return to civilian life. Contemporaneously, the global war on terror has relied extensively on scientists and healthcare personnel to assist in enhanced interrogation techniques, calling into question the ethical and legal obligations of these professionals.

At the same time, scientific and technological advances have expanded the realm of warfare into cyberspace and have catapulted the military biomedical complex into the realm of human enhancement. As the military candidly states, one of its primary goals is to exploit the life sciences to create warfighters with superior physical, physiological, and cognitive abilities. The demand for super soldiers has never been greater. Projects include pharmaceuticals and medical devices that enhance learning, memory, and warfighter performance. The military also has emphasized integrating neuroscience into military operations. Although biomedical enhancements and new technologies may provide warfighters with unique competitive advantages, they also come with special risks.

As this book details in Part V, the current legal and ethical frameworks should be restructured to recalibrate regulatory levers that apply to military science, broaden the health and social safety net for service members and veterans, expand the ethical duties of military researchers and physicians, and reconfigure the state secrets privilege and other governmental immunities. I group these recommendations under a concept that I call *jus in praeparatione bellum*—justice in war preparations—a normative framing which is grounded in just war theory but also draws upon principles of international relations found within the foreign policy doctrines of realism and liberal internationalism.

For the military biomedical complex, *jus in praeparatione bellum* assesses ethical and legal elements related to military medicine, preventive health measures, and war-related biomedical innovations. As with other areas within just war theory, *jus in praeparatione bellum* is a doctrine of restraint and responsibility. It acknowledges that war is a unique human activity, presumes that war is sometimes necessary, and recognizes that war preparations are perpetual. *Jus in praeparatione bellum* endeavors to draw a line between just and unjust means of preparing for war. As long as one rejects the maxim *inter arma enim silent leges* (in times of war, laws are silent), an inquiry into just and unjust conduct is essential. War is a high-stakes endeavor that is morally dubious and immensely challenging, but a nation can promote its self-interest while maintaining principles of justice and human dignity.

Note

1. Quoted in Stockholm International Peace Research Institute (SIPRI), *The Problem of Chemical and Biological Warfare: The Rise of CB Weapons*, vol. 1 (1971), 232.

PART I
1775–1917
Maintaining a Healthy Fighting Force

1
Military Medicine During the American Revolutionary War

Introduction

On July 21, 1775, General George Washington wrote to the Continental Congress to advocate for creating a military medical department. "I could wish it was immediately taken into consideration," Washington implored, "as the lives and health of both officers and men so much depend on a due regulation of this department."[1] Six days later, Congress established a Hospital Department to manage military medical affairs.[2] The department's beginnings were austere: it was plagued by underfunding, supply shortages, mismanagement, and infighting among physicians and leaders. These limitations hindered the provision of quality healthcare, while dreadful camp conditions exacerbated the spread of disease.

Notwithstanding the significant impediments, military exigencies and an entrepreneurial spirit nurtured several wartime medical innovations, chief among them a new system for battlefield triage, sanitary measures for military installations, and a force-wide system for smallpox variolation. Variolation is a technique that predates vaccination but is premised on an analogous method—intentionally infecting a healthy person with a virus to build immunity. The ethical and operational implications of compelled variolation received considerable attention during the war: debate centered on balancing military objectives, individual rights, religious beliefs, and public health goals.

Wartime Health: Challenges and Innovations

Colonial-era medicine bears little resemblance to today's comprehensive healthcare enterprise. By the late eighteenth century, there were few colonial medical schools, and only a small number of physicians had studied medicine in Europe: of the more than 3,500 physicians in the American colonies, fewer than 400 had earned a medical degree.[3] There were no laws governing who could practice medicine, a gap that sometimes was exploited by charlatans who feigned medical expertise.[4] Apprenticeships were the primary pedagogical method, and doctors relied heavily on medical handbooks to guide their practice.[5]

Although physicians typically could reset fractures, amputate limbs, and dress ulcers, among other procedures, they commonly utilized practices such as bloodletting, emetics, laxatives, and enemata.[6] Of these, bloodletting was widely viewed as a panacea for a variety of illnesses and injuries, including burns and fractures.[7] These medical practices—which reflected the contemporary field of medicine—typically offered little respite to ailing patients. According to Benjamin Rush, a renowned

America's Military Biomedical Complex. Efthimios Parasidis, Oxford University Press.
 DOI: 10.1093/9780199351473.003.0002

physician and civic leader who signed the Declaration of Independence: "Fatal experience has taught the people of America that a greater proportion of men have perished with sickness in our armies than have fallen by the sword."[8]

Battles during the Revolutionary War were infrequent and short, and soldiers spent considerable time in camp or in transit between positions.[9] The majority of wartime medical care was provided by physicians assigned to local regiments, most of whom were appointed by local militia.[10] The colonies had relied on a militia system for more than 100 years prior to the Revolutionary War: each colony's militia maintained its own laws and culture, served as a local police force, and provided armed support during battles with Native Americans and rival European settlers. Colonial governments provided little funding for local militia, and throughout the Revolutionary War, communication between militia was limited because it was difficult to send messages across long distances. In large part, it was a system of self-reliance based on citizen-soldiers, with little centralized control from leaders of the Continental Army.[11]

Regimental physicians served under a regimental commanding officer. They were obligated to follow the chain of command, but their day-to-day practice carried on largely without oversight from medical officers or other physicians.[12] Across regiments, unsanitary camp conditions contributed to poor health among soldiers, and the quality of care varied in relation to the local physician's expertise.[13] As General Washington explained to Congress during the autumn of 1776, the wretched conditions motivated some soldiers to seek fraudulent medical waivers: "The regimental surgeons I am speaking of, many of whom are very great rascals, countenancing the men in sham complaints to exempt them from duty, and often receiving bribes to certify indispositions, with a view to procure discharges or furloughs."[14]

Deplorable management and persistent quarrels among physicians and department leaders further burdened wartime medical care. Benjamin Church was the first director of the Hospital Department, and his term was mired in controversy. Church was a distinguished Boston physician who had occupied a high position with the Sons of Liberty, a secret society that promoted colonial interests and fought against British taxation. Despite his civilian success and reputation, he was a fractious and inefficient leader in the Hospital Department, and within weeks of his appointment he was arrested "for carrying on a traitorous correspondence with the enemy."[15] Church asked his mistress to deliver an encrypted letter to the British that expressed his devotion to the Crown and detailed Continental Army positions. En route, she was intercepted by a former lover, who dutifully reported the incident to colonial authorities. A military tribunal found Church guilty of treason and imprisoned him. Shortly after his conviction, he was deported to the West Indies, though the vessel he boarded reportedly was lost at sea.[16]

Following Church's short and disastrous tenure, Congress appointed John Morgan as the Hospital Department's new leader. Morgan served during the French and Indian War and was a pioneer in medical education—he held the first medical professorship in the colonies. He co-founded a medical school at the College of Philadelphia (known today as the University of Pennsylvania).[17] He inherited a chaotic, undersupplied department that was overcrowded with sick and injured soldiers.[18] Additionally, resentment was growing between physicians serving in general hospitals and those serving with regiments. Supplies to regimental physicians flowed from general hospitals, and general hospital physicians often hoarded medical supplies rather than apportioning their stockpiles to assist injured soldiers on the front lines. Meanwhile, regimental physicians controlled the decision to transfer an injured soldier to a general hospital, and no standard protocols governed such transfers.[19] Within this framework, several factors impacted soldier health, including physician quality, competencies of commanding officers, and access to food, medical supplies, clothing, and shelter.[20] A weakness in any area could have disastrous health effects. Morgan sought to implement regulations to govern these interactions, but the daunting challenges proved insurmountable.[21]

In turn, Morgan was replaced by William Shippen. Morgan and Shippen cofounded the Philadelphia medical school, but the two men were bitter rivals who constantly jockeyed for prestige.[22] Wartime health challenges continued, and Shippen resigned amid accusations (set forth by Morgan and others) that he underreported deaths and misappropriated medical supplies. After Shippen's departure, John Cochran took over the department. Cochran served during the French & Indian War and volunteered to serve in the Continental Army without pay. Of the four Hospital Department leaders, he was the only one who had not earned a medical degree. His tenure was less contentious than that of his predecessors, and he served as director until the end of the war.[23]

Although the four directors were accomplished civilian physicians, the first three were unable to effectively lead the Hospital Department. This was partly because the department was created after the war began, with no established infrastructure for training, command structure, policy implementation, and supply management. The lack of rank for military physicians complicated these efforts. Although soldiers were obligated to obey orders from commanding officers, doctors were outside the chain of command, and their orders could be freely ignored. Rudimentary medical techniques and inadequate resources were also key factors: regiments often relied on civilians to provide supplies and abodes for treatment of the injured, and many physicians abandoned their posts because they went months without pay. As officials later acknowledged, these shortcomings allowed formidable talents to wither within an unsound institution.[24]

Despite grave health challenges, resource limitations, relentless infighting, and administrative inefficiencies, several medical advancements emerged from the war. One centered on preventive health measures. In 1777, Rush published a 14-page pamphlet, *Directions for Preserving the Health of Soldiers*, which outlined health protocols and underscored the importance of regular exercise, cleanliness, and a sound diet. The first Army Regulations, published in 1779, adopted many of Rush's recommendations.[25]

The war also spurred a new method for battlefield triage. Initially, the Continental Army did not have an established system for removing and treating wounded soldiers. Expanding upon British and European models, the Army created an echelon system of evacuation and medical care. Each regiment had its own hospital, typically a log cabin, barn, or tent. Injured soldiers were transported from the battlefield to a regimental hospital, often by being carried in a wheelbarrow. Those seriously ill or wounded were transferred to a general hospital, with "flying" hospitals—structures that could be moved easily, such as tents—serving as intermediaries.[26]

Notwithstanding the Army's new triage system and Rush's preventive health guidelines, disease spread rapidly, and wounds often became infected, leading to death or amputation (which also had a high mortality rate). Given the conditions, working in hospitals was extremely hazardous.[27] Of the estimated 25,000 deaths suffered by the Americans during the Revolutionary War, approximately 90 percent were caused by disease, with smallpox, dysentery, and typhoid fever being the most common ailments.[28]

Smallpox ravaged the Continental Army, killing thousands of soldiers and leaving countless others blind or otherwise permanently injured.[29] As public health expert Donald Hopkins succinctly explains, smallpox was widely feared because of "the suddenness and unpredictability of its attack, the grotesque torture of its victims, the brutality of its lethal or disfiguring outcome, and the terror that it inspired."[30] Coupled with high fever and malaise, painful pustules covered one's body, and pustules inside the mouth made it difficult to eat and drink. There was no treatment for the disease.[31] As one report explained, throughout the world, individuals infected with smallpox often were "abandoned by their nearest relatives and friends, as persons doomed by divine wrath to irrevocable death."[32] Fatality rates ranged from 20 percent to 60 percent.[33]

Smallpox was more common in Britain than in the colonies, and a majority of British soldiers were immune to the disease because they had survived an infection. The British knew that the Americans were particularly susceptible to smallpox and were rumored to have deliberately spread the disease in Boston and other cities. Some British officers and deserters corroborated the reports of elementary biological warfare. It was not the first time that the British used smallpox to debilitate a foe: British soldiers attempted to infect their adversaries during the French and Indian War, sometimes by distributing blankets used by sick individuals.[34] There was no smallpox vaccine at the time, though there existed a controversial method of inoculation known as variolation.

Smallpox Variolation

First practiced by the Chinese in the fifteenth century, variolation had spread to India, Britain, Europe, North America, and elsewhere by the early eighteenth century. The method involved intentionally exposing a person to a mild smallpox case, typically by scraping a pustule of an infected individual and placing the diseased skin into an open wound.[35] Figure 1.1 depicts a developing smallpox pustule and medical instruments used for inoculation.[36] The individual undergoing variolation was monitored for a few weeks as their immune system attempted to fight off the virus. It was a risky procedure: variolation mortality rates ranged from 2 percent to 20 percent, many individuals suffered permanent injuries such as blindness, and individuals undergoing variolation were contagious and had to be quarantined.[37] Due to the risks, variolation was prohibited in several colonies, though it had gained general acceptance in Britain, Europe, and New England.[38]

Washington survived a smallpox infection that he acquired during a journey to Barbados in 1751, and knew firsthand how devastating the disease could be.[39]

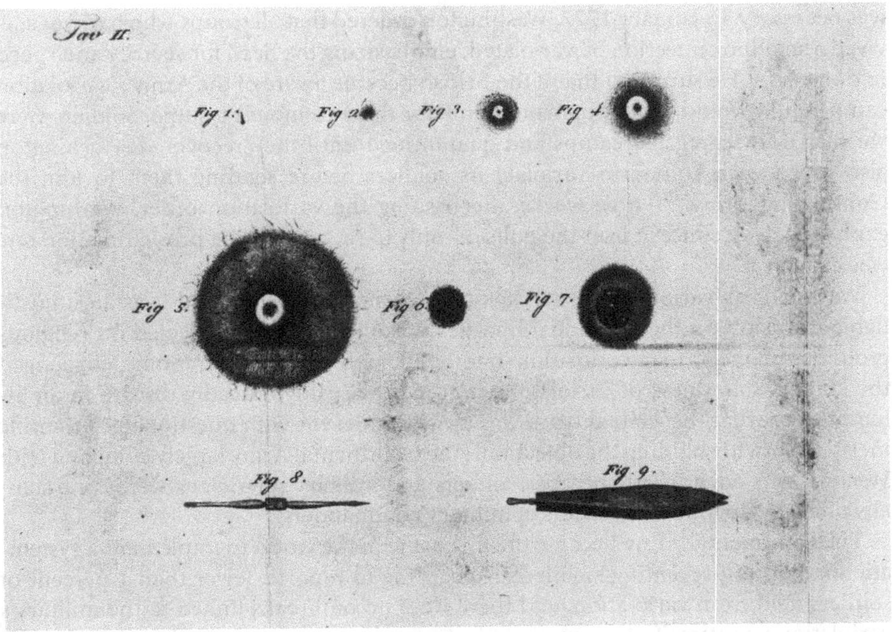

Figure 1.1 An engraving depicting a developing smallpox pustule (figs. 1–7) and medical instruments used for inoculation (figs. 8–9).
Image from the National Library of Medicine.

Initially, he decided not to variolate his troops.[40] Apart from the chance of dying during the procedure, in the midst of variolation soldiers would be unfit for battle and potentially could trigger a smallpox outbreak. Transmission of other diseases during variolation—such as tuberculosis or syphilis—also was a concern.[41]

Rather than permit the procedure, Washington prohibited variolation and ordered that soldiers who became infected with smallpox be isolated from their comrades.[42] He warned, "any disobedience to this order will be most severely punished."[43] The Continental Congress supported Washington's decision, and doctors caught variolating soldiers were fined and punished.[44]

Despite Washington's edict and the prosecution of those defying his order, many soldiers secretly variolated themselves, and some physicians offered to variolate soldiers surreptitiously for a fee. The custom of purchasing smallpox scabs for variolation was called "buying the smallpox."[45] This was dangerous, as inoculated soldiers fell ill and often spread smallpox to others. Some died during the procedure. By the summer of 1776, smallpox continued to proliferate despite Washington's policy against variolation. Fear of smallpox debilitated the Continental Army, and recruiting new soldiers became increasingly difficult. The adverse health conditions brought the Army to the brink of a mutiny.[46]

The toll on human life and the likelihood that smallpox could prevent the Continental Army from winning the war convinced Washington that a new policy was necessary. In January 1777, Washington ordered that all troops who had not survived a smallpox infection be variolated, emphasizing the need for secrecy and speed in doing so.[47] He surmised that if the British became aware of the Army's variolation campaign, it would be an opportune time for them to mount a surge. Soldiers were variolated in segregated camps and quarantined until they recovered. Washington also urged each colony to variolate its soldiers before sending them to join the Continental Army.[48] Three weeks after issuing the variolation order, Washington expressed deep concern over the policy,[49] only to revert to being pro-variolation one week later.[50]

Washington's variolation order was controversial. Many viewed it as an unjustifiable intrusion on the individual liberties of soldiers. Others objected on religious grounds—namely, that variolation interfered with God's will.[51] Some challenged the clinical soundness of variolation, arguing that using a deadly disease in an attempt to prevent the same disease was a risky endeavor with questionable scientific merit.[52] Notwithstanding the objections, the Continental Army largely complied with Washington's edict, partly because soldiers and military physicians were contractually obligated to follow the orders of military commanders.[53]

The Continental Army became the first army in the world to implement a systematic method of preventing smallpox. According to reports, fewer than 1 percent of soldiers died from variolation, and there were no outbreaks linked to the military's variolation program. To be sure, these facts may be inaccurate. Medical records were scarce, and mortality and morbidity rates could not be verified.[54] Moreover, many variolated soldiers reported that "very little care and attention" were provided while under variolation and quarantine, and some contemporaries noted that not all deaths from variolation were recorded.[55] In the end, however, smallpox immunity resulted in a healthier fighting force, contributing to the Army's ability to defeat the British.

Summary

The trials and tribulations of the Hospital Department during the Revolutionary War paralleled the Continental Army's scramble against Britain, a mighty empire with superior resources and military experience. The department was created shortly after hostilities commenced, and there was little time to consider the administrative structure and build medical supply stockpiles. While the department's leaders had illustrious civilian careers, they had limited experience in wartime medical affairs. They often were preoccupied with securing their legacy rather than spearheading the development of a robust department. Varying standards of medical care in the field, coupled with contentious relationships between regimental and general hospital physicians, contributed to poor health outcomes.

Notwithstanding mismanagement and resource limitations, the Continental Army created preventive health guidelines, established an innovative method for battlefield triage, and implemented a smallpox variolation program. Although one now speaks with the benefit of hindsight, the success of the variolation directive remains a testament to the benefits that may flow from a risky but well-reasoned integration of biomedical techniques into military strategy. At the same time, the arguments set forth by critics of Washington's policy highlight fundamental interests implicated when a military commander compels a subordinate to submit to a medical treatment they otherwise might refuse. As will be discussed in the following chapter, these wartime experiences framed the military medical establishment in the newly established nation.

Notes

1. George Washington, Letter to John Hancock, July 21, 1775.
2. *Journal of the Continental Congress* (July 27, 1775), in *Library of Congress: Journals of the Continental Congress 1774-1789*, vol. II (Washington, DC: Government Printing Office, 1905), 209–11.
3. John T. Greenwood and F. Clifton Berry, *Medics at War* (2005), 3; Francis R. Packard, *History of Medicine in the United States*, vol. 1 (1963), 273.
4. Packard, *History of Medicine*, vol. 1, 273.
5. Greenwood and Berry, *Medics*, 3; William G. Rothstein, *American Physicians in the 19th Century* (1985), 34–35.
6. Rothstein, *American Physicians*, 45–54; Bernard Rostker, *Providing for the Casualties of War* (2013), 62.
7. Rothstein, *American Physicians*, 45–54.
8. Benjamin Rush, *Directions for Preserving the Health of Soldiers* (1777), 3.
9. Rostker, *Providing*, 244.
10. Harvey E. Brown, *The Medical Department of the United States Army from 1775 to 1873* (1873), 3–6; John Duffy, *The Healers* (1979), 74–76.
11. Allan R. Millett, Peter Maslowski, and William B. Feis, *For the Common Defense* (2012), 1–43; Gian Gentile, Jameson Karns, Michael Shurkin, and Adam Givens, *The Evolution of U.S. Military Policy from the Constitution to the Present*, vol. 1 (2019), 8–9.
12. Regimental Surgeon Warrant Form (1775), reprinted in Brown, *The Medical Department*, 5.
13. Brown, *The Medical Department*, 6.
14. George Washington, Letter to the President of Congress, September 24, 1776.
15. Brown, *The Medical Department*, 8.
16. John Bakeless, *Turncoats, Traitors, and Heroes* (1959), 9–23; Brown, *The Medical Department*, 8–11.
17. Brown, *The Medical Department*, 10–12; P. M. Ashburn, *A History of the Medical Department of the United States Army* (1929), 14.
18. Brown, *The Medical Department*, 11–14.
19. Brown, *The Medical Department*, 30–31; Packard, *History of Medicine*, vol. 1, 544–45.
20. Duffy, *The Healers*, 76–77.
21. Packard, *History of Medicine*, vol. 1, 544–47.
22. Richard H. Shryock, *Medicine and Society in America, 1660-1860* (1960), 31.
23. Ashburn, *A History*, 20.
24. Greenwood and Berry, *Medics*, 2.
25. Friedrich Wilhelm Steuben, *Regulations for the Order and Discipline of the Troops of the United States* (1779).
26. Greenwood and Berry, *Medics*, 1–2.
27. Greenwood and Berry, *Medics*, 5.

28. Howard H. Peckham, *The Toll of Independence: Engagements & Battle Casualties of the American Revolution* (Chicago: University of Chicago Press, 1974); Rostker, *Providing*, 60–62; Brown, *The Medical Department*, 11–14.
29. Brown, *The Medical Department*, 11–14.
30. Donald R. Hopkins, *The Greatest Killer* (2002), 3.
31. Donald A. Henderson, "The Eradication of Smallpox—An Overview of the Past, Present, and Future," *Vaccine* 29S (2011): D7–D9.
32. Joseph Jones, "Spurious Vaccination," *Nashville Journal of Medicine and Surgery* (1867): 8.
33. Stefan Riedel, "Edward Jenner and the History of Smallpox and Vaccination," *BUMC Proceedings* 18 (2005): 21–25.
34. Elizabeth A. Fenn, *Pox Americana* (2001), 88–90, 131–33.
35. Riedel, "Edward Jenner."
36. Luigi Sacco, *Osservazioni Pratiche sull'uso Del Vajuolo Vaccino* (Milan: Nella Stamperia Italiana e Francese, 1801), Table 2.
37. Fenn, *Pox Americana*, 32–33; Jeanne E. Abrams, *Revolutionary Medicine* (2013), 28–29.
38. Fenn, *Pox Americana*, 39–40; Riedel "Edward Jenner."
39. Fenn, *Pox Americana*, 13–15.
40. Abrams, *Revolutionary Medicine*, 54–57.
41. Riedel "Edward Jenner."
42. Abrams, *Revolutionary Medicine*, 54–57.
43. George Washington, *General Orders* (May 20, 1776).
44. Fenn, *Pox Americana*, 70–71, 88–89.
45. Fielding H. Garrison, *Notes on the History of Military Medicine* (1922), 159.
46. Brown, *The Medical Department*, 27.
47. George Washington, Letter to William Shippen, January 6, 1777.
48. Fenn, *Pox Americana*, 80–103.
49. George Washington, Letter to William Shippen, January 28, 1777.
50. George Washington, Letter to William Shippen, February 6, 1777.
51. Edmund Massey, *A Sermon against the Dangerous and Sinful Practice of Inoculation* (London, 1722).
52. Fenn, *Pox Americana*, 36–42.
53. Brown, *The Medical Department*, 5.
54. Fenn, *Pox Americana*, 263–66.
55. Quoted in Fenn, *Pox Americana*, 80–82, 95–98.

2

Revolution to Civil War

Medical Struggles and Innovations

Introduction

The success of the scrappy and resourceful revolutionaries galvanized a postwar jubilance that swept across the nascent American republic. A new political democracy was emerging that valued freedom of thought and shunned traditional European concepts of aristocracy, imperialism, and militarism. Rather than maintain a large national army and a centralized medical department, the United States decided to rely primarily on local militia and their respective physicians.[1] A large European-style army was unaffordable and unpopular, and American leaders recognized that the Atlantic Ocean served as a buffer to protect against encroaching adversaries. As George Washington remarked after the Revolutionary War, "A large standing Army in time of Peace hath ever been considered dangerous to the liberties of a Country."[2] Washington also cautioned against permanent military alliances and sought to minimize US political connections with foreign nations.[3]

Washington and other American leaders envisaged a small national army that would set standards that militia would adopt, but a top-down managerial framework did not materialize. In the context of military medicine, adherence to preventive health measures and the quality of medical care varied dramatically across militia.[4] This fragmented system contributed to poor health outcomes, particularly during the War of 1812, and in 1818 Congress attempted to address the concerns by creating a permanent military medical department.

Although a permanent department brought organizational improvements, public scorn toward the medical profession intensified as quack medicines proliferated and physicians often were unable to provide care that ameliorated health concerns. A common perspective from the time held that "if all the medicines were thrown into the sea it would be so much better for mankind and so much worse for the fishes."[5] It would take decades for these sentiments to shift.

Notwithstanding the challenges, throughout the nineteenth century, the military was at the forefront of advancements in medical practice and medical research. As these fields expanded, the line between practice and research became increasingly amorphous, and it was not always clear whether a therapy was administered primarily to treat a patient or to gather general medical knowledge. Within the burgeoning research environment, soldiers, slaves, and the disabled were viewed as convenient patient populations upon which to experiment with novel treatments. Few contemporaries raised ethical or legal concerns.

In the eight decades between revolution and civil war, the military medical establishment wrestled incessantly with the clinical, operational, moral, and societal implications of the aforementioned dilemmas. When the Civil War began, high casualty

America's Military Biomedical Complex. Efthimios Parasidis, Oxford University Press.
 DOI: 10.1093/9780199351473.003.0003

rates strained the medical departments of the Union and Confederacy, both of which were understaffed and underresourced. The Union Army responded by instituting transformative improvements: battlefield triage was reformulated, medical personnel tracked patient health, and publications detailed best practices for physicians and medical staff. These innovations contributed to a healthier fighting force and later would influence military and civilian medicine across the globe.

Military Medicine from Revolution to the War of 1812: Precursors to a Permanent Military Medical Department

After the Revolutionary War ended in 1783, the Continental Army disbanded, the Hospital Department dissolved, and local regiments were responsible for maintaining the health of their units.[6] Although the Continental Congress established a national standard for war pensions, the federal government did not have adequate resources to fulfill its obligations, and it was reluctant to impose a tax to cover the gap. Seeking to salvage some value from their pensions, many beneficiaries sold their rights to speculators at discounts as low as 12.5 cents on the dollar. In 1791, the government agreed to honor its commitments—much to the benefit of the speculators—though some beneficiaries did not receive payment until 1828.[7]

Peacetime in the new nation was short-lived. Battles with Native Americans were frequent, intensifying with the southern and westward expansion of the nation. America also faced threats from France, Spain, and Britain, each of which maintained control over certain lands on the continent and sought to limit American commerce.[8] America's relationship with France soured as the United States increased trade with Britain, balked at French demands for loans to support its military, and refused to uphold its obligations under existing treaties to defend Caribbean islands under French control.[9] During the French Revolution (1789–1799), France issued an order that sanctioned the seizure of American merchant ships, and many vessels were plundered and confiscated by French officials or privateers. The impact on US commerce was significant, diplomatic efforts to resolve the conflict were unsuccessful, and anti-French sentiment increased across America.[10]

The possibility of war with France was imminent, and the United States mobilized its Army and created a Department of Navy. George Washington was recalled from retirement to command the Army, and James Craik, Washington's personal physician, led a reestablished wartime Hospital Department.[11] The conflict became known as the Quasi-War with France. Although neither country declared war, by 1798, the Navy was fighting the French in the Caribbean. The British offered support, but President John Adams refused, seeking to avoid a full-scale confrontation. The French Revolution ended with a *coup d'etat* led by Napolean Bonaparte, and by 1800, the threat of war between France and America had subsided.[12] In turn, the Hospital Department disbanded and healthcare responsibilities again fell to local regiments.[13]

Thereafter, the Navy saw increased action as it patrolled the seas to protect American merchants, fighting Barbary Pirates and combatting other global threats to US commerce.[14] Not only did Navy medical personnel typically receive less pay than their Army counterparts, sailors also self-funded a portion of their healthcare needs.

The US government deducted twenty cents per month from sailor pay and deposited the proceeds into a Marine Hospital Fund. Over the next three decades, the funds were used to build Navy hospitals in Brooklyn, Philadelphia, Norfolk, and elsewhere. Meanwhile, Navy medical personnel published best practices that detailed protocols for diet, sanitation, clothing, ventilation, and isolating sick sailors.[15]

At the turn of the century, Britain and France were engaged in near continuous conflict, and each nation attempted to restrict the other from trading with America. The British also encouraged Native American uprising within the United States, and the British Royal Navy often removed American seamen from merchant ships and forced them to serve on behalf of Britain—a practice known as impressment. Estimates place the number of Americans seized by the British between six thousand and ten thousand.[16] In 1812, in the wake of what many called British insults, the United States declared war against Britain.

When the War of 1812 began, the US military was ill-prepared. There was no medical department, and medical stockpiles were inadequate to address wartime needs.[17] Nine months into the war—after American forces suffered significant casualties from battle and disease—Congress resurrected the Hospital Department and appointed James Tilton as its leader.[18] Three decades earlier, during the Revolutionary War, Tilton witnessed firsthand the depletion of the Army caused by unsanitary conditions and disease.[19] He was determined not to repeat the experience. Tilton helped lead efforts to implement preventive health measures and create a health information reporting system. As to the latter, the goal was to build a storehouse of medical knowledge from which evidence-based policies could be formulated, though few physicians complied with the requirements.[20]

During the war, variolation was prohibited and smallpox vaccination was required. The vaccine was developed in 1796 by Edward Jenner, an Englishman, and by 1812, it had become a generally accepted method of smallpox prevention.[21] Jenner had honed his vaccine by experimenting on what contemporaries characterized as "pauper children," inoculating them and then exposing the children to smallpox to test vaccine effectiveness.[22] When British ships first brought the inoculations to the United States during the early 1800s, some cities conducted analogous experiments to evaluate the vaccine independently.[23] While there were risks to vaccination, including transmission of other diseases such as syphilis and tuberculosis, the risks posed by the vaccine were less than those of variolation.[24]

From the perspective of military medicine, the War of 1812 is noteworthy for several reasons: it marked the first time the military instituted a mandatory vaccination program, and the first time the military endeavored to use data to track and study disease and injuries. The smallpox vaccine successfully reduced the number of soldiers who contracted smallpox. Still, of the approximately twenty thousand casualties during the war, disease was the cause of nearly three out of four deaths. Malaria, dysentery, pneumonia, and typhoid fever were common ailments.[25]

The Hospital Department was disbanded after the war, though some officials recognized that a permanent department was necessary to maintain experienced medical staff and adequate medical stockpiles.[26] These sentiments were representative of a shift in military policy, which recognized the weaknesses of the militia system and sought to expand the national army.[27] In 1818, Congress created a permanent Army

Medical Department led by a Surgeon General.[28] Joseph Lovell, the department's first Surgeon General, issued *Regulations for the Medical Department*, which set forth the responsibilities of medical officers, established a single Medical Corps, and elucidated the command structure within the department.[29] The organizational advancements were significant, but their impact on warfighter health was minimal in light of the deplorable state of the field of medicine.

The Medical Profession Through the Mid-Nineteenth Century

Despite organizational improvements in military medical affairs, throughout America the practice of medicine was lamentable. There were no enforceable standards for medical education or physician licensure, and the demand for physicians far exceeded supply. Opportunistic physicians and private medical schools exploited this predicament—a person could attend lectures for three months and receive a medical degree. The caliber of training varied considerably, and some schools sold fake diplomas.[30] Public resentment toward physicians and medical schools heightened, not only because of the proliferation of unskilled doctors but also because of the practices of many schools. For example, schools routinely paid "bodysnatchers" to dig up graves for bodies to use during anatomy lessons. Medical schools across the country were ransacked and torched as mobs stormed buildings to reclaim the bodies of lost loved ones.[31]

Physicians relied excessively on bloodletting, opium, calomel (a mercury-based compound with devastating side effects), and other medicines and practices that afforded little to no benefits to patients.[32] As one example, excessive bloodletting hastened the death of George Washington in 1799.[33] In light of the inability of physicians to treat disease effectively, the popularity of medical sects increased. These sects included homeopaths, minister-physicians, and Thomsonians. Thomsonians—who relied primarily on steam baths and herbal remedies, and admonished bloodletting and the use of drugs such as mercury and opium—flourished in homesteader communities and were at the forefront of a popular revolt against standard medical practices.[34]

Coupled with the rising popularity of medical sects, the "patent medicine" industry prospered. Contrary to their name, these medicines were not patented and were sold without regulatory oversight. Proprietors created secret nostrums that were marketed and sold directly to the public with extravagant and unsupported claims, such as cures for malaria and skin diseases. Many concoctions were comprised simply of alcohol and spices, while others included morphine, cocaine, or other toxic substances. The industry used its wealth and political connections to stall regulation and influence media content. The quack medicine industry thrived amid a *laissez-faire* political economy that heralded American entrepreneurship and shunned health and consumer protection regulations.[35]

Although there was fierce competition among medical sects and within the patent medicine industry, as a general matter, public confidence in the field of medicine was quite low.[36] There was scant evidence of effectiveness among the various practices and therapies, which served to discredit the profession at large. A group of doctors created the American Medical Association in 1847 to help build the prestige of the profession, but for decades, the organization was largely unsuccessful in bringing about meaningful changes to medical education and physician licensure.[37] Within the dire state of the medical profession, however, the military was at the vanguard of medical care and medical research.

The Military at the Forefront of Medical Care and Research

During the mid-nineteenth century, the military was at the forefront of medical care and research. In an attempt to ensure that its doctors were of high quality, in 1834, the Medical Corps established an entrance examination to test whether budding military physicians were well versed in the theory and practice of medicine.[38] During the 1840s, the Army began collaborating with universities and scientific institutions to conduct medical research.[39] The Army also created reporting mechanisms to collect and study patient health outcomes. These efforts led to the publication of volumes of studies over the next half-century.[40]

Military-led endeavors also helped build a culture of research throughout the field of medicine. During the first half of the nineteenth century, more than one hundred medical journals were founded.[41] Physicians would document novel cases and results from experiments with new therapies; much of the research was conducted on soldiers. In one well-documented case from 1833 involving a soldier with a gunshot wound to his stomach, the preeminent Army surgeon William Beaumont studied gastric digestion by injecting the patient with various experimental substances. On several occasions, the patient ran away from the hospital due to intense pain from the experiments, only to be brought back for further study.[42] Apart from research on injured soldiers, some physicians purchased slaves solely for purposes of experimentation, and medical journals and physician memoirs routinely discussed the use of slaves in medical experiments. These practices did not raise ethical quandaries within the profession.[43]

Contemporaneous with the proliferation of medical research, the concept of manifest destiny drove territorial expansion and resulted in frequent battles with Native Americans, European colonialists, and Mexico.[44] Despite organizational improvements since creating a permanent Army Medical Department in 1818, overcrowded and unsanitary camps were typical.[45] During the Mexican-American War (1846–1848), the quality of medical care was hindered by poorly trained medical personnel, lackluster preparation for wartime health needs, and inadequate medical resources.[46] As a report from a senior physician stationed in Puebla, Mexico, explained, many soldiers "are not only filthy but covered with vermin."[47] Even when best practices had been established, soldiers often ignored health recommendations due to their mistrust of doctors.[48]

The battlefield experiences underscored the importance of maintaining enforceable preventive health protocols, trained medical personnel, adequate medical stockpiles, and an efficient health administration command. Peacetime preparations were essential for a smooth transition to wartime needs. To help address the concerns, in 1847, Congress enacted legislation that provided Medical Corps officers with military rank, fair pay, and an equal command level with officers in other departments.[49] As military historians John Greenwood and Clifton Berry explain, "after 72 years, Medical Corps officers had finally gained an unquestioned right to command within their department and full status and privileges of officers of the U.S. Army."[50] These new measures were tested with the onset of the Civil War.

Military Medicine During the American Civil War

When the Civil War commenced in 1861, neither the Union nor the Confederacy had experience mobilizing a large army. Regiments reported for duty and found few or no provisions for housing, food, and healthcare. There was little funding for military medicine, partly because lawmakers and military leaders mistrusted physicians. Most military doctors had little experience in battlefield medicine, and Union and Confederate medical departments were disorganized, poorly managed, and undersupplied.[51] As historian Colonel P. M. Ashburn wrote, medical leaders primarily "were old men, most of them veterans of the War of 1812, honored relics of a distant past, in which their minds lingered."[52] Wartime conditions were grim: makeshift medical clinics were established in churches and other buildings, bandages and dressings often were reused, and diseases such as measles, mumps, and chickenpox spread rapidly in camps. Surgeons frequently improvised, creating operating tables from two barrels and a door as moaning, injured men waited in line for amputations, a pile of arms and legs beside the table.[53] Figure 2.1 provides an example of a Civil War first aid station.

To support the wartime health needs of Union soldiers, in 1861 Congress created the Sanitary Commission, a civilian relief agency that sought to emulate the success of Florence Nightingale and her British colleagues during the Crimean War (1853–1856).[54] The organization, which included thousands of American women, inspected field medical clinics, helped staff medical facilities, and trained and deployed thousands of nurses. The group also advocated for significant reforms in military medicine.[55] Although Congress and President Abraham Lincoln supported the Sanitary Commission, many Army leaders opposed its efforts, and the two entities maintained a strained relationship. The Sanitary Commission had no official authority to enforce its recommendations, though some Army field commanders valued the Commission's advice and implemented its policies.[56] Coupled with the creation of the Sanitary Commission, in 1863, Congress established the National Academy of Sciences to address Civil War needs. The group comprised distinguished scientists, engineers, and health professionals, and the organization's mandate was to "investigate, examine, experiment, and report upon any subject of science or art" when called upon by the government.[57]

In light of ongoing wartime medical struggles, one year into the Civil War Lincoln reorganized the Army Medical Department. Over the objections of senior Army leaders, including Secretary of War Edwin Stanton, Lincoln promoted a young physician, William Hammond, to the position of Surgeon General.[58]

Lincoln's choice was controversial, but it proved transformative. Hammond was a well-respected physician and medical researcher, a prolific scholar, and an articulate

Figure 2.1 Wounded soldiers, some of whom are amputees, at a first aid station near Fredericksburg, Virginia, during the US Civil War.
Photograph from the National Library of Medicine

teacher. A visionary with a dynamic personality, Hammond worked tirelessly to improve the Army Medical Department and reorganize the existing systems of field medicine, medical supply, and hospitals.[59] Hammond welcomed the support of the Sanitary Commission and updated entrance examinations to ensure the proficiency of military doctors.[60] He placed leading physicians on examination boards and emphasized the importance of high standards for entry.[61] Physicians also served as key counsel to commanding officers, helping to determine camp placement and organization, dietary requirements, and triage management. The doctors had two chains of command: one to the regimental commanding officer and another that extended to the Surgeon General.[62]

Hammond hired and promoted physicians and officers based on merit rather than seniority or connections, and delegated to these individuals the authority to lead their respective missions.[63] He instituted a robust medical records system and ordered that physicians collect medical specimens and document the nature of wounds, the impact of treatment, and the results of autopsies.[64] This work resulted in a meticulous six-volume treatise, published between 1870 and 1888, titled *The Medical and Surgical History of the War of the Rebellion*.[65] The European medical community regarded this monumental series as America's first major contribution to academic medicine.[66] Separate books that focused on naval medicine also were produced.[67]

Coupled with the scholarly publications, Hammond periodically issued pamphlets that provided guidance on a variety of wartime medical matters, including best practices for administering medical care and collecting health data.[68] A key member of Hammond's team was Jonathan Letterman, often referred to as the "Father of Battlefield Medicine." Letterman issued preventive health guidelines and worked to ensure that medical units were well-staffed and supplied. He improved the procurement process, helped create a triage system, and investigated fraudulent government contractors.[69]

Battlefield triage and large military hospitals also were hallmark Civil War innovations. Field clinics typically were small, makeshift outfits. Each regiment maintained several ambulance wagons: some transported medical supplies, others injured soldiers. Once an injured soldier was stabilized on the battlefield, a wagon transferred the patient to a local clinic where further care could be provided. Hospital trains and ships, each of which could hold thousands of patients, relocated the injured to general hospitals in larger cities. The Union Army established 192 general hospitals, some containing more than 3,000 beds, with a total capacity of over 118,000 beds. The Union Army also created convalescence camps that provided soldiers with a halfway house once they were fit enough to leave the hospital but not yet fit to return to battle. The convalescence camps also provided healthcare for transient soldiers, discharged soldiers, and returning prisoners of war.[70]

Initially, Union Army commanders disapproved of the system, arguing that it would "increase the expenses and immobility of our army . . . without any corresponding advantages."[71] Indeed, at the outset, the network was inefficient and poorly managed, and wagon drivers often were accused of greed and thievery.[72] Yet, over time the system proved to be successful in saving lives and limiting the spread of disease. By the summer of 1864, the Union Army had employed approximately eight hundred ambulances that facilitated quick and careful removal of battlefield casualties.[73] To help handle the flow of patients, the Union Army created a health record system that followed patients from the battlefield to general hospitals.[74] As the Sanitary Commission remarked after the war, the Union Army's hospital and triage system was "one of the noblest triumphs of the war . . . on a scale unprecedented in history, not only in their vastness, but in their fulfillment of all the requirements of humanity and science."[75] Military and civilian medical units throughout the world later would emulate the Union Army's model.

The Civil War also served as the genesis of several clinical innovations, particularly in pain management and surgery. Union Army physicians were encouraged to conduct medical research alongside their clinical and preventive health responsibilities.[76] Anesthesia is one example. While anesthesia was developed in the 1840s, it had yet to gain widespread acceptance, and many physicians believed it contributed to poor health outcomes. To administer anesthesia, military physicians would soak a rag with ether or chloroform and place the rag over the nose and mouth of an injured soldier. The soldier would become unconscious, though not completely so, and the physician would proceed with medical treatment such as surgery or an amputation. More than sixty thousand amputations were performed during the war, representing 75 percent of all surgeries. Amputations were quick procedures, often lasting less than ten minutes. However, injured soldiers sometimes would wait in queue for days because of the large number of injuries that required amputating a limb. More than 30 percent of amputees died, most from postoperative infections. Due to water scarcity, at times, surgeons would not wash their hands for two days or more, contributing to the spread of infections.[77]

While wartime advancements were significant, many physicians protested the administration of novel therapies, remarking that such treatment was contrary to sound medical judgment. Even more controversial was medical experimentation. In one example, three renowned physicians experimented upon patients in a military hospital to study phantom limb and nerve damage. The doctors administered thousands of experimental injections of morphine and atropine, and later published their findings.[78] Others observed and documented the psychological impact of war, noting that soldiers sometimes experienced unbearable shock to the "nervous system" that caused "raving and violent" mania.[79] In these studies, it is unclear if the physicians provided notice or obtained consent from the patients. Medical research expanded as the war progressed, and injured soldiers afforded physicians a patient base to test and hone new treatments.

Apart from medical experiments, serious complications sometimes arose with established treatments. For example, although smallpox vaccination was mandated by both the Union and Confederacy, reports of "spurious" vaccinations—where the vaccine would cause death or disability, introduce other diseases, such as syphilis, or fail to confer immunity—were commonplace.[80] In light of the reports, many soldiers refused to be vaccinated. The resistance troubled officials and physicians, who feared a smallpox outbreak. One article, published in the *Nashville Journal of Medicine and Surgery*, noted: "So common had accidents become after vaccination, and so strong was the prejudice growing, both in the army and amongst citizens against its employment."[81] As with smallpox variolation during the Revolutionary War, individuals opposed mandatory vaccinations not only because of the potential adverse health effects, but also for religious reasons and under principles of autonomy and individual liberty.[82]

Notwithstanding the preventive health and clinical innovations, as with wars of the past, during the Civil War, disease caused more deaths than combat. Of the 3 million soldiers who fought during the war, more than 600,000 died, 67 percent from disease. The Union Army registered over six million cases of disease, and the wartime disease mortality rate was more than five times higher than during peacetime. When compared to previous wars, the percent drop in deaths from disease (which was 90 percent during the Revolutionary War and 75 percent during the War of 1812) was not due to improved medical care but rather reflected how deadly combat had become because of new weapons such as shrapnel shells and the "repeating" rifle, which could fire seven shots in thirty seconds.[83] Each side treated hundreds of thousands of gunshot wounds, and 500,000 soldiers were permanently injured.[84] These figures most likely underestimate the actual numbers: many records were destroyed, and soldiers often did not report their ailments to avoid interactions with physicians, who were still generally deemed to be incompetent. It likewise was typical for the sick or injured to go home rather than report to a military physician.[85]

Looking beyond wartime needs, Surgeon General Hammond set forth a vision for the future of military medicine, including a permanent ambulance corps, an Army Medical School, an Army Medical Museum, a central laboratory, and permanent military hospitals throughout the country. The recommendations were well-received, but it took forty years to implement the proposals.[86]

Despite Hammond's transformative impact, at the behest of Secretary of War Stanton—who had opposed Hammond's initial appointment—Hammond was court-martialed for improper requisition of supplies and conduct unbecoming an officer. Many at the time viewed the court-martial as a politically motivated vendetta. Although Hammond vigorously refuted the charges, he was dismissed from the Army in August 1864.[87] Hammond continued to fight for his reputation, and a Senate committee report from 1879 found that Hammond's dismissal was based on "disproved" and "trifling, if not frivolous" charges.[88]

After the war, Congress mandated that the government provide veterans with medical care and disability benefits, but did not allocate sufficient resources to meet the demand. By 1870, the backlog of veterans waiting for treatment was three thousand, rising to nine thousand the following year.[89] Compounding the concerns was rampant abuse of alcohol, morphine, and opiates. While these drugs were frequently administered during the war to alleviate pain and psychological ailments—which were sometimes referred to as "acute nostalgia" or "nervousness"—they often led to addiction.[90] After the war, hundreds of thousands of veterans were addicted to opium, and morphine addiction became known as "army disease."[91] To help address health and societal concerns, the government created a system of soldiers' homes to assist war veterans.[92]

The large number of benefits claims, many of which were fraudulent, overwhelmed the War Department. The costs were astronomical. In 1880 war pensions accounted for 21 percent of the federal budget, rising to over 41 percent by 1893. By 1898, approximately 750,000 veterans were receiving a pension. Former Confederate states provided benefits for their veterans, many of whom believed that accepting federal pensions would undermine their pride and sense of honor. In 1958, Congress afforded Confederate veterans and their families with benefits that matched those of their Union counterparts. The last Civil War veteran died in 1959, and the last person to receive a Civil War pension died in 2020.[93]

Summary

In the three decades following the Revolutionary War, military medicine floundered amid underfunding and substantial variation in quality due to reliance on local militia to provide healthcare and enact preventive health measures. The establishment of a permanent military medical department in 1818 ushered in some organizational improvements. However, the quality of care was still lacking owing to meager standards for entry into the medical profession and the absence of regulatory oversight for apothecaries and medicinal products. Thereafter, the military took several steps that placed it at the vanguard of medical care and research: the armed forces set educational standards for physician entry, established reporting systems to collect health data and track health outcomes, and collaborated with universities to conduct medical research.

Historians often remark that the Civil War marked the beginning of the modern era of military medicine. In large part, this characterization is due to significant innovations in battlefield triage and preventive health, organizational improvements in health data collection, and a growing consensus that a hospital was a place where one could regain health rather than where one goes to die. Wartime innovations also furthered a cultural shift within the field of medicine that increasingly placed more value on the academic study of disease and injury. America was a fractured and battered nation by the conclusion of the Civil War, but the fields of medicine and research were on the verge of a magnificent expansion.

Notes

1. Brown, *The Medical Department*, 70–72; Greenwood and Berry, *Medics*, 5.
2. George Washington, *Sentiments on a Peace Establishment* (May 1, 1783).
3. George Washington, *Farewell Address* (September 17, 1796).
4. Gentile et al., *The Evolution*, 9–11.
5. Quoted in Richard H. Shryock, *The Development of Modern Medicine* (1936), 304.
6. Greenwood and Berry, *Medics*, 5.
7. Rostker, *Providing*, 62–66.
8. Gentile et al., *The Evolution*, 9; Monica Duffy Toft and Sidita Kushi, *Dying by the Sword* (2023), 40–42.
9. Toft and Kushi, *Dying*, 40–42.
10. Donald R. Hickey, "The Quasi War: America's First Limited War, 1798-1801," *The Northern Mariner/Le Marin du Nord* 18, no. 3-4 (July–October 2008): 67–77.
11. Greenwood and Berry, *Medics*, 5, 11.
12. Millett et al., *For the Common Defense*, 89–95.
13. Greenwood and Berry, *Medics*, 5.
14. "Barbary Wars, 1801-1805 and 1815-1816," Milestones in the History of U.S. Foreign Relations, U.S. Department of State, Office of the Historian (accessed March 5, 2024).
15. Greenwood and Berry, *Medics*, 11–13.
16. J. C. A. Stagg, *The War of 1812* (2012), 28.
17. Mary C. Gillett, *The Army Medical Department, 1775-1818* (1981), 148; Millett et al., *For the Common Defense*, 95–107.
18. Stanhope Bayne-Jones, *The Evolution of Preventive Medicine in the United States Army, 1607-1939* (1968), 77–78; Greenwood and Berry, *Medics*, 6–7.
19. Bayne-Jones, *The Evolution*, 48, 77–78.
20. Mary C. Gillett, *The Army Medical Department, 1818-1865* (1987), 34.
21. Bayne-Jones, *The Evolution*, 75 76.
22. C. F. Nichols, *Vaccination* (1902), 6–10.
23. Henry B. Shafer, *The American Medical Profession, 1783 to 1850* (1936), 110–11.
24. Stefan Riedel, "Edward Jenner and the History of Smallpox and Vaccination," *BUMC Proceedings* 18 (2005): 21–25.
25. Spencer C. Tucker, ed., *The Encyclopedia of the War of 1812: A Political, Social, and Military History* (Santa Barbara, CA: ABC-CLIO, 2012); Packard, *History of Medicine*, vol. 1, 628.
26. Brown, *The Medical Department*, 84; Greenwood and Berry, *Medics*, 8.
27. Gentile et al., *The Evolution*, 33–36.
28. An Act Regulating the Staff of the Army, ch. 61, 3 Stat. 426 (April 14, 1818).

29. Adjutant and Inspector General's Office, *Regulations for the Medical Department* (1818).
30. Duffy, *The Healers*, 172–77; Shryock, *The Development*, 251; Rothstein, *American Physicians*, 85–100; Donald E. Konold, *A History of American Medical Ethics, 1847-1912* (1962), 14–31.
31. Duffy, *The Healers*, 168–69; Norman L. Cantor, *After We Die: The Life and Times of the Human Cadaver* (Washington, DC: Georgetown University Press, 2010), 249–50.
32. Konold, *A History*, 1–7; Rothstein, *American Physicians*, 49–52.
33. Harriet A. Washington, *Medical Apartheid* (2006), 28.
34. Duffy, *The Healers*, 30–41, 109–28.
35. Konold, *A History*, 14–31.
36. Shryock, *Medicine and Society*, 144–45; Duffy, *The Healers*, 109–28; Rothstein, *American Physicians*, 63–65.
37. Rothstein, *American Physicians*, 76–79, 114–21.
38. Packard, *History of Medicine*, vol. 1, 622–23.
39. Gillett, *The Army Medical Department, 1818-1865*, 78.
40. Brown, *The Medical Department*, 210–14; Bayne-Jones, *The Evolution*, 82–84.
41. Shafer, *The American Medical Profession*, 182.
42. Ronald L. Numbers, "William Beaumont and the Ethics of Human Experimentation," *Journal of the History of Biology* 12, no. 1 (Spring 1979): 113–35.
43. Shryock, *The Development*, 177; Washington, *Medical Apartheid*, 29–30, 57.
44. Millett et al., *For the Common Defense*, 108–41.
45. Ashburn, *A History*, 55–59; Greenwood and Berry, *Medics*, 9–10.
46. Brown, *The Medical Department*, 186–89; Greenwood and Berry, *Medics*, 9–10; Richard V. N. Ginn, *The History of the U.S. Army Medical Service Corps* (1997), 8.
47. Report of R. S. Satterlee, Senior Surgeon, 1st Division, U.S. Army (July 5, 1847), reprinted in Brown, *The Medical Department*, 186–88.
48. Brown, *The Medical Department*, 186–89; Greenwood and Berry, *Medics*, 9–10; Ashburn, *A History*, 55–59.
49. An Act to Raise for a Limited Time an Additional Military Force, and for Other Purposes, 29th Cong., Chap. VIII (February 11, 1847).
50. Greenwood and Berry, *Medics*, 9–10.
51. Duffy, *The Healers*, 206–27.
52. Ashburn, *A History*, 68.
53. Packard, *History of Medicine*, vol. 1, 640–41; Duffy, *The Healers*, 206–27.
54. William Q. Maxwell, *Lincoln's Fifth Wheel* (1956), 4–8.
55. Rostker, *Providing*, 77-84.
56. Charles J. Stille, *History of the United States Sanitary Commission, Being the General Report of Its Work During the War of the Rebellion* (1866), 509.
57. An Act to Incorporate the National Academy of Sciences, 37th Cong., 12 Stat. 806 (March 3, 1863).
58. John T. Greenwood, "Hammond and Letterman: A Tale of Two Men Who Changed Army Medicine," *Landpower Essay* (June 2003): 1–6.
59. Ashburn, *A History*, 74–86; Greenwood, "Hammond and Letterman."
60. Shauna Devine, "Examined at the University of Pennsylvania: Dr. Fulton, His Professional Milieu, and Military Medicine 1862-1864," in *Civil War Medicine*, edited by Robert D. Hicks (2019), 265–69; Mark D. Miller, "William A. Hammond: Restoring the Reputation of a Surgeon General," *Military Medicine* 152, no. 9 (September 1987): 452–57.
61. Devine, "Examined," 265–69.
62. Hicks, *Civil War Medicine*, 25–26.
63. Greenwood, "Hammond and Letterman."

64. Ashburn, *A History*, 74–86; Rostker, *Providing*, 88–89.
65. *The Medical and Surgical History of the War of the Rebellion*, 2 vols. in 6 parts (1870–1888).
66. Rostker, *Providing*, 89.
67. Packard, *History of Medicine*, vol. 2, 712–15.
68. Hicks, *Civil War Medicine*, 35.
69. Jonathan Letterman, *Medical Recollections of the Army of the Potomac* (1866); Greenwood, "Hammond and Letterman"; Garrison, *Notes*, 174–76.
70. Rostker, *Providing*, 84–88, 252; Gillett, *The Army Medical Department, 1818-1865*, 227–28.
71. P. H. Watson, Assistant Secretary of War, to William A. Hammond, Surgeon General (August 29, 1862), reprinted in *The Medical and Surgical History of the War of the Rebellion*, vol. II, part III (1883), 933–34.
72. Gillett, *The Army Medical Department, 1818-1865*, 191–93.
73. Ginn, *The History*, 15–16.
74. Greenwood and Berry, *Medics*, 29–30.
75. Quoted in Stille, *History*, 95.
76. Hicks, *Civil War Medicine*, 28.
77. Gillett, *The Army Medical Department, 1818-1865*, 283–87; Rostker, *Providing*, 93.
78. Gillett, *The Army Medical Department, 1818-1865*, 283–87.
79. Thomas T. Ellis, *Leaves from the Diary of an Army Surgeon* (1863), 71, quoted in Hicks, *Civil War Medicine*, 35.
80. Joseph Jones, "Spurious Vaccination," *Nashville Journal of Medicine and Surgery* (1867): 3–134.
81. Jones, "Spurious Vaccination," 4.
82. Jones, "Spurious Vaccination," 8–9; Nichols, *Vaccination*, 4–10, 35–45
83. Rostker, *Providing*, 91; Duffy, *The Healers*, 206–7.
84. Vincent J. Cirillo, *Bullets and Bacilli* (2004), 24; Julius Bonello, "Civil War Medicine," *The Surgical Technologist* (January 2000): 10–17.
85. Duffy, *The Healers*, 206–7.
86. Ashburn, *A History*, 74–75.
87. Miller, "William A. Hammond."
88. Quoted in Robert S. Henry, *The Armed Forces Institute of Pathology* (1964), 30.
89. Mary C. Gillett, *The Army Medical Department, 1865-1917* (1995), 21–25.
90. Gillett, *The Army Medical Department, 1865-1917*, 4, 48–49; Gillett, *The Army Medical Department, 1818-1865*, 7; John M. Kinder, *Paying with Their Bodies* (2015), 23–24.
91. John Frederick Logan, "The Age of Intoxication," *Yale French Studies* 50 (1974): 89.
92. Rostker, *Providing*, 99–106.
93. Gillett, *The Army Medical Department, 1865-1917*, 21–25; Rostker, *Providing*, 99–109; Alex Fox, "The Last Person to Receive a Civil War Pension Dies at Age 90," *Smithsonian Magazine*, June 8, 2020.

3
A Postbellum Renaissance in Military Medicine and Research

Introduction

Throughout the nineteenth century, physicians endured persistent scrutiny for their inability to treat disease effectively. During hearings in the 1840s to consider medical practice regulations, a state senator from New York lamented, "The people of this state have been bled long enough in their bodies and pockets."[1] After the Civil War, the president of Harvard University, Charles Eliot, remarked: "It is fearful to think of the ignorance and incompetence of most American doctors who have graduated at American schools. They poison, maim, and do men to death in various ways, and are unable to save life or preserve health."[2] These sentiments led individuals to eschew medical care and disregard preventive health measures.[3] As a Georgia lawmaker explained, public health measures were primarily viewed as just another "trick of the doctors."[4] These were not merely wartime concerns.

The reputation of the medical profession changed substantially during the late nineteenth and early twentieth centuries. The public marveled as scientists uncovered the role of viruses and bacteria in disease. Research flourished, and medical advancements helped nurture a healthier and more prosperous society. The military played a central role in these endeavors—forming research teams, collaborating with universities, and publishing volumes of influential studies. During these developments, some physicians raised ethical concerns regarding human experimentation and called for the adoption of research ethics guidelines. Lawmakers and the American Medical Association rebuffed the movement, and ethical analysis was left to the discretion of individual scientists. At a societal level, ethical transgressions were tolerable so long as research harms fell on the few and research benefits flowed to the many.

Contemporaneous with the scientific advancements, a resource-strapped military struggled to accommodate the healthcare needs of military personnel. The plight of service members during the Spanish-American War highlighted the deficiencies and led to public hearings and significant institutional reforms in military medical affairs. Coupled with the structural reforms, military medical research expanded during the early twentieth century, and military researchers conducted cutting-edge work on infectious diseases of particular import to the armed forces.

A Scientific Revolution

By the end of the nineteenth century, Europe was in the midst of a scientific revolution. Charles Darwin's groundbreaking work on evolution and human origins shook the foundation of intellectual inquiry and inspired copious examination of natural phenomena.

America's Military Biomedical Complex. Efthimios Parasidis, Oxford University Press.
 DOI: 10.1093/9780199351473.003.0004

Many new or improved instruments—including the microscope, clinical thermometer, stethoscope, X-ray, and hypodermic needle—facilitated innovative studies. Pioneering research by scientists such as Louis Pasteur, Ignaz Semmelweis, Joseph Lister, and Dmitri Ivanovsky led to significant insights regarding bacteria and viruses. European researchers discovered the causal agents for malaria, cholera, tuberculosis, typhoid fever, tetanus, gonorrhea, and dysentery. New clinical methods and preventive measures followed, such as sterilizing wounds and surgical equipment, eliminating disease-carrying insects, and implementing protocols to ensure a clean water supply.[5]

In Europe, the field of medicine was evolving at an exponential pace, and a new subfield of bacteriology emerged.[6] Initially, many US physicians questioned the germ theory of disease, characterizing the concept as a "medical fashion" or "bacteriomania."[7] American physicians maintained their traditional methods and were slow to reject the long-held miasmatic theory, positing that bad air (miasma) from rotting organic matter was the root cause of disease. Medical sects continued to jostle for patients and prestige, medical schools retained lax standards, and the government and private sector allocated little funding to basic science, which stunted the American research enterprise.[8] Quinine, calomel, opium, and alcohol-based tonics continued to be widely utilized. Hypodermic administration of opium became commonplace, notwithstanding the recognition that it led to addiction. Although physicians often complained of nighttime calls from patients who required additional opium shots, their prescribing habits changed little. By the early twentieth century, the United States imported twenty times as much opium per capita as Germany, Italy, and Austria-Hungary. In an attempt to free themselves from opium overuse, some physicians turned to cocaine for a variety of ailments, including colds.[9]

Despite early skepticism regarding the germ theory of disease, scientific developments that radiated from Europe began transforming medicine in the United States. The public began to recognize that disease was not a divine punishment but rather was rooted in natural causes that could be studied and addressed. In 1893, Johns Hopkins Medical School set a new paradigm for American medical education by incorporating scientific exploration into medical theory, practice, and research. That same year, the Army Medical School was established, a landmark moment in the professionalization of military medicine and research. The Navy Medical School was formed in 1902 and became a leading teaching and research center. By 1903, dozens of medical schools nationwide had followed their lead.[10]

The Army and Navy medical colleges provided budding physicians with a general medical curriculum supplemented with instruction specific to military needs. Many physicians gained expertise in other fields, such as bacteriology, immunology, pharmacology, and botany. Across the country, in addition to their clinical work as healthcare providers, military physicians played an instrumental role in medical research and the development of preventive health measures.[11]

Scientific Progress and Human Experimentation

Proving that bacteria and viruses caused disease required experiments with humans. As bioethicist Susan Lederer details in her groundbreaking book *Subjected to Science*, the "use of human beings to confirm that a microbe caused a particular disease or to demonstrate the mode of transmission was a harsh legacy of the germ theory of disease."[12] Lederer documented scores of studies where scientists intentionally infected research subjects with gonorrhea, syphilis, or other pathogens. Some researchers experimented on disabled children, while others utilized infants, dying patients, or adults with mental illness.

In one study conducted in 1895, renowned physicians George Sternberg and Walter Reed tested smallpox vaccine immunity at orphanages in Brooklyn. The children were inoculated with experimental vaccines, and the scientists investigated smallpox immunity and vaccine effectiveness over time.[13] Sternberg was the Army Surgeon General and an American pioneer in bacteriology, and Reed was a well-respected bacteriologist and professor at the Army Medical School. In other experiments to test vaccine effectiveness, physicians injected children with smallpox virus after inoculating the children.[14]

The methods were not unusual for the era, as medical research often involved experimentation on impoverished children, particularly those who received free healthcare in public or charitable hospitals.[15] As Lederer explained, Americans generally suspected "that indigent patients would become the unwitting subjects of therapeutic experimentation in life and the objects of dissection after death."[16] These practices were not solely an American phenomenon. Physicians in Vienna, Berlin, Paris, and throughout Europe often experimented on their patients, particularly the poor. As renowned physician John Finney explained in his memoir, a common perspective among German doctors during the late nineteenth and early twentieth centuries was that a "patient was something to work on, interesting experimental material, but little more."[17] A Swedish physician practicing in a charity hospital explained that using children as research subjects was preferable, as typically, it was cheaper than obtaining animals for experimentation.[18]

Scientific experiments on humans were not conducted in an ethical vacuum. In 1865, the eminent French physiologist Claude Bernard published *An Introduction to the Study of Experimental Medicine*, wherein he inquired into the limits of human experimentation, remarking that principles of "medical and surgical morality" dictate that one should never perform "on man an experiment which might be harmful to him to any extent, even though the result might be highly advantageous to science, i.e., to the health of others."[19] Bernard recognized that "Instruction comes only through experience" but stated that experimentation is morally justifiable only if there is a chance of benefit to the subject.[20] Bernard further admonished the common practice of experimenting on criminals and the imprisoned, "granting them pardon in the exchange. Modern ideas of morals condemn such actions," he explained.[21] Given his stature in Europe, Bernard's work was well-known in American medical circles.[22]

Writing in 1886 and building from Bernard's work, renowned Boston physician Charles Francis Withington explained that physicians had an obligation to obtain consent from individuals when research involved discomfort and had no therapeutic benefit.[23] The work of Bernard and Withington helped frame the movement against "human vivisection." Led in part by animal rights groups that had lobbied against using animals in research, human vivisection referred to using humans in nontherapeutic medical experiments.[24] During a meeting of the American Humane Association in 1897, physician Albert Leffingwell lambasted scientists engaged in nontherapeutic medical research, dubbing the work "scientific assassination" that exploited vulnerable patients, many of whom were newly arrived immigrants detained in quarantine wards.[25]

By the early twentieth century, an ethical distinction had been recognized between the administration of novel or experimental therapies that could benefit the patient and those that could not. As to the former, experimentation was largely deemed reasonable as long as the patient had been fully informed and provided consent.[26] In a comprehensive work published in 1914, Leffingwell provided a detailed ethical code to help researchers evaluate whether an experiment was justifiable. Leffingwell highlighted the need to respect patient autonomy, promote human rights, and obtain consent after disclosure of research goals, risks, and benefits.[27]

Notwithstanding the scholarly writings, implementation of ethical protocols was haphazard, and enforcement was sparse.[28] As a general matter, the field of medical ethics was less concerned with moral imperatives and more focused on building the prestige of the medical profession, including a prohibition on "case stealing" and encouraging a custom of not testifying against other physicians in malpractice suits.[29] By the early twentieth century, the field of medical ethics in the United States was not grounded in the writings of Bernard, Withington, and Leffingwell, but rather primarily was framed by Thomas Percival's 1803 treatise, *Medical Ethics*, which comprised a series of maxims related to physician etiquette and trade union rules. Contrary to the book's title, the treatise did not examine moral principles in medical care and research.[30] Percival's trade guidelines heavily influenced the bylaws of medical societies throughout America.[31]

In 1916, the American Medical Association (AMA)—the nation's largest physician group—debated whether to establish ethical guidelines for research involving humans. Six years earlier, the AMA created ethical guidelines for animal experimentation, and medical schools across the country adopted the rules. The AMA refused to amend its code of ethics to include protections for humans. During the debate on the proposal, physicians balked at any intrusion upon their independence and were adamant in their ability to act judiciously and do no harm.[32]

Apart from the AMA discussions, Congress and various state legislatures considered laws to protect human research subjects, but none resulted in an adopted rule.[33] The AMA would not amend its code of ethics to include protections for human subjects until 1946—in large part due to pressure from American prosecutors from the Nuremberg Doctors' Trial, who wanted to point to established ethics principles in their case against German researchers for atrocities committed during World War II. As detailed later in this book, the 1946 AMA guidelines did not have the force of law, and a federal statute outlining legal protections for human subjects would not be enacted until decades after World War II.

Despite a lack of legislation or codified ethical guidelines, American court cases from the early 1900s underscored the physician's obligation to obtain patient consent.[34] In a landmark case from 1914 that still is taught in law schools, jurist Benjamin Cardozo wrote: "Every human being of adult years and sound mind has a right to determine what shall be done with his own body."[35] The sole exception, Cardozo explained, is when a person is unconscious and needs emergency medical care.[36]

Although court cases opened a pathway for legal recourse, nonconsensual research continued. There was not a clear line between medical care and experimentation, particularly if a medical technique had the potential to benefit the individual patient. A code of silence among physicians made prosecution extremely difficult. Furthermore—given the prevalent paternalistic relationship between doctor and patient—physicians generally did not feel the need to explain their actions to patients, and patients did not feel empowered to ask. This confluence of factors nurtured a scientific culture that afforded broad discretion to physician-researchers and resulted in the exploitation of patients, most of whom were socially marginalized due to their race, ethnicity, health condition, or socioeconomic status.

Reorganization of Military Medical Affairs

Notwithstanding significant advances in medicine and research during the late nineteenth century, inadequate funding and mismanagement led to a military medical department ill-prepared to tackle wartime challenges. The US experience during the Spanish-American War (1898) is illustrative. The war commenced amid the Cuba Libre movement, where island revolutionaries sought freedom from Spanish colonial rule. America viewed Cuba as a strategic military hub and maintained significant business interests on the island, including the export of 90 percent of Cuban sugar.[37] On February 15, 1898, a massive explosion rocked the *USS Maine*, which was stationed in Havana Harbor, killing 266 sailors. Despite contradictory reports as to the cause, American media outlets largely blamed the Spanish, and public opinion favored a declaration of war against Spain.[38] War advocates rallied around the cry, "Remember the *Maine*! To hell with Spain!"[39] The war marked a watershed moment for US military policy, as American leaders decided to take a more active role in foreign policy and global affairs.[40]

Congress appropriated $50 million for war funding, $20,000 of which was allotted to the Army Medical Department. This paltry sum—a mere 0.04 percent of the war budget—was unsettling given that the department was already grossly underequipped.[41] As the Army grew from 28,000 to 275,000, Congress did not authorize an increase in medical personnel. The prewar ratio of medical personnel to soldiers was 35:1, rising to 348:1 during the war.[42] Military hospitals were overcrowded and understaffed, and available medical supplies were often misappropriated.[43] According to Theodore Roosevelt, who led the Rough Riders, a volunteer cavalry that fought during the war, "the condition of the wounded in the big field hospitals in the rear was so horrible, from the lack of attendants as well as of medicines, that we kept all the men we possibly could at the front."[44] Figure 3.1 depicts wounded soldiers being loaded onto horse-drawn ambulances during the Spanish-American War.

The war lasted less than twelve weeks, during which the United States suffered 2,308 deaths, less than 400 of which were caused by battle.[45] To provide context, the ratio of deaths from disease and battle, which was 2:1 during the Civil War, ballooned to 6:1 during the Spanish-American War. The ratio was exponentially higher, and it occurred in an age of medical enlightenment. Dreadful sanitary measures

Figure 3.1 Wounded soldiers being loaded onto horse-drawn ambulances during the Spanish-American War.
Photograph from the National Library of Medicine.

exacerbated the spread of disease: in particular, typhoid fever ravaged warfighters, accounting for over 20,000 documented cases and 87 percent of disease-related wartime deaths. Typhoid fever is a debilitating illness that spreads quickly and can leave an individual unfit for duty for two months or more, causing enormous strain on the fighting force.[46]

By 1898, the medical community knew the causal agent of typhoid fever, that the disease is transmitted via contact with infected feces, and how to implement effective preventive measures. Yet, commanding officers and soldiers largely neglected to implement, enforce, and abide by preventive measures such as boiling water, properly disposing of raw sewage, and disinfecting kitchen and hospital supplies. Many commanders viewed sanitary measures as medical fads that had no place in the rugged life of a soldier.[47] One commanding officer explained: "You have got to get camp fevers with camp experience just as much as a child gets teeth."[48] In some instances, commanding officers threatened medical officers with court-martial if they disclosed the extent of typhoid outbreaks in camps.[49]

Despite the dreadful camp conditions, the mortality rate of wounded warfighters was 5 percent, the lowest in American history. In large part, this was due to antiseptic surgical techniques and the diagnostic use of X-rays to pinpoint bullets and bone fractures. Taken together, camps on American soil were far more deadly than Cuban battlefields.[50]

Despite the brevity of the war and the relatively small number of battle-related casualties, the war motivated significant institutional changes.[51] The public accused the War Department of incompetence and neglect, and President William McKinley appointed a commission to "investigate all charges of criminal neglect of the soldiers."[52] Chaired by Major General Grenville Dodge, the commission took 3,800 pages of testimony. It produced an eight-volume report wherein it concluded that the government was unprepared for wartime medical needs and did not provide adequate funding for personnel and resources.[53] The commission also determined that military leaders underestimated the health risks and downplayed the extent of disease-related casualties.[54] Among its recommendations, the report called for an increase of commissioned medical officers, the creation of a reserve medical corps, financial allocations for a robust reserve medical supply, and the streamlining of military health administration.[55]

Thereafter, Surgeon General Sternberg established a Typhoid Board to investigate the causes of wartime failings and create policies to correct them. The board placed primary responsibility on commanding officers—who had a duty to enforce camp sanitation measures to maintain a healthy fighting force—and recommended that preventive medicine be integrated into officer training at military schools. The new courses, which were adopted in 1905 and offered via a new Department of Military Hygiene at West Point, were intended to introduce preventive health concepts and reinforce the commanding officer's duty to cooperate with medical officers and take affirmative steps to protect the health of their subordinates. The Typhoid Board also outlined measures to help ensure proper disposal of raw sewage and sanitary measures in food preparation, mess halls, and camp hospitals.[56]

An organizational revolution in military medicine followed.[57] Field service regulations were codified, and the Army Medical Department regularly updated its policies to align with medical best practices. Between 1901 and 1911, the Nurse Corps, Dental Corps, and Medical Reserve Corps were established, each comprising prominent civilian health professionals.[58] The military created new general hospitals nationwide and increased medical supply stockpiles. Warehouse depots were built in New York, San Francisco, Saint Louis, and Washington, DC, and in foreign locations such as Manila, Philippines. The depots in New York and San Francisco could hold supplies sufficient for 100,000 soldiers for six months. A critical component of the reorganization was an emphasis on centralized control and close coordination between the various military branches.[59]

As military medicine was being refined, so was the entire medical establishment. The AMA played a vital role in the modernization of medical school curricula and worked to eliminate schools viewed as diploma mills. Between 1905 and 1915, the number of medical schools decreased from 160 to 96; those remaining were of a higher quality.[60] The government increased medical and scientific research funding, and wealthy industrialists, including John D. Rockefeller and Andrew Carnegie, created medical research institutes.[61] The US Public Health Service was established in 1912 to investigate "the diseases of man and conditions influencing the propagation and spread thereof."[62] These efforts set the stage for a remarkable expansion of military medicine and research.

Advancing Military Medical Research

Military medical research burgeoned during the early twentieth century. The military built new laboratories, encouraged research in the Army and Navy Medical Schools, and created a series of medical boards to study infectious diseases.[63] Surgeon General Sternberg, a leading bacteriologist, played an influential role in spearheading the developments, which included scores of studies related to malaria, dysentery, dengue fever, plague, and other diseases.[64] Among the studies, the work of the Yellow Fever Commission is particularly noteworthy.

Yellow fever was responsible for more than 100,000 American deaths between 1793 and 1901.[65] This feared pestilence was called "yellow jack" because of the yellow quarantine flags flown on ships where cases were prevalent. There was no known cause or cure for the disease.[66] An epidemic in 1793 devastated Philadelphia, the nation's capital. Approximately 10 percent of the population died, and nearly half the city's inhabitants fled, including President George Washington and the entire national government.[67] The disease also negatively impacted military personnel during the Civil War and the Spanish-American War.[68]

In 1900, Major Walter Reed led a team of researchers to study the transmission of yellow fever. A physician who taught bacteriology at the Army Medical School and worked for the New York Board of Health, Reed previously had collaborated with Sternberg on smallpox vaccine research, including research on children. Reed and his physician colleagues—James Carroll, Jesse Lazear, and Aristides Agramonte—were eager to test the hypothesis of Cuban physician Carlos Juan Finlay, who posited that a specific species of mosquito transmitted yellow fever. Malaria had already been proven to be mosquito-borne, and mosquito control measures helped decrease the spread of the disease. Yet, Finlay had been unable to prove his theory and was ridiculed for his hypothesis.[69]

Reed's yellow fever research required two core components: mosquitos and human participants. The researchers pledged to be research subjects, a common practice in the nineteenth and early twentieth centuries. Reed also requested permission from local authorities to solicit volunteers from Cuba.[70] Newspapers in Havana deplored the use of humans as research subjects, but the Spanish consul in Cuba approved Reed's request, primarily because of Reed's assurances that he would use the utmost care in protecting and compensating volunteers.[71]

Reed drafted contracts for the research participants, written in English and Spanish, and each person provided written consent after being informed about the study's purpose, protocols, and risks. Volunteers were paid $100 in gold for participating in the research and another $100 if they became ill. To provide context, $100 in 1900 equates to approximately $3,670 in 2024, and in 1900, the median annual income in the United States was $437.[72] Initially, Cubans volunteered enthusiastically, mainly because of the generous compensation and since most people found Reed's hypothesis ridiculous.[73] According to local reports, those not selected "almost wept."[74] The enthusiasm ended abruptly once volunteers began to fall sick with yellow fever.[75]

Sixteen of the thirty men who volunteered for the study were American soldiers. According to some accounts, American volunteers refused compensation, indicating that their decision to volunteer was "solely in the interest of humanity and the cause of science."[76] There is also evidence that military leaders were hesitant to set a precedent whereby service members received compensation for participation in military research due to concerns that research would become too costly if compensation became the norm.[77]

Reed was widely criticized at medical conferences, where physicians contended that he was unduly endangering humans.[78] Publicly, he defended the research protocols, but privately, he expressed great concern to Sternberg when a research subject fell ill, stating: "Should he die, I shall regret that I ever undertook this work. The responsibility for the life of a human being weighs upon me very heavily at the present, and I am dreadfully melancholic."[79] These remarks suggest that Reed understood the complex moral calculus involving nontherapeutic medical research.

Twenty-two of the thirty volunteers developed yellow fever—remarkably, no volunteer died from the disease despite an expected death rate of 20–40 percent.[80] During the research, however, Lazear succumbed to yellow fever, apparently due to a bite from a stray mosquito in the lab, possibly due to self-experimentation.[81]

In the end, Reed and his colleagues concluded that yellow fever was mosquito-borne and recommended preventive measures such destroying mosquito breeding grounds. Their work benefited military and civilian populations worldwide. After the implementation of mosquito control measures, cases of yellow fever decreased significantly, and outbreaks became increasingly rare.[82]

The Yellow Fever Commission stands out not only for its research findings but also for its methods—in particular, the consent and compensation components. The impetus for the consent forms is unclear. Research scandals from the 1890s may have played a role: in one case involving the development of a syphilis vaccine, a medical professor in Prussia infected four children and three prostitutes with syphilis without their consent. The public was outraged once newspapers broke the story, and in 1900, the Prussian government instituted consent requirements for human experiments.[83]

Perhaps more relevant were the remarks of William Osler—a leading physician of his day and one of the first professors at Johns Hopkins Medical School. In 1897, Osler characterized nonconsensual yellow fever experiments conducted by Giuseppe Sanarelli, an Italian scientist, as "criminal."[84] He argued that obtaining "full consent" before experimentation or utilization of new therapies was necessary, adding: "We have no right to use patients entrusted to our care for the purpose of experimentation unless direct benefit to the individual is likely to follow."[85] Osler acknowledged that the pursuit of science had led to "regrettable transgressions" but also noted that "once this limit is transgressed, the sacred cord which binds physician and patient snaps instantly."[86]

Although the protocols of the Yellow Fever Commission were commendable, not all research during the era was conducted with volunteers, and sometimes coercive tactics were employed. In the Philippines, US scientists solicited research subjects from American soldiers stationed abroad, promising increased pay and favorable assignments.[87] In one study from 1904 involving an experimental killed-bacteria typhoid vaccine, ten of twelve research subjects contracted typhoid, though none died. Posttrial experiments revealed that the vaccines contained live typhoid bacilli, and the Army blocked publication of the results due to fear of negative publicity.[88]

Between 1908 and 1910, after further research, the Army offered its personnel the ability to be inoculated with a newly developed typhoid vaccine. Over twelve thousand volunteered, and the results were astonishing: the prevalence of typhoid among the inoculated was 0.4 percent, ten times lower than those who were not immunized. In 1911, immunization became compulsory, a factor that contributed greatly to low rates of typhoid fever during World War I.[89]

In 1906, following the Philippine-American War of 1899–1902, American researchers experimented on prisoners in Bilibid prison, where the US military held Filipino revolutionaries. In one experiment, Colonel Richard Strong, who later would become a Harvard professor, induced beriberi disease in Filipino prisoners of war to study whether a diet rich in vitamin B could stave off the disease, which has devastating effects such as paralysis, mental disturbance, and heart failure. At least one prisoner died during the experiment.[90] In a separate study, thirty-four prisoners received a contaminated cholera vaccine;[91] in another, an experimental vaccine comprised of bubonic plague and cholera serum was tested on ten prisoners, each of whom died.[92] Senator Jacob Gallinger—a New Hampshire physician who previously had introduced legislation to establish guidelines for research with human subjects—requested an investigation into the research. During the review, Secretary of War William Howard Taft indicated that no prisoner participated against their will. However, some questioned whether the prisoners could refuse participation due to coercion or fear of negative repercussions.[93]

The military's robust research enterprise brought about transformational changes in public health. State and local governments increasingly created public health boards and instituted public health laws.[94] Construction of the Panama Canal, a decade-long project that commenced in 1904, was made possible by mosquito control measures initiated under the leadership of Colonel William Crawford Gorgas, an Army physician.[95] Years earlier, the French abandoned their efforts to build the canal after more than 20,000 workers died of mosquito-borne illnesses. Between 1881 and 1889, the French lost 33 percent of their workforce annually due to yellow fever, whereas the ratio dropped to 2 percent for the Americans.[96]

Summary

The germ theory of disease galvanized clinical innovations and groundbreaking preventive health measures, advancements that helped propel the reputation of the medical and scientific communities. In the public's eye, not only could scientists help control the spread of disease, but their work also contributed to societal and economic progress by facilitating the development of large infrastructure projects. Society's embrace of public health and infectious disease research largely overlooked ethical concerns regarding human experimentation. Lawmakers and the AMA dismissed calls to formulate research guidelines, indicating that researchers could exercise sound moral judgment. The decision to implement safeguards remained with individual scientists, but only a small number adopted disclosure, consent, or compensation protocols.

Contemporaneously, Congress increased funding for military medical readiness, the War Department adopted significant organizational changes to medical departments, and officer training was restructured to emphasize the importance of preventive health.[97] As military historian Mary Gillett explains, by 1913, "the Medical Department may well have been prepared to meet the demands of a conflict like the one in which it had most recently been involved."[98] However, Gillett adds, the department "was not ready for a conflict on a scale never before encountered in the course of human history."[99]

Notes

1. *New York Medical Society Proceedings, 1844-1846* (1846), 71, quoted in Shafer, *The American Medical Profession*, 210.
2. Quoted in John S. Haller, *American Medicine in Transition, 1840-1910* (1981), 218–19.
3. Gillett, *The Army Medical Department, 1865-1917*, 4.
4. Quoted in Shryock, *The Development*, 222.
5. Gillett, *The Army Medical Department, 1865-1917*, 93–94.
6. Ashburn, *A History*, 148.
7. E. T. Tibbets, *Medical Fashion in the Nineteenth Century; Including a Sketch of Bacteriomania and the Battle of the Bacilli* (1884); Susan E. Lederer, *Subjected to Science* (1997), 3.
8. Duffy, *The Healers*, 104–6, 228–46.
9. Rothstein, *American Physicians*, 188–97; Fitz Hugh Ludlow, "What Shall They Do to Be Saved?," *Harper's Magazine*, August 1867.
10. Rothstein, *American Physicians*, 290–97.
11. Gillett, *The Army Medical Department, 1865-1917*, 318–41; Duffy, *The Healers*, 228–46.
12. Lederer, *Subjected to Science*, 3.
13. George M. Sternberg and Walter Reed, "Report on Immunity Against Vaccination Conferred Upon the Monkey by the Use of the Serum of the Vaccinated Calf and Monkey," in *Transactions of the Association of American Physicians*, vol. X (Philadelphia: Dornan, 1895), 57–69.
14. Lederer, *Subjected to Science*, 4.
15. Lederer, *Subjected to Science*, 4–7.
16. Lederer, *Subjected to Science*, 6.
17. J. M. T. Finney, *A Surgeon's Life* (1940), 127.
18. *Human Vivisection: Foundlings Cheaper than Animals* (Washington, DC: Humane Society, 1901), cited in Lederer, *Subjected to Science*, 51.
19. Claude Bernard, *An Introduction to the Study of Experimental Medicine* (1865), 101.
20. Bernard, *An Introduction*, 101–2.
21. Bernard, *An Introduction*, 101.
22. Lawrence J. Henderson, "Introduction," in Bernard, *An Introduction*, v–xii.
23. Charles Francis Withington, *The Relation of Hospitals to Medical Education* (1886), 16–18.
24. Lederer, *Subjected to Science*, xiv–xv, 27.
25. Albert Leffingwell, "Scientific Assassination," in *Report of the Proceedings of the Twenty-first Annual Convention of the American Humane Association* (Fall River, MA: American Humane Society, 1897), 37–40.
26. Albert Leffingwell, *An Ethical Problem* (1914), 290–91.
27. Leffingwell, *An Ethical Problem*, 322–25.
28. Shafer, *The American Medical Profession*, 220–25; Konold, *A History*, 75.
29. Konold, *A History*, 48–51.
30. Shryock, *The Development*, 260; Rothstein, *American Physicians*, 82–84.
31. Shafer, *The American Medical Profession*, 148–54, 220–21; Konold, *A History*, 9–13.

32. Lederer, *Subjected to Science*, 73–74.
33. Lederer, *Subjected to Science*, 61, 73–75.
34. *See, e.g., Mohr v. Williams*, 95 Minn. 261 (1905); *Pratt v. Davis*, 224 Ill. 300 (1906); *Schloendorff v. Society of New York Hospital*, 211 N.Y. 125 (1914).
35. *Schloendorff*, 211 N.Y. at 129–30.
36. *Schloendorff*, 211 N.Y. at 129–30.
37. Louis A. Pérez, *Cuba: Between Reform and Revolution*, (Oxford: Oxford University Press, 1995), 149; Cirillo, *Bullets and Bacilli*, 6.
38. Cirillo, *Bullets and Bacilli*, 7.
39. "Feb. 15, 1898: U.S. Battleship Maine Explodes in Havana Harbor," *The New York Times*, February 15, 2012.
40. Sean Zeigler, Alexandra Evans, Gian Gentile, and Badreddine Ahtchi, *The Evolution of U.S. Military Policy from the Constitution to the Present*, vol. 2 (2019), 5–24.
41. Ashburn, *A History*, 162–68.
42. Ginn, *The History*, 20.
43. Ashburn, *A History*, 162–68.
44. Theodore Roosevelt, *The Rough Riders* (1899), 179.
45. Enrique Chaves-Carballo, "Clara Maass, Yellow Fever and Human Experimentation," *Military Medicine* 178, no. 5 (May 2013): 557–62.
46. Cirillo, *Bullets and Bacilli*, 29–33, 83.
47. Cirillo, *Bullets and Bacilli*, 57, 83–87.
48. *Report of the Commission Appointed by the President to Investigate the Conduct of the War Department in the War with Spain*, vol. 5 (1899), 1781 (testimony of Major General Joseph C. Breckinridge).
49. Cirillo, *Bullets and Bacilli*, 83–86.
50. Cirillo, *Bullets and Bacilli*, 30, 57, 83–84.
51. Rostker, *Providing*, 246.
52. *Report of the Commission*, vol. 1, 107.
53. *Report of the Commission*, vol. 1, 188–89.
54. Gillett, *The Army Medical Department, 1865-1917*, 193–95; Cirillo, *Bullets and Bacilli*, 100–3; Ashburn, *A History*, 174–76.
55. *Report of the Commission*, vol. 1, 188–89.
56. Cirillo, *Bullets and Bacilli*, 72–79, 131–35.
57. Gillett, *The Army Medical Department, 1865-1917*, 313.
58. Greenwood and Berry, *Medics*, 55–58; Gillett, *The Army Medical Department, 1865-1917*, 318–29; Garrison, *Notes*, 186.
59. Gillett, *The Army Medical Department, 1865-1917*, 313–14, 339–40.
60. Ginn, *The History*, 64.
61. Duffy, *The Healers*, 228–46.
62. An Act to Change the Name of the Public Health and Marine-Hospital Service to the Public Health Service, to Increase the Pay of Officers of Said Service, and for Other Purposes, Pub. L. No. 265, 62nd Cong., 37 Stat. 309 (August 14, 1912).
63. Bayne-Jones, *The Evolution*, 123–46.
64. Gillett, *The Army Medical Department, 1865-1917*, 285–93.
65. Rostker, *Providing*, 117.
66. J. Gordon Frierson, "The Yellow Fever Vaccine: A History," *Yale Journal of Biology and Medicine* 83 (2010): 7–85.
67. Eve Kornfeld, "Crisis in the Capital: The Cultural Significance of Philadelphia's Great Yellow Fever Epidemic," *Pennsylvania History: A Journal of Mid-Atlantic Studies* 51, no. 3 (July 1984): 189–205.

68. K. David Patterson, "Yellow Fever Epidemics and Mortality in the United States, 1693-1905," *Social Science and Medicine* 34, no. 8 (April 1992): 855–65; John Duffy, "Yellow Fever in the Continental United States During the Nineteenth Century," *Bulletin of the New York Academy of Medicine* 44, no. 6 (June 1968): 687–701.
69. Gillett, *The Army Medical Department, 1865-1917*, 191, 239–41; Ashburn, *A History*, 267.
70. John R. Pierce, "'In the Interest of Humanity and the Cause of Science': The Yellow Fever Volunteers," *Military Medicine* 168, no. 11 (November 2003): 857–63.
71. Gillett, *The Army Medical Department, 1865-1917*, 242.
72. CPI Inflation Calculator (accessed February 29, 2024); Gilson Willets, *Workers of the Nation: An Encyclopedia of the Occupations of the American People and a Record of Business, Professional and Industrial Achievement at the Beginning of the Twentieth Century*, vol. 2 (New York: P. F. Collier and Son, 1903), 1048.
73. Pierce, "'In the Interest of Humanity'"; Gillett, *The Army Medical Department, 1865-1917*, 242.
74. Quoted in William B. Bean, "Walter Reed and the Ordeal of Human Experimentation," *Bulletin of the History of Medicine* 51, no. 1 (Spring 1977): 87.
75. Gillett, *The Army Medical Department, 1865-1917*, 242.
76. Quoted in Pierce, "'In the Interest of Humanity.'"
77. Jonathan D. Moreno, *Undue Risk* (1999), 19–20.
78. Chaves-Carballo, "Clara Maass."
79. Walter Reed, "Letter to General Sternberg" (January 31, 1901), quoted in Bean, "Walter Reed," 88.
80. Pierce, "'In the Interest of Humanity.'"
81. Chaves-Carballo, "Clara Maass"; Packard, *History of Medicine*, vol. 2, 1113–15.
82. Bayne-Jones, *The Evolution*, 134–35; Ashburn, *A History*, 265–67.
83. Moreno, *Undue Risk*, 19–21.
84. William Osler, Discussion on George M. Sternberg, "The Bacillus Icteroides (Sanarelli) and Bacillus X (Sternberg)," in *Transactions of the Association of American Physicians*, vol. XIII (Philadelphia, PA: Dornan, 1898), 71.
85. William Osler, "The Evolution of the Idea of Experiment in Medicine," *Transactions of the Congress of American Physicians and Surgeons* (1907): 7.
86. Osler, "The Evolution," 7.
87. Gillett, *The Army Medical Department, 1865-1917*, 285–93.
88. Cirillo, *Bullets and Bacilli*, 123–25.
89. Cirillo, *Bullets and Bacilli*, 123–25.
90. Ashburn, *A History*, 276–77; M. H. Pappworth, *Human Guinea Pigs* (1967), 61.
91. Lederer, *Subjected to Science*, 75–76.
92. Gillett, *The Army Medical Department, 1865-1917*, 293.
93. Lederer, *Subjected to Science*, 75–76.
94. Rothstein, *American Physicians*, 310–11.
95. William Crawford Gorgas, *A Few General Directions with Regard to Destroying Mosquitoes, Particularly the Yellow-Fever Mosquito* (Washington, DC: Government Printing Office, 1904); William Crawford Gorgas, *Sanitary Work on the Isthmus of Panama During the Last Three Years* (New York: William Wood & Company, 1907).
96. Cirillo, *Bullets and Bacilli*, 118–19.
97. Gillett, *The Army Medical Department, 1865-1917*, 324–25.
98. Gillett, *The Army Medical Department, 1865-1917*, 341.
99. Gillett, *The Army Medical Department, 1865-1917*, 341.

4
Military Medicine During World War I

Introduction

When war in Europe erupted in 1914, President Woodrow Wilson pledged neutrality and attempted to broker a truce between the warring nations.[1] Wilson noted that the "people of the United States are drawn from many nations, and chiefly from the nations now at war."[2] He further stated, "Every man who really loves America will act and speak in the true spirit of neutrality, which is the spirit of impartiality and fairness and friendliness to all concerned."[3] Although, for years, the American public predominantly favored isolationism and nonintervention, public opinion tilted toward war against Germany following a series of actions. Key among them were Germany's offer of alliance with Mexico, whereby Germany promised to help Mexico regain land lost during the Mexican-American War, and its relentless submarine attacks, including the sinking of American commercial liners and the British vessel *Lusitania*; 128 American civilians were among the 1,198 killed.[4] In his war message to Congress on April 2, 1917, Wilson explained: "The day has come when America is privileged to spend her blood and her might for the principles that gave her birth and happiness and the peace which she has treasured. God helping her, she can do no other."[5]

War preparations were extensive: the government instituted a draft and established hundreds of agencies and committees to assist in war efforts. The Committee on Medicine, created by the National Defense Act of 1916, spearheaded medical mobilization.[6] The department introduced pre-deployment health screenings, enacted wartime preventive health measures, and established a robust healthcare infrastructure to address combat-related injuries. For the first time in its history, the military dedicated significant resources to assess and treat mental illness. The health and organizational advancements were instrumental in addressing wartime health concerns, and later would become embedded features of the military biomedical complex.

The War to End All Wars

Shortly after World War I (WWI) commenced, H. G. Wells published *The War That Will End War*, a collection of articles wherein he argued that defeating German militarism was the only way to end armed conflicts between nations.[7] The phrase would be shortened to "the war to end war" and later would evolve to "the war to end all wars," but the key sentiment remained constant: military engagement was a moral imperative that would bring about world peace and prosperity.

When WWI began, domestically the United States was in the midst of the Progressive Era, a period during which the government assumed an increased role in regulating monopolistic companies such as banks and railroads, instituted workplace safety and child labor laws, and bolstered consumer protection regulations.

America's Military Biomedical Complex. Efthimios Parasidis, Oxford University Press.
 DOI: 10.1093/9780199351473.003.0005

Progressivism, whose proponents included leaders from both parties—Republicans (Theodore Roosevelt, William Howard Taft) and Democrats (Woodrow Wilson)—was a social and political movement that centered on defeating corruption and minimizing waste and inefficiencies. It also sought to address poverty and exploitation that stemmed from industrialism and urbanization. A central component of Progressivism was revitalizing democracy through election reform and targeting miscreant elected officials and political machines.[8]

America's reluctance to enter WWI was partly due to its isolationist stance, which had widespread public support across the country.[9] Also influential was Wilson's belief that mediation and rational discourse could reconcile the concerns of adversaries. When Wilson's mediation efforts proved fruitless, he saw US military engagement as the only way to bring about peace, a position supported by many Progressivists.[10] The notion that war was a means of achieving peace represented a fundamental historical shift in US military policy and foreign affairs.

While some Progressivists were pacifists, others viewed war as necessary if one side is irrational and armed conflict could contribute to a liberal world order.[11] Progressivists optimistically believed that rational thought, free trade, and technological innovation could address societal and political disputes. They viewed war as a public institution that could be eliminated if nations were economically intertwined and subject to binding resolutions of an international body of dispute resolution. These were the motivations behind the League of Nations, which was created by the Paris Peace Conference that ended WWI.[12]

Despite Wilson's role in creating the League of Nations, the United States never joined the organization, primarily due to antipathy to further entrenchment in European affairs and a desire to avoid joining an institution that might impede American independence.[13] The League of Nations struggled to fulfill its mandate and was reluctant to impose sanctions on nations that violated its principles. Several countries (including Germany, Japan, and Italy) resigned from the group in the early 1930s. The League of Nations dissolved during World War II, and after the war, its assets and properties were transferred to the newly created United Nations.[14]

Another key aspect of Progressivism was utilizing scientific and technological expertise to reform government, education, industry, and society.[15] This extended to using technological innovations to create military advantages.[16] Automobiles and airplanes were integrated into war strategy, and America's preeminent contemporary inventor, Thomas Edison, provided the military with a series of inventions that sought to enhance US warfighting abilities.[17] In addition to the technological advancements, new medical knowledge facilitated pragmatic efforts to structure medical mobilization, build the military's healthcare infrastructure, and address wartime health concerns. WWI also marked the first time that physicians applied their expertise beyond military medical affairs and into projects to develop weapons of mass destruction.

Medical Mobilization and Healthcare Infrastructure

Upon entering WWI, the United States mobilized more than four million service members, half of whom deployed to Europe. The Army Medical Department grew enormously during the war—from under 1,000 personnel in April 1917 to more than 350,000 by November 1918. The number of physicians increased from 444 to 31,530, and record numbers of dentists, nurses, and veterinarians were deployed. About one in four American physicians served in the Army.[18] Universities, industry, and civilian organizations helped implement preventive health measures, staff military hospitals, and manufacture and coordinate medical supplies. The military created a potpourri of divisions to coordinate these efforts. Several medical specialties maintained their own departments, such as Internal Medicine, Orthopedic Surgery, Physical Reconstruction, and Roentgenology. Although, for the most part, the compartmentalization provided organizational benefits to the military medical enterprise, inefficiencies sometimes resulted from duplicative efforts and communication lapses.[19]

Along with administrative restructuring, the battlefield triage system was modernized to incorporate innovations such as motorized vehicles. Under the new system, stretcher-bearers and motorized ambulances removed injured soldiers from the battlefield. Ambulances were stationed about one mile to the rear of the front line, and a constant circulation of vehicles helped transport the wounded from the field to unit hospitals. Hospital trains were used for larger distances and could carry up to 1,500 people, though transit often was hindered by poor conditions and the constant threat of shelling or air attack. Figure 4.1 depicts wounded soldiers arriving by train at a triage station in the French countryside.

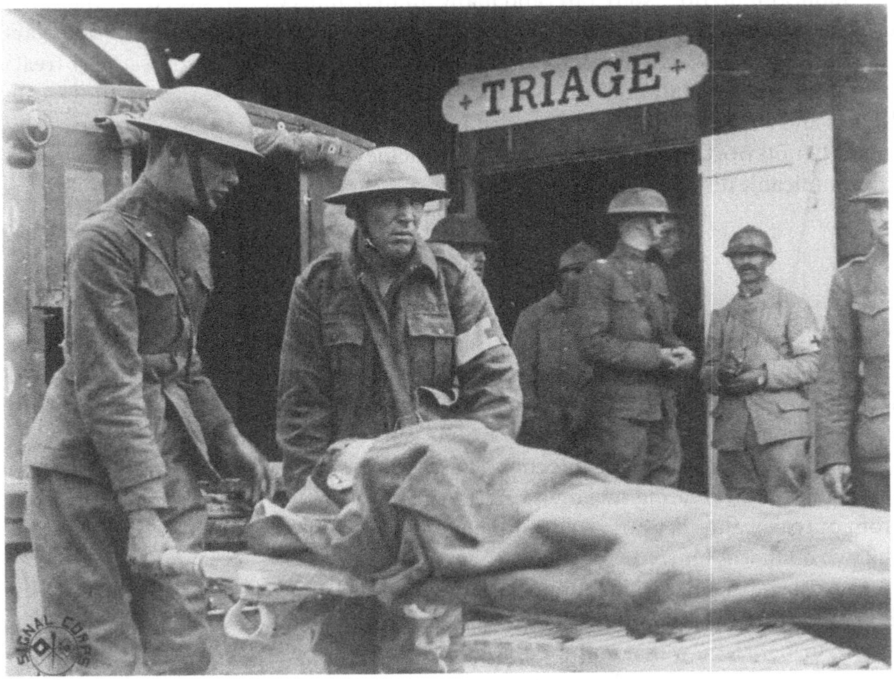

Figure 4.1 Wounded WWI soldiers arriving at a triage station near Suippes, France.
Photograph from the Otis Historical Archives, National Museum of Health and Medicine.

To reduce the time between injury and care, surgical units were stationed close to the front line, and field hospitals triaged the injured and performed emergency operations. Evacuation hospitals, which were much better equipped than field hospitals, were stationed twelve to twenty miles back from the line.[20] Despite a comprehensive healthcare infrastructure, grueling trench warfare overburdened medical personnel, and the extensive use of chemical weapons created horrid conditions.

Notwithstanding the challenges, for the first time in American military history, the number of deaths from disease was less than those from battle. Of the 106,378 deaths during the war, 51,417 were from disease. This was largely due to medical breakthroughs developed in the decades prior to the war, most notably the use of anesthesia and antiseptics. Also important were extensive preventive health measures, such as water purification, sewage treatment, and rodent extermination. The medical department controlled mosquito-borne diseases by draining and oiling marshes near camps, and they deloused service members to help prevent the spread of typhus. In addition, military officials inspected farms to ensure a safe food supply and conducted unannounced inspections to monitor the quality of the food served at hospitals and camps.[21]

As with previous wars, WWI inspired several significant medical and technological innovations. Military physicians developed new and improved methods of reconstructive surgery and wound treatment to deal with the unrelenting trench warfare and the large number of shell wounds. Although bone and joint injuries were common, amputations were rare, and many injured limbs were restored to functional usefulness, often through newly developed modes of physical therapy. The first wartime blood transfusions were accomplished, neurosurgery emerged as a new treatment for head trauma, and X-rays were frequently used to diagnose injuries and locate bullets and shrapnel.[22] Amid the myriad innovations that emerged from the war, particularly noteworthy accomplishments were achieved in the fields of infectious and communicable diseases, and mental health.

Infectious and Communicable Diseases

The military dedicated extensive resources to combat infectious and communicable diseases. In addition to mandating vaccines to protect against smallpox and typhoid fever, the military employed bacteriologists, epidemiologists, and other specialists to monitor disease and institute preventive measures.[23] The work was divided primarily between the Division of Infectious Diseases and Laboratories and the Division of Sanitation. In creating two divisions, the goal was to distinguish medical research from public health measures to be implemented in the field, though the line between the two often was blurred.[24]

The role of the medical laboratory in the wartime health enterprise grew rapidly. By 1917, every hospital was expected to have a lab that could examine specimens of blood, urine, feces, water, and food. Major facilities were also equipped to track disease, test water for bacteria or other contaminants, and make vaccines. Some locations maintained central laboratories that could conduct more elaborate studies, including research with animals.[25]

By the fall of 1918, an influenza pandemic enveloped the globe—nearly five hundred million people (about one-third of the world's population) contracted the virus, which had a mortality rate of 10–20 percent and was dubbed the "greatest medical holocaust in history."[26] The strain of the virus was particularly troubling because it was most lethal in healthy adults. War hastened the spread of the virus and may have caused mutations that increased the virus's lethality. American camps and hospitals were overwhelmed with flu-stricken service members, and nearly half of disease-related deaths among American military personnel—more than twenty-four thousand people—were due to flu or flu-related complications.[27]

The success of smallpox and typhoid fever vaccines brought hope that immunizations could help prevent other diseases, including influenza, meningitis, and pneumonia. However, vaccine development progressed slowly. Even when a new vaccine was created—such as the pneumococcal pneumonia vaccine—its value was limited because it only protected against one of the many strains of pneumonia. In turn, measures such as hygiene, sanitation, and quarantine were more successful preventive health strategies.[28]

Syphilis and gonorrhea also were significant wartime health concerns. At some camps, one in three men had contracted a sexually transmitted infection, and across US forces, more than 1.7 million days of duty were lost due to venereal diseases.[29] Not only were officials concerned about the impact of sexually transmitted infections on the ability of American troops to fight, but they also feared the consequences once service members returned home. For example, gonorrhea leads to sterility and causes blindness in newborns. As military historian Mary Gillett explains, the "possibility that thousands of young men might return from their military service to spread the fruits of immorality to their innocent families and thereby to undermine society produced a nationwide battle against venereal disease . . . unprecedented in magnitude and intensity."[30]

Military and civilian organizations emphasized education and the use of prophylaxis. The Public Health Service embarked on research and a public outreach campaign within the United States, while in the war theater, some commanders precluded service members from entering brothels. Those who contracted a sexually transmitted infection were questioned about their contacts, and failure to use prophylaxis constituted neglect of duty. Medical officers were required to report infected military personnel to the camp commander, who determined whether the person should be reprimanded. The measures were successful, and the rate of sexually transmitted infections dropped 300 percent.[31]

Mental Health

WWI also marked the first time in US history that the armed forces dedicated substantial resources to address wartime mental health concerns. Over the previous 140 years, mental health received scant attention compared to physical battle wounds.[32] During the Revolutionary War, some Continental Army physicians documented an illness they termed "nostalgia," which "caused much distress, and unfitted a large portion of the army for duty."[33] Civil War physicians occasionally observed cases of warfighters suffering from "shock" to their "nervous system."[34] Treatment typically included rest or the administration of opium.[35] By the early twentieth century, psychiatry and psychology were disciplines emerging in importance and public acceptance. During war preparations, a WWI advisory committee determined that the military was not equipped to handle mental health concerns on the scale that would be encountered in wartime. Following the report, the Army created a Neuropsychiatry Division that spearheaded the development of diagnostic procedures and treatments.[36]

Leading American psychiatrists and psychologists, along with the American Psychological Association, assisted in the war efforts.[37] Although psychological testing was a new discipline with questionable scientific merit, the military used the exams to evaluate recruits and assign service duties. Individuals who exhibited signs of mental instability were flagged for additional evaluation. This included alcoholics, drug addicts, psychopaths, and "grotesque liars," as well as individuals with epilepsy, multiple sclerosis, or advanced syphilis. The military also included homosexuals as a mental illness category. Commanders worried that war "neuroses" might result in instances where a soldier's fear, homesickness, or instinct for self-preservation would prevail over military values of "duty, honor, and discipline."[38] By February 1918, no less than 1 percent of recruits were deemed unfit for psychiatric reasons.[39]

The evaluations also served as an intelligence test. At the time, the Army's service requirement equated to the cognitive ability of a ten-year-old, though recruits with lower scores periodically were admitted. The military sometimes provided educational courses to individuals whose difficulties were linked to a lack of formal education.[40] The military also used psychiatric evaluations to scrutinize conscientious objectors, many of whom were described as "strict vegetarians" and "great readers, especially of socialistic literature."[41] Officials worried about releasing conscientious objectors from military duty due to concerns that the men might unduly influence public opinion outside the military, such as becoming the "rankest sort of propagandists."[42] Keeping conscientious objectors in the military, and subjecting them to military law, was the only way to keep them "from becoming a public menace."[43]

Another motivation behind the tests was to reduce government spending on pensions and healthcare for service-related mental health concerns. The military also did not want to invest in training men who might be unfit for duty. Warfighters suffering from what was then deemed "insanity" reached 12:1,000 during WWI, a rate twelve times higher than the civilian population.[44] Later termed "shell shock" because "the conditions were associated with the central nervous system and the shock of exposure to the strain of battle under new conditions of warfare," military officials and mental health experts feared "a new type of casualty which might threaten most seriously the manpower of armies."[45]

Treatment during the war was minimal: many service members briefly rested and were fed, then sent back to the front line or evacuated to the rear if their condition was dire.[46] The high rate of mental illness was primarily caused by grueling trench warfare and relentless gas attacks. Treatment often included drugs, including cocaine and morphine, which were provided to service members to help calm their nerves and boost their performance.[47]

Summary

America maintained a comprehensive healthcare infrastructure during WWI due largely to organizational improvements throughout the armed forces and advancements in medicine and research from the late nineteenth and early twentieth centuries. Innovations in medical care and public health contributed to positive health outcomes and significantly decreased disease-related casualties. During the war, these efforts were particularly noteworthy in the fields of infectious diseases and mental health.

Notwithstanding the salutary achievements, the indelible legacy of WWI lies in a shadowy corner, captured succinctly by Wilfred Owen's well-known poem, *Dulce et Decorum Est.* Owen—an Englishman who died in combat during the war—vividly portrayed the horrors of gas attacks, concluding his poem with an expression of disenchantment with war: "The old Lie: *Dulce et decorum est, Pro patria mori*" (It is sweet and fitting to die for one's country).[48] The advent of gas warfare cast immeasurable harm and fear among combatants, and significantly expanded the scope of military medical affairs.

Notes

1. Woodrow Wilson, *Proclamation on Neutrality* (August 4, 1914); Woodrow Wilson, *Message on Neutrality* (August 19, 1914).
2. Wilson, *Message on Neutrality.*
3. Wilson, *Message on Neutrality.*
4. Sean Zeigler et al., *The Evolution*, 63–66.
5. Woodrow Wilson, *War Message to Congress* (April 2, 1917).
6. Bayne-Jones, *The Evolution*, 147–51.
7. H. G. Wells, *The War That Will End War* (1914).
8. Maureen A. Flanagan, *America Reformed: Progressives and Progressivisms, 1890s-1920s* (New York: Oxford University Press, 2006); William A. Link and Susannah J. Link, eds., *The Gilded Age and Progressive Era: A Documentary Reader* (Oxford: Wiley Blackwell, 2012).
9. Kendrick A. Clements, "Woodrow Wilson and World War I," *Presidential Studies Quarterly* 34, no. 1 (March 2004): 65.
10. Woodrow Wilson, *Address of the President of the United States to the Senate* (January 22, 1917).
11. Joe E. Decker, "The Progressive Era and the World War I Draft," *OAH Magazine of History* 1, no. 3–4 (Winter–Spring 1986): 15–16.
12. John J. Tierney, "For America 'The War to End War' Was Just the Beginning," *Brown Journal of World Affairs* 21, no. 1 (Fall–Winter 2014): 219–29.
13. Henry Cabot Lodge, "Constitution of the League of Nations," 57 Cong. Rec. 4520, 4520-30 (February 28, 1919) (statement of Henry Cabot Lodge and ensuing discussion and debate).
14. Leland M. Goodrich, "From League of Nations to United Nations," *International Organization* 1, no. 1 (February 1947): 3–21.
15. David M. Kennedy, "The Progressive Era," *The Historian* 37, no. 3 (May 1975): 453–68.
16. Barton C. Hacker, "Military Institutions, Weapons, and Social Change: Toward a New History of Military Technology," *Technology and Culture* 35, no. 4 (October 1994): 768–834.
17. Patrick Coffey, *American Arsenal* (2014), 3–5, 12–25.

18. Ginn, *The History*, 37–38.
19. Ashburn, *A History*, 305–6.
20. Greenwood and Berry, *Medics*, 68–75; Ashburn, *A History*, 341–43.
21. Ginn, *The History*, 66.
22. Thomas Helling, *The Great War and the Birth of Modern Medicine* (2022); Ashburn, *A History*, 354.
23. Mary C. Gillett, *The Army Medical Department, 1917-1941* (2009), 145–49.
24. Bayne-Jones, *The Evolution*, 153–54.
25. Gillett, *The Army Medical Department, 1865-1917*, 391–92, 413.
26. Quoted in Joseph Waring, *A History of Medicine in South Carolina: 1900-1970* (Columbia: South Carolina Medical Association, 1971), 33.
27. Gillett, *The Army Medical Department, 1917-1941*, 145; Ginn, *The History*, 38.
28. Gillett, *The Army Medical Department, 1917-1941*, 145–51.
29. Gillett, *The Army Medical Department, 1917-1941*, 145–51; Ashburn, *A History*, 337.
30. Gillett, *The Army Medical Department, 1917-1941*, 33–34.
31. Duffy, *The Healers*, 237–38; Ashburn, *A History*, 336–37; Gillett, *The Army Medical Department, 1917-1941*, 33–34, 149–51.
32. Gillett, *The Army Medical Department, 1917-1941*, 68–69, 134.
33. Quoted in Brown, *The Medical Department*, 14.
34. Thomas T. Ellis, *Leaves from the Diary of an Army Surgeon* (1863), 71, quoted in Hicks, *Civil War Medicine*, 35.
35. Gillett, *The Army Medical Department, 1865-1917*, 4, 48–49; Gillett, *The Army Medical Department, 1818-1865*, 7; Kinder, *Paying*, 23–24.
36. Surgeon General's Office, *The Medical Department of the United States Army in the World War: Neuropsychiatry*, vol. 10 (1929); Gillett, *The Army Medical Department, 1917-1941*, 36, 68–69.
37. Ginn, *The History*, 74–77.
38. Surgeon General's Office, *The Medical Department*, 511–12.
39. Gillett, *The Army Medical Department, 1917-1941*, 37, 68–69, 119–20, 134–37.
40. Gillett, *The Army Medical Department, 1917-1941*, 136–37.
41. George McPherson, "Neuro-Psychiatry in Army Camps," *Boston Medical and Surgical Journal* 181 (November 20, 1919): 607.
42. McPherson, "Neuro-Psychiatry," 607.
43. McPherson, "Neuro-Psychiatry," 607.
44. Gillett, *The Army Medical Department, 1917-1941*, 134–35, 376–77.
45. Surgeon General's Office, *The Medical Department*, 1–2.
46. Ashburn, *A History*, 341.
47. Lukasz Kamienski, *Shooting Up* (2016), 94–103.
48. Wilfred Owen, *Dulce et Decorum Est* (1920).

PART II
1917–1946
Beyond Disease Prevention and Medical Care

5
Chemical Warfare
Expanding the Scope of Military Medical Affairs

Introduction

The deployment of chemical weapons during World War I (WWI) devastated warfighters and altered the path of military medicine and research. Dating back to the Revolutionary War, military medical affairs in the United States focused exclusively on disease prevention and medical care. Gas warfare expanded the scope of military medicine beyond these areas and into the realm of weapons development.

Coupled with gas defense research—which required human experiments to test gas masks, protective clothing, and skin ointments to treat gas injuries—physicians and scientists conducted chemical weapons research to create toxic substances for deployment on the battlefield. Integrating weapons development into the military biomedical complex created clinical and ethical challenges for military physicians and researchers, and opened military personnel to new sources of service-related health risks. This shift accelerated after a ferocious German offensive at Ypres, Belgium.

Ypres Gas Attacks

On April 22, 1915, a sunny spring day with a slight breeze from the north, German soldiers opened the valves of 5,730 cylinders filled with pressurized liquid chlorine that were hidden near Allied troops.[1] A greenish-yellow cloud of chlorine gas formed slowly, becoming a five-foot-high dense cloud that gently moved over French and Algerian soldiers positioned nearby. The troops had no protective equipment and no training in chemical warfare. Within minutes, mayhem ensued. Chlorine gas strips the lining of the bronchial tubes and lungs, creating a fluid that fills the lungs and causes frothing from the mouth. Thousands of Allied soldiers were clutching their throats, gasping for air. The Germans could hardly believe their great success.[2]

In one day, and without suffering any casualties, the Germans blew open a four-mile-wide hole in a position that the Allies had held for months. Within thirty-six hours, the Germans struck again, this time sending a chlorine gas cloud barreling over Canadian forces.[3] Figure 5.1 illustrates the vast area of land enveloped by a gas attack.

A French colonel caught in a German offensive described individuals "running hither and thither like madmen, crying loudly for water, spitting blood, some even rolling on the ground in a desperate struggle to breathe."[4] A Canadian private recounted that entire divisions abandoned the line, throwing away their rifles as they ran, "only to fall writhing on the ground, clutching at their throats, tearing open their shirts in a last struggle for air, and after a while ceasing to struggle and lying still

America's Military Biomedical Complex. Efthimios Parasidis, Oxford University Press.
© Efthimios Parasidis 2025. DOI: 10.1093/9780199351473.003.0006

Figure 5.1 WWI poison gas attack near Flanders, Belgium.
Photograph from the National Archives and Records Administration.

while a greenish foam formed over their mouths and lips."[5] With two gas attacks, the Germans killed five thousand Allied soldiers and wounded ten thousand; more than half of the wounded were sent home due to gas-related disabilities.[6]

Germany's deployment of chlorine gas at Ypres was the first successful use of chemical weapons during WWI. Previously, reports indicated that French troops tossed gas grenades at German soldiers, German soldiers launched gas shells toward British positions, and German forces fired gas shells at the Russians. In each case, however, no ill effects were detected.[7] Germany's Ypres gas attacks initiated a relentless chemical arms race that reoriented the path of military medicine and research across the globe.

Proliferation of Gas Warfare in Europe

Less than two decades prior to the Ypres attacks, Germany and twenty-five other nations pledged under the 1899 Hague Convention "to abstain from the use of projectiles the object of which is the diffusion of asphyxiating or deleterious gases."[8] The agreement represented a multinational attempt to ban the deployment of chemical and biological weapons, though means of enforcement and penalties for violation of the agreement were not specified. In part, the Hague Convention was a response to the carnage of many new war technologies, such as machine guns and mobile heavy artillery. As experienced during the American Civil War (1861–1865), the Crimean War (1853–1856), and the Franco-Prussian War (1870–1871), the new methods of war devastated combatants and civilians and caused environmental and economic catastrophes.[9]

Building on previous doctrines—including the Lieber Code (adopted by the Union Army during the American Civil War to establish rules of land warfare), the 1868 St. Petersburg Declaration (a multinational treaty that renounced the use of certain exploding projectiles), and the 1874 Brussels Declaration (which set forth rules of land warfare)—the delegates at the 1899 Hague Convention sought to curtail limitless war and pass moral judgment on various means of warfare, including chemical and biological weapons. During the second half of the nineteenth century, scientists examined how best to use available chemical agents—such as cyanide, hydrochloric acid, and chlorine—as weapons of war.[10] Although the use of poison gases in war can be traced to ancient India, Greece, and China, among others,[11] advancements in chemical sciences made chemical weapons much more potent. Among their goals, the Hague delegates endeavored to prevent the exploitation of scientific advancements for military uses.

After the Ypres attacks, Germany claimed that it did not violate the Hague Convention because it did not use projectiles but instead released poison gas from cylinders.[12] It was a dubious interpretation that belied the spirit of the agreement. Indeed, a separate provision promulgated at the 1899 convention banned all "poison or poisoned arms."[13] Nevertheless, German media outlets supported the government's specious position, adding that gas warfare "is not only permissible by international law, but is an extraordinarily mild method of war."[14]

Regardless of the legality of Germany's gas warfare attacks, the Allied scientific community assembled its greatest minds to unravel what the Germans were using and how to counter it. Allied nations commenced chemical weapons programs, and within months, Britain and France initiated gas attacks against Germany. A chemical weapons tug-of-war ensured, and Europe marched precipitously toward unlimited chemical warfare.[15] As journalists Robert Harris and Jeremy Paxman explained, "virtually every leading chemist in Britain was at work on some aspect of gas warfare," and "there was no time to worry about ethics."[16] Dozens of British labs experimented with more than 150,000 chemical compounds to advance the nation's gas weapons arsenal. In 1916, Britain's chemical weapons program was centralized in a new facility at Porton Down, which employed more than one thousand scientists and military personnel.[17] Still active today, Porton Down remains a preeminent center for chemical and biological warfare.

Germany maintained a wartime poison gas advantage due to its superior chemical industry, which worked closely with military leaders to assist in war efforts.[18] Leading German companies—such as Bayer and BASF—were intimately involved in producing gas weapons. Bayer employees celebrated their work by painting a giant mural in company headquarters that depicted the manufacture of poison gas and shells being filled with toxic substances.[19] Fritz Haber, a leading German scientist who was director of the Kaiser Wilhelm Institute and helped build Germany's gas arsenal, famously remarked: "During peacetime a scientist belongs to the world, but during wartime he belongs to his country."[20]

Haber's wife, Clara, also a chemist, abhorred her husband's work, deeming it a perversion of the ideals of science. After a party that celebrated Haber and the Ypres attacks, Clara walked into her garden with her husband's pistol and shot herself in the heart, dying instantly. Hours later, Haber boarded a train to supervise a gas attack against Russian troops; he did not return for his wife's funeral.[21] In a wartime manifesto, Haber, Max Planck, and ninety-one other leading German scientists defended their work, refuting claims that gas weapons were inhumane or violative of international law.[22] A few German scientists disagreed, including Albert Einstein, then a professor at Berlin University, who characterized the new weapons as "an axe in the hand of a pathological criminal."[23]

Haber's commitment to chemical warfare was undeterred, though, after the war, he dressed in disguise and fled to Switzerland to avoid prosecution. In 1919, he was awarded the Nobel Prize in Chemistry for an ammonia synthesis process used to create fertilizer.[24] For the Nobel committee, Haber's gas weapons work was not disqualifying. This, perhaps, is not surprising since Alfred Nobel, who established the prize, was the inventor of dynamite and other weapons of war. Haber was proud of his gas warfare work and kept a framed photograph of the first gas cloud drifting toward Ypres.[25] As Haber explained: "In no future war will the military be able to ignore poison gas. It is a higher form of killing."[26]

During World War II, Haber was forced to leave Germany because he was Jewish. However, poison gases that he helped develop—including a lethal cyanide gas dubbed Zyklon B—were used in concentration camp mass killings.[27] The Fritz Haber Institute, a German research center in Berlin established in 1911, is still active today. In a 2011 publication celebrating the institute's centennial, German scientists heralded the institute's "central role in the German development of chemical warfare during World War I."[28]

Meanwhile, as protective gas masks were honed during the war, scientists produced new and more toxic substances that could penetrate the masks. Germany took chemical warfare to new heights with the introduction of mustard gas: a blister agent whose effects—which include painful blisters, chemical burns, respiratory ailments, and temporary blindness—begin hours after exposure. On July 12, 1917, Germans loaded mustard gas into artillery shells and launched them at Allied forces who, once again, were positioned near Ypres.[29]

This time, however, no gas formed after the shells exploded—instead, Allied soldiers observed an oily substance that smelled like garlic or mustard. Hours later, soldiers experienced insufferable eye pain and vomited uncontrollably; painful blisters appeared all over their bodies. Chronic health conditions plagued those who had not endured a slow and agonizing death. Land exposed to mustard gas can remain contaminated for months, and bleach is the only known decontaminate. In one instance, a mustard gas attack during the winter of 1917 poisoned soldiers in the spring once the ground thawed. Due to its uniquely devastating properties, mustard gas was dubbed the "King" of the war gases.[30]

By August 1918, the warring nations were deploying chemical weapons frequently, each justifying its use under principles of military necessity and national security. Chemical warfare also was used for war propaganda, to build animosity against the enemy.[31] Winston Churchill likened his nation's chemical weapons arsenal to a beehive stocked with "the pure essence of slaughter" with "diabolical schemes for killing men on a vast scale by machinery or chemistry."[32] In many respects, it was war without limits.

Throughout WWI, Germany led in most areas of gas warfare—not only were the Germans the first to use chlorine, phosgene, and mustard gas, they were the first to integrate gas attacks into tactical operations.[33] Of the three chemicals, phosgene killed nearly everything it came into contact with, including humans, animals, insects, and vegetation.[34] Overall, thirty different chemicals were used in combat during the war. For their part, American leaders viewed gas weapons as merely one of the many deplorable aspects of war.[35]

Development of America's Chemical Warfare Program

Of the major WWI belligerents, the United States was the sole nation at the 1899 Hague Convention that dissented to the ban on the use of poison gas during war. America did not support limitations on the creation of new weapons of war, nor did it view death by poison gas any worse than other wartime horrors, such as choking on seawater after a submarine attack.[36] Notwithstanding America's refusal to ratify the agreement, initially, the United States demonstrated little interest in gas warfare. This was mainly because the military was already overcommitted and underfunded.[37]

The first American agency to conduct gas research was the Bureau of Mines, which had extensive experience in mine gases. In February 1917, the director of the Bureau of Mines wrote to the chairman of the Military Committee of the National Research Council and offered the bureau's services in developing a gas warfare program.[38] Within weeks the bureau was charged with conducting "investigations into noxious gases, generation, and antidote for same, for war purposes."[39] By May 1917, the bureau was collaborating on gas weapons research with twenty-one universities and three private companies, including Harvard, Princeton, Yale, American University, Ohio State, MIT, Johns Hopkins, and General Electric. The nation's leading chemists were eager to use their expertise to help develop America's chemical warfare program.[40] In August 1917, America's chemical warfare efforts—including defensive and offensive measures—were transferred to the newly formed Division of Gas Defense, which was housed within the Surgeon General's Office. This represented a watershed moment in military medicine since, until then, the Army Medical Department had not been responsible for developing weapons.[41]

The question of whether the medical department should create chemical weapons was hotly contested from the beginning. British experts sought to exclude medical officers from participating in gas weapons development since it was viewed as a combatant responsibility.[42] Several top American officials shared this view, and the Division of Gas Defense was removed from the medical department less than one year after its inception. In June 1918, all responsibilities were transferred to the newly formed Chemical Warfare Service, even though the Army Medical Department contested the move because it did not want to lose control of this important and well-funded division.[43]

Thereafter, the Chemical Warfare Service established a medical division staffed with physicians and researchers. Although medical personnel still were involved in chemical weapons development, they were not under the umbrella of the Army Medical Department. They thus were not directly responsible for providing medical care to military personnel. As the work of the Chemical Warfare Service and Army Medical Department evolved, however, the boundary between weapons research and the practice of medicine proved difficult to maintain.[44] Nevertheless, the efforts underscored an important inquiry that remains today: What are the practical and normative boundaries between health professionals involved in weapons development and those providing medical care?

The American chemical warfare program was headquartered at Edgewood Arsenal, Maryland, a $40,000,000 complex with more than five hundred buildings, thirty-six miles of railroad tracks, and its own power plant. At Edgewood, a staff of 1,900 examined thousands of substances for potential use as chemical weapons. Equipped to produce 200,000 chemical bombs and shells per day, Edgewood was a massive military-scientific endeavor unparalleled in size and scale.[45] During WWI, the United States produced more than ten thousand tons of chemical weapons, much of which was sold to Britain and France.[46]

The Chemical Warfare Service collaborated with leading universities, including Harvard, Yale, the University of Chicago, the University of Wisconsin, the University of Michigan, and MIT.[47] The work was important, challenging, and dangerous, and sometimes caused residual harm to unsuspecting civilians. According to George Temple, an Army engineer who participated in chemical warfare studies at American University: "At the end of the day the camp personnel, their clothes impregnated by gas, would pile into trolleys. As the trolley cars neared the downtown area, civilians began to board them. Soon they were all sneezing or crying, depending upon the type of gas the soldiers had been working with."[48]

An accidental explosion at a Washington, DC, chemical weapons facility gassed a residential neighborhood, including a former US senator and his wife, who were sitting on their porch and were impacted by the chemical mist.[49] As Temple surmised, "More men were killed by gas on the experimental side than in actual use."[50] Decontamination efforts on the American University campus continued into the 1990s and early 2000s; during this period there were several instances of construction workers who suffered severe skin and eye injuries after digging up rusted bombs and old lab equipment that contained chemical elements used during WWI research.[51]

Notwithstanding substantial investments in chemical warfare before entering WWI, the United States was relatively unprepared to mount or defend against gas attacks once it entered the war. Although American gas medical officers stationed in Europe underwent a four-day training program held at the *École de Gaz* at the University of Paris School of Pharmacy, most American service members received little training in gas warfare. Moreover, the first twenty-five thousand masks that the Americans shipped to Europe were defective and could be penetrated by gas in less than one minute. Frequently, American service members purchased—at their own expense—high-quality French gas masks. American-made masks that provided adequate protection, as depicted in Figure 5.2, were not available until the fall of 1918. Meanwhile, given the unique concerns raised by gas casualties and the precautions

Figure 5.2 American soldiers during WWI, equipped with the latest gas masks.
Photograph from the National Archives and Records Administration.

needed to treat gas victims, specialized gas hospitals were established in the war theater.[52]

American officials understood that a robust defense strategy requires offensive capabilities. In other words, to effectively combat actual or anticipated chemical warfare, the armed forces must have the ability to deploy chemical weapons. Even if the weapons were not used, the ability to use them might disincentivize a gas attack from an adversary. Moreover, gas defenses could not be validated unless they were tested against the chemicals they sought to protect against. Taken together, as a practical matter, the United States had to create, test, and maintain chemical weapons to defend against chemical attacks.

During the war, America's gas scientists—which included a research team at Harvard—achieved a major breakthrough with the creation of lewisite. A blister agent more lethal than mustard gas, lewisite causes immediate and intense eye pain, skin irritation, nausea, vomiting, tightness of breath, and chest pain. At a secret facility located near Cleveland, Ohio, hundreds of scientists and staff worked to produce lewisite bombs, and the employees were threatened with court-martial if they revealed any details of the chemical weapons work.

Lewisite bombs were tested in fields where goats and dogs were tied to stakes; researchers watched in awe as the animals struggled to handle the violent effects of the chemical.[53] One of the lead lewisite researchers, Harvard scientist James Conant, wrote: "To me, the development of new and more effective gases seemed no more immoral than the manufacture of explosives and guns."[54] The first batch of lewisite was en route to Europe when the armistice was signed in November 1918. Rather than return the noxious chemicals to America, the ship was sunk at sea.[55]

The Aftermath

Chemical weapons were responsible for more than 1.3 million injuries and 90,000 deaths during WWI, though some experts posit the actual numbers were significantly higher. Russia alone had 500,000 casualties from chemical warfare. Of British service members, chemical weapons wounded more than 180,000 and killed more than 6,000. France and Germany also suffered over 200,000 gas casualties each. Among the German casualties was a twenty-nine-year-old corporal, Adolf Hitler, who was temporarily blinded after a mustard gas attack near Ypres.[56]

American forces were particularly devastated by the use of chemical weapons; more than seventy thousand service members were admitted to a military hospital for a gas injury, amounting to more than 31 percent of all US battle injuries and 9 percent of deaths.[57] Mustard gas was responsible for two in five of these gas casualties. The American casualties were demoralizing because the United States was in a position of advantage since the nation entered the war after gas weapons had been introduced and some defense measures had already been created.[58] In large part, American casualties were due to inadequate preparations for gas warfare.

The casualty figures—though shocking—do not capture the colossal psychological impact of gas warfare. Fear of gas attacks drained the strength and willpower of warfighters. According to one eyewitness account, "Gas shock was as frequent as shell shock."[59] In addition, countless service members experienced "gas fright," whereby a person would develop gas poisoning symptoms even if they were not gassed.[60]

Summary

Innovations utilized during WWI included machine guns, barbed wire, chemical weapons, tanks, fighter planes, communication with telephone and wireless devices, and motorized ambulances, cars, and trucks.[61] Of these, General John Pershing, commander of American forces during WWI, highlighted gas warfare as the technological advancement that caused him deep concern.[62] Pershing's remarks were prescient, as a worldwide proliferation of chemical weapons marked the decades following WWI. Equally disruptive was the paradigm shift instilled by gas warfare, which expanded the scope of military medical affairs beyond disease prevention and medical care to encompass weapons development. A new military biomedical complex was emerging.

Notes

1. Charles E. Heller, *Chemical Warfare in World War I* (1984), 7–10.
2. Robert Harris and Jeremy Paxman, *A Higher Form of Killing* (2002), 3–5.
3. Harris and Paxman, *A Higher Form of Killing*, 3–5.
4. Jean Jules Henri Mordacq, *Le drame de l'Yser: La surprise de gaz, Avril 1915* (Paris: Editions des Portiques, 1933), 62–63, quoted in Diana Preston, *A Higher Form of Killing* (2015), 102.
5. Quoted in James McWilliams and James Steel, *Gas!: The Battle for Ypres, 1915* (St. Catharines, Canada: Vanwell Publishing, 1985), 106.

6. Harris and Paxman, *A Higher Form of Killing*, 5–6.
7. Frederic J. Brown, *Chemical Warfare* (1968), 6–7; Thomas J. Craughwell, *The War Scientists* (2011), 177–78.
8. Declaration on the Use of Projectiles the Object of Which is the Diffusion of Asphyxiating or Deleterious Gases, The Hague (July 29, 1899); Seymour M. Hersh, *Chemical and Biological Warfare* (1968), 4–5.
9. Jeanne Guillemin, *Hidden Atrocities* (2017), 1–3.
10. Guillemin, *Hidden Atrocities*, 1–3.
11. Adrienne Mayor, *Greek Fire, Poison Arrows, and Scorpion Bombs* (2022).
12. "German Defense of Gas," *Times*, June 28, 1915, 10, cited in SIPRI, vol. 1., 232–33.
13. Hague Convention with Respect to the Laws and Customs of War on Land, July 29, 1899, art. 23.
14. "German Defense of Gas," *Times*, June 28, 1915, 10, quoted in SIPRI, vol. 1, 232–33.
15. Ulf Schmidt, *Secret Science* (2015), 27–28.
16. Harris and Paxman, *A Higher Form of Killing*, 23.
17. Harris and Paxman, *A Higher Form of Killing*, 23.
18. Brown, *Chemical Warfare*, 6.
19. Preston, *A Higher Form of Killing*, 83–86.
20. Quoted in Preston, *A Higher Form of Killing*, 1.
21. Preston, *A Higher Form of Killing*, 82, 87, 112–13.
22. Professors of Germany, "To the Civilized World," *The North American Review* 210, no. 765 (August 1919): 284–87.
23. Albert Einstein, Letter to Henrich Zangger (December 1917), quoted in Albrecht Folsing, *Albert Einstein: A Biography*, trans. Ewald Osers (New York: Penguin Books, 1998), 399–402.
24. Harris and Paxman, *A Higher Form of Killing*, 34.
25. Craughwell, *The War Scientists*, 9, 129–39.
26. Quoted in Preston, *A Higher Form of Killing*, 1.
27. Craughwell, *The War Scientists*, 186–87.
28. Bretislav Friedrich, Dieter Hoffmann, and Jeremiah James, "One Hundred Years of the Fritz Haber Institute," *Angewandte Chemie* 50, no. 43 (October 17, 2011): 10022–49.
29. Harris and Paxman, *A Higher Form of Killing*, 26–27.
30. Heller, *Chemical Warfare*, 14; Harris and Paxman, *A Higher Form of Killing*, 26–29.
31. Brown, *Chemical Warfare*, 12–17.
32. Winston Churchill, *The World Crisis: 1916-1918*, vol. II (New York: Charles Scribner's Sons, 1927), 208.
33. Heller, *Chemical Warfare*, 22.
34. Harris and Paxman, *A Higher Form of Killing*, 21.
35. Brown, *Chemical Warfare*, 17, 38.
36. Heller, *Chemical Warfare*, 3–4.
37. Brown, *Chemical Warfare*, 22–23.
38. Van H. Manning, *War Gas Investigations* (Washington, DC: Government Printing Office, 1919), cited in Leo P. Brophy and George B. Fisher, *The Chemical Warfare Service* (1958), 3–4.
39. NRC Record of Meeting (April 3, 1917), quoted in Brophy and Fisher, *The Chemical Warfare Service*, 4.
40. Thomas I. Faith, *Behind the Gas Mask* (2014), 14–15.
41. Gillett, *The Army Medical Department, 1917-1941*, 69–73.
42. Gillett, *The Army Medical Department, 1917-1941*, 69–73.
43. Ginn, *The History*, 78.

44. Gillett, *The Army Medical Department, 1917-1941*, 69–73.
45. Harris and Paxman, *A Higher Form of Killing*, 35.
46. Ginn, *The History*, 77.
47. Moreno, *Undue Risk*, 38–39.
48. "Secret, Deadly Research: Camp AU Scene of World War Training Trenches, Drill Field," *The American University Eagle* (January 15, 1965), quoted in Moreno, *Undue Risk*, 39.
49. Coffey, *American Arsenal*, 30–37.
50. "Secret, Deadly Research," quoted in Moreno, *Undue Risk*, 39.
51. Coffey, *American Arsenal*, 41–42.
52. Heller, *Chemical Warfare*, 35–67; Brown, *Chemical Warfare*, 29.
53. Coffey, *American Arsenal*, 30–37.
54. James B. Conant, *My Several Lives* (1970), 49–50.
55. Institute of Medicine, *Veterans at Risk* (1993), 26.
56. Harris and Paxman, *A Higher Form of Killing*, 13, 18, 34.
57. Rostker, *Providing*, 128.
58. Heller, *Chemical Warfare*, 34, 67.
59. Quoted in Harris and Paxman, *A Higher Form of Killing*, 18.
60. Brown, *Chemical Warfare*, 36–37.
61. Rostker, *Providing*, 123.
62. John Pershing, *Final Report of General John J. Pershing, Commander-in-Chief American Expeditionary Forces* (1920), 75–77.

6
The Military Biomedical Complex During the Interwar Period

Introduction

In the two decades between world wars—an era now commonly referred to as the interwar period—the military biomedical complex expanded its pre-World War I (WWI) medical research programs. It established new divisions that built off wartime chemical weapons research. Considerable efforts also were dedicated to demobilization and attending to the healthcare needs of millions of returning WWI veterans. Although the nation was elated by the Allied victory, a euphoria that carried into the 1920s, the positive sentiment dissolved following the 1929 market crash and the ensuing economic depression. A debilitated economy curtailed the health system for service members and veterans, which already had been resource-strapped due to the downsizing of the defense budget after WWI. Public backlash to subpar treatment of veterans eventually led to the creation of the Veterans Administration, a department charged with coordinating and providing healthcare and benefits for veterans.

On a global scale, throughout the 1920s, nations worked to reconstruct damaged communities and rebuild trust among nations. The League of Nations was formed to help mediate conflicts, and several treaties were enacted which sought to limit certain methods of warfare. Among these international agreements were attempts to instill a global norm against the deployment of poison gases and other deleterious substances. Yet, rather than curtailing development in these fields, the doctrines had the perverse effect of encouraging clandestine research and stockpiling. In the United States, military-civilian chemical warfare collaborations increased significantly, and the pursuit of military science enveloped the nation's factories and college campuses. A global arms race followed, with several nations utilizing chemical weapons during battles and colonial exploits. Each nation justified its actions with crafty legal arguments that exposed the fragility of existing weapons bans: national security, however defined by each country, was the consummate trump card.

Demobilization and Veterans Healthcare

Demobilization after WWI was a colossal undertaking. The Army Medical Department administered exit health examinations to three million service members, two million of whom were returning from Europe.[1] The department offered healthcare and benefits for service-related harms, though inadequate resources hindered the department's ability to meet the demand. Over 930,000 veterans applied for disability benefits, more than 200,000 of whom were permanently disabled.[2] In 1921, a committee appointed by President Warren Harding and chaired by General Charles

America's Military Biomedical Complex. Efthimios Parasidis, Oxford University Press.
 DOI: 10.1093/9780199351473.003.0007

Dawes reported: "It cannot be too strongly emphasized that the present deplorable failure on the part of the government to properly care for the disabled veterans is due in large part to an imperfect organization of government effort."[3]

Five months after the committee's scathing report, Harding created the Veterans' Bureau, an independent agency that reported directly to the president.[4] Veterans' benefits increased slowly over the years, and in 1930, President Herbert Hoover signed an executive order that created the Veterans Administration (VA) to consolidate and coordinate healthcare and benefits for veterans.[5] Thereafter, veterans continued to protest for more robust benefits, staging a large demonstration on the steps of the US Capitol (see Figure 6.1). During the summer of 1932, thousands of veterans and their supporters occupied government buildings and set up encampments in parks throughout Washington, DC. Violent clashes with police followed, and the US Army, under orders from President Hoover, used tanks and tear gas to evict protestors from government property. Local hospitals were overwhelmed with wounded protestors, several of whom died during the clashes.[6]

Initially, the VA extended benefits and healthcare coverage to include non-service-related health conditions, but the American Medical Association (AMA) vigorously opposed the expansion, characterizing it as socialist and unworkable. Behind the scenes, doctors were concerned that government-sponsored healthcare would cut into their profits.[7] President Franklin D. Roosevelt vetoed certain veterans benefits, stating in 1933 that "the mere wearing of a uniform in the war years does not entitle a

Figure 6.1 WWI veterans and their supporters hold a demonstration on the steps of the US Capitol during the summer of 1932 to advocate for veterans benefits.
Photograph from the Library of Congress.

veteran to a pension from his government for a disability incurred after his period of service has ended."[8] Roosevelt contended that state and local authorities, not the federal government, should bear responsibility for non-service-related health concerns. This was the first time since 1820 that the federal government revoked benefits for veterans.[9]

After extensive lobbying from veterans' organizations, in 1934, Congress overrode Roosevelt's veto and restored most of the benefits, including access to the VA for non-service-related health needs.[10] Despite the legal shift, Congress did not allocate adequate funding, and veterans with non-service-related health conditions were treated only insofar as facilities and resources were available.[11] This predicament epitomized what some scholars call "hollow government syndrome," which occurs when Congress affords powers and responsibilities to an agency but does not provide funding to meet the obligations.[12] Despite the resource limitations, the new benefits marked a key expansion in healthcare policy for veterans.

As had been the case throughout America's military history, Congress decreased funding for the military after WWI. The peacetime downsizing of the armed forces brought a substantial curtailment of funding for military medicine.[13] As Surgeon General Merritte Ireland explained, nearly a decade after WWI, "the Medical Department is less well prepared for field service than before the war with Germany."[14] The budgetary shortfalls were exacerbated by the Great Depression, which resulted in a fighting force that, according to military historian Russell Weigley, by the late 1930s, "may have been less ready than at any time in its history to handle combat."[15] The decreased funding for the armed forces and military medicine impacted military medical research, though significant advancements still occurred during the interwar period.

Medical Research During the Interwar Period

The military research enterprise that blossomed during the early twentieth century continued to expand during the interwar period, though the work was often completed on a shoestring budget due to funding shortfalls and the domestic backlash caused by the Great Depression. Researchers tracked the health of veterans and analyzed information from wartime observations on disease and injuries. During the 1920s, medical boards were established in the Philippines to examine tropical diseases such as malaria, dengue fever, and parasitic infections.[16]

American service members stationed in the Philippines often served as research subjects. However, protocols from Reed's yellow fever research in Cuba—which included disclosure of risks, compensation for participation, and compensation for research-related harms—were not widely adopted. At the time, the AMA continued to refuse to enact ethical guidelines for research with human subjects, and no state or federal statute afforded specific protections for research participants. As was the case before WWI, during the interwar period, untoward research practices were common throughout the United States. Poor and marginalized segments of society were particularly prone to exploitation. As one example, children with cognitive and developmental disabilities were test subjects in risky medical research. During the studies, which took place at institutions and orphanages throughout New Jersey and Pennsylvania, dozens of children were inoculated with an experimental measles vaccine and then exposed to measles to test vaccine efficacy.[17]

Within this regulatory gap, the military enacted guidelines that afforded some protections to research subjects in military studies. In 1925, the Army established a policy that required the use of volunteers in research, and in 1932, the Navy instructed that, for some studies, "all subjects should be informed volunteers, that the detailed protocol be approved in advance, and that every precaution be taken to prevent accidents."[18] Analogous policies were not commonly found among civilian research organizations. Although the extent to which researchers adhered to the military guidelines is unclear, the policies illustrated that military leaders understood the need to address research subject vulnerability.

Military research during the interwar period led to several scientific advances, including a better understanding of dengue fever and the development of a new rabies vaccine. A new field of aviation medicine also emerged, which included research to determine best practices for clothing, equipment, and the health of airmen.[19] Studies also examined the impact of partial asphyxiation on aviators, an important inquiry given the primitive oxygen masks of the era and the lack of pressurized cabins. Other tests examined the "physical and temperamental deterioration" of pilots.[20]

Meanwhile, books, plays, and films glamorized the breakthroughs of military medicine. A 1926 bestseller, Paul de Kruif's *Microbe Hunters*, described the work of Reed's Yellow Fever Commission. As bioethicist Jonathan Moreno explains, the book "charmingly," but "somewhat inaccurately . . . introduced generations of lay readers to the wonders of public health microbiology."[21] The work of Reed and his colleagues was further glamorized in a 1934 Broadway play, *Yellow Jack*, which starred James Stewart and later was released as a film. According to one critic: "Graphic scenes of emotion, humor, and laboratory ritual unroll new understanding of how old and new schools clash, how doctors and plain men sacrifice lives for other lives, what spiritual force it takes to ask such sacrifices."[22]

Expansion of Chemical Weapons Research and Development

Alongside the groundbreaking medical research conducted during the interwar period and increased recognition of the need for safety and consent protocols in medical research, the United States vastly expanded the scope of its chemical warfare program. This expansion involved substantial debate within the military on whether chemical weapons should remain an integral component of America's weapons arsenal. It also paralleled American efforts to enact international doctrines to limit the proliferation of chemical and biological weapons.

In the years following the conclusion of WWI, Army leaders questioned the role of the Chemical Warfare Service within the military establishment.[23] Some were skeptical of integrating science and technology into combat, while others characterized the group as "a corps of college professors."[24] General John Pershing, who commanded the American Expeditionary Force on the Western Front during WWI, favored the continuation of gas research "because we cannot trust the other fellow."

On the morality of chemical warfare, Pershing remarked: "I cannot see very much difference myself in methods of killing."[25]

The American chemical industry, along with the Chemical Warfare Service, used its money and political influence to lobby for an expansion of chemical weapons research and development. Lobbyists crafted a nationwide campaign to magnify the dangers of chemical warfare, ghostwriting editorials and sponsoring speakers at conferences, veterans groups, and Chambers of Commerce.[26] An address given to the Rotary Club in Flint, Michigan, is illustrative. During the talk, Dow Chemical Vice President William Hale proclaimed that gas was "the most effective weapon of all time" and "the most humane ever introduced into war by man."[27] Hale concluded his talk by calling for import tariffs, which would represent a new "battle cry . . . 'To Hell with all German imports!' "[28]

The lobbying campaign proved successful. Import restrictions and tariffs were enacted, and the Chemical Warfare Service's place within the military was cemented.[29] According to Lieutenant Colonel Frederic Brown, who worked in the Office of the Special Assistant to the Joint Chiefs of Staff for Counter-Insurgency and Special Activities during the 1960s, the ability of the Chemical Warfare Service to establish itself as a separate service "was a remarkable example of pressure-group activity conducted outside of the normal constraints of the military bureaucracy."[30]

The Chemical Warfare Service and chemical industry also downplayed the long-term effects of gas poisoning. Due to lingering adverse health effects from the influenza pandemic of 1918–1920, physicians could not pinpoint whether respiratory conditions were caused by the flu or exposure to chemical weapons. In fact, some researchers—including the chief of the Chemical Warfare Service Medical Division—conducted studies to evaluate potential therapeutic uses of nonlethal doses of chlorine gas to treat whooping cough, bronchitis, pneumonia, and influenza. During one public relations event, President Calvin Coolidge was locked in a gas chamber and "treated" with chlorine gas for forty-five minutes a day for three straight days. The researchers claimed the treatment successfully rid Coolidge of his cold, though subsequent studies questioned the effectiveness of chlorine gas therapy.[31]

To further expand societal uses of chemical agents—particularly in the realm of law enforcement—the Chemical Warfare Service recommended that chemical weapons be used as a method of crowd control and as a means of executing prisoners sentenced to death. Thereafter, tear gas entered the arsenals of law enforcement agencies, and twelve states adopted laws that permitted execution by poison gas.[32]

Although the military was determined to maintain an active chemical weapons program, debate continued regarding the structure and organization of the Chemical Warfare Service within the military.[33] In 1920, the War Department recommended that "no troops of the Chemical Warfare Service should be assigned within the Army as combat organizations, but that this service should carry on research and development."[34] The following year, the policy shifted to permit "training and instruction in chemical warfare, both offensive and defensive" for officers of all branches of the military.[35] In turn, the Chemical Warfare Service was assigned a continuing combat mission.[36]

The intramilitary debate also was influenced by the Department of State, which was working to create an international ban on chemical weapons. During the 1922

Washington Arms Conference, the United States proposed a ban on the use of chemical weapons in war.[37] The treaty decreed that the ban should "be universally accepted as a part of international law binding alike the conscience and practice of nations."[38] Following the conference, all work of the Chemical Warfare Service was limited to defensive measures.[39] As one contemporary commentator wrote in April 1922, whereas "pious resolutions of this sort, passed in time of peace, undoubtedly express the conscience of mankind . . . In the hands of a war-mad humanity such a rule is likely to prove brittle."[40] Indeed, the 1922 treaty was never ratified by France, a principal signatory, and thus, the provisions of the agreement never came into effect.[41] Thereafter, the United States abandoned its defense-only policy for chemical weapons.

Throughout the world, the interwar years were a period of chemical weapons proliferation. By the mid-1920s, Britain, Italy, Spain, France, Germany, Japan, and the Soviet Union maintained chemical weapons programs, and several nations used chemical weapons during combat. Britain utilized chemical weapons in 1919 against Afghans near India and during Middle East conflicts in the 1920s. According to one British official, speaking in the context of chemical warfare in India:

> On the question of morality . . . gas has been openly accepted as a recognized weapon for the future, and there is no longer any question of stealing an unfair advantage by taking an unsuspecting enemy unaware. Apart from this, it has been pointed out that tribesmen are not bound by the Hague Convention and they do not conform to its most elementary rules.[42]

The French and Spanish deployed chemical weapons in Morocco in 1925, and chemical weapons were widely used between 1919 and 1921 during the Russian Civil War.[43]

A doctrinal crescendo was achieved with the Geneva Protocol of 1925, which prohibited the "use in war of asphyxiating, poisonous or other gases, and of all analogous liquids, materials, or devices."[44] Passed under the rubric of the League of Nations, the United States led the enactment of the Geneva Protocol, which sought to internationalize the principles outlined in the 1922 Washington Arms Treaty.[45]

In 1928, forty-three nations, including the United States, signed a separate peace accord, the Kellogg-Briand Pact, which condemned and renounced "recourse to war for the solution of international controversies" and mandated "settlement or solution of all disputes or conflicts of whatever nature or of whatever origin they may be . . . by pacific means."[46] As scholars have detailed, the Kellogg-Briand Pact served as a regional security agreement and an attempt to increase US participation in European affairs.[47]

These peace accords were largely driven by a desire to end war, to fulfill the Progressivist idea of mediation of international disputes, and to help build a stable international order and economy. Notwithstanding the noble goals, the accords did little to curb wars of aggression or limit the global proliferation of chemical and biological weapons. Even US Secretary of State Frank Kellogg (one of the principal authors of the Kellogg-Briand Pact, the other being French Foreign Minister Aristide Briand) supported the continuation of America's chemical weapons program, remarking that "I have never seen any proposal seriously advanced by any government to provide that national preparation for the use of and for defense against chemical warfare, if such warfare should be used by an enemy contrary to treaty engagements, should be abolished or curtailed in the slightest."[48]

The Geneva Protocol, which applies to chemical weapons and "bacteriological methods of warfare," was not binding upon signatories until individual governments ratified. Many nations either refused to ratify the protocol or did so after adding restrictions. For example, Britain, France, and Canada ratified the protocol on the condition that the agreement would be binding only against nations that ratified as well, and that chemical or biological weapons could be used to retaliate in kind.[49] Although the protocol affirms that wartime uses of biological or chemical weapons have "been justly condemned by the general opinion of the civilized world," as a practical matter the accord amounted to a ban on first use of chemical or biological weapons. The protocol did not prohibit research and development of chemical or biological weapons, nor did it prohibit stockpiling.

The United States signed the Geneva Protocol in 1925, but the Senate Foreign Relations Committee held a closed-door meeting during which it decided that the United States would not ratify the treaty.[50] The Chemical Warfare Service, veterans groups, and the chemical industry lobbied extensively against ratification. Lobbyists and military officials argued that gas was likely to be used in future wars and that a robust chemical weapons program was necessary to maintain a balance of power.[51] US leaders also sought to protect American strategic interests and the nation's armaments industry, and cautioned against what many viewed as European disarmament to safeguard against another war in Europe.[52] As will be detailed in chapter 11, ratification of the Geneva Protocol by the United States would not come until 1975.

Across the globe, one unintended consequence of the Geneva Protocol was to drive weapons research underground. During the late 1920s, Germany and the Soviet Union began collaborating on top-secret chemical weapons research—this was despite the fact that the two powers fought against each other just a decade earlier, and the 1919 Treaty of Versailles banned the importation and manufacture of "asphyxiating, poisonous or other gases" in Germany for all time.[53] Notwithstanding the Treaty of Versailles and Germany's ratification of the 1925 Geneva Protocol, Germany maintained a secret chemical weapons program throughout the interwar period. Meanwhile, the British conducted secret field tests with chemical weapons—including phosphorus bombs and poison gases—in Scotland, Wales, India, Australia, and the Middle East.[54] In 1934, British scientists in India examined whether "Indian skin" was more susceptible to blister agents than "European skin."[55]

In the United States, ambivalence toward chemical warfare research and development persisted throughout the interwar period. During the 1930s, General Douglas MacArthur characterized a chemical weapons arsenal as an insurance policy, noting that "I am personally more or less indifferent to the retention or abolition of gas. I do not believe it has ever been a vital factor or ever will be. It is hard to control, of little effectiveness against trained troops and lends itself to vicious exploitation of civilian populations."[56] For their part, Presidents Hoover and Roosevelt strongly opposed chemical weapons.[57] In 1937, Roosevelt vetoed a bill that would change the name of the Chemical Warfare Service to the Chemical Corps. Although the War Department supported the name change, Roosevelt remarked:

> I have a far more important objection to this change of name. It has been and is the policy of this Government to do everything in its power to outlaw the use of chemicals in warfare. Such use is inhuman and contrary to what modern civilization should stand for. I am doing everything in my power to discourage the use of gases and other chemicals in any war between nations. While, unfortunately, the defensive necessities of the United States call for study of the use of chemicals in warfare, I do not want the Government of the United States to do anything to aggrandize or make permanent any special bureau of the Army or the Navy engaged in these studies. I hope the time will come when the Chemical Warfare Service can be entirely abolished. To dignify this Service by calling in the "Chemical Corps" is, in my judgment, contrary to a sound public policy.[58]

Notwithstanding Roosevelt's stern and perhaps aspirational statement, America's chemical warfare program grew during the interwar period and expanded significantly under Roosevelt during World War II (WWII).

As chemical weapons research established a solid footing within the military, cities were bombarded with recruitment posters that sought to entice Americans to join the Chemical Warfare Service.[59] The department's central facility remained at Edgewood Arsenal in Maryland, though thirteen manufacturing facilities across the country were equipped to produce and test mustard gas, phosgene, and other toxic substances.[60] Research and development laboratories spanned from coast to coast, often in collaboration with leading universities such as MIT, Harvard, Princeton, Cornell, Johns Hopkins, and Yale.

The Army Medical Department collaborated with the Chemical Warfare Service on chemical weapons research, and tactical training programs were created whereby service members would practice offensive and defensive maneuvers.[61] Although there were practical reasons for this integration, the institutional line between medical care and weapons development blurred during these endeavors. It also was dangerous work with occupational hazards, causing countless illnesses and injuries such as chronic coughs, blistering skin, damaged vision, and cognitive difficulties.

Meanwhile, the deployment of chemical weapons continued into the 1930s: the Japanese utilized chemical weapons against the Chinese, and the Italians deployed chemical weapons during their invasion of Abyssinia. As to the latter, at least fifteen thousand Abyssinians were killed or injured by chemical weapons, representing one-third of Abyssinian casualties during the war.[62] According to firsthand reports, Italy's mustard gas attacks drenched the countryside, poisoned food and water supplies, and caused immeasurable pain to civilians, including children, who ate poisoned food and walked barefoot along contaminated pastures.[63] The Italians claimed that their actions did not violate the Geneva Protocol, despite being a first strike with chemical weapons, because of "other illegal acts of war" committed by the Abyssinians.[64] Ethiopia petitioned for redress to the League of Nations, but no meaningful action was taken to address Italy's use of chemical weapons.[65]

In 1936, German scientist Gerhard Schrader discovered a particularly toxic substance during his work on insecticides. At the time, German law required inventions with military potential to be kept secret and reported to the government, and Schrader dutifully complied. He discovered what became known as Tabun—named after the German word for "taboo"—a lethal substance that is colorless, practically odorless, and can be ingested into the human body via inhalation or contact with skin.[66] Humans or animals exposed to Tabun, which is classified as a nerve agent, will vomit and froth at the mouth, their pupils will shrink to a little dot, limbs will twitch and jerk, and, within minutes, they will suffer convulsions and die.

About one year after discovering Tabun, Schrader and his colleagues discovered a new and more lethal substance. They dubbed it Sarin, an acronym derived from their last names (*S*chrader, *A*mbros, *R*üdriger, and van der L*in*de).[67] Within months of the discovery of Sarin, Germany invaded Poland, commencing what was to become WWII. Three weeks into the new war, Adolf Hitler delivered a speech in which he warned his enemies: "The moment may come when we use a weapon which is not yet known and with which we could not ourselves be attacked . . . It is to be hoped that no one will then complain in the name of humanity . . . I need not remind you that I have given orders to keep to the rules of war, but those in the West need not imagine that it must always be so."[68]

Enter Biological Weapons

Whereas chemical weapons are created using synthetic compounds, with biological weapons, the goal is to weaponize naturally occurring pathogens (such as viruses and bacteria) or toxins (poisonous substances produced by living organisms). Biological warfare has ancient roots. For more than two thousand years, various armies used animal cadavers to pollute enemy drinking water.[69] The Scythians, a large Eurasian group of nomadic tribes, were master warfighters who dipped their arrows into decomposing cadavers or blood mixed with manure.[70] The Tatars (fourteenth century) catapulted the corpses of plague victims onto enemy positions.[71] While many of these tactics were unsuccessful, the weaponization of biological agents accelerated significantly during the first half of the twentieth century as the germ theory of disease revealed that biological agents could be adopted as viable weapons of war. The work centered on isolating pathogens and constructing mechanisms by which they could be deployed.

Allegations of biological warfare surfaced during and after WWI. The Germans reportedly poisoned water wells with cholera, attempted to spread plague bacteria in Russia, and contemplated the use of zeppelins to drop plague bombs on British citizens. During WWI, a group of German-American saboteurs created liquids laced with anthrax and glanders for use in biological warfare. The saboteurs infiltrated three American ports (Baltimore, Norfolk, and Newport News), killing thousands of horses and mules that were bound for Europe to assist in Allied war efforts.[72] A top secret American report found accounts of German biological warfare as "confirmed and undoubted."[73]

Biological warfare research accelerated during the interwar period, and some historians contend that the 1925 Geneva Protocol motivated a biological weapons arms race. Japan, for example, created a biological weapons program in 1932 after Japanese army major Shiro Ishii returned from a trip to Europe "convinced that biological weapons were an effective means of fighting a war . . . otherwise the statesmen at Geneva would not have gone to the trouble of banning them."[74] Ishii and his team conducted biological weapons research in occupied Manchuria, which Japan invaded in 1931.[75]

Ishii's reasoning mirrored that of the British, who, in 1934, began a biological weapons program in part because of international initiatives aimed at banning toxic weapons. The British built a secret biological weapons laboratory at Porton Down, the country's center for chemical weapons that was created during WWI. In 1940, Canada began biological weapons research: at the University of Toronto, Canadian researchers examined whether infectious bacteria, such as typhoid and salmonella, could be mixed with sawdust and sprayed from an airplane. By 1941, several additional nations had established biological warfare programs, including Germany, Italy, France, Poland, and the Soviet Union.[76] The United States, as will be discussed in succeeding chapters, did not commence biological weapons research until after WWII began. Across the globe, biological warfare research and development proliferated during and after WWII.

Summary

In the two decades between world wars, the military biomedical complex balanced its traditional healthcare responsibilities with chemical warfare research and development. Post-WWI efforts sought not only to rebuild the global social order and infrastructure, but also to establish internationally recognized rules to mediate conflicts and set limits to battle. The measures included the creation of the League of Nations and the enactment of treaties that elucidated proper and improper conduct during war, such as guidelines to curtail the use of chemical and biological weapons. In practice, however, the international treaties and institutions had little impact on weapons proliferation and deployment of chemical weapons during military missions. Moral and ethical quandaries were subsumed by the geopolitical advantages of maintaining and using these powerful weapons. The worldwide quest for scientific and military superiority carried into the next world war.

Notes

1. Gillett, *The Army Medical Department, 1917-1941*, 417–29.
2. Kinder, *Paying*, 5.
3. *To Establish a Department of Health: Hearings before the Subcommittee on Reorganization of the Senate Committee on Government Operations*, 82nd Cong. 181 (1952) (quoting Dawes Committee Report).
4. An Act to Establish a Veterans' Bureau and to Improve the Facilities and Service of Such Bureau, and Further to Amend and Modify the War Risk Insurance Act, Pub. L. No. 47, 42 Stat. 147 (August 9, 1921).
5. Executive Order 5398, 38 C.F.R. 1.1 (July 21, 1930).

6. "Bonus Expeditionary Forces March on Washington," National Park Service (accessed June 11, 2024).
7. Duffy, *The Healers*, 303.
8. Quoted in Mitchel B. Wallerstein, "Terminating Entitlements: Veterans' Disability Benefits in the Depression," *Policy Sciences* 7 (1976): 177.
9. Rostker, *Providing*, 169–73.
10. "House, 310 to 72, Overrides Roosevelt's Veteran Veto; Close Senate Vote Likely," *The New York Times*, March 28, 1934; "Senate Overrides Veterans Veto, Crippling Whole Economy Program; 29 Democrats Desert the President," *The New York Times*, March 29, 1934.
11. Rostker, *Providing*, 172–73.
12. Peter B. Hutt, "The State of Science at the Food and Drug Administration," *Administrative Law Review* 60, no. 2 (Spring 2008): 432.
13. Gillett, *The Army Medical Department, 1917-1941*, 471–82.
14. War Department, *Annual Report of the Surgeon General* (Washington, DC: U.S. Government Printing Office, 1926), 15.
15. Russell F. Weigley, *History of the United States Army* (1984), 402.
16. Gillett, *The Army Medical Department, 1917-1941*, 503–4.
17. Allen M. Hornblum, Judith L. Newman, and Gregory J. Dober, *Against Their Will* (2013), 49.
18. Bureau of Medicine & Surgery, Letter to Secretary of the Navy (April 5, 1932), cited in Charles W. Shilling, "History of the Research Division, Bureau of Medicine and Surgery, U.S. Department of Navy," Attachment D to Advisory Committee on Human Radiation Experiments, Washington, DC Meeting, vol. 3 (June 13–14, 1994).
19. Gillett, *The Army Medical Department, 1917-1941*, 496–504.
20. Ginn, *The History*, 76–77.
21. Moreno, *Undue Risk*, 17.
22. "Again 'Yellow Jack,'" *American Journal of Public Health* 24 (1934): 651.
23. Brown, *Chemical Warfare*, 72–92.
24. *Army Reorganization: Hearings before the House Committee on Military Affairs*, 66th Cong. 53 (1919) (testimony of General Peyton March).
25. *Reorganization of the Army: Hearings before the Subcommittee of the Senate Committee on Military Affairs*, 66th Cong. 1644 (1919) (testimony of General John Pershing).
26. Brown, Chemical Warfare, 57–61; Coffey, *American Arsenal*, 38–42.
27. *Munitions Industry: Hearings before the Special Senate Committee Investigating the Munitions Industry*, 73rd Cong. 2565 (1935) (Exhibit 914).
28. *Munitions Industry*, 73rd Cong. 2568 (Exhibit 914).
29. Brown, *Chemical Warfare*, 59–61.
30. Brown, *Chemical Warfare*, 82.
31. Faith, *Behind the Gas Mask*, 90–102.
32. Faith, *Behind the Gas Mask*, 90–102.
33. Brophy and Fisher, *The Chemical Warfare Service*, 18–22.
34. Lassiter Board Report (July 8, 1920), quoted in Brown, *Chemical Warfare*, 88.
35. War Department General Orders No. 42 (August 17, 1921): 2.
36. Brown, *Chemical Warfare*, 92.
37. Brophy and Fisher, *The Chemical Warfare Service*, 18–22; Brown, *Chemical Warfare*, 92–96; Faith, *Behind the Gas Mask*, 104–6.

38. Treaty Relating to the Use of Submarines and Noxious Gases in Warfare, Washington, DC, art. 5 (February 6, 1922).
39. Brown, *Chemical Warfare*, 92–96.
40. Roland Hugins, "The Washington Arms Conference," *The Open Court* 36, no. 4 (April 1922): 199.
41. Brophy and Fisher, *The Chemical Warfare Service*, 20.
42. C. H. Foulkes, Letter to War Office (November 5, 1919), quoted in Harris, and Paxman, *A Higher Form of Killing*, 45–46.
43. Harris and Paxman, *A Higher Form of Killing*, 45–46; Nick McCamley, *Secret History of Chemical Warfare* (2006), 40.
44. Protocol for the Prohibition of the Use of Asphyxiating, Poisonous or Other Gases, and of Bacteriological Methods of Warfare, Geneva, art. 5 (June 17, 1925).
45. Brown, *Chemical Warfare*, 98-103.
46. General Treaty for the Renunciation of War (Kellogg-Briand Pact), Paris (August 27, 1928).
47. Melvyn P. Leffler, *Safeguarding Democratic Capitalism* (2017), 104–13.
48. Frank Kellogg, Letter to Charles Parsons, 68 Cong. Rec. 366 (December 13, 1926).
49. John Norton Moore, "Ratification of the Geneva Protocol on Gas and Bacteriological Warfare: A Legal and Political Analysis," *Virginia Law Review* 58, no. 3 (March 1972): 452–54.
50. Hersh, *Chemical and Biological Warfare*, 6–7.
51. Brown, *Chemical Warfare*, 103–7.
52. Leffler, *Safeguarding Democratic Capitalism*, 95–96.
53. Treaty of Peace with Germany (Treaty of Versailles), art. 171 (June 28, 1919).
54. Harris and Paxman, *A Higher Form of Killing*, 46–49.
55. Schmidt, *Secret Science*, 144.
56. Douglas MacArthur, Letter to Brig. Gen. G. Simmonds (February 26, 1932), quoted in Brown, *Chemical Warfare*, 140.
57. Brown, *Chemical Warfare*, 123–25.
58. Franklin D. Roosevelt, Letter to the Senate on Chemicals in Warfare (August 4, 1937).
59. Harris and Paxman, *A Higher Form of Killing*, 35.
60. Jonathan B. Tucker, *War of Nerves* (2006), 89.
61. Brown, *Chemical Warfare*, 128–29.
62. Harris and Paxman, *A Higher Form of Killing*, 48–52.
63. Jeffery K. Smart, "History of Chemical and Biological Warfare," in *Medical Aspects of Chemical and Biological Warfare*, edited by Frederick R. Sidell, Ernest T. Takafuji, and David R. Franz (Washington, DC: Borden Institute, 1997), 34–35.
64. Quoted in McCamley, *Secret History*, 44.
65. McCamley, *Secret History*, 44–45.
66. Harris and Paxman, *A Higher Form of Killing*, 55–57; Coffey, *American Arsenal*, 149.
67. Harris and Paxman, *A Higher Form of Killing*, 55–57.
68. War Cabinet Memorandum by the Secretary of State for Foreign Affairs, "Herr Hitler's Speech at Danzig on September 19," Britannic Majesty's Government (September 20, 1939).
69. Mayor, *Greek Fire*, 93–103.
70. Mayor, *Greek Fire*, 66–78.
71. Mayor, *Greek Fire*, 119–21.
72. Craughwell, *The War Scientists*, 216–18; Harris and Paxman, *A Higher Form of Killing*, 76; Sheldon H. Harris, *Factories of Death* (1994), 160.
73. Colonel William Creasy, Report to Secretary of Defense Ad Hoc Committee (February 24, 1950), quoted in Harris and Paxman, *A Higher Form of Killing*, 76.

74. Quoted in Harris and Paxman, *A Higher Form of Killing*, 77.
75. George W. Christopher, Theodore J. Cieslak, Julie A. Pavlin, and Edward M. Eitzen, "Biological Warfare: A Historical Perspective," *Journal of the American Medical Association* 278, no. 5 (August 6, 1997): 412–17.
76. Wendy Barnaby, *The Plague Makers* (2002), 7–78; Harris and Paxman, *A Higher Form of Killing*, 83–85; Stephen Endicott and Edward Hagerman, *The United States and Biological Warfare* (1998), 27.

7
World War II
Transformational Developments in Military Medicine and Research

Introduction

During World War II (WWII), nearly every segment of society was engulfed by war efforts. Car manufacturers produced airplanes, shipping companies constructed military vessels, and pharmaceutical firms developed medical products for wartime health needs. A sprawling healthcare system supported American military personnel across the globe, and the drive to weaponize atomic energy, toxic chemicals, and biological agents situated the entire scientific enterprise in a high-stakes race for military superiority. At the peak of the battle, over 89 percent of the federal budget was dedicated to national defense, and 37.5 percent of the gross domestic product was war-related.[1] United as a nation, Americans worked with a deep resolve to promote military preparedness and the creation of weapons of mass destruction.

Wartime needs inspired innovations in healthcare operations, medical treatment, and preventive health. The government also created a new, civilian-led military science agency that maintained a robust agenda in medical research and weapons development. Funding for military-industry collaborations was plentiful; universities dedicated laboratories to further wartime research, and faculty applied their expertise to military matters. The line between military and civilian science largely disappeared. Both on and off the battlefield, Americans embraced the disruptive awe of military science.

In the course of this work, the ethical implications of research with human subjects received little attention. Scientists exposed countless individuals to lethal pathogens, chemical weapons, and radiological harms, and researchers themselves were subject to dangerous occupational hazards. Many research subjects were coerced into experiments or were unwitting participants. Others were told they would be prosecuted for espionage if they revealed the existence of the studies. Amid the most expansive war in the history of humankind, national security and wartime exigencies cast aside the legal and ethical concerns.

Mobilization and Medical Innovations

Mobilization of American service members began after Germany's march through Europe, the shocking fall of France in June 1940, and Vichy leader Henri-Philippe Pétain's armistice pact with Adolf Hitler. Congress authorized an increase in the size of the armed forces, federalized the National Guard, and instituted America's first

America's Military Biomedical Complex. Efthimios Parasidis, Oxford University Press.
 DOI: 10.1093/9780199351473.003.0008

peacetime draft. By December 7, 1941—the day Japan bombed Pearl Harbor—the Army Medical Department was supporting 1.6 million soldiers.[2] During the war, Navy medical personnel staffed fifteen hospital ships and dozens of other hospitals and medical clinics, while the US Army Air Force practiced aviation medicine, ran aeromedical evacuations, and spearheaded a national program in aviation research.[3] More than sixteen million Americans served during WWII, and at the height of the war, 61 percent of men between the ages of 18 and 36 were in uniform.[4]

As with World War I (WWI), during WWII, the United States sought to exclude from service individuals with mental or emotional health concerns.[5] Between 1942 and 1943, more than 1.2 million men were excluded for "mental derangement." This equated to a rejection rate of nearly 12 percent and was the most common reason for an exclusion. Notwithstanding the screenings, mental health ailments during WWII more than doubled when compared to WWI, and discharge rates due to mental illness were significantly higher. Furthermore, since officials thought they had screened out the health concerns, there was a shortage of psychiatrists, and military hospitals were not adequately equipped to care for mental health conditions.[6] More than 500,000 men were discharged for psychiatric reasons. At times, the number of psychiatric discharges exceeded the number of recruits.[7]

In an attempt to help address the challenges, medical personnel administered questionable treatments such as hypnosis and pharmaceuticals with unproven efficacy and significant risks. One commonly used drug was sodium pentothal, a barbiturate general anesthetic that was thought to relieve a person's mental anguish enough so that they could return to battle.[8] The powerful drug did little to assist service members and had significant side effects. After the war, it was used for a variety of other purposes, such as medically induced comas, truth serums to assist in interrogations, and lethal injections for executing prisoners. Warfighters also were routinely provided amphetamines to address fatigue and mental anguish, and to enhance performance. Colloquially referred to as bennies, the drugs were used by Allied and Axis combatants, though often they led to addiction and other adverse health consequences.[9]

Although mental health concerns were poorly anticipated and addressed, a comprehensive triage system of care helped preserve the fighting force in the field. Medics played an integral role throughout the war by providing frontline emergency care. An organized system that utilized ambulances, trains, ships, field surgical units, hospital ships, base hospitals, and other clinical stations linked treatment from the front lines to hospitals in the United States.[10] The creation of portable surgical hospitals was also an important WWII innovation. Portable hospitals were particularly useful in the South Pacific, where it was difficult to evacuate injured service members from remote islands.[11]

Of wartime clinical innovations, blood transfusions and infection control measures were transformative. Research during the interwar period led to the development of an effective method of transfusion that utilized plasma and whole blood, which increased survival rates for service members who suffered significant blood loss or were undergoing surgery. In addition, the use of antibiotics grew exponentially during the war. Warfighters were provided an envelope with a shaker top containing sulfa powder and were instructed to sprinkle the powder on open wounds to help heal and prevent infection. By 1944, the administration of sulfa drugs decreased substantially due to the introduction of penicillin, which was more effective and had fewer side effects. A wonder drug created by British scientists, penicillin revolutionized treatment and was widely used to treat wounds, sexually transmitted diseases, and other bacterial infections, saving countless lives.[12]

Civilian specialists in biochemistry, immunology, bacteriology, infectious diseases, and public health worked with medical officers to establish preventive health policies, collect and analyze data, and conduct laboratory research. Wartime preventive health measures included mosquito control and chemical warfare detection kits for food and water supplies.[13] Immunizations for smallpox, tetanus, typhoid fever, and yellow fever were mandatory for military personnel. Those who refused a vaccine were subject to disciplinary proceedings, including the potential for court-martial before a military court.[14]

Of the inoculations, the yellow fever vaccine was controversial. The vaccine was under development and had not been licensed by American regulators. Nevertheless, weeks after the attack on Pearl Harbor, US officials feared a biological attack with yellow fever and ordered mandatory immunizations for all active-duty military personnel. After the vaccine campaign began, it was discovered that approximately 427,000 doses were contaminated with Hepatitis B, a virus that causes jaundice, liver failure, and cancer. During vaccine manufacturing, the pharmaceutical company mistakenly used contaminated human blood that harbored the dangerous virus.[15]

Approximately 330,000 service members contracted Hepatitis B; 51,000 were hospitalized because of disease-related complications, and no less than 100 died.[16] The yellow fever vaccine mandate was halted in April 1942 due to the adverse health impact but resumed a few months later.[17] Apart from significant vaccine-induced injuries, it marked the first time that an unlicensed vaccine was a requirement of service.

The medical, legal, and ethical concerns of mandatory vaccination with an unlicensed vaccine were bypassed under the rubric of force protection. In the rush to utilize a vaccine that was believed to be safe and effective (despite not being approved by regulators) against a perceived threat (biological warfare), military leaders ended up compromising military missions by creating a health crisis that impacted a significant proportion of the fighting force. The vaccine mandate also negatively impacted morale, as the yellow fever vaccine crisis planted seeds of doubt as to whether the military health system maintained adequate safeguards to protect service members from adverse health risks from mandatory prophylaxis.

A New Model for Military Research

Coupled with a comprehensive healthcare enterprise to assist military personnel, WWII ushered in a new era for military science. Although military and civilian scientists collaborated on medical research and weapons development for decades, the military primarily initiated and directed these efforts. A new model emerged during the summer of 1940 when an elite group of civilian scientists approached President Franklin D. Roosevelt with a novel idea. The group was led by Vannevar Bush, an engineering professor and president of the Carnegie Institution of Washington, a nonprofit organization that maintained a substantial budget for scientific research.

Previously, Bush was dean of MIT's School of Engineering and vice president of the university.

The cadre believed that the military had been ineptly synthesizing American scientific knowledge and that an alternative to the slow-moving military bureaucracy was necessary to promote the progress of military science. Bush and his colleagues contended that academic and industry scientists were best suited to lead wartime research efforts because they nurture a culture of creativity and entrepreneurship rather than abide by a rigid and politically controlled system that stunts growth by punishing mistakes and favoring seniority. The group also believed that, since military leaders were not sufficiently acquainted with modern scientific advancements, they were unable to fully contemplate how science might address military needs.[18]

According to Irvin Stewart, one of Bush's colleagues, this predicament called for "letting men who knew the latest advances in science become more familiar with the needs of the military in order that they might tell the military what was possible in science so that together they might assess what should be done."[19] As Bush added:

> It is the primary responsibility of the Army and Navy to train the men, make available the weapons, and employ the strategy that will bring victory in combat. The Armed Services cannot be expected to be experts in all of the complicated fields which make it possible for a great nation to fight successfully in total war. There are certain kinds of research—such as research on the improvement of existing weapons—which can best be done within the military establishment. However, the job of long range research involving application of the newest scientific discoveries to military needs should be the responsibility of those civilian scientists in the universities and in industry who are best trained to discharge it thoroughly and successfully. It is essential that both kinds of research go forward and that there be the closest liaison between the two groups.[20]

In a one-page document prepared for Roosevelt, Bush outlined the contours of a civilian-led research center—with its own funding and a direct report to the president—that would spearhead scientific work to further national security. Within ten minutes, Roosevelt approved the plan.[21]

Roosevelt created the National Defense Research Committee (NDRC) and tapped Bush as its leader. Only two of the eight NDRC members were military men (an Army and Navy representative).[22] As Bush later wrote in his memoir, the establishment of the NDRC was "an end run, a grab by which a small company of scientists and engineers, acting outside established channels, got hold of the authority and money for the program of developing new weapons."[23]

One important aspect of the NDRC's funding scheme was that it entered into contracts directly with universities and covered research overhead costs. Previously, professors would obtain research grants, and the university was responsible for covering the costs of facilities and other research-related matters. In light of the new funding model—which brought significant revenue to academic institutions for constructing, maintaining, and staffing laboratories and research centers—universities encouraged their faculty to pursue NDRC grants. The link between academia and the military grew stronger, and faculty across the country increasingly dedicated their efforts to military science.[24]

By the spring of 1941, the NDRC sought more government funding and government approval to expand the committee's mandate beyond weapons research to include weapons development and medical research. Initially, Bush did not want to include medical research under the NDRC's umbrella, but his position changed once he learned that the American Medical Association (AMA) was lobbying for control of such an agency. Bush—like many of his contemporaries—did not trust the AMA for a variety of reasons, including the group's vigorous opposition to a national health insurance program and its attempts to squash the development of group medical practices by refusing to provide care in hospitals that sought to create such programs. For years, the Department of Justice prosecuted the AMA and local medical societies for price-fixing, unlawful restraints on trade, and other antitrust violations.[25]

Ever the savvy politician, Bush cautioned Roosevelt against entrusting military medicine to the AMA. Bush and his colleagues then drafted an executive order to create a new agency, the Office of Scientific Research and Development (OSRD). Roosevelt signed Executive Order 8807 on June 28, 1941—among its provisions, the order placed NDRC within OSRD and created several other OSRD subgroups, including the Committee on Medical Research (CMR).[26] Bush resigned his position at NDRC and Roosevelt appointed him the leader of OSRD, which now controlled a massive enterprise encompassing weapons research, weapons development, and medical research.[27]

Of the OSRD subgroups, the CMR worked closely with established centers such as the Public Health Service, the National Research Council, and the National Academy of Sciences. The CMR also subsumed several military research divisions, including the Chemical Warfare Service.[28] Throughout the war, the CMR entered into approximately five hundred research contracts with more than one hundred institutions, including universities, research centers, and hospitals.[29]

At the time, no federal or state regulations governing safety and consent protocols for research with human participants existed. Nor had medical societies, such as the AMA, established ethical guidelines. In 1942, the CMR attempted to fill the gap by instituting a research policy:

> Human experimentation is not only desirable, but necessary in the study of the problems of war medicine which confront us. When any risks are involved, volunteers only should be utilized as subjects, and these only after the risks have been fully explained and after signed statements have been obtained which shall prove that the volunteer offered his services with full knowledge and that claims for damage will be waived. An accurate record should be kept of the terms in which the risks involved were described.[30]

Although the CMR policy recognized the need for research protocols to help protect research participants, adherence to the policy was inconsistent. As discussed in chapter 8, this would haunt American prosecutors during the Nuremberg Doctors' Trial held after WWII.

With respect to the wartime research, as bioethicist Susan Lederer details: "In order to solve pressing military medical problems, the CMR committed funds for nontherapeutic research involving orphans, the retarded, prisoners, and the mentally ill . . . moral qualms about using these populations in nontherapeutic studies faded in the harsh light of wartime necessity."[31] Penicillin research was one of the CMR's most successful projects, though it sometimes involved experimentation on unsuspecting patients at university or state hospitals.

The CMR also dedicated substantial resources to combatting malaria, a devastating disease caused by parasites transmitted by mosquitos. Warfighters were ordered to take one Atabrine pill per day to suppress the risk of contracting malaria, but they often ignored the mandate because of harsh side effects that included skin discoloration, nausea, vomiting, and diarrhea. For some, the drug caused more serious adverse reactions such as hallucinations, psychosis, and liver damage, and rumors spread that Atabrine caused sterility. Medical officers raised concerns about the drug's risks, but military and government leaders campaigned to increase adherence to the Atabrine mandate by downplaying the risks and overstating the benefits. Despite the availability of Atabrine and the deployment of mosquito control measures, malaria devastated military personnel throughout the war. Half a million American service members contracted the disease, with those in the South Pacific hit particularly hard. At times, three in five were sick with malaria.[32]

In light of the significant health concerns, a CMR research division focused exclusively on antimalarial treatments. The researchers worked with the Public Health Service and leading companies, including Abbott Laboratories, American Cyanamid, Dow Chemical, DuPont, Eli Lilly, Merck, Parke-Davis, and Sharp & Dohme. The CMR sponsored hundreds of studies with dozens of scientists from top universities, including the California Institute of Technology, Johns Hopkins, Harvard, the University of Chicago, Ohio State University, and UCLA. New antimalarial drugs were tested on patients at state hospitals and psychiatric institutions in Georgia, Illinois, Massachusetts, Michigan, New York, Tennessee, South Carolina, and elsewhere. In a typical study, a scientist would surreptitiously inject a research subject with a serum that contained malaria and then administer one of several experimental drugs. For years, the research was kept secret and research subjects were unaware that they were part of an experiment.[33] In written research reports, the research subjects often were dehumanized, referred to by the scientists as "clinical material" or "material."[34] At the time, few scientists acknowledged the ethical concerns with nontherapeutic malaria research.[35]

Potentially therapeutic malaria studies involved more complex ethical questions. Malaria therapy was a medically accepted procedure for patients with end-stage syphilis, providing benefits in up to 50 percent of cases. Although physicians were not sure how the therapy worked, the induction of high fevers (caused by malaria) was deemed a "miracle cure," notwithstanding a mortality rate of up to 15 percent from the treatment.[36] In one case, a psychiatrist-researcher asked for written documentation broadly defining therapeutic care, noting that such a letter "would be most important should any legal question arise" concerning the experiments.[37] A lead researcher provided the exculpatory letter, even though it expanded therapeutic care beyond that which was medically accepted for the time.[38] During the war, penicillin was found to cure syphilis, and malaria therapy was no longer an acceptable treatment option, thus erasing the pseudo-clinical loophole for research.

Antimalarial drugs also were tested on healthy civilians and on service members who had contracted malaria and were being treated in military hospitals. Researchers sought men of military age who would agree to be infected with malaria and then be exposed to experimental treatments. This included studies on hundreds of medical students at Ohio State University, and on prisoners throughout the country. As to the latter, research at the Atlanta Federal Penitentiary was paradigmatic. The warden of the facility initially denied the request to use inmates as test subjects, not out of concern for their health but because prison labor brought $2 million a year to the prison's coffers: the prisoners had been making mattresses for the military and mailbags for the US Postal Service. Eventually, the warden acquiesced when researchers explained the dire need for antimalarial drugs for service members.[39]

The prisoners were told of the potential side effects and that they would need to remain in the study for up to eighteen months. In return, each inmate would receive $50, access to a physician for all their medical needs, six months taken off their sentence, and a letter of appreciation added to their file, which, they were told, might be taken into account by the parole board. In Atlanta, five hundred prisoners signed up, later to be joined by thousands more who were offered similar incentives at prisons across America.[40]

As the war was winding down, the government allowed certain aspects of the malaria research to be publicly disclosed.[41] A profile was published in the June 4, 1945, issue of *Life* magazine that included photographs of prisoners with jars of malaria-infected mosquitos placed over their stomachs and arms. The article noted that, in some cases, malaria was "allowed to progress considerably" before treatment, but assured readers that "prisoners are not pardoned or paroled for submitting to infection."[42] As with other articles in prominent outlets such as *The New York Times*, the *Life* piece highlighted that the work was conducted on volunteers doing their part to further America's war efforts.[43]

Coupled with studies involving psychiatric patients, medical students, and prisoners, CMR researchers also experimented on volunteers from the Civilian Public Service, where twelve thousand pacifists supported war efforts as farmers, as orderlies, and in other noncombatant positions. Conscientious objectors volunteered for studies involving treatments for malaria and hepatitis, saltwater studies to test survival strategies for shipwrecked service members, and starvation studies to investigate the revival of malnourished combatants.[44] Once the work was revealed to the public, military officials closely monitored publications that discussed the research and sought to influence public opinion.[45] In one case, a Philadelphia newspaper was urged to "adopt a moderate tone" and avoid the term "guinea pigs," which military health officials characterized as "undignified, hackneyed, threadbare and hardly appropriate."[46]

The OSRD's expansive research enterprise drew on leading scientists and universities to further America's national security goals. The science was new, exciting, and developing rapidly, and government funding for projects was plentiful. At the same time, the creation of the OSRD and the new research model that was implemented throughout the country further eroded the line between civilian and military science.

This was particularly true in research related to atomic, chemical, and biological warfare.

Atomic Weapons Research

Within weeks of Germany's invasion of Poland, Albert Einstein wrote a letter to President Roosevelt that highlighted the recent discovery of nuclear fission and the possibility of "extremely powerful bombs of a new type." Einstein—who had admonished the development and use of chemical weapons during WWI—posited that atomic weapons "could be achieved in the immediate future" and recommended that Roosevelt help "speed up the experimental work" by funding atomic energy research.[47] Einstein's letter set in motion a series of events that would lead to the Manhattan Project.

The OSRD coordinated atomic weapons research, which included the construction of atomic bombs and radiological warfare. As to the latter, a National Academy of Sciences report from May 1941 described the "production of violently radioactive materials . . . carried by airplanes to be scattered as bombs over enemy territory."[48] By the spring of 1943, when the viability of an atomic bomb was unclear, military leaders considered poisoning the German food supply with radioactive elements to cause slow, progressive injuries. Resurrecting arguments that Germans used to justify the deployment of chemical weapons during WWI, American officials reasoned that radiological weapons were a more humane form of warfare.[49]

As with OSRD's medical research, atomic weapons research harnessed the brightest minds in industry and academia. Labs and research centers were created throughout the United States, yet officials recognized that atomic innovations were more likely to result if scientists worked together in one location. In turn, physicist Robert Oppenheimer—who led the Manhattan Project facility at Los Alamos—recruited scientists from across the country. As Oppenheimer remarked, although initially, the "notion of disappearing into the New Mexico desert for an indeterminate period and under quasi-military auspices disturbed a good many scientists, and the families of many more," in the end, "most of those with whom I talked came to Los Alamos."[50] As Oppenheimer further detailed:

> Almost everyone realized that this was a great undertaking. Almost everyone knew that if it were completed successfully and rapidly enough, it might determine the outcome of the war. Almost everyone knew that it was an unparalleled opportunity to bring to bear the basic knowledge of art and science for the benefit of his country. Almost everyone knew that this job, if it were achieved, would be a part of history. This sense of excitement, of devotion and of patriotism in the end prevailed.[51]

Eight Nobel Prize winners worked as Manhattan Project scientists,[52] and the money and prestige associated with atomic research were astonishing. According to one Manhattan Project scientist: "It was fantastic—we could buy any piece of machinery or equipment, and you never had to justify it."[53] The Manhattan Project had its own funding source, and all project details were considered top secret. Even congressional leaders were kept in the dark.

Between 1942 and 1946, the Manhattan Project employed more than 130,000 people at a cost of approximately $2 billion (which equates to about $38 billion in 2024). While atomic bombs were designed and built at Los Alamos, other facilities included production reactors and plutonium extraction canyons at Hanford, Washington, and an 800-meter-long uranium enrichment facility at Oak Ridge, Tennessee. Some scientists firmly believed there was a moral imperative to conduct the work, while others struggled with the morality of dedicating their efforts to weapons of mass destruction.[54] As Edward Teller—a physicist who was an integral member of the Manhattan Project—stated: "If the scientists in the free countries will not make weapons to defend the freedom of their countries, then freedom would be lost."[55]

Researchers understood that radiation exposure was likely to be quite dangerous, even though "the deleterious effects of radiation could not be seen or felt and the results of over-exposure might not become apparent for long periods after such exposure."[56] Although internally, Manhattan Project physicians warned of radiation dangers from occupational exposure and fallout from test explosions, they also worked with military leaders to downplay the adverse effects, often due to fear of litigation.[57]

One Manhattan Project physician recalled that the military was "interested in having a usable bomb and protecting security. The physicists were anxious to know whether the bomb worked or not and whether their efforts had been successful . . . radiation hazards were entirely secondary."[58] In 1943, scientists noted, "Never before has so large a collection of individuals been exposed to so much radiation."[59] According to the Final Report of the President's Advisory Committee on Human Radiation Experiments, which was published in 1995, rather than disclose the risks and take appropriate mitigation measures, officials were concerned that "word of death or toxic hazard could leak out to the surrounding community and blow the project's cover."[60]

The first atomic bomb test explosion—code-named Trinity—took place on July 16, 1945, just over three years after the creation of the Manhattan Project and one day before the commencement of the Potsdam Conference, where Harry Truman, Joseph Stalin, and Winston Churchill met to decide the postwar fate of Germany, which had agreed to unconditional surrender on May 8, 1945. In the weeks prior to the test blast, several meetings considered "the medico-legal aspects" of radiation hazards. Military lawyers required radiation readings before and after the blast and signed affidavits that attested to their accuracy so the documents could be used in court proceedings. Manhattan Project physicians sought to distance themselves from potential legal liability, and wrote memos that warned of potential radiation harms to people and the environment.[61] As one physician, Louis Hempelmann, recalled: "All my memos were put in the waste basket."[62] As he further detailed, the radiation exposure limits that the doctors had deemed to be "safe" during the Trinity test were "awfully arbitrary . . . We didn't know what the hell we were doing . . . and we were just hoping that the situation wouldn't get terribly sticky."[63]

Although post-blast readings found radiation levels near homes higher than that deemed to be safe, no one was evacuated. According to reports, the fallout from the Trinity blast reached a campsite where twelve girls were sleeping. During the day, the girls frolicked in the summer "snow"—white ash that was warm to the touch. Ten of the twelve girls died of cancer before turning thirty. Studies later revealed that the radiation exposure from the Trinity blast was more than eight hundred times higher than that deemed safe.[64] As Hempelmann remarked: "No one really wanted to pursue the radiation possibilities for fear of getting involved in litigation."[65]

Four hours after the Trinity explosion, the *USS Indianapolis* sailed under San Francisco's Golden Gate Bridge with the Little Boy bomb in its hull. On July 26, 1945, the ship arrived at Tinian, a small island in the South Pacific, and discharged its valuable cargo. Four days later, the Japanese torpedoed the *Indianapolis*, and within twelve minutes, the vessel sank.[66] On the morning of August 6, 1945, the Enola Gay took flight, and the world witnessed the first attack with an atomic weapon. Just before takeoff, a chaplain offered a prayer to God "to be with those who brave the heights of Thy heaven and who carry the battle to our enemies."[67]

Shortly after the Hiroshima bombing, a White House press release lauded the attack as "the greatest achievement of organized science in history."[68] Secretary of War Henry Stimson remarked: "No praise is too great for the unstinting efforts, brilliant achievements, and complete devotion to national interest of the scientists of this country. Nowhere else in the world has science performed so successfully in time of war."[69]

Although most researchers celebrated the bombing, some expressed remorse. One of the chief architects of the Manhattan Project, Leo Szilard—a physicist who conceived of the nuclear chain reaction, was a co-inventor with Enrico Fermi on a patent for a nuclear reactor, and drafted the letter for Albert Einstein's signature that eventually led to the creation of the Manhattan Project—called the bombing of Hiroshima "one of the greatest blunders of history."[70] After the war, Oppenheimer admitted "a very great sense of revulsion and of wrong."[71] Months later, Einstein would write that the "unleashed power of the atom has changed everything save our mode of thinking, and thus we drift towards unparalleled catastrophe."[72]

Years later, President Dwight Eisenhower recalled a pre-Hiroshima discussion regarding the use of atomic weapons:

> I listened, and I didn't volunteer anything because, after all, my war was over in Europe and it wasn't up to me. But I was getting more and more depressed just thinking about it. Then he [Secretary of War Stimson] asked for my opinion, so I told him I was against it on two counts. First, the Japanese were ready to surrender and it wasn't necessary to hit them with that awful thing. Second, I hated to see our country be the first to use such a weapon. Well . . . the old gentleman got furious. And I can see how he would. After all, it had been his responsibility to push for all the huge expenditure to develop the bomb, which of course he had a right to do, and was right to do. Still, it was an awful problem.[73]

Eisenhower echoed these thoughts during his first inaugural address, where he warned that "science seems ready to confer upon us, as its final gift, the power to erase human life from this planet."[74]

Notwithstanding the sheer devastation that fell upon Hiroshima, Japan refused to surrender unconditionally. Three days later, the United States dropped the Fat Man atomic bomb on Nagasaki, and within days, unconditional surrender was achieved. As Truman wrote in his private diary: "It is certainly a good thing for the world that Hitler's crowd or Stalin's did not discover this atomic bomb. It seems to be the most terrible thing ever discovered, but it can be made the most useful."[75]

Chemical Weapons Research

Alongside atomic weapons research, chemical weapons research expanded during WWII. The budget of the Chemical Warfare Service grew exponentially—from $2 million in 1940 to $60 million in 1941, reaching $1 billion in 1942 (which, combined, equates to about $20.2 billion in 2024).[76] In 1939, the Chemical Warfare Service employed 1,000 officers and 1,300 civilians, rising to more than 66,000 officers and 25,000 civilians between 1942 and 1944.[77] The United States established thirteen new chemical weapons facilities, including a fifteen thousand-acre complex at Arkansas's Pine Bluff Arsenal, which employed ten thousand at its peak, and the Rocky Mountain Arsenal in Colorado, which employed three thousand. A massive test site was created at Dugway Proving Ground in Utah that was forty times the size of Britain's Porton Down; among its structures, the facility included replicas of German and Japanese homes.[78]

American scientists examined thousands of compounds to find those that were highly toxic, available in large quantities, nonflammable, noncorrosive, and stable enough to be stored and transported. The work was conducted in collaboration with several leading companies, including Monsanto, DuPont, and American Cyanamid. The most desirable chemicals had a density heavier than air so that the compound would linger over the targeted area for an extended time.[79] The goal was to contaminate a large area efficiently and effectively.

Since the 1922 Washington Arms Treaty was never ratified and the United States did not ratify the 1925 Geneva Protocol, the government claimed that the restrictions on chemical and biological weapons did not legally bind the United States. War Department field manuals published in 1934 and 1940 underscored that "the United States is not a party to any treaty, now in force, that prohibits or restricts the use in warfare of toxic or non-toxic gases, or of smoke or incendiary materials."[80] In 1940, before the United States officially entered the war, it was secretly selling chemical weapons to Britain; the poison gases were manufactured by American companies and shipped in vessels registered to foreign countries.[81] When America entered the war, the United States was equipped to mount and defend against a chemical attack.

The OSRD led wartime chemical weapons research, which included tests with human subjects. During the war, more than sixty thousand service members participated in chemical weapons research, at least four thousand of whom were exposed to mustard gas or lewisite.[82] Many of the studies were conducted in collaboration with leading universities such as the University of Chicago, Harvard, and Cornell. Some of the studies examined prophylactic ointments and gas masks, while others sought to determine whether race or skin complexion impacted one's susceptibility to mustard gas. For example, scientists at Cornell posited that non-white people had "thicker skin," making them less sensitive to mustard gas.[83] In 1944, the Army provided the

Chemical Warfare Service with forty Japanese-American service members to study variations in response to blistering agents between Asians and Caucasians.[84]

Experiments with mustard gas bomb cases were conducted behind Harvard Stadium.[85] In other tests, service members stood in a field wearing various levels of protective clothing as low-flying airplanes sprayed mustard gas or other toxic chemicals.[86] Military officials also ordered subordinates to crawl through fields saturated with mustard gas. According to Rollins Edwards, a participant in the mustard gas experiments: "It took all the skin off your arms . . . You do what they tell you to do and you ask no questions."[87] Speaking in 2015, Edwards indicated that he did not refuse or question the work, since "defiance was unthinkable . . . especially for black soldiers."[88] For more than seventy years his skin was constantly irritated and flaking, a painful daily reminder of the wartime experiments.[89]

Field studies with gas weapons also were conducted outside America's borders. On an island off the coast of Panama, American bombers dropped more than two hundred mustard gas bombs that caused concentrations "sufficient to cause 100 per cent casualties from severe blistering and systemic poisoning among masked but otherwise unprotected troops."[90] One hour later, "fully protected" American troops engaged in mock warfare over the field.[91] As reports detailed, many sustained gas-related injuries, and "there were moments when panic or mass hysteria seemed close to the surface."[92]

In addition to the field experiments, researchers conducted what they called "man-break" tests whereby service members were locked in gas chambers and exposed to noxious chemicals.[93] In many tests, service members would remain in the chambers inundated with mustard gas or lewisite for up to four hours; thereafter, they were ordered to wear their clothes and protective gear for up to twenty-four hours.[94] Some men endured experiments in gas "chambers either every day or every other day until they developed moderate to intense erythema," which entails skin rashes and irritation.[95]

At times, officers working on the experiments recruited service members under false pretenses, and when the men would report for duty, they would be ordered into a gas chamber, much like that depicted in Figure 7.1. The experience of seventeen-year-old Nathan Schnurman, who was recruited to test summer uniforms for the Navy, is illustrative. Schnurman arrived at the test facility, where he was issued a gas mask and specialized clothing and ordered into a gas chamber so the Navy could examine the effectiveness of the protective equipment.[96] The chamber could not be unlocked from the inside. As Schnurman recalled, "I looked up at the ceiling and saw dark yellow oily mist rolling in."[97] The mist penetrated his mask, and he pleaded to be let out of the chamber. The researchers refused—Schnurman vomited into his mask, suffered a heart attack, and fainted.[98]

A Naval Research Laboratory report noted that for men who "did not cooperate fully" in the man-break experiments, "a short explanatory talk, and, if necessary, a slight verbal 'dressing down' has always proven successful."[99] As the report further stated: "There has not been a single instance in which a man has refused to enter the gas chamber."[100] Commanding officers threatened men with sanctions that included court-martial and prison time. Some officials ordered "malingerers and psychoneurotics" to the gas chambers as a form of corporal punishment.[101]

Figure 7.1 Test subjects enter a gas chamber as part of the US military's mustard gas experiments.
U.S. Army photograph.

As their name suggests, the man-break tests were grueling and resulted in severe research-related injuries, sometimes even death. At times, the gas levels were equivalent to those reported on WWI battlefields. Research participants experienced severe eye injuries and skin blisters on nearly every part of their body. Exposure to mustard gas also causes blindness, intense vomiting, internal and external bleeding, and damage to the lungs and respiratory system. Many suffered long-term health effects that included cancer, asthma, and psychological disorders. On top of their research-related injuries, the men were told that they would be prosecuted under the Espionage Act if they disclosed the true reason for their ailments. This led to misdiagnoses and insufficient medical care after the war.[102]

For decades the US government refused to acknowledge the existence of the studies or provide injured service members with compensation or long-term healthcare. Schnurman filed a lawsuit in 1979, but a court dismissed his case pursuant to the *Feres* doctrine's legal immunities that shield the government against claims for service-related harms.[103] In 1990, he led a group of men who decided to mount a public campaign that detailed the experiments. One year later, the government admitted to using service members in the research. The government also revealed that it did not fully disclose safety risks or obtain informed consent from the men, and that individuals may have suffered adverse health effects as a result of their participation in the studies. In turn, the government offered compensation and medical treatment for research-related harms.[104]

Following the government's admissions, the Veterans Administration (VA) asked the Institute of Medicine (IOM) to investigate the mustard gas experiments. During its investigation, the IOM found that "an atmosphere of lingering secrecy still existed in the Department of Defense," including "a picture of abuse and neglect that was impossible for the committee to ignore."[105] The IOM expressed dismay that the Department of Defense (DoD) opted not to conduct any epidemiological studies of individuals exposed to the chemical weapons, either as test subjects or from occupational exposure as weapons production workers or war gas handlers. The IOM also concluded that exposure to the noxious substances was causally related to health conditions that included respiratory cancers, skin cancers, leukemia, chronic respiratory diseases, bone marrow depression, psychological disorders, and more. The IOM recommended that the VA create a program to identify and provide treatment to individuals exposed to chemical weapons and to conduct follow-up studies to examine the health impact of exposure.[106]

Notwithstanding the IOM report—and despite assurances from the VA that surviving veterans from the mustard gas experiments would be contacted and provided appropriate care—an investigative report published in 2015 revealed that the VA did not contact 85 percent of the approximately four thousand surviving research participants. For the 610 veterans who were contacted, the VA simply sent a letter in the mail. Although the VA later claimed that incomplete medical records caused the lackluster outreach, within two months an investigative journalist and her team were able to locate more than 1,200 veterans using the VA's records.[107]

According to the 2015 report, many veterans who attempted to obtain medical care or compensation described "an unending cycle of appeals and denials as they struggled to get government benefits for mustard gas exposure. Some gave up out of frustration."[108] Consider the experience of Charlie Cavell, who was nineteen years old at the time of the experiments: "We weren't told what it was . . . until we actually got into the process of being in that room and realized, wait a minute, we can't get out of here."[109] After one hour of mustard gas exposure in the gas chamber, half the service members were let out, though Cavell and five others were told to remain; after a second hour in the chamber, Cavell was released but told he must continue wearing his gas-saturated uniform. Officials also told him that if he revealed what had happened to him, he would be dishonorably discharged and incarcerated in a military prison at Fort Leavenworth, Kansas.[110]

As he stated in 2015: "They put the fear of God in just a bunch of young kids."[111] Cavell's claims for his skin cancer and respiratory ailments were denied, despite the fact that he actually obtained medical records that detailed his exposure to mustard gas. Following the 2015 investigative report, the VA reconsidered Cavell's application, though the agency required that he complete a reevaluation before any funds could be disbursed.[112] Just months before his death, and nearly seventy years after the experiments, the VA finally approved Cavell's claim.[113]

America's chemical warfare experiments were similar to those conducted by other Allied Powers. For example, Britain conducted hundreds of experiments utilizing service members as human subjects, including gas chamber tests and studies involving anti-gas ointments, equipment, and clothing. The tests caused short- and long-term adverse health consequences and, in some cases, death. Many of the studies were conducted at Porton Down, Britain's center for chemical warfare, in collaboration with scientists from leading British universities, including the University of Oxford. Similar to their American counterparts, British service members were legally precluded from disclosing the existence of, or their participation in, the research.[114] In analyzing the work of Porton Down scientists, historian Ulf Schmidt explains: "Without malicious intent, they frequently breached established standards of professional medical conduct. Theirs was a wartime emergency requiring urgent and decisive action, they believed, one in which there was little time to worry about ethics."[115]

The Allied Powers considered the deployment of chemical weapons during the war. In late spring of 1942, Churchill announced that "we shall treat the unprovoked use of poison gas against our Russian ally exactly as if it were used against ourselves, and if we are satisfied that this new outrage has been committed by Hitler we will use our great and growing air superiority in the West to carry gas warfare on the largest possible scale far and wide against military objectives in Germany."[116] Within weeks Roosevelt issued an analogous warning against Japan, should it continue its chemical attacks against China or commence attacks against other allied nations.[117]

In July 1944, Churchill wrote to a key advisor: "It may be several weeks or even months before I shall ask you to drench Germany with poison gas, and if we do it, let us do it one hundred percent. In the meanwhile, I want the matter studied in cold blood by sensible people and not by that particular set of psalm-singing uniformed defeatists which one runs across now here now there." Churchill further explained:

> It is absurd to consider morality on this topic when everybody used it in the last war without a word of complaint from the moralists or the Church. On the other hand, in the last war the bombing of open cities was regarded as forbidden. Now everybody does it as a matter of course. It is simply a question of fashion changing as she does between long and short skirts for women. . . . I shall of course have to square Uncle Joe and the President; but you need not bring this into your calculations at the present time.[118]

Sir John Dill, a high-ranking British military official, added: "At a time when our National existence is at stake, when we are threatened by an implacable enemy who himself recognizes no rules save those for expediency, we should not hesitate to adopt whatever means appear to offer the best chance of success."[119]

On the eve of the D-Day invasion, both Roosevelt and Churchill feared that Germany would respond with chemical or biological warfare. In preparation, the nations stockpiled a sixty-day supply of chemical bombs that could be used to mount two four-hundred-bomber attacks that would dispense hundreds of tons of mustard gas and phosgene.[120] In the spring of 1943, Germany doubled its Sarin and Tabun production, and during the summer of 1943, a German attack on US merchant ships stationed in Italy released mustard gas, which was secretly stored in the cargo of one of the ships, causing severe injuries to sailors and medical professionals who handled patients without knowing they were exposed to the noxious substance. Of the 628 individuals hospitalized due to the exposure, 83 died; American documents related to the incident were not declassified until 1959.[121]

American leaders also approved a preemptive gas attack on Tokyo and other Japanese cities. At the time, the American public supported chemical warfare: one study found that 40 percent of Americans favored a gas attack if it would reduce American casualties, and a separate study found support at 70 percent. Following the war, the secret document that authorized the gas attack on Japan was altered to add the word "retaliatory," and both the original and altered documents were kept classified until 1995.[122]

Biological Weapons Research

WWII also marked the commencement of America's biological warfare program. In 1942 the National Academy of Sciences issued a secret report that evaluated biological warfare threats and highlighted the potential use of biological agents to harm humans, livestock, and agriculture. Shortly thereafter, Secretary of War Stimson wrote to President Roosevelt: "Biological Warfare is, of course, 'dirty business' but in light of the committee's report, I think we must be prepared. And the matter must be handled with great secrecy as well as great vigor." Stimson called for the establishment of a civilian agency that would secretly evaluate biological warfare threats and the creation of biological weapons. "Entrusting the matter to a civilian agency," Stimson noted, might help assure the public that the United States was not contemplating offensive use of biological weapons. "To be sure," Stimson added, "Offensive possibilities should be known to the War Department" since they are "not beyond the bounds of possibility any more than they are in the field of gas attack for which the Chemical Warfare Service of the War Department is prepared."[123]

Within days Roosevelt approved the plan. The program was given the innocuous title of the War Research Service and was housed in a civilian outpost, the Federal Security Agency. George Merck, founder of Merck & Company, ran the program, and its scientific director was Ira Baldwin, a professor of bacteriology from the University of Wisconsin. Merck was a leading figure in the pharmaceutical and vaccine industries; among his company's products were the first smallpox vaccine produced in the United States and penicillin G, one of the first antibiotics. Just as the chemical industry lobbied during the interwar period for an expansive chemical warfare program, the pharmaceutical and vaccine industries touted the benefits of biological warfare research during WWII.[124]

As biological warfare research expanded, military officials recognized the moral quandaries of employing physicians in the work. In a series of meetings and reports dating back to the summer of 1941, officials suggested that the work of physicians in the Army Medical Corps and the Surgeon General's Office be limited to defensive measures while the Chemical Warfare Service would be charged with offensive protocols. Although the Chemical Warfare Service also employed physicians, military and government leaders viewed the dichotomy as an acceptable moral compromise because physicians with clinical responsibilities were not conducting research into the offensive use of biological weapons.[125]

Between 1942 and 1945, the United States invested more than $40 million in its biological weapons program (which equates to approximately $760 million in 2024).

The main research center was located at Camp Detrick in Frederick, Maryland, primarily because of its proximity to Washington, DC and Edgewood Arsenal, the headquarters of the Chemical Warfare Service. Research and development sites also were established at Horn Island, Mississippi, Dugway Proving Ground, Utah, and Vigo, Indiana. Researchers studied the weaponization of biological agents such as anthrax, plague, yellow fever, botulinum toxin, glanders, tularemia, typhus, and encephalitis. Using gas chambers, scientists analyzed the impact of these agents on animals and explored the use of beetles, fleas, and other insects as disease-carrying weapons.[126]

Army and Navy physicians cooperated on biological weapons research, and by the end of WWII, approximately four thousand people were working on two hundred projects in collaboration with dozens of private companies and universities, including Columbia, Cornell, Harvard, Johns Hopkins, Northwestern, Ohio State, Stanford, Notre Dame, Vanderbilt, and the universities of California, Chicago, Illinois, Pennsylvania, and Wisconsin.[127] As Merck explained: "All known pathogenic agents were subjected to thorough study and screening by scientists of the highest competence."[128] By the spring of 1944, Camp Detrick had produced 5,000 anthrax-filled bombs, and the Vigo plant was equipped to make 500,000 anthrax bombs a month.[129]

In 1944, Roosevelt liquidated the War Research Service and placed the biological weapons program under the control of the War Department, which housed the program within the Chemical Warfare Service.[130] The biological weapons research in the Chemical Warfare Service was furthered by the Office of Strategic Services (OSS), a department Roosevelt created in 1942 that was the precursor to the Central Intelligence Agency (CIA). OSS's leader, General William "Wild Bill" Donovan, promised Roosevelt that OSS would be staffed with young officers who were "calculatingly reckless with disciplined daring, who are trained for aggressive action."[131] Roosevelt recognized the wartime benefits of the intelligence unit, but he was concerned that the OSS would evolve into an American Gestapo. In turn, he ordered his chief White House military aide to conduct a secret investigation into OSS's operations.[132] More than thirteen thousand agents served in OSS during the war, including ivory-tower intellectuals, corporate executives, scientists, lawyers, movie directors, and mafia hitmen. The work was shrouded in secrecy, funding was plentiful and largely unregulated, and several agents stole substantial amounts of cash that was intended to support foreign assets.[133]

Among its myriad projects, the OSS researched biological warfare. Donovan selected Stanley Lovell, a scientist and businessman from Boston, to spearhead OSS's clandestine projects, including biological weapons research.[134] Donovan asked Lovell to think outside the box and create "every subtle device and every underhanded trick to use against the Germans and Japanese."[135] In one project, the group created goat dung laced with debilitating diseases that could be scattered by aircraft over enemy villages; in another, Lovell's team shipped botulinum toxin pills to prostitutes in occupied China with instructions to slip the pills to unsuspecting Japanese clients.[136] Botulinum toxin is among the most toxic substances known to man—its effects begin as a nasty bout of food poisoning but will kill three in five of those infected.

As with atomic and chemical weapons, there were significant occupational and research-related hazards in biological weapons research. At one point the head of the

Chemical Warfare Service considered whether the military could secretly dispose of the remains of civilians or service members in the event of research-related deaths.[137] Thereafter, the Judge Advocate General of the Army "decreed that by establishing a restricted military area at Camp Detrick, deceased personnel might be placed in a hermetically sealed metal casket and interred by military personnel in the area, without disclosing by certificate, report or statement the nature or cause of death."[138]

Throughout the war, America collaborated with Britain and Canada on biological weapons research, and Britain asked the United States to manufacture weapons-grade anthrax and botulinum toxin.[139] Following the war, Britain's anthrax experiments on Gruinard Island, a small islet off the northwest coast of Scotland, received considerable attention. Anthrax exposure, even in very small amounts, can cause skin ulcers, blood poisoning, intense coughing, difficulty breathing, and death. The tests at Gruinard—which included bombing areas staged with sheep and other animals—confirmed the lethality of anthrax and the ease at which it can spread and lie dormant in the land. Gruinard Island remained contaminated with anthrax for over forty years; every four hundred yards around the perimeter of the island, a sign advised of the contamination and prohibited entry. In 1986, the island was decontaminated with formaldehyde and seawater, and in 1990 British officials visited the island, declared it safe, and removed the signs.[140]

Research and development of biological weapons accelerated after reports indicated that Japan dispersed plague-infested fleas in China.[141] British officials ordered a full-scale review of their biological weapons program, urging that "It should take the form of a thorough and practical examination of the military factors involved, and should ignore ethical and political considerations."[142] A subsequent report concluded that biological weapons "could probably make a material change in the war situation before the end of 1945."[143] On whether anti-crop or anti-livestock agents should be used against the Japanese, American officials reasoned that the United States "is entitled to deprive the enemy of food and water, and to destroy the sources of supply in his fields."[144] The Allies also prepared plans to deploy biological weapons against Germany, estimating that if anthrax were used, it might kill half of the inhabitants and contaminate cities for decades. Reports alleged that the Allies used biological agents to assassinate German officials, though the claims were unsubstantiated.[145]

By 1945, some Allied leaders viewed atomic and biological weapons as "complementary," while others characterized biological weapons "as very humane indeed by comparison with atom bombs."[146] America's biological weapons program was not publicly revealed until January 3, 1946, when the government published a report authored by Merck. Merck's report warned of the dangers of biological weapons, confirmed the existence of an American biological weapons program, and indicated that the program served both to deter biological attacks and to arm the United States with biological weapons. The report characterized biological weapons as "possible in many countries, large and small, without vast expenditures of money or the construction of huge production facilities."[147] The report also noted that biological warfare research could be conducted "under the guise of legitimate medical or bacteriological research."[148] As Merck concluded: "Work in this field, born of the necessity of war, cannot be ignored in time of peace."[149]

Summary

For more than 140 years—from the American Revolutionary War up to the beginning of WWI—the fulfillment of military medicine's mandate to ensure a healthy fighting force encompassed work aimed at disease prevention and treatment for injured service members. A new path was forged during the first half of the twentieth century, as physicians and researchers began assisting in the development of chemical, biological, and atomic weapons. This transformation occurred over three tumultuous decades, during which the United States fought two world wars and endured the Great Depression.

As distinguished physician and Brigadier General Stanhope Bayne-Jones explained, incorporating weapons development into military medicine created a "Janus-like configuration," whereby roles and fiduciary duties of military physicians and researchers conflicted in irreconcilable ways.[150] According to Bayne-Jones, who was dean of Yale Medical School and Deputy Chief of the Preventive Medicine Service in the Office of the Surgeon General during WWII, the incongruous duties generated "a serious moral and practical problem" and caused "a somewhat schizophrenic behavior of the Army Medical Department."[151] Despite the quandaries, the front lines of science were marching on. National security was the ultimate driving force, a fact that would encumber American prosecutors during postwar tribunals at Nuremberg.

Notes

1. Christopher J. Tassava, "The American Economy During World War II," *Economic History Association*, February 10, 2008; Stephen Daggett, "Costs of Major U.S. Wars," *Congressional Research Service* (June 29, 2010): 2.
2. Gillett, *The Army Medical Department, 1917-1941*, 536.
3. Greenwood and Berry, *Medics*, 95–98.
4. Rostker, *Providing*, 175.
5. Gillett, *The Army Medical Department, 1917-1941*, 564.
6. Rostker, *Providing*, 179, 200–13.
7. Dave Grossman, *On Killing* (2009), 43.
8. Rostker, *Providing*, 204–7.
9. Kamienski, *Shooting Up*, 104–42.
10. Rostker, *Providing*, 192–98.
11. Greenwood and Berry, *Medics*, 103.
12. Greenwood and Berry, *Medics*, 88–89; Rostker, *Providing*, 190–92.
13. Ginn, *The History*, 173–74.
14. Martin Furmanski, "Unlicensed Vaccines and Bioweapon Defense in World War II," *Journal of the American Medical Association* 282, no. 9 (September 1, 1999): 822.
15. Furmanski, "Unlicensed Vaccines"; Leonard B. Seeff, "Yellow Fever Vaccine-Associated Hepatitis Epidemic During World War II," in Institute of Medicine, *Epidemiology in Military and Veteran Populations: Proceedings of the Second Biennial Conference, March 7, 1990* (Washington, DC: National Academy Press, 1991), 9–17.

16. Furmanski, "Unlicensed Vaccines"; Paul A. Lombardo, "'Of Utmost National Urgency': The Lynchburg Colony Hepatitis Study, 1942," in *In the Wake of Terror*, edited by Jonathan D. Moreno (2003), 3–13.
17. Frank L. Smith, *American Biodefense* (2014), 97.
18. Irvin Stewart, *Organizing Scientific Research for War* (1948), 3–6.
19. Stewart, *Organizing*, 6.
20. Vannevar Bush, *Science: The Endless Frontier* (1945), 34.
21. Vannevar Bush, *Pieces of the Action* (1970), 36.
22. Bush, *Pieces*, 36–37.
23. Bush, *Pieces*, 31–32.
24. Bush, *Pieces*, 38–39.
25. Bush, *Pieces*, 42–47; Patricia Spain Ward, "The Medical Antitrust Case of 1938-1943," *American Studies* 30, no. 2 (Fall 1989): 123–53.
26. Executive Order 8807 (June 28, 1941), 6 Fed. Reg. 3207 (July 2, 1941).
27. Bush, *Pieces*, 42–47.
28. Moreno, *Undue Risk*, 40.
29. Bush, *Science*, 53.
30. A. N. Richards, Letter to J. E. Moore, October 31, 1942 ("Revision of Dr. Richards' letter of October 9, 1942"), quoted in Final Report of the President's Advisory Committee, *The Human Radiation Experiments* (1996), 54.
31. Lederer, *Subjected to Science*, 140.
32. Peter J. Weina, "From Atabrine in World War II to Mefloquine in Somalia: The Role of Education in Preventive Medicine," *Military Medicine* 163, no. 9 (September 1998), 635–39; Karen M. Masterson, *The Malaria Project* (2014), 13, 260.
33. Masterson, *The Malaria Project*, 161–62, 299.
34. James Shannon, Letter to Dr. Andrus (October 26, 1942), quoted in Masterson, *The Malaria Project*, 163.
35. Masterson, *The Malaria Project*, 272.
36. Gretchen Vogel, "Malaria as Lifesaving Therapy," *Science* 342 (November 8, 2013): 686.
37. Alf Alving, Letter to Robert Loeb (June 3, 1944), quoted in Masterson, *The Malaria Project*, 272.
38. Masterson, *The Malaria Project*, 272.
39. Masterson, *The Malaria Project*, 223, 230–32, 266–72.
40. Masterson, *The Malaria Project*, 230–32.
41. Masterson, *The Malaria Project*, 327.
42. "Prison Malaria," *Life*, June 4, 1945, 43–46.
43. Masterson, *The Malaria Project*, 327.
44. Masterson, *The Malaria Project*, 266–72.
45. Moreno, *Undue Risk*, 22.
46. Stanhope Bayne-Jones, Letter to G. D. Fairbairn (February 16, 1945), quoted in Moreno, *Undue Risk*, 22.
47. Albert Einstein, Letter to F. D. Roosevelt (August 2, 1939); Albert Einstein, Letter to Henrich Zangger (December 1917), quoted in Albrecht Folsing, *Albert Einstein: A Biography*, trans. Ewald Osers (New York: Penguin Books, 1998), 399–402.
48. Arthur H. Compton, Letter to Frank B. Jewett, "Report of National Academy of Sciences Committee on Atomic Fission" (May 17, 1941): 2.
49. Final Report, *The Human Radiation Experiments*, 325.
50. J. Robert Oppenheimer, Letter to Maj. Gen. K. D. Nichols, General Manager, AEC (March 4, 1954).
51. Oppenheimer, Letter to Nichols.

52. James L. Nolan, *Atomic Doctors* (2020), 10.
53. Stanton Cohn, quoted in Eileen Welsome, *The Plutonium Files* (1999), 210.
54. Richard Rhodes, *The Making of the Atomic Bomb* (2012), 1, 331–36.
55. Edward Teller, *Energy from Heaven and Earth* (1979), 145.
56. Manhattan District Program, "Medical Program" (December 31, 1946), quoted in Final Report, *The Human Radiation Experiments*, 6.
57. Nolan, *Atomic Doctors*, 2.
58. James Nolan, Letter to Richard Newcomb (August 12, 1957), quoted in Nolan, *Atomic Doctors*, 11.
59. Robert Stone, "Health Radiation and Protection" (May 10, 1943), quoted in Final Report, *The Human Radiation Experiments*, 6.
60. Final Report, *The Human Radiation Experiments*, 6.
61. Nolan, *Atomic Doctors*, 47–48.
62. Quoted in Nolan, *Atomic Doctors*, 48.
63. Quoted in Nolan, *Atomic Doctors*, 50.
64. Nolan, *Atomic Doctors*, 50–57.
65. Quoted in Nolan, *Atomic Doctors*, 56.
66. Rhodes, *The Making*, 678; Nolan, *Atomic Doctors*, 62–68.
67. Quoted in Rhodes, *The Making*, 704.
68. Harry Truman, Statement by the President Announcing the Use of the A-Bomb at Hiroshima (August 6, 1945).
69. Henry Stimson, Statement of the Secretary of War (August 6, 1945).
70. Leo Szilard, Letter to Gertrud Weiss (August 6, 1945), quoted in Rhodes, *The Making*, 735.
71. Quoted in Rhodes, *The Making*, 570.
72. Quoted in "Atomic Education Urged by Einstein; Scientist in Plea for $200,000 to Promote New Type of Essential Thinking," *The New York Times*, May 25, 1946.
73. "Ike on Ike," *Newsweek*, November 11, 1963, quoted in Rhodes, *The Making*, 688.
74. Dwight D. Eisenhower, Inaugural Address (January 20, 1953).
75. Harry S. Truman, "Diary on the Atomic Bomb" (1945).
76. Tucker, *War of Nerves*, 89.
77. Schmidt, *Secret Science*, 138–39.
78. Harris and Paxman, *A Higher Form of Killing*, 118–19.
79. Leo P. Brophy, Wyndham D. Miles, and Rexmond C. Cochrane, *The Chemical Warfare Service* (1959), 49–50; McCamley, *Secret History*, 91–92.
80. War Department, *Basic Field Manual*, Vol. VII, Part Two, *Rules of Land Warfare* (1934), 8; War Department, *Field Manual 27-10: Rules of Land Warfare* (1940), 8.
81. Harris and Paxman, *A Higher Form of Killing*, 117.
82. IOM, *Veterans at Risk*, 1.
83. Marion B. Sulzberger et al., "Skin Sensitization to Vesicant Agents of Chemical Warfare," *Journal of Investigative Dermatology* 8 (1947): 365–93; Susan L. Smith, "Mustard Gas and American Race-Based Human Experimentation in World War II," *Journal of Law, Medicine and Ethics* 36, no. 3 (Fall 2008): 517–21.
84. Schmidt, *Secret Science*, 142.
85. Coffey, *American Arsenal*, 93–94.
86. Smith, "Mustard Gas."
87. Quoted in Caitlin Dickerson, "Secret World War II Chemical Experiments Tested Troops by Race," *NPR*, June 22, 2015.
88. Quoted in Dickerson, "Secret World War II Chemical Experiments."
89. Dickerson, "Secret World War II Chemical Experiments."
90. Chemical Corps Association, *The Chemical Warfare Service in World War II* (1938), 38.

91. Chemical Corps, *The Chemical Warfare Service*, 39.
92. Chemical Corps, *The Chemical Warfare Service*, 39.
93. Senate Committee on Veterans' Affairs Staff Report, *Is Military Research Hazardous to Veterans' Health?: Lessons Spanning Half a Century* (Washington, DC: US Government Printing Office, 1994), 18–19.
94. IOM, *Veterans at Risk*, 36.
95. IOM, *Veterans at Risk*, 39.
96. *Schnurman v. United States*, 490 F. Supp. 429 (E.D. Va. 1980).
97. Quoted in Patrick Cockburn, "US Navy Tested Mustard Gas on its Own Sailors," *The Independent*, March 14, 1993.
98. Cockburn, "US Navy."
99. Navy Research Laboratory, *Report on Chamber Tests with Human Subjects* (December 22, 1943), 23.
100. Navy Research Laboratory, *Report on Chamber Tests*, 23.
101. IOM, *Veterans at Risk*, 67–68.
102. IOM, *Veterans at Risk*, vii, 1–8, 81–213.
103. *Schnurman*, 490 F. Supp. at 429.
104. IOM, *Veterans at Risk*, v–vi, 1–3.
105. IOM, *Veterans at Risk*, vi–vii.
106. IOM, *Veterans at Risk*, 3–8.
107. Caitlin Dickerson, "The VA's Broken Promise to Thousands of Vets Exposed to Mustard Gas," *NPR*, June 23, 2015.
108. Quoted in Dickerson, "The VA's Broken Promise."
109. Quoted in Dickerson, "The VA's Broken Promise."
110. Dickerson, "The VA's Broken Promise."
111. Quoted in Dickerson, "The VA's Broken Promise."
112. Dickerson, "The VA's Broken Promise."
113. Caitlin Dickerson, "WWII Veteran, Who Fought to Expose Secret Mustard Gas Experiments, Dies," *NPR*, May 30, 2016.
114. Schmidt, *Secret Science*, 126–29, 257.
115. Schmidt, *Secret Science*, 127.
116. Winston Churchill, Broadcast Report on the War (May 10, 1942).
117. Franklin Roosevelt, Statement on Japanese Use of Poison Gas (June 5, 1942).
118. Winston Churchill, Memorandum (July 6, 1944), quoted in Harris and Paxman, *A Higher Form of Killing*, 129–30.
119. John Dill, "The Use of Gas in Home Defence" (June 15, 1940), quoted in Harris and Paxman, *A Higher Form of Killing*, 111–12.
120. Tucker, *War of Nerves*, 64–65.
121. Coffey, *American Arsenal*, 153–57; Jennet Conant, *The Great Secret* (2020).
122. Brown, *Chemical Warfare*, 287–89; Thomas Allen and Norman Polmar, "Poisonous Invasion Prelude," *Pittsburgh Post-Gazette*, August 4, 1995.
123. Henry Stimson, Letter to President Roosevelt (April 29, 1942), quoted in Harris and Paxman, *A Higher Form of Killing*, 97–98.
124. Harris and Paxman, *A Higher Form of Killing*, 98; Endicott and Hagerman, *The United States and Biological Warfare*, 27–33.
125. Endicott and Hagerman, *The United States and Biological Warfare*, 27–33.
126. Harris and Paxman, *A Higher Form of Killing*, 98–100.
127. Endicott and Hagerman, *The United States and Biological Warfare*, 27–33.
128. George W. Merck, *Biological Warfare: Report to the Secretary of War* (January 3, 1945): 3.
129. Harris and Paxman, *A Higher Form of Killing*, 104–6.

130. Smith, *American Biodefense*, 39.
131. Quoted in Central Intelligence Agency, "Some Aspects of the Activities of the Office of Strategic Services Operational Groups in World War II" (1945): 5.
132. Tim Weiner, *Legacy of Ashes* (2007), 5–9.
133. Max Hastings, *The Secret War* (2016), 284–303; Douglas Waller, *Wild Bill Donovan* (2011), 5–6, 71–73.
134. Waller, *Wild Bill Donovan*, 101–3.
135. Stanley P. Lovell, *Of Spies and Stratagems* (1963), 17.
136. Harris and Paxman, *A Higher Form of Killing*, 204–7.
137. Moreno, *Undue Risk*, 44.
138. Rexmond C. Cochrane, "Biological Warfare Research in the United States," *History of the Chemical Warfare Service in World War II*, vol. 2 (November 1947): 176–77.
139. Smith, *American Biodefense*, 42.
140. Harris and Paxman, *A Higher Form of Killing*, 72–76; "Britain's Anthrax Island," *BBC News*, July 25, 2001.
141. Harris and Paxman, *A Higher Form of Killing*, 82–83.
142. Vice Chiefs of Staff, "Instructions to the Joint Planning Staff" (July 16, 1944), quoted in Harris and Paxman, *A Higher Form of Killing*, 132.
143. Joint Planning Staff, "Military Considerations Affecting the Initiation of Chemical and Other Special Forms of Warfare" (July 27, 1944), quoted in Harris and Paxman, *A Higher Form of Killing*, 135.
144. Maj. Gen. Myron Cramer, Judge Advocate General, to Secretary of War, "Destruction of Crops by Chemicals" (March 5, 1945), quoted in Barton Bernstein, "Origins of the U.S. Biological Warfare Program," in *Preventing a Biological Arms Race*, edited by Susan Wright (1990), 19.
145. Harris and Paxman, *A Higher Form of Killing*, 91–96, 106–8.
146. Quoted in Harris and Paxman, *A Higher Form of Killing*, 108.
147. Merck, *Biological Warfare*, 6.
148. Merck, *Biological Warfare*, 6.
149. Merck, *Biological Warfare*, 6.
150. Bayne-Jones, *The Evolution*, 157–58.
151. Bayne-Jones, *The Evolution*, 157–58.

8
Justice at Nuremberg
Establishing Principles of Research Ethics

Introduction

In November 1943, the Soviet Union, the United Kingdom, and the United States issued a joint declaration proclaiming that the Allied Powers would pursue Germans responsible for wartime atrocities "to the uttermost ends of the Earth and will deliver them to their accusers in order that justice be done."[1] Two years later, the Allies fulfilled their promise, creating an international tribunal to prosecute German political leaders, military officials, industrialists, and physicians for war crimes and crimes against humanity. Held at the Palace of Justice in Nuremberg—a city regarded as the ceremonial home of the Nazi party—prosecutors presented copious evidence of genocide, mass sterilizations, and gruesome experiments in concentration camps, including inoculations with dubious vaccines, chemical weapons research, and scores of other studies.

In their defense, the Germans contended that each country must take actions that further its national security, a position that harkens back to Cicero's maxim that laws are silent in times of war. The Germans also analogized their conduct to untoward research in America, including studies conducted in prisons and psychiatric institutions, and highlighted that Germany's sterilization program was modeled on that pursued in the United States and sanctioned by American courts. The defendants further noted that their conduct did not violate any standards or norms because there was no written legal or ethical code governing research with human subjects. American prosecutors at the Nuremberg tribunal scrambled to provide convincing counterarguments, even going as far as to present false testimony regarding research ethics protocols in the United States.

At the conclusion of the proceedings, the Nuremberg tribunal set forth ten principles governing research with human subjects. Commonly referred to as the Nuremberg Code, scientists in America and other Allied nations largely ignored the protocols, characterizing the guidelines as ethical rules for barbarians. Applying the principles to their own work, they reasoned, would unnecessarily restrict researcher independence and hinder the progress of science. This moral dichotomy became embedded in American medicine and research for decades.

Germany's Wartime Research Enterprise

German universities and industries were global leaders in medicine and science during the first half of the twentieth century. The German government fostered civilian-military partnerships by providing grants for war-related projects in weapons

America's Military Biomedical Complex. Efthimios Parasidis, Oxford University Press.
 DOI: 10.1093/9780199351473.003.0009

development and medical research. Leading German companies—including Bayer, IG Farben, and Siemens—embraced the business of war, though they often structured shell corporations in an attempt to create a liability shield should postwar claims arise.[2] To further encourage robust collaborations, the German government assured physicians and scientists that they would receive full legal immunity if they participated in war preparations.[3]

Throughout Germany's research enterprise, hundreds of physicians and industrialists utilized concentration camp prisoners in forced labor and scientific experiments. The prisoners were commonly referred to as "test persons" in lab notes, scientific publications, and government reports.[4] The legion of experiments using test persons was conducted on the premise that the work would benefit German warfighters and the nation.[5] The experiments also furthered Germany's eugenics movement to create a "master race" of Aryans.

At the concentration camp in Dyhernfurth, scientists studied the effects of mustard gas, nerve agents, and other toxic chemicals by locking test persons in gas chambers for various lengths of time.[6] German researchers examined skin decontamination agents and protective clothing, often by conducting field tests whereby test persons were forced to crawl over land contaminated with chemical agents.[7] Other research subjects were treated as "human canaries" by being locked in train cars or munitions depots that were loaded with chemical weapons.[8] At Raubkammer, where a team of German scientists known as Group X was based, Allied forces discovered a display of organs from animals exposed to nerve agents and four thousand photographs of humans exposed to chemical weapons.[9] Many research subjects were photographed several times a day to examine the progression of injuries, and physicians dissected the bodies of test persons to examine the impact of chemical weapons on the lungs and other organs.[10]

As depicted in Figure 8.1, German physicians also used concentration camp prisoners to study experimental drugs and vaccines, including treatments for malaria, spotted fever, and typhus.[11] During these studies, test persons were intentionally infected with pathogens and then administered an experimental medical product. In other studies, researchers induced sepsis and examined the efficacy of novel therapies.[12] At the Gross-Rosen concentration camp, German scientists conducted biological warfare experiments on Russian prisoners of war.[13] German physicians often maintained live human "carriers"—test persons intentionally infected with a disease solely to provide fresh blood for additional studies.[14] At Dachau and Auschwitz, physicians examined "chemical methods for the abrogation of the will" by administering mescaline, barbiturates, and other hallucinogenic substances in an attempt to elicit information from prisoners of war.[15]

Josef Mengele's concentration camp experiments were particularly repugnant. In collaboration with other German physicians and university professors, including scientists at the renowned Kaiser Wilhelm Institute, Mengele—who was the chief physician at the Auschwitz concentration camp—conducted genetic studies and biological warfare experiments on twins, many of whom were children and infants. In one line of experiments, a twin would be injected with a pathogen; if the inoculated twin died, their sibling would be killed, and autopsies would be performed on both to compare their organs. In other studies, Mengele and his team experimented with dubious

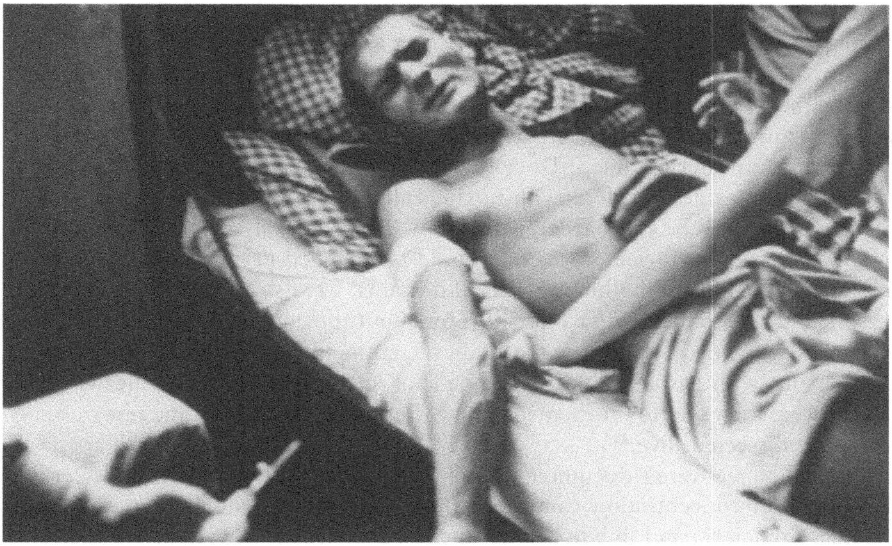

Figure 8.1 Test subject of a Nazi medical experiment conducted at the Dachau concentration camp in Germany.
Photograph from the National Archives and Records Administration.

fertility methods in an effort to create twins with Nordic features such as blue eyes and blond hair. Mengele hoped his research would lead to a prestigious academic appointment after the war. As Germany neared defeat, however, Mengele fled to South America, where Nazi sympathizers in Argentina, Brazil, and Paraguay helped him evade prosecution for war crimes.[16]

Germany's war research enterprise also included high-altitude tests whereby test persons parachuted from altitudes up to forty-six thousand feet without oxygen and fifty-nine thousand feet with oxygen. Coupled with open-air studies, test persons were exposed to altitudes as high as sixty-nine thousand feet without oxygen in pressurized chambers. Freezing experiments were conducted to evaluate protective clothing and methods of restoring body warmth. At Dachau, test persons were immersed in tubs filled with ice and cold water measuring as low as 36.5° F. Researchers monitored body temperatures, which sometimes would drop to 79.5° F, and explored several revival techniques, including hot baths, blankets, lying next to a naked person, and sex. Scores of test persons died during the studies, and autopsies were performed to examine the impact of high altitudes and cold temperatures on the human body. The experiments were conducted to further aviation medicine and devise techniques for rescuing warfighters who had been stranded in cold water for an extended period.[17]

German medical students practiced their surgical skills on live test persons, including surgery on stomachs, gallbladders, spleens, and throats.[18] Bone grafting procedures were conducted, often by intentionally breaking the bones of test persons with a hammer. Physicians amputated limbs from living prisoners and attempted transplants to injured warfighters. Coupled with unimaginable suffering, many died during the exercises.[19]

German physicians would test antibiotics by inflicting wounds on test persons and stuffing the wounds with wood shavings and cloth laced with bacteria; these studies were structured to mirror battlefield injuries.[20] Other experiments analyzed methods for making seawater potable.[21] Doctors tested dubious methods of sterilization, including intrauterine injections with noxious chemicals and castration desks where men would sit and be bombarded with X-rays. Physicians also injected dye into the eyes of test persons to see if eye color could be changed permanently.[22] Lab notes and specimens, including human eyes and brains, were shared with German professors and researchers, often shipped throughout the country with the label "War Materials—Urgent."[23] Following the studies, German officials killed countless ailing test persons with gasoline injections, one shot to the head, or other means—their bodies were then sent to the crematorium.[24] Before being killed, some research subjects were dissected alive.[25]

The Allies recovered documents implicating more than two hundred German physicians in concentration camp experiments, many of whom were professors, leading physicians, and public health officials. Postwar investigations revealed that hundreds more participated in the studies, with no documented objections from German medical societies.[26] As an article in a German medical journal detailed in 1939: "Removing this Jewish infection, which could lead to a national disease and to the death of the people, is also a duty of our health leadership."[27] According to historian Robert Proctor: "Contrary to post-war apologies, doctors were never forced to perform such experiments. Physicians volunteered—and in several cases, Nazi officials actually had to restrain overzealous physicians from pursuing even more ambitious experiments."[28] War was an engine of innovation for the Germans, who pursued science at any price and rewarded the production of newfound knowledge. Results from the experiments were presented to military research groups and at scientific conferences.[29] Some researchers received professional accolades via formal awards ceremonies from prestigious German institutes.[30]

The full extent of the German research enterprise may never be known. As Allied forces were gaining momentum and battling their way through Germany, German officials issued orders that researchers burn their lab notebooks—in turn, thousands of documents that detailed research methods and findings were destroyed. IG Farben, a large and influential German company with deep ties to the Nazi regime, is estimated to have burned millions of sheets of paper. Not all German scientists complied with the order, reasoning that Allied leaders might value the scientific information.[31] As detailed in chapter 9, this gamble paid dividends, as postwar geopolitics included a scramble to acquire Germany's scientific trade secrets and the experts who created them.

The Nuremberg Doctors' Trial

Between December 9, 1946, and August 19, 1947, the Nuremberg tribunal heard the Medical Case, commonly referred to as the Nuremberg Doctors' Trial. Twenty-three people were named as defendants, but as historian Ulf Schmidt explains, "the entire German medical profession was on trial—their moral integrity and scientific reputation. It was a scenario that had been totally unthinkable in pre-war Europe: German medical science had led the world, particularly in the fields of physiology, biochemistry, surgery, and public health."[32] Of the hundreds of physicians who conducted concentration camp experiments, the named defendants were largely at the top of the German medical hierarchy.[33]

US Brigadier General Telford Taylor was the lead attorney for the prosecution. In his opening statement, Taylor remarked: "The defendants in this case are charged with murders, tortures and other atrocities committed in the name of medical science."[34] During the next eight months, the prosecution introduced into evidence more than fourteen hundred documents and testimony from eighty-five witnesses, including medical ethics experts and concentration camp survivors.[35]

The Germans defended their actions on the grounds of national security, arguing that the studies were structured to elicit important information to benefit the military and nation. The defendants also rationalized their use of concentration camp prisoners. According to Fritz ter Meer, a member of the board of directors of IG Farben, who admitted using prisoners in the development of nerve agents, "no harm had been done to the KZ [concentration camp] inmates as they would have been killed anyway and were thus offered a chance of survival."[36] Ter Meer also highlighted what he characterized as the "humanitarian purpose" of the research: to develop therapies to save German lives.[37]

Pointing to research in the United States—in particular, malaria studies spearheaded by the Office of Scientific Research and Development (OSRD)—the German defendants contended that experimentation on prisoners was not per se unlawful.[38] To bolster their arguments, the defendants also analogized their conduct to nonconsensual research in Britain, the Netherlands, and France.[39] Karl Brandt—Adolf Hitler's personal physician and a leader in Germany's medical and research establishment—defended the studies, stating that "experiments on human beings are absolutely essential" and "have been conducted as long as any scientific efforts have been made in medicine."[40] He noted that the critical question is "whether the experiment is important or unimportant?[41] As Brandt further explained, "Do you think that one can obtain any worthwhile fundamental results without a definite toll on lives?"[42]

Regarding eugenics and mass sterilizations, the defendants indicated that the United States served as Germany's model. Prior to WWII, American philanthropic organizations—including the Carnegie Institution and Rockefeller Foundation—funded eugenics research not only at American universities that included Yale, Harvard, Princeton, and Stanford, but also at institutions in Germany.[43] In 1913, former President Theodore Roosevelt wrote that society has "no business to permit the perpetuation of citizens of the wrong type."[44] Several states maintained offices that promoted eugenic policies, and by 1931, thirty states had enacted eugenics laws that permitted forced sterilization among individuals deemed unfit to reproduce, which primarily included people with physical abnormalities or mental illness.[45] In 1927, the US Supreme Court held that Virginia's compulsory sterilization law was a legitimate exercise of a state's constitutional right to promote public health and welfare: the court analogized forced sterilization to vaccination laws.[46] By the mid-1930s, eugenics was being taught at hundreds of high schools and colleges—including Harvard, Johns Hopkins, Princeton, Northwestern, Stanford, the University of Chicago, Columbia, Cornell, and the University of Wisconsin—and government officials had sterilized more than twenty thousand Americans against their will.[47] Leading American medical journals, prominent physicians, and public health leaders routinely praised eugenics policies.[48]

Germany's eugenic movement was codified with the 1933 Law for the Prevention of Progeny with Hereditary Disease, which permitted compulsory sterilization and other public health measures. By 1945, German officials forcibly sterilized more than 400,000 people who were deemed "incurably sick" with alleged genetic disorders or hereditary abnormalities, including the disabled and those with mental illnesses, as well as individuals from ethnicities or religions that they deemed inferior, such as Jews, Poles, Russians, and Roma. German officials used the eugenics movement to rationalize the mass killing of more than ten million civilians, including infants and children.

The defendants' arguments during the Nuremberg trial raised questions of whether the Americans were also guilty of war crimes or crimes against humanity, and if not, why. Accordingly, an integral aspect of the case rested on the prosecution's ability to distinguish the illegality of the defendants' conduct. Initially, the prosecution argued that the acts of the German scientists violated German law.[49] In 1931, the German government established requirements that included obtaining consent from research participants. The law was motivated by a medical scandal from the previous year whereby seventy-five children died following a nonconsensual tuberculosis vaccine experiment. After Hitler was sworn in as Chancellor in 1933, however, the country suspended civil liberties, including the consent requirements. Thus, for the Nuremberg prosecutors, the 1931 regulations were a legal dead end.[50]

The prosecution next appealed to the professional duties of physicians and researchers. As Taylor proclaimed: "Whatever book or treatise on medical ethics we may examine, and whatever expert on forensic medicine we may question, will say that it is a fundamental and inescapable obligation of every physician under any known system of law not to perform a dangerous experiment without the subject's consent."[51] Yet, Taylor could not point to any law in any nation that codified this obligation.[52]

To succeed in the case, Taylor was forced to elucidate a clear distinction between ethical and unethical experimentation. To do so he turned to a team of medical experts that included Leo Alexander and Andrew Ivy. Alexander was an Austrian-Jewish psychiatrist who fled to the United States in the mid-1930s and later joined the Army Medical Department; Ivy was a physician, professor of physiology, and vice president of the University of Illinois.[53]

During the Nuremberg trial, Ivy returned to the United States to help construct a record of research ethics protocols. He contacted Illinois Governor Dwight Green and urged him to establish a committee to investigate the ethics of America's wartime malaria research, a component of which occurred in Illinois prisons.[54] Ivy was the man who approved the malaria research program at Stateville Prison in Illinois.[55]

Dubbed the Green Committee, Ivy served as chair and recruited prominent American physicians to join him. As historian Jon Harkness explains, however, the Green Committee had yet to meet when Ivy returned to Nuremberg to continue with his testimony. Thereafter, Ivy lied under oath when he stated that the Green Committee had met and issued its conclusions. Ivy further lied when he testified that there was "no connection" between the establishment of the Green Committee and the Doctors' Trial.[56]

During cross-examination, Ivy admitted that the "conclusions" he testified to were "in my own mind" and that the Green Committee was established to help assist the prosecution during the Doctors' Trial.[57] Ivy further admitted that principles of research ethics in the United States were not adopted by the American Medical Association (AMA) until December 28, 1946, nineteen days after the Doctors' Trial commenced. In fact, Ivy submitted the list of principles to the AMA and urged the group to adopt them in order to assist the prosecution.[58] The principles did not even exist in written form prior to the commencement of the trial.[59]

Coupled with the prosecution's challenges in demonstrating legal and ethical doctrines in force at the time of Germany's wartime experiments, there were political tensions among the Allied Powers regarding the scope of the trial. For example, the Soviets sought the death penalty for Hans Fritzsche, a key member of Germany's biological weapons program who advocated for biological warfare against the Allies. However, the British and Americans recognized that their biological warfare programs were more advanced than Germany's, and feared exposure during the trial. To avoid this predicament, the Americans and British refused to prosecute Fritzsche, and he was acquitted.[60]

The Nuremberg judges were in a difficult position. They were responsible for adjudicating the atrocious conduct of the German scientific enterprise without calling into question common research practices in Allied nations. The tribunal was asked to provide justice in light of war crimes and crimes against humanity, but also to be mindful of the precedent that would be set if laws and ethical codes were retroactively applied. Specific to the Medical Case, the judges endeavored to balance the continuing need for medical experimentation with fundamental notions of liberty, justice, and autonomy.

To synthesize these aims, the tribunal turned to a series of research ethics principles that Ivy and Alexander had prepared as part of their testimony. These principles served as the foundation for the Nuremberg Code, which is reprinted in Table 8.1.

As a practical matter, the Nuremberg Code was forward-looking—an attempt to guide physicians and researchers on the ethical and legal boundaries of research with human subjects.[61] As Secretary of War Henry Stimson explained: "We can understand the law of Nuremberg only if we see it for what it is—a great new case in the book of international law, and not a formal enforcement of codified statutes."[62] The ten principles focus on the need to balance risks and benefits and obtain informed consent from research participants. Elements of fraud, deceit, coercion, or undue influence are strictly forbidden.

In the end, the Nuremberg tribunal found fifteen of the twenty-three defendants guilty; seven of the fifteen were sentenced to death, five received life sentences, and the remaining three received prison terms ranging from ten to twenty years.[63] None of the defendants expressed remorse or acknowledged that their actions were unlawful or unethical.[64] Standing on the gallows, just moments before being hanged, Karl Brandt proclaimed: "It is no shame to stand on this scaffold. I served my Fatherland as others before me."[65]

Table 8.1 THE NUREMBERG CODE

1. The voluntary consent of the human subject is absolutely essential. This means that the person involved should have legal capacity to give consent; should be so situated as to be able to exercise free power of choice, without the intervention of any element of force, fraud, deceit, duress, overreaching, or other ulterior form of constraint or coercion; and should have sufficient knowledge and comprehension of the elements of the subject matter involved as to enable him to make an understanding and enlightened decision. This latter element requires that before the acceptance of an affirmative decision by the experimental subject there should be made known to him the nature, duration, and purpose of the experiment; the method and means by which it is to be conducted; all inconveniences and hazards reasonably to be expected; and the effects upon his health or person which may possibly come from his participation in the experiment. The duty and responsibility for ascertaining the quality of the consent rests upon each individual who initiates, directs or engages in the experiment. It is a personal duty and responsibility which may not be delegated to another with impunity.
2. The experiment should be such as to yield fruitful results for the good of society, unprocurable by other methods or means of study, and not random and unnecessary in nature.
3. The experiment should be so designed and based on the results of animal experimentation and a knowledge of the natural history of the disease or other problem under study that the anticipated results will justify the performance of the experiment.
4. The experiment should be so conducted as to avoid all unnecessary physical and mental suffering and injury.
5. No experiment should be conducted where there is an a priori reason to believe that death or disabling injury will occur; except, perhaps, in those experiments where the experimental physicians also serve as subjects.
6. The degree of risk to be taken should never exceed that determined by the humanitarian importance of the problem to be solved by the experiment.
7. Proper preparations should be made and adequate facilities provided to protect the experimental subject against even remote possibilities of injury, disability, or death.
8. The experiment should be conducted only by scientifically qualified persons. The highest degree of skill and care should be required through all stages of the experiment of those who conduct or engage in the experiment.
9. During the course of the experiment the human subject should be at liberty to bring the experiment to an end if he has reached the physical or mental state where continuation of the experiment seems to him to be impossible.
10. During the course of the experiment the scientist in charge must be prepared to terminate the experiment at any stage, if he has probable cause to believe, in the exercise of the good faith, superior skill, and careful judgment required of him, that a continuation of the experiment is likely to result in injury, disability, or death to the experimental subject.

Post-Nuremberg Research Ethics

The international community paid little attention to the Nuremberg Code. As historian Ulf Schmidt explains, the Code was "mostly hidden away from the public eye, known to a few experts and institutions, and applied by even fewer medical innovators."[66] Physician and bioethicist Jay Katz—who graduated from Harvard Medical

School in 1949, worked in an Air Force hospital, and joined Yale University in 1953—summarized the response of the American medical community: "It was a good code for barbarians but an unnecessary code for ordinary physician-scientists."[67] After the war, newspapers and magazines published articles that provided details of research using "human guinea pigs," including research involving experimental vaccines and ingestion of radioactive drugs. Journalists portrayed human subjects as thrill-seekers contributing to noble studies and the progress of science. Few mentioned the Nuremberg Code or raised any ethical concerns with research protocols.[68]

Notwithstanding the prevailing view that the Nuremberg Code was largely irrelevant, scientists around the globe feared that their research would be hindered should the Code's principles be widely implemented. Physicians in the United States and the United Kingdom lobbied for scientific autonomy and opposed codified research protections. Leading scientific publications, such as the *British Medical Journal*, supported a framework that deferred to the individual conscience of the researcher, arguing that government intervention would lead to a totalitarian takeover of medicine by the state.[69]

Meanwhile, in 1947, representatives of medical societies from thirty-two countries formed the World Medical Association (WMA), seeking to maintain the honor of the medical profession, promote world peace, and improve public health. Despite the noble mission, the WMA lobbied vigorously for medicine's monied interests. During its first meeting, the association received a $250,000 donation from American healthcare institutions and pharmaceutical companies (equivalent to $3.5 million in 2024), a move that sought to solidify the WMA's pro-industry, anti-consumer agenda. As many war-torn nations sought to create a healthcare system where individuals could access medical care regardless of their socioeconomic status, the WMA declared it "unethical" for physicians to provide care absent complete independence from the government.[70]

The WMA did not adopt the Nuremberg Code in 1947 or at any time. Rather, in 1964, the group published its own set of research ethics principles in a document titled the Declaration of Helsinki. As ethicists have detailed, the Declaration of Helsinki is less rigorous than the Nuremberg Code and affords significant discretion to the individual judgment of researchers.[71] Since the WMA is a special interest lobby group, nations have no obligation to adopt WMA policies, nor must courts give deference to WMA principles. Over the years, however, the Declaration of Helsinki has been widely viewed as an important framework for research ethics; since 1964, it has been updated nine times, most recently in 2013.[72]

The American research establishment largely dismissed the Nuremberg Code—the AMA did not adopt the Code, and the US government did not codify the Nuremberg principles into federal law. Rather than implement protections for research subjects, government and military officials sought legal avenues for indemnifying scientists in the event of research-related harms to research subjects. During the late 1940s and early 1950s, the military sometimes utilized release forms that summarized study protocols and risks, and protected the Department of Defense (DoD) from liability. Researchers at universities who worked on military studies sought equivalent protections, and in 1952 Congress passed a law that provided indemnification for military research.[73] This new law provided legal cover for scientists in the event of injury or death to human subjects enrolled in experiments.[74]

In the United States, the military was the first to seriously consider implementation of protections for research subjects. The debate within military channels lasted for years and involved several high-level DoD committees. In a classified memo from the fall of 1952, the Armed Forces Medical Policy Council issued a resolution that called for the adoption of the Nuremberg Code. The council's legal adviser, Stephen Jackson, supported the resolution, adding that the United States would be deemed hypocritical should it modify, or fail to adopt, the Code.[75]

The council's decision was controversial. As bioethicist Jonathan Moreno explains, many Pentagon officials opposed "any written policy that threatened to restrict human experiments for national security needs, or that questioned the moral integrity of physicians and commanding officers and their ability to make tough ethical calls."[76] The Pentagon's Committee on Medical Sciences argued that "human experimentation within the field of medical sciences has [been], in years past, and is at present governed by an unwritten code of ethics administered informally by fellow workers in the field [and] is considered to be satisfactory." A written policy, according to the committee, would "cause more harm than good" and "would focus unnecessary attention on the legal aspects of the subject."[77]

Other officials took a more measured approach. For instance, the chairman of the DoD's Medical Sciences Committee was reluctant to adopt a written policy but did not oppose it. Despite some pushback, Jackson and other DoD attorneys had the support of Anna Rosenberg, an influential Pentagon official. Rosenberg recommended that the consent language be modified to require that "consent be expressed in writing before at least one witness," a requirement not found in the Nuremberg Code.[78] The lawyers and the chair of the medical council concurred with her suggestion.[79]

Shortly after President Dwight Eisenhower entered the White House in January 1953, George Underwood—the Director of the Executive Office of the Secretary of Defense—sent a memorandum to a deputy secretary under Charles Wilson, the former chairman of General Motors who had just entered office as Eisenhower's Secretary of Defense.[80] In a letter dated February 5, 1953, Underwood summarized the ongoing debate regarding research protections and presented Wilson with a draft policy. Underwood stated that the policy contains "conditions which will govern the use of human volunteers in experimental research in the field of atomic, biological and/or chemical warfare" and noted that "There is no DOD policy on the books which permits this type of research."[81] Meanwhile, the Armed Forces Medical Policy Council "strongly recommended" that Wilson institute a policy modeled on the Nuremberg Code that would govern research conducted by, or on behalf of, the military.[82]

Within three weeks, Wilson memorialized the policy. On February 26, 1953, Wilson issued a memorandum to the Secretaries of the Army, Navy, and Air Force wherein he indicated that the DoD had officially adopted the Nuremberg Code. The memo, which was drafted by Jackson, stated:

> Based upon a recommendation of the Armed Forces Medical Policy Council, that human subjects be employed, under recognized safeguards, as the only feasible

> means for realistic evaluation and/or development of effective preventive measures of defense against atomic, biological or chemical agents, the policy set forth below will govern the use of human volunteers by the Department of Defense in experimental research in the fields of atomic, biological and/or chemical warfare.[83]

Thus, at a time when civilian medical and research enterprises refused to adopt the Nuremberg principles for their own work, the US military took the lead in instituting research protections. Indeed, the new policy added protections that went beyond those enumerated in the Nuremberg Code, including that (1) consent to experimentation must be in writing, to which the Nuremberg Code must be affixed, and must be in the presence of one witness; (2) the number of volunteers must be as little as necessary; (3) prisoners of war must be excluded from experiments; and (4) researchers must obtain the approval of the secretary of their service prior to commencing a study.[84]

Wilson's memo was classified as top secret, perhaps due to the subject matter of human experimentation for atomic, biological, and chemical warfare. The memo was shared with the Joint Chiefs of Staff and the DoD's Research and Development Board, though it is unclear how widely the memo was circulated within the ranks of each service. The secretary of each service was responsible for disseminating Wilson's memo, but the top-secret classification limited distribution, which in turn hindered implementation of the policy. Although the Secretary of the Army drafted an unclassified version that applied to military researchers and civilian research contractors, it is not clear how widely the protocols were adhered to. The Navy and Air Force likewise adopted policies that afforded some protections for research subjects, but these policies were not always known or followed by researchers. Moreover, some scientists simply ignored the policy, believing that it unduly restricted military research.[85]

The peculiarities surrounding Wilson's memo remain a mystery, though implementation failures may be linked to the haphazard notice of the policy to researchers and widespread resistance within military research circles.[86] As will be detailed in Part III of this book, the call for research protections dissipated as the Cold War intensified, and the drive for scientific superiority was intimately linked with military excellence and geopolitical dominance. A 1975 report from the Army inspector general found that there had been a "startling . . . lack of consistency in the interpretations" of the Wilson memo.[87] Whether this circumvention was deliberate, negligent, or due to a lack of knowledge of the standards remains unclear.[88] As the Advisory Committee on Human Radiation Experiments concluded in its 1995 final report: "It is difficult to determine why these requirements were applied to some activities and not to others."[89]

Summary

Germany's wartime research enterprise was on trial at Nuremberg, but the proceedings highlighted a global deficiency in research protections for human subjects. The medical and scientific communities could have used Nuremberg as a springboard to pursue a culture change in research ethics, but instead elected to pigeonhole German science as uniquely heinous and characterize ethical protocols as an unnecessary hindrance on researcher autonomy and the progress of science. In the United States, the military took the lead in adopting and strengthening the Nuremberg principles, though the policies were inconsistently applied. As the quest for scientific superiority burgeoned during the Cold War, the demand for research ethics protections largely fell silent.

Notes

1. The Tripartite Conference in Moscow, Moscow Declaration on Atrocities (November 1, 1943).
2. Gerald L. Posner and John Ware, *Mengele* (1986), 21; Harris and Paxman, *A Higher Form of Killing*, 57.
3. Alexander Mitscherlich and Fred Mielke, *Doctors of Infamy* (1949), 93.
4. Mitscherlich and Mielke, *Doctors of Infamy*, 8.
5. Mitscherlich and Mielke, *Doctors of Infamy*, 77.
6. Tucker, *War of Nerves*, 51; Mitscherlich and Mielke, *Doctors of Infamy*, 75–80.
7. Harris and Paxman, *A Higher Form of Killing*, 62; Schmidt, *Secret Science*, 84.
8. Tucker, *War of Nerves*, 51.
9. Harris and Paxman, *A Higher Form of Killing*, 60–62; Linda Hunt, *Secret Agenda* (1991), 13.
10. Mitscherlich and Mielke, *Doctors of Infamy*, 78.
11. Mitscherlich and Mielke, *Doctors of Infamy*, 42–51; Masterson, *The Malaria Project*, 308–19; Robert N. Proctor, "Nazi Doctors, Racial Medicine, and Human Experimentation," in *The Nazi Doctors and the Nuremberg Code*, edited by George J. Annas and Michael A. Grodin (1992), 26.
12. Mitscherlich and Mielke, *Doctors of Infamy*, 68–71; Masterson, *The Malaria Project*, 314.
13. Posner and Ware, *Mengele*, 59.
14. Mitscherlich and Mielke, *Doctors of Infamy*, 45.
15. Quoted in Norman Ohler, *Blitzed* (2017), 209–11.
16. Eva Mozes-Kor, "The Mengele Twins and Human Experimentation: A Personal Account," in *The Nazi Doctors*, edited by Annas and Grodin, 55–56; Posner and Ware, *Mengele*, 31–41, 94–181; Miklos Nyiszli, *Auschwitz* (2011), 17, 58–59.
17. Mitscherlich and Mielke, *Doctors of Infamy*, 4, 12–33.
18. Masterson, *The Malaria Project*, 311–14.
19. Mitscherlich and Mielke, *Doctors of Infamy*, 55–57, 64–65.
20. Posner and Ware, *Mengele*, 79.
21. Mitscherlich and Mielke, *Doctors of Infamy*, 33–41.
22. Proctor, "Nazi Doctors," 26.
23. Posner and Ware, *Mengele*, 34–41.
24. Mitscherlich and Mielke, *Doctors of Infamy*, 63; Posner and Ware, *Mengele*, 28; Nyiszli, *Auschwitz*, 67.
25. Leo Alexander, "Medical Science Under Dictatorship," *New England Journal of Medicine* 241, no. 2 (July 14, 1949): 39–47.
26. Pappworth, *Human Guinea Pigs*, 185; Mitscherlich and Mielke, *Doctors of Infamy*, x–xi.
27. *Medical Journal for Lower Saxony* (1939), quoted in Ohler, *Blitzed*, 18.
28. Proctor, "Nazi Doctors," 26.
29. Mitscherlich and Mielke, *Doctors of Infamy*, 20–33.
30. Posner and Ware, *Mengele*, 79.

31. Tucker, *War of Nerves*, 69–74.
32. Ulf Schmidt, *Justice at Nuremberg* (2006), 3.
33. Mitscherlich and Mielke, *Doctors of Infamy*, xvii.
34. Telford Taylor, "Opening Statement of the Prosecution, December 9, 1946," in *The Nazi Doctors*, edited by Annas and Grodin, 67.
35. George Annas and Michael A. Grodin, "Introduction," in *The Nazi Doctors*, edited by Annas and Grodin, 3–11.
36. Quoted in Tucker, *War of Nerves*, 94.
37. Quoted in Tucker, *War of Nerves*, 94.
38. Masterson, *The Malaria Project*, 326–29.
39. Schmidt, *Justice at Nuremberg*, 206–8.
40. Karl Brandt, Nuremberg Medical Case Trial Transcript (February 4, 1947): 2375–78.
41. Karl Brandt, Nuremberg Medical Case Trial Transcript (February 6, 1947): 2566.
42. Quoted in John Marks, *The Search for the "Manchurian Candidate"* (1979), 10.
43. Stefan Kuhl, *The Nazi Connection* (1994).
44. Theodore Roosevelt, Letter to Charles B. Davenport (January 3, 1913).
45. Kenneth L. Garver and Bettylee Garver, "Eugenics: Past, Present, and Future," *American Journal of Human Genetics* 49 (1991): 1109–18.
46. *Buck v. Bell*, 274 U.S. 200 (1927).
47. Hornblum et al., *Against Their Will*, 35–37.
48. Dozens of such articles are referenced in Paul A. Lombardo, "'Ridding the Race of His Defective Blood'—Eugenics in the *Journal*," *New England Journal of Medicine* 390, no. 10 (March 7, 2024): 869–72.
49. Schmidt, *Justice at Nuremberg*, 164–71.
50. Moreno, *Undue Risk*, 64–65.
51. Taylor, "Opening Statement," 89.
52. Schmidt, *Justice at Nuremberg*, 222–26, 233–43.
53. Schmidt, *Justice at Nuremberg*, 34–66, 134, 222–26, 233–43.
54. Moreno, *Undue Risk*, 76–77.
55. Masterson, *The Malaria Project*, 328.
56. Jon M. Harkness, "Nuremberg and the Issue of Wartime Experiments on US Prisoners: The Green Committee," *Journal of the American Medical Association* 276, no. 20 (November 27, 1996): 1672–75.
57. Quoted in Harkness, "Nuremberg," 1674.
58. Andrew C. Ivy, Nuremberg Medical Case Trial Transcript (June 13, 1947): 9168–70.
59. Schmidt, *Justice at Nuremberg*, 233–36.
60. Harris and Paxman, *A Higher Form of Killing*, 87.
61. Jay Katz, "The Nuremberg Code and the Nuremberg Trial: A Reappraisal," *Journal of the American Medical Association* 276, no. 20 (November 27, 1996): 1662–66.
62. Henry L. Stimson, "The Nuremberg Trial: Landmark in Law," *Foreign Affairs* 25, no. 2 (January 1947): 179–89.
63. Alexander Mitscherlich and Fred Mielke, "Seven Were Hanged," in *The Nazi Doctors*, edited by Annas and Grodin, 105–7.
64. Mitscherlich and Mielke, *Doctors of Infamy*, 146–48; Pappworth, *Human Guinea Pigs*, 185; Telford Taylor, *The Anatomy of the Nuremberg Trials* (1992), 607–11.
65. Quoted in Mitscherlich and Mielke, "Seven Were Hanged," 106.
66. Schmidt, *Justice at Nuremberg*, 264.
67. Jay Katz, "The Consent Principle of the Nuremberg Code: Its Significance Then and Now," in *The Nazi Doctors*, edited by Annas and Grodin, 228.
68. Final Report, *The Human Radiation Experiments*, 86–87.

69. Schmidt, *Justice at Nuremberg*, 264–67.
70. Schmidt, *Justice at Nuremberg*, 264–67.
71. Katz, "The Nuremberg Code"; Schmidt, *Justice at Nuremberg*, 282–83.
72. World Medical Association, *Declaration of Helsinki: Ethical Principles for Medical Research Involving Human Subjects* (accessed March 14, 2024).
73. An Act to Facilitate the Performance of Research and Development Work by and on behalf of the Departments of the Army, the Navy, and the Air Force, and for Other Purposes, Pub. L. 82-557, 66 Stat. 725 (July 16, 1952).
74. Final Report, *The Human Radiation Experiments*, 55–56.
75. Final Report, *The Human Radiation Experiments*, 56–64.
76. Moreno, *Undue Risk*, 168.
77. F. Lloyd Mussells, Executive Director, Committee on Medical Sciences, Memorandum to Floyd L. Miller, Vice Chairman, Research and Development Board, DoD (November 12, 1952), quoted in Final Report, *The Human Radiation Experiments*, 57.
78. Stephen Jackson, Assistant General Counsel, Office of the Secretary of Defense, Memorandum to Melvin Casberg, Chairman of the Medical Policy Council (October 22, 1952), quoted in Moreno, *Undue Risk*, 167.
79. Moreno, *Undue Risk*, 157–69.
80. Moreno, *Undue Risk*, 170–73.
81. George V. Underwood, Director of the Executive Office of the Secretary of Defense, Memorandum to Deputy Secretary Foster (February 5, 1953), quoted in Moreno, *Undue Risk*, 171.
82. Melvin Casberg, Chairman, Armed Forces Medical Policy Council, Memorandum to the Secretary of Defense (January 13, 1953), quoted in Final Report, *The Human Radiation Experiments*, 58.
83. Charles E. Wilson, "The Use of Human Volunteers in Experimental Research" (February 26, 1953), reprinted in Final Report, *The Human Radiation Experiments*, 59–60.
84. Wilson, "The Use of Human Volunteers."
85. Final Report, *The Human Radiation Experiments*, 56–64, 104–6; Moreno, *Undue Risk*, 175–78.
86. Schmidt, *Justice at Nuremberg*, 278–91.
87. Army Inspector General Report, *Use of Volunteers in Chemical Agent Research* (July 21, 1975), 45–46.
88. Moreno, *Undue Risk*, 178–84.
89. Final Report, *The Human Radiation Experiments*, 500.

PART III

1946–1991

Proliferation of the Military Research Enterprise

9
The Spoils of War
Exploiting the German and Japanese Research Enterprises

Introduction

World War II (WWII) could have been the war to end all wars. Six years of global combat took the lives of more than 21 million warfighters and 48 million civilians.[1] Genocide by the Germans claimed over 11 million lives, more than half of whom were Jewish and more than 1 million of whom were children.[2] Atomic bombs annihilated Hiroshima and Nagasaki. Famines ravaged countless communities, and the task of reconstruction loomed large for war-torn nations. In the context of military science, the United States flexed its atomic muscle, chemical weapons were stockpiled by Axis and Allied powers, and Japan's expansive biological weapons program underscored the centuries-old realization that disease can sway the balance of power in war.

In the wake of unprecedented carnage that illustrated man's limitless brutality, leading nations exited WWII with a ferocious drive to embolden their respective militaries. Although, historically, postwar periods in the United States were characterized by a decrease in military size and spending, the race for world hegemony pushed post-WWII America into an extraordinary state of scientific and military proliferation, particularly since the atomic bomb demonstrated that scientific knowledge can translate to military and geopolitical power. According to Major General Hugh Knerr, a US deputy commander in Europe: "Occupation of German scientific and industrial establishments has revealed the fact that we have been alarmingly backward in many fields of research. If we do not take the opportunity to seize the apparatus and the brains that developed it and put the combination back to work promptly, we will remain several years behind while we attempt to cover a field already exploited. Pride and face-saving have no place in national insurance."[3]

Knerr's sentiment captured the essence of the postwar scramble to capitalize on the scientific riches of Germany and Japan. The US government recruited Nazi scientists to work in key areas of military science, granted Japanese researchers immunity from war crimes prosecution in exchange for biological warfare secrets, and offered significant funding and support to American industries and universities for research that furthered military goals. These endeavors dovetailed with a grand strategy developed by America's foreign policy elite—characterized as "armed primacy" and a "pointillist empire" by historians Stephen Wertheim and Daniel Immerwahr, respectively—whereby the United States did not need to occupy entire nations but would promote its interests by establishing a US-led global financial system and maintaining American military bases throughout the world.[4] This framework contributed to American dominance by rewarding allies and suppressing military, economic, and ideological threats. As Cold War fears intensified in the decades following WWII, research

America's Military Biomedical Complex. Efthimios Parasidis, Oxford University Press.
 DOI: 10.1093/9780199351473.003.0010

flourished across the country, largely unhindered by legal or regulatory guidelines and with little attention to moral or ethical concerns.

Operation Paperclip

Operation Paperclip was a secret program that brought more than sixteen hundred German scientists to the United States after WWII to work in chemical and biological warfare, aviation and space medicine, rocket development, and other areas of military science.[5] The code name Paperclip arose because American intelligence officers would place a paperclip on the dossier of promising German scientists.[6] According to leading officials—including President Harry Truman, FBI Director J. Edgar Hoover, and military brass—the potential benefits of collaborating with Nazi scientists outweighed the scandalous notion of partnering with key members of the German war machine.[7] The program remained active for decades, with Nazi scientists embedded in the American research enterprise into the 1970s and beyond.[8]

The precursor to Operation Paperclip was Project Overcast, a program approved in July 1945—just weeks after the Allied victory in Europe—whereby the United States employed Nazi scientists to assist in the ongoing war with Japan.[9] The idea for Project Overcast developed as American intelligence agents uncovered Germany's wartime scientific secrets. Throughout the war, agents from the Army, Navy, Chemical Warfare Service, and Office of Strategic Services (OSS) worked to uncloak Germany's military science programs—by capturing and interrogating German researchers, copying documents, and stealing equipment and materials. In the course of their work, the intelligence officers discovered that Germany was conducting gruesome experiments in concentration camps.[10] Notwithstanding the shocking revelations, within the first six months of Project Overcast, 160 Nazi scientists were working for the United States.[11] Vannevar Bush—wartime director of the Office of Scientific Research and Development (OSRD)—welcomed the Germans, marveling at those he described as "intellectual giants of Nobel Prize stature."[12]

Four of the twenty-three defendants at the Nuremberg Doctors' Trial had been supporting the Americans under Operation Paperclip, a fact that was kept secret for forty years.[13] Karl Krauch—a leading figure in Germany's chemical weapons program, who advocated for the deployment of nerve agents against the Allies—was offered a Paperclip contract while incarcerated at Nuremberg.[14] Some Paperclip scientists were found guilty at Nuremberg, while others were acquitted.[15] It later was revealed that American officials suppressed evidence of war crimes for scientists harbored under Paperclip. One of the acquitted was Kurt Blome—a physician and biological warfare expert who later worked in America's biological weapons program.[16]

During the Doctors' Trial, American prosecutors proclaimed that Germany's wartime "experiments revealed nothing which civilized medicine can use," though they posited that the research "may be useful to criminals everywhere and there is no doubt that they may be useful to a criminal state."[17] After the trial, and during the same week that some Nazi scientists were hanged and their ashes were discarded into a river, American officials conducted high-level meetings to discuss which Nazis to bring to America and how best to integrate them into the military biomedical complex.[18]

Some officials considered the exploitation of German science as a component of reparations.[19]

National security and the progress of science usurped moral concerns of employing war criminals and utilizing data derived from concentration camp experiments. A secret 1945 report from the US Naval Technical Mission in Europe noted that the information "may contribute materially to the present knowledge of physiology and through practical application may be the means of saving lives." The report advocated for use of data from concentration camp experiments, indicating that "a moral responsibility exists to make available the information gained through the sacrifice of lives and sufferings of the prisoners who served as experimental subjects." Use of such data, the report added, "is not intended to condone to any degree whatsoever the violations of the Oath of Hippocrates and the flouting of humanitarian principles."[20]

American media outlets—including *The New York Times* and *Life*—eventually learned of some aspects of Operation Paperclip and published articles that revealed Nazi scientists were living and working in the United States.[21] The War Department did not deny the allegations but rather orchestrated a public relations campaign that heralded the scientific genius of the Germans and assured the public that the foreigners were thoroughly screened.[22] Undersecretary of War Robert Patterson publicly declared that "no scientists who are alleged war criminals are brought to the United States."[23]

Behind the scenes, however, Patterson warned that the Germans are our "enemies and it must be assumed they are capable of sabotaging our war effort. Bringing them to this country raises delicate questions, including the strong resentment of the American public, who might misunderstand the purpose of bringing them here and the treatment accorded them."[24] Patterson's fears were not unfounded. During the Nuremberg Doctors' Trial, American officials asked Paperclip scientists who were not named as defendants to translate German documents recovered from concentration camps—the documents were deemed integral to the prosecution, but many disappeared after being placed in German hands. Despite the known risks of employing Nazi scientists in key military projects, lax surveillance plagued the Paperclip program, including unmonitored travel, mail, and phone calls. It later was uncovered that some Paperclip scientists maintained secret postboxes, while others had thousands of dollars deposited into their bank accounts from unnamed sources. Some Paperclip scientists were caught disclosing classified American secrets to the Soviet Union and other nations, while others were suspected of sabotage, particularly in the area of rocket development.[25]

Notwithstanding the precarious arrangement with the Nazi scientists, American officials worked quickly to round up more German researchers and disperse them throughout the United States. Fearful that the Soviets would reach the scientists first, American operatives appeared in the homes of German university professors at all hours of the day and night and informed the scientists that they had no more than twenty-four hours to pack their belongings.[26] According to John Gimbel, a member of the American occupation force in postwar Germany, who later conducted twenty years of research on Operation Paperclip: "They were asked to come voluntarily, but those who asked what would happen if they refused were told that force would be used or that they would be arrested."[27] The scientists could bring their families and were promised jobs, lab facilities, housing, and furniture.[28]

Scientists, their teams, and their dependents were exempted from immigration requirements, and upon their arrival in the United States, they were put to work in military and civilian facilities. Under a parallel secret project code-named National Interest, the Central Intelligence Agency (CIA) and military encouraged universities and industry to hire Paperclip scientists. These institutions—which included Lockheed Martin, the University of North Carolina, Boston University, the University of Texas, and Washington University in St. Louis, among others—were grateful for the opportunity to hire foreign scientists at salaries significantly lower than those commanded by Americans.[29]

America was one of several nations scavenging Germany's scientific enterprise. The British instituted their own version of Paperclip—Project Matchbook—whereby German scientists were solicited to assist with Britain's weapons programs.[30] Fritz Haber, Germany's World War I chemical weapons mastermind, was reportedly granted refuge in England.[31] Australia and Canada each imported dozens of Nazi scientists, and others went to work in France, Argentina, and other nations.[32] The Soviet Union amassed a wide range of Nazi documents, materials, and researchers from biological weapons labs and nerve agent factories—acquisitions that were particularly important since the Soviets had yet to create an atomic bomb.[33]

Several German scientists—lab notebooks in hand—voluntarily turned themselves over to American officials, confident they would be welcomed into the United States.[34] Others conducted interviews with several Allied nations in order to negotiate favorable terms. One of the most coveted was renowned rocket scientist Werner von Braun, a Nazi officer.[35] The rocket factory where von Braun worked utilized slave labor from the Dora concentration camp. As bioethicist Jonathan Moreno explains: "No matter how absorbed the scientists might have been in their work, none could have been unaware of the corpses of men who had dropped at their workstations from exhaustion, or those who were left hanging in the factory for days as a warning to other prisoners."[36]

After negotiating with several countries, von Braun and his team selected the United States because America had offered the best deal. A memorandum identifying von Braun as a security threat was removed from his file. Falsification of von Braun's record was not an aberration: American officials systematically altered Paperclip files and, at times, concealed the true identities of scientists in order to facilitate a transfer to the United States. Several Paperclip scientists were identified as security risks and war criminals, and others were known to have conducted horrific concentration camp experiments. Although war crimes, torture, and murder were insufficient reasons to reject a German scientist, "Communist affiliations or inclinations" resulted in disqualification. The military justified its actions by arguing that national security interests outweighed any legal or ethical concerns.[37]

Prominent individuals and groups protested against Operation Paperclip, including Albert Einstein, Eleanor Roosevelt, the American Jewish Congress, and the Federation of American Scientists.[38] Telford Taylor—who led the prosecution during the Doctors' Trial—was incensed, stating that the Paperclip program represented "a blow to the principles of international law and concepts of humanity for which we fought the war."[39] Notwithstanding some criticism from contemporaries, as historian Ulf Schmidt explains, for the most part the moral dilemma of simultaneously prosecuting and exploiting German scientists "exists mainly in the retrospective analysis."[40] Even Leo Alexander—a key medical ethics adviser during the Doctors' Trial, who provided a conceptual framework for the Nuremberg Code—recommended at least one Nazi scientist for Operation Paperclip.[41] During the 1950s, as scientists convicted of war crimes were released from prison, several were recruited to the United States under Paperclip and placed in civilian and military positions.[42]

From the perspective of military science, the recruitment of Germany's finest scientists paid massive dividends for the United States. With assistance from German researchers, the United States created a plethora of chemical and biological weapons, including Sarin nerve agent bombs and weaponized bubonic plague. Several Nazi scientists earned military and civilian awards, one was appointed the first director of the Kennedy Space Center in Florida, and another had a government building named after him.[43] Von Braun became a celebrated member of America's space program: it was not until after a man walked on the moon that he publicly confirmed that he had been a Nazi officer.[44]

Contemporaneous with Operation Paperclip and exploitation of the German scientific enterprise, the US military and government worked behind the scenes to capitalize on a treasure trove of knowledge from Japan's expansive biological warfare program.

Commandeering Japan's Biological Warfare Program

By the mid-1940s, Japan maintained the world's preeminent biological weapons program. Led by physician and army general Shiro Ishii—a decorated military officer and public health official who held leadership positions at Japan's Water Purification Bureau and the Epidemic Prevention Department of the Kwantung Army—Japanese scientists investigated the weaponization of anthrax, cholera, plague, typhoid, typhus, tetanus, botulinum toxin, smallpox, brucellosis, glanders, tuberculosis, tularemia, tick encephalitis, and other agents. In addition to engineering biological weapons to harm humans, Japanese researchers studied how best to poison water, crops, and livestock.[45] Funding for Japan's biowarfare endeavors rivaled that of the Manhattan Project.[46]

The program was headquartered at the Pingfan Institute, a massive complex that employed 3,000 and maintained 150 buildings on 1,000 acres.[47] When locals inquired as to what the Japanese were building at Pingfan, they were told the complex was a lumber mill; Japanese researchers privately joked that people were the logs. From that point, Japanese scientists often referred to human test subjects as *maruta*, the Japanese word for logs.[48] According to one Japanese researcher stationed at the biowarfare unit: "At that time, the general thinking at the unit was that it was necessary to sacrifice three *maruta* in order to save one hundred Japanese soldiers."[49]

Apart from Pingfan, biological and chemical weapons research also was conducted at eighteen satellite locations.[50] As with military science enterprises in other nations, leading Japanese physicians, public health officials, industry experts, and university researchers dedicated their efforts to promote war-related projects.[51] Coupled with biowarfare experiments, other projects included examining human responses to extreme conditions, such as high-voltage electric shocks, oxygen deprivation, and extreme cold to study anti-frostbite techniques. Japanese researchers also attempted blood transfusions by replacing human blood with horse blood.[52] Figure 9.1 depicts Japanese scientists engaged in a biological warfare test in Jilin Province, China.[53]

Japanese scientists conducted field tests whereby they would tie humans to stakes and explode a bomb laced with a biological agent—researchers then would measure the potency of the agent at various distances.[54] In other studies, humans were placed in gas chambers and exposed to toxic pathogens. Japanese researchers fed botulinum toxin to prisoners of war, injected humans with bacteria and viruses, and studied experimental vaccines.[55] In some instances, Japanese researchers dissected living humans to examine the impact of biological agents on the human body. Children and infants were subjected to some of the studies.[56]

Most of the experiments took place in Manchuria while it was under Japanese occupation.[57] The majority of test subjects were Chinese, though some reports indicated that the Japanese also experimented on prisoners of war from America, Australia, Britain, and the Soviet Union.[58] The Japanese maintained meticulous notes and assembled archives of film and specimens that frequently dehumanized research subjects, referring to them as "experimental materials" or "Manchurian monkeys."[59]

Japan's biological weapons work was not limited to the lab or field tests. The Japanese poisoned over one thousand wells and sprayed anthrax and glanders as

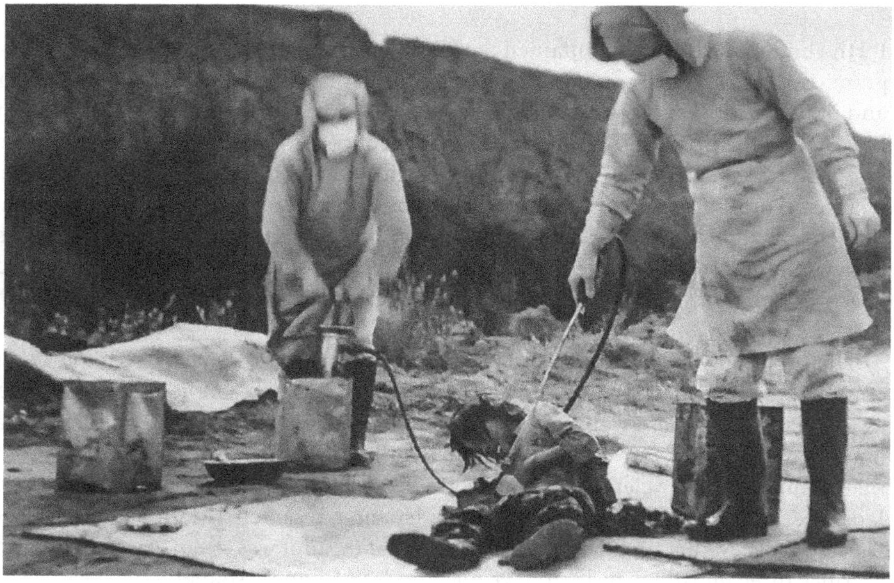

Figure 9.1 Japanese scientists from Unit 731 engage in a biological warfare test in Jilin Province, China.

Photograph from the Jilin Provincial Archives.

they drove through Chinese farms. In other attacks, low-flying airplanes would drop grain and rice mixed with millions of plague-infested fleas. Unsuspecting rats would carry the fleas throughout Chinese cities, and several Chinese communities suffered disease outbreaks. Unable to contain the spread of disease, however, the bioweapons also killed thousands of Japanese service members.[60] Despite his support for America's biowarfare program, President Franklin D. Roosevelt characterized Japan's biological attacks as an "inhumane form of warfare."[61]

The full extent of Japan's biological warfare research may never be known. In August 1945, as Soviet forces were nearing Pingfan, Japanese researchers destroyed the entire complex, burned incriminating documents, and killed any surviving research subjects. Near one biological weapons facility, Soviet troops discovered a shallow grave with thousands of Chinese and Mongolians, including children. Meanwhile, Ishii and his colleagues pledged mutual silence to protect themselves and their homeland.[62]

Unlike postwar Germany, which was divided among the four Allied powers, postwar Japan was entirely under US supervision. Given their success in exploiting Germany's military science enterprise, American officials pondered what deal they could strike to take advantage of Japan's biological weapons program. Notwithstanding his pledge of silence, Ishii himself proposed that the Americans hire him, stating: "In the preparation for the war with Russia, I can give you the advantage of my 20 years [of] research and experience." Ishii added that he could "write volumes" regarding the "strategic and tactical employment" of biological weapons, and that he had researched "the best agents to be employed in various regions and in cold climates."[63]

American diplomats and military officials debated the risks and benefits of a deal with Ishii.[64] Characterizing biological weapons as "demoralizing, silent, and insidious," the War Department's Military Intelligence Division warned that France, Britain, and the Soviet Union could initiate large-scale biological attacks within years. As the secret report asserted, the United States had to keep pace and be armed "to strike the enemy's populated areas."[65] A secret cable transmitted to Washington on May 6, 1947, read: "Ishii states that if guaranteed immunity from 'war crimes' in documentary form for himself, his superiors, and subordinates, he can describe program in detail."[66] A separate secret memorandum candidly explained that the "value to U.S. of Japanese BW data is of such importance to national security as to far outweigh the value accruing from war crimes prosecution."[67] Army officials added: "Naturally, the results of these experiments are of the highest intelligence value."[68]

Although US leaders broadly supported expanding America's biowarfare program, they also feared a biological weapons arms race with the Soviet Union. Officials likewise recognized that the public would be appalled should the Ishii deal become public, particularly since America was leading the prosecution against German researchers who conducted analogous wartime experiments. The public relations risks moved to the fore once American media outlets exposed Operation Paperclip. Since the government successfully mitigated the negative impact of the public disclosure of Operation Paperclip—by promoting a sanitized, and largely false, version of the program—it was not unreasonable to conclude that a deal with Ishii, should it become public, also could be whitewashed.[69]

In the end, the Americans determined that the potential benefits of gaining intimate access to Japan's biological warfare apparatus outweighed the risks—another example of the United States prioritizing the pursuit of military science at any price. American officials interviewed Ishii and dozens of his subordinates, each of whom received full immunity in exchange for their cooperation.[70] In secret meetings over the course of several months, Ishii and his team provided hundreds of autopsy reports and specimens from five hundred humans who had been experimented upon.[71] An examination of the specimens revealed causes of death to include cholera, mustard gas, and plague, among others.[72] Other reports detailed Japan's use of chemical weapons against China, including dozens of attacks with mustard gas, tear gas, suffocating gases, and other agents.[73]

As the United States was secretly brokering a deal with Ishii and his biological warfare comrades, American leaders publicly joined the call for "stern justice" for Japanese war criminals and participated in the International Military Tribunal for the Far East (IMTFE), which prosecuted more than 5,700 Japanese defendants for war crimes and crimes against humanity. The trials—which took place between 1946 and 1951—were held in Tokyo and fifty-two other venues, and involved prosecutors and judges from Australia, Britain, Canada, China, France, India, the Netherlands, New Zealand, the Philippines, the Soviet Union, and the United States.[74]

Although top-secret American reports concluded that Japan's wartime experiments were similar to those condemned as war crimes at Nuremberg, Ishii and most of his associates were not prosecuted.[75] As one secret report noted, Japan's biological weapons research "is the only known source of data from scientifically controlled experiments showing the direct effect of BW agents on man." The report concluded that, since raising war crimes charges "would completely reveal such data to all nations," disclosure "must be avoided in the interests of defense and national security of the U.S."[76] A Soviet prosecutor at the IMTFE proposed submission of war crimes charges for Japan's medical atrocities, but American prosecutors at the IMTFE opposed the proposal and the charges were never brought.[77]

The USSR was the sole country to prosecute Japanese officers for biological warfare, although the trial occurred outside the purview of the IMTFE. Between December 25 and 31, 1949, the Soviets conducted the Khabarovsk War Crime Trials, where twelve Japanese military officials were tried and convicted of war crimes. Ishii and most of his colleagues were not identified as defendants, though their names appeared in the evidentiary record.[78] One Japanese researcher who testified during the trial stated that American war prisoners were used to study "the immunity of Anglo-Saxons to infectious diseases."[79]

During the trial, Soviet media reported that America was conducting biological warfare research and protecting Japanese war criminals.[80] The Americans denied the allegations and attempted to discredit Khabarovsk, dubbing it a show trial and publicly asserting that there was no evidence the Japanese had used human beings in biological warfare research.[81] Rebutting the Americans, the Soviets published a 535-page English translation of the Khabarovsk proceedings, complete with witness testimony and copies of Japanese documents that revealed key aspects of Japan's biowarfare program.[82]

Decades later, during congressional hearings held in the 1980s, Frank James, a veteran and WWII prisoner of war (POW), testified that he and other POWs were experimented upon by the Japanese. James further stated that some American POWs died during the experiments, that Japanese scientists removed specimens from their bodies after conducting autopsies, and that the specimens "were placed into containers and marked with the POW's number" and then "taken away by the Japanese medical group."[83] Warren "Pappy" Whelchel, another POW, described in detail the biological warfare experiments that he was coerced into while being held captive by the Japanese.[84] Greg Rodriquez testified that his father was experimented on by the Japanese and suffered research-related injuries, but that his father could not obtain adequate healthcare or veterans benefits for his health ailments due to a lack of records to substantiate that the health conditions were service-related.[85] As Rodriquez further stated: "The lives of the American FEPOW's [Far East Prisoners of War] experimented on by Unit 731 at Mukden were forfeited in the name of national security.[86]

The United States received valuable information from the immunity deal with Ishii and his comrades. The cadre prepared a comprehensive report on human experiments with biological weapons, and a separate report on biological and chemical agents to harm crops and livestock. Ishii and his team reproduced more than eight thousand slides containing tissue samples from humans and animals exposed to toxic agents. Many of the slides were from research subjects who died from inhalation anthrax, which America was keenly interested in weaponizing.[87] As one secret US report candidly stated: "Such information could not be obtained in our own laboratories because of scruples attached to human experimentation."[88]

In a classified memo, American officials acknowledged that the pact with Ishii "might later be a source of serious embarrassment for the United States."[89] In turn, the United States kept the deal secret for over thirty years.[90] National security and the progress of science superseded legal, moral, or ethical concerns. Public discussion of the Ishii deal would arise following a series of articles written by journalist John Powell, whom at one point the United States charged with sedition for his work to uncover and publish details of America's biowarfare program and the deal with Ishii and his team. During the 1980s and 1990s, other investigators confirmed Powell's revelations, and some Japanese biowarfare veterans publicly expressed remorse for their actions.[91] As historian Sheldon Harris, who spent a decade researching the Japanese biological weapons program, succinctly concluded, "American scientists hungered after forbidden fruit."[92]

Postwar Expansion of America's Research Enterprise

Operation Paperclip and the Japanese biowarfare pact were valuable initiatives that helped propel the military biomedical complex, but more important was the widespread adoption of military science as a primary driving force behind America's postwar research enterprise. On November 17, 1944, President Roosevelt wrote to his top scientific advisor, Vannevar Bush, asking for advice on how to convert the wartime OSRD into a postwar national research program.[93] Responding to Roosevelt's request in July 1945, Bush prepared a report, *Science: The Endless Frontier*, wherein he outlined a comprehensive model for scientific research and advocated for the creation of a permanent, independent, apolitical, national research foundation that would be civilian-controlled but would maintain close ties with the military. In Bush's vision, the foundation would receive funding from Congress to support medical research at hospitals and universities, industrial research for commercial applications, and military research to further weapons development.[94]

The report—which was drafted with guidance from four subcommittees consisting of fifty advisers from the highest ranks of academia and industry, including the presidents of Harvard, MIT, Johns Hopkins, Bell Telephone, and Polaroid—emphasized the importance of a robust national research enterprise and noted that "it has become clear beyond all doubt that scientific research is absolutely essential to national security."[95] The Secretaries of War and Navy agreed with the assessment, adding that "war is increasingly total war, in which the armed services must be supplemented by active participation of every element of civilian population."[96] Among its recommendations, the report advocated for corporate tax breaks for research and development, an overhaul of patent laws, an expanded science curriculum in public schools, and university scholarships for students with a strong aptitude for science.[97] The proposals contributed to a post-WWII militarization of American society.

In 1947, Congress passed a law that operationalized many of the report's core proposals, including the creation of the National Science Foundation (NSF).[98] President Truman vetoed the bill, primarily because it did not provide the president with the authority to appoint the NSF's leaders.[99] In 1950, Truman signed a revised version of the bill that afforded him the power to appoint the NSF director and a twenty-four-member governing board.[100] Nine years later, during the administration of President Dwight D. Eisenhower, Congress amended the law to grant the governing board the power to delegate its funding authority.[101] Although the shift did not eliminate the influence of politics within the NSF, it placed day-to-day funding decisions in the hands of NSF scientists rather than political appointees.[102]

Contemporaneously, the United States reorganized its military and intelligence agencies. Via the National Security Act of 1947, Congress created a Department of Defense (DoD) that maintained separate services for the Army, Navy, and Air Force. The new law also created the National Security Council (NSC), which advises the president on national defense and foreign affairs, and the Central Intelligence Agency (CIA).[103] Within months of the new framework, the military asked the CIA to assist in covert operations, including paramilitary activities. CIA attorneys advised against collaborating with the military on covert operations, indicating that the National Security Act did not provide legal authority for such activities because they were tenuously related to intelligence.[104]

The military then lobbied Truman for an expansion of the CIA's mandate, and in 1948 the National Security Council issued Directive 10/2, which granted intelligence officials—via a new Office of Special Projects—broad authority to engage in covert operations such as paramilitary activities, espionage, political propaganda, sabotage, assisting guerrilla movements, and more.[105] The Office of Special Projects merged with the CIA in 1952, forming a new combined directorate, the CIA Clandestine Services. The intent, as detailed in a secret CIA report, was for the CIA Clandestine Services "to operate as independently of the other offices of CIA as efficiency would permit."[106]

Two years later, President Eisenhower commissioned a special study group to prepare a CIA report on the use of covert operations to support America's geopolitical priorities.[107] In the top-secret report, the committee advocated for "an aggressive covert, psychological, political and paramilitary organization" that would be "more effective, more unique and, if necessary, more ruthless than that employed by the enemy."[108] The group further advised that "no one should be permitted to stand in the way of the prompt, efficient and secure accomplishment of this mission."[109] Speaking in reference to the Soviet Union, the report indicated that "It is now clear that we are facing an implacable enemy whose avowed objective is world domination by whatever means and at whatever cost. There are no rules in such a game. Hitherto acceptable norms of human conduct do not apply."[110] This philosophy guided covert operations and intelligence activities throughout the Cold War, including collaborations with industry and academics on scientific research.

In public and private institutions throughout the country, military science was a core national priority. In 1950, as the Korean War commenced and the Cold War was heating up, President Truman ordered that all federal agencies, including research institutes, focus their resources on matters of national security.[111] That same year Congress passed the Defense Production Act, which authorized the President to require that private businesses accept and prioritize government contracts for materials deemed necessary for national defense.[112]

In 1952, an internal National Institutes of Health (NIH) planning memo explained that the NIH "will not wait for formal requests by the armed forces . . . to undertake research which NIH staff knows to be of urgent military and civilian defense significance." The NIH went on to list research priorities to "begin immediately," which included "work directly related to biological warfare, shock, radiation injury and thermal burns."[113] This work further entrenched the symbiotic relationship between the civilian and military research establishments.

Coupled with these military science activities, in 1956 the DoD created the Defense Science Board to provide guidance on military science pursuits. Still active and influential today, the board is housed within the Pentagon and is composed of industry and academic leaders, as well as former high-ranking government and military officials. During the early stages of these scientific developments, several university scientists raised concerns about academic freedom and the militarization of research. For example, in 1946, Cornell physicist Philip Morrison lamented that the university research environment was becoming "narrow, national, and secret." Morrison warned that "American science will appear to the world as the armorer of a new and more frightful war."[114]

Military science work accelerated exponentially following the Soviet Union's successful 1957 launches of Sputnik 1 and Sputnik 2, the first artificial satellites to circle the Earth. These Soviet achievements—combined with a televised US launch disaster whereby a Vanguard rocket lifted a few feet off the ground, fell back, and then blew up—shocked and motivated US leaders into action. Fearing that the USSR was far superior in several aspects of military science, America summoned all of its technological and scientific resources.[115] The government brought together leaders from academia and industry and established advisory committees and research groups that specialized in key areas of military science.

Within months of the Sputnik launches, the United States created the Advanced Research Projects Agency (ARPA)—later renamed the Defense Advanced Research

Projects Agency (DARPA)—to conduct classified research that could reshape the frontiers of technology and military science. Initially, the military and intelligence community mistrusted ARPA and feared that the agency would siphon resources and scientific talent, but eventually, both would come to view ARPA as an integral component of the military biomedical complex.[116]

Publicly, President Eisenhower called for a "Science for Peace" program, but privately, his administration worked fervently to expand military science and create technologically superior weapons of war.[117] One group, known as JASON and funded by ARPA, brought together leading scientists to tackle some of the government's toughest technological challenges.[118] Throughout the Cold War, the government and military also increasingly relied on the RAND Corporation, a nonprofit institute created in 1948 to assist in translating theoretical concepts from the social and physical sciences into applied science for military and national security purposes. RAND (which stands for Research and Development) is a distinguished policy think tank, and several of its members have served in high-ranking roles in government and the military.[119] Still active today, JASON and the RAND Corporation are frequently relied upon to advise the government on pressing issues in military science, while DARPA is the military's preeminent research agency, responsible for revolutionary innovations that have been utilized in military and civilian applications such as the computer mouse, drones, GPS, and the Internet.

Federal funding for research grew exponentially: NIH grants increased from $85,000 in 1945 to $200 million in 1959 (which included $45 million for in-house research at the NIH's hospital, the NIH Clinical Center), rising to $1.5 billion by 1970.[120] NSF funding grew from $16 million in 1956 to $159 million in 1960 and $280 million in 1970.[121] Supplementing these investments were millions of dollars in research funding from the Public Health Service, Atomic Energy Commission, Veterans Administration (VA), CIA, Army, Navy, and Air Force.

By the early 1960s, the federal government funded two-thirds of all scientific research and development in the United States, including 60 percent of industry research and 70 percent of university research.[122] This massive funding portfolio afforded the government wide latitude to dictate what was researched and by whom. In practical effect, since much of the funding was devoted to areas of interest to military science, the intellectual and economic powerhouses of America continued to be devoted to war preparations. Between WWII and the mid-1970s, this included government funding for thousands of research projects involving human subjects.[123] Notwithstanding the research ethics principles set forth in the Nuremberg Code, the studies were conducted with few legal protections for research participants.

Development of Legal Guidelines for Human Experimentation

As research was proliferating throughout the country, policies governing research with human subjects applied in very few instances.[124] In 1953, the military issued a policy governing atomic, biological, and chemical weapons research that was modeled on the Nuremberg Code, though as detailed in chapter 8 the policy was sparingly enforced. That same year, the NIH instituted guidelines for studies conducted internally at the NIH Clinical Center.[125] The NIH guidelines required consent from research participants and "group consideration" of research that "deviated from acceptable medical practice or involved unusual hazard."[126] These terms left ample room for researcher discretion. More importantly, the NIH policy did not apply to the majority of NIH projects, which were projects funded by the NIH but conducted externally at universities, hospitals, or research institutes.[127] Given the very limited reach of the DoD and NIH policies, after WWII and into the early 1960s, for the majority of studies conducted across the country—in hospitals, universities, prisons, clinics, and corporate labs—the decision whether to implement safeguards was left to individual institutions and their researchers.[128]

Some additional protections for research subjects were established during the early 1960s. In 1962, the Army issued Regulation 70-25, titled *Use of Volunteers as Subjects of Research*, which detailed guidelines, such as voluntary consent and written approval of a study after a risk-benefit assessment, in instances where Army volunteers "are deliberately exposed to unusual or potentially hazardous conditions."[129] The guidelines defined such conditions in narrow terms—specifically, instances where risk is "beyond the normal call of duty."[130] The rules also explicitly exempted research and activities related to military training, field testing, or clinical studies "conducted by the Army Medical Service for the benefit of patients."[131] As a practical matter, the research guidelines were applicable in very few Army endeavors.

Also in 1962, the Food and Drug Administration (FDA) instituted new research guidelines, sparked by a public health tragedy caused by thalidomide, a drug that was first marketed in Europe but was also used by pregnant women in the United States to combat nausea. As the agency was evaluating thalidomide, Frances Kelsey, an FDA scientist, pressed the manufacturer for more data due to her concerns that the drug caused peripheral neuropathy in those who ingested the medicine. At the time, it was common for pharmaceutical companies to provide American physicians with free samples of experimental drugs and pay physicians to collect data on patients who were administered the therapies. Physicians typically did not inform their patients that the drugs were experimental and that their health records were being shared with drug developers.[132] In line with this common practice, the American women who received thalidomide were not informed that they were participating in a study. During the ongoing research, scientists discovered that ingestion of thalidomide during pregnancy was correlated with a significant increase in birth defects—by 1961, the drug was linked to birth defects in more than ten thousand children.[133] Many children were born with phocomelia, a shortening or absence of limbs.

The thalidomide calamity led to significant changes in protocols governing drug approval and pharmaceutical research.[134] Via the 1962 Kefauver-Harris Amendments to the Federal Food, Drug, and Cosmetic Act (FDCA), drug companies were now required to prove that their product was effective in order to earn FDA approval.[135] Under the original provisions of the FDCA, which were established in 1938, drug companies only were responsible for providing evidence that their product was safe.[136] The changes to the drug approval process called for more studies involving human participants, and the 1962 amendments outlined new protections for research subjects. At the time, some commentators warned that, due to the new FDA regulations, "imaginative research will be stifled and that new therapeutics will be dangerously delayed."[137]

The new guidelines included the need to obtain consent and inform research participants of the investigational status of the drug, though the guidelines only applied to pharmaceutical research conducted for purposes of FDA review.[138] The FDA did not issue regulations that detailed how to obtain consent or what information should be disclosed, and the consent requirements contained several exceptions. For example, consent was not required if the researcher, in their own judgment, determined that obtaining consent was not feasible or contrary to the best interests of the research subject.[139] Fundamentally, the consent requirement served as a mechanism by which institutions and researchers obtained written legal waivers for research-related harms rather than a means for protecting research subjects and providing meaningful information to potential research participants.

Notwithstanding their limited reach, the 1962 Army Regulation and FDA guidelines, coupled with the 1953 DoD and NIH policies, were important first steps in the codification of protections for research participants. By emphasizing consent, informational disclosures, and an independent assessment of risks and benefits, the guidelines acknowledged that individual scientists should not be afforded full discretion to determine whether research is ethical and the circumstances under which research should be conducted. Yet, since the guidelines were rarely enforced and contained exemptions that dispensed with many of the safeguards, the policies afforded few real-world protections for research subjects. The safety gaps—coupled with a professional ethos that blended ambition, paternalism, and hubris—resulted in a research culture that routinely exploited the poor, disenfranchised, or otherwise vulnerable individuals. This was not limited to military science, but rather was an aspect of the entire US research enterprise in the decades following WWII.

Within this environment, institutionalized children were widely viewed as quintessential research subjects: they lived in a controlled setting, could be intensely monitored, were shunned by the public, and did not have agency or authority to push back against the scientific pursuits of the adults in the room. During the summer of 1947, as American prosecutors presented their closing arguments at the Nuremberg Doctors' Trial, researchers from the University of Pennsylvania and other colleges conducted experiments at Pennsylvania's Pennhurst State School and Hospital, an institution for children with mental and physical disabilities. Among the myriad studies, scientists laced the children's milk with "pooled feces" to conduct a "challenge study" to test the effectiveness of an experimental hepatitis vaccine.[140] In a vaccine challenge study, a research subject is inoculated with an experimental vaccine and then exposed to the disease to test the vaccine's effectiveness.

Leading physicians and vaccine pioneers, including Jonas Salk, commonly sought out institutionalized children to test their medical products. As Salk candidly explained during his quest to obtain child research subjects: "I have investigated the local possibilities for such an experiment and find . . . there are institutions for hydrocephalics and other similar unfortunates. I think we may be able to obtain permission for a study."[141] Challenge studies were conducted on children at several institutions, and included tests with experimental vaccines for polio, hepatitis, influenza, meningitis, mumps, rubella, and measles.[142] Howard Howe, a polio researcher and physician at Johns Hopkins, published an article in 1952 wherein he described studies that compared the "antibody response of the chimpanzee and the child." Howe characterized the latter as "low grade idiots or imbeciles with congenital hydrocephalus, microencephaly, or cerebral palsy."[143]

Between 1956 and 1971, researchers conducted a slew of experiments at New York's Willowbrook State School, which housed thousands of children with cognitive disabilities. The researchers infected children with hepatitis, measles, mumps, and rubella to study the progression of the illnesses and test new vaccines. In a hepatitis study conducted at Willowbrook, a research team led by New York University (NYU) physician Saul Krugman fed children, aged three to eleven, chocolate milk laced with human feces collected from individuals suffering from hepatitis and jaundice. Children were provided the milk concoction at various intervals, such as six months and a year after inoculation, in order to hone the vaccine formula and analyze whether long-term immunity could be achieved. In the midst of the research, parents who sought to have their children enrolled at Willowbrook were told that admission was possible only if they agreed to allow the researchers access to their child for experimentation.[144] By making such offers, researchers and Willowbrook administrators took advantage of desperate parents with few options for their children with disabilities.

Following the Willowbrook studies, countless children suffered research-related harms from the inducement of hepatitis and other diseases, including abdominal pain, vomiting, and liver inflammation, among other ailments. Meanwhile, Krugman and his colleagues were praised in medical journals for their findings and methods. The work led to important new insights, such as the discovery of two types of hepatitis, A and B, and contributed to the development of prophylactic vaccines.[145] The editors of the *Journal of the American Medical Association* heralded Krugman's "carefully planned and executed study" and his "judicious use of human beings," while an article in the *New England Journal of Medicine* noted that Krugman's study "was sanctioned by the authorities of the New York State Department of Mental Hygiene and by the Armed Forces Epidemiological Board, Office of the Surgeon General."[146] The medical and research establishments did not raise legal or ethical concerns with the Willowbrook studies.

In addition to challenge studies to test experimental vaccines, at orphanages and mental institutions across the country, researchers subjected children to a litany of tests to analyze the risks and benefits of experimental pharmaceuticals and novel consumer products such as enriched bread, fluoride toothpaste, and refined sugar. At New Jersey's State Colonies for the Feebleminded, researchers conducted ringworm studies where they would scrape the scalp of children, aged six to ten, and rub ringworm into the wound to examine the progression of the disease and test investigational ointments.[147]

Physicians also conducted psychiatric research on children, including honing techniques such as electroconvulsive therapy and lobotomy. At Creedmoor State Hospital in New York, renowned child psychiatrist Lauretta Bender led a team that analyzed whether LSD could assist children with schizophrenia, autism, or other mental ailments. Bender and her colleagues administered LSD to children, aged six to twelve, over a period of weeks or months, and in some cases for more than a year. In many instances, administration of hallucinogenic drugs was supplemented with electroconvulsive therapy or other psychiatric practices. The extent to which parents understood or consented to the experiments is unknown.[148] Oftentimes, researchers and administrators saw the experiments as a way for the children to pay back a debt to society for the care they received in state institutions.

In addition to pediatric institutions, prisons were a key source of research subjects. A captive population with a dire need for money, prisoners often were willing to accept a meager stipend for participation in risky research. Prisoners needed the money for bail, to support their families, and to pay for everyday items that their jailer did not provide, such as toothpaste, soap, and writing materials.[149] After the Nuremberg Doctors' Trial, the United States was unique in that it continued to maintain an extensive program of prison experiments.[150]

Studies conducted from the 1950s through the 1970s at Philadelphia's Holmesburg prison are illustrative. At Holmesburg, with funding from corporations and the government, physicians and researchers from the University of Pennsylvania and other institutions tested a variety of experimental drugs and commercial products, including diet pills, eye drops, sunscreen, and detergents.[151] One distinguished dermatologist, Albert Kligman, referred to Holmesburg as "an anthropoid colony" ideal for human experimentation. Kligman marveled at the opportunity to conduct research on prisoners: "All I saw before me were acres of skin. It was like a farmer seeing a fertile field for the first time."[152] As Kligman later explained: "I began to go to the prison regularly, although I had no authorization. It was years before the authorities knew that I was conducting various studies on prisoner volunteers. Things were simpler then. Informed consent was unheard of. No one asked me what I was doing. It was a wonderful time."[153]

Prisoners were paid for their participation, ranging on average from $10 to $300 per study. Army experiments to test LSD and other psychotropic drugs were particularly enticing, paying up to $1,500. The cash was lucrative by prison standards, where inmates typically received fifteen cents a day to make shoes or knit clothes, but it was far less than that demanded by research volunteers outside prison walls.[154] According to William Robb, an inmate who participated in the studies, the payments were "a king's ransom for an inmate trying to raise bail money."[155] In one experiment, researchers conducted burn tests to examine new healing ointments; in another study, scientists performed scalp transplants in an attempt to rejuvenate hair growth by moving hair from the back of a person's head to the front. During a "sweat test," prisoners were placed in an overheated chamber, after which their armpit glands were cut and examined. Researchers also dropped Johnson & Johnson baby shampoo into the eyes of prisoners to test which formula was least irritable.[156]

For decades the majority of early-stage pharmaceutical research—which involved risky toxicity studies—was conducted on prisoners. Hundreds of studies exposed inmates to bacteria, viruses, or fungi, and then examined investigational drugs, ointments, or other therapies. In an experiment involving ringworm, prisoners wore infected boots continuously for a week. In other studies, scientists injected viruses—including wart viruses, vaccinia, herpes simplex, and herpes zoster—on various parts of a prisoner's body, including the forearm, palm, back, face, scalp, and penis.[157] As Allen Hornblum explains in his groundbreaking book, *Acres of Skin*, "Holmesburg

Prison became one of postwar America's largest, nontherapeutic, human research factories."[158]

In 1966, the FDA raised concerns with Kligman's practices and notified thirty-three drug companies that Kligman was no longer an FDA-accepted investigator due to his repeated failures to comply with research guidelines.[159] Within a month, the agency reversed its decision, citing assurances from Kligman that he would follow the rules.[160] News reports indicated that pharmaceutical firms and the University of Pennsylvania placed political pressure on the FDA, and several FDA officials—including Frances Kelsey—publicly remarked that it was unusual for the FDA to grant such a quick reinstatement.[161]

When details of the prison experiments reached mainstream media outlets, one reporter likened Kligman to a Nazi physician. "I'm Jewish," he responded, "it struck me as ludicrous and incredible that I'd be compared to that."[162] Kligman professed that he recognized and considered the ethical implications of his work: "We had an ethical problem. How much right do you have to cause risk to a prisoner in medical tests from which he has no direct benefit?... We pay him to lend us his body for some time. But we pre-determine whether a test is dangerous, and the prisoner has to depend on our judgment."[163] During the mid-1970s, as research guidelines were adopted to afford prisoners with some legal protections, Kligman remarked that it was "a very good case of the triumph of the do-gooders," whom he defined as liberals, lawyers, and prison reformers.[164] He destroyed all his research notes in 1974, just as lawsuits against him began to heat up.[165]

One of the most hazardous Holmesburg experiments was a collaboration between Dow Chemical and the University of Pennsylvania that examined dioxin, a component of Agent Orange, which Americans were deploying extensively in Vietnam. Scientists placed dioxin directly onto the backs of inmates to study adverse health effects.[166] The university released Dow Chemical from liability in the event of research-related harms, and required that inmates waive all legal claims in exchange for the modest payment they received to take part in the study. The comprehensive legal waiver was distilled to one sentence: "I, the undersigned, also hereby give my permission to the hospital, laboratories or others to perform medical and other tests on me: the hospital, laboratories, etc. or the prison are not to be held responsible in any way for any complications or untoward results that may arise."[167] The waiver was so broad that it encompassed injuries caused by negligence, recklessness, and intentionally harmful conduct. Figure 9.2 depicts one prisoner who volunteered for a study.

A report that revealed details of the dioxin experiment led to significant public backlash, and in 1981, the Environmental Protection Agency (EPA) initiated an investigation and offered medical monitoring for prisoners who were subjected to the study.[168] Dioxin was known to cause cancer and other significant health ailments, but the EPA's efforts were largely a façade: the EPA placed the burden on inmates to contact the agency and prove their exposure.[169] As one internal EPA memo explained, "I hope everybody involved in this remembers that the purpose is to calm things down—not stir them up."[170]

Of the forty inmates who wrote to the EPA, the agency only interviewed eighteen. Of those interviewed, since the dioxin records were destroyed by Dow Chemical and the University of Pennsylvania, the EPA concluded it could not confirm that

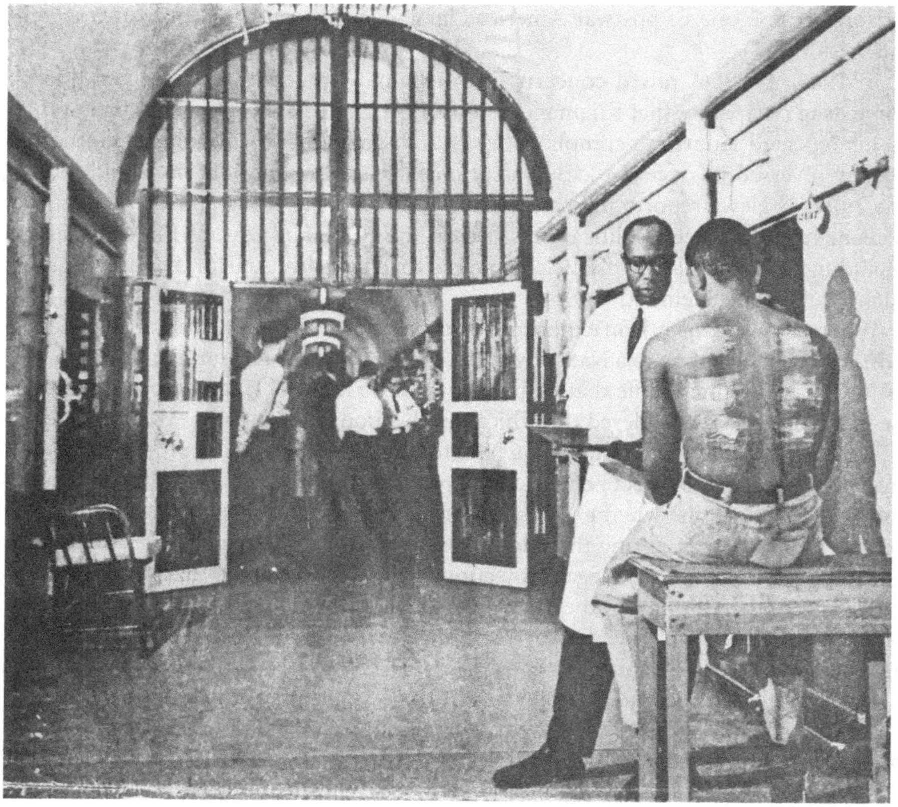

Figure 9.2 A prisoner volunteering for a medical experiment at Holmesburg Prison speaks with a medical administrator.
Photograph reproduced with permission from the Special Collections Research Center, Temple University Libraries, Philadelphia, PA.

the inmates were exposed to dioxin and thus refused to conduct medical monitoring. The inmates then turned to the courts, filing lawsuits against Dow Chemical, the University of Pennsylvania, the researchers, and the City of Philadelphia (which ran the prison).[171] Some plaintiffs settled, though at least one lost on a legal technicality because the judge ruled that he took too long to file the lawsuit.[172] In another case, a court ruled against an inmate who took part in the studies on the grounds that the inmate consented to the research and there was no evidence that prison officials acted with malice or inflicted cruel or unusual punishment.[173] The City of Philadelphia did not issue a formal apology for the experiments until 2022.[174]

The experiments at Holmesburg exemplified post-WWII research in prisons. Throughout the 1950s, prisoners in Georgia and Illinois were test subjects for new malaria treatments, while inmates in Ohio were administered experimental vaccines for influenza and the common cold. Inmates also were test subjects for Albert Sabin's oral polio vaccine, which is known to cause polio in a percentage of people who are inoculated because the vaccine uses attenuated strains of live polio virus. At Sing Sing prison in New York, prisoners provided muscle tissue and blood samples to NYU researchers for an arthritis study, while others were injected with syphilis to examine the progression of the disease. At New Jersey's only female prison, two hundred women took part in a study to test a vaccine to protect against hepatitis—all contracted the disease. By 1974, approximately 75 percent of all FDA-approved drugs utilized prison research, and many drug companies maintained laboratories and clinics near prisons.[175]

Of the hundreds of prison studies conducted across the country, one of the most controversial involved injecting live cancer cells into healthy adults to study how the body reacts to the cells. The NIH-funded study took place in Ohio—a collaboration between researchers at Ohio State University and Manhattan's Sloan-Kettering Institute for Cancer Research. After the injections, part of the inmate's forearm would be surgically removed for further examination. The project was lauded in mainstream media outlets, including *The New York Times*, as a breakthrough in oncology research involving humanitarian contributions by prisoners.[176] The *Times* article stated that the experiment bears "no great risk" and that "there is nothing but admiration for men who are regarded as enemies of society yet are fully aware of the contribution that they can make to medicine."[177]

One of the Sloan-Kettering researchers from the Ohio study, Chester Southam, led an analogous experiment in 1963 at the Jewish Chronic Disease Hospital in New York, a facility that provided long-term care for elderly patients. Southam wanted to compare the immunological response of healthy subjects (from the prison experiment) with debilitated individuals housed at the hospital. Of the twenty-two hospital patients who had live cancer cells injected into their bodies, most were not told of the study, and many were incompetent to provide consent.[178]

Ethical concerns about exploiting vulnerable populations received little attention in medical circles. Such experiments were not unusual for the era, nor were they deemed unethical or dishonorable by prevailing medical and scientific standards. Scientists viewed associations with prisons, institutions, and hospitals as springboards to government research grants and lucrative contracts with pharmaceutical companies. Young academics working toward tenure were hard-pressed not to fall in line. As scientists vigorously competed with one another to hone new vaccines, pharmaceuticals, and medical techniques, mistreatment of human subjects was commonly viewed as an oversight, not a problem endemic to research.[179]

A small number of contemporary scientists voiced ethical concerns. In his seminal 1959 article, *Experimentation in Man*, Henry Beecher, a physician at Massachusetts General Hospital, argued that it was "unethical and immoral to carry out potentially dangerous experiments without the subject's knowledge and permission."[180] As Beecher added: "Any classification of human experimentation as 'for the good of society' is to be viewed with distaste, even alarm. Undoubtedly, all sound work has this as its ultimate aim, but such high-flown expressions are not necessary, and have been used within living memory as cover for outrageous ends."[181]

Beecher's remarks echoed calls from anti-vivisectionists from the early 1900s, which were discussed in chapter 3. In 1962, as ethical objections began to trickle into reputable scientific journals, the NIH commissioned a study of human subject research policies at university medical schools. Of the fifty-two universities that responded, few had written guidelines, and a majority opposed peer review of research protocols, preferring to leave research decisions to the discretion of individual faculty members.[182]

Shortly thereafter, a 1963 anthology published by Boston University's Law-Medicine Research Institute included several articles that highlighted the expansion of clinical research after WWII and the need for more robust ethical guidelines.[183] The anthology noted, "There are no statutes and no reported cases dealing directly with clinical investigation," but "medical research activity, like all activities in our society, is subject to common-law principles of general applicability."[184] In 1964, a report circulated within the NIH that warned of "untoward events" that would "rudely shake" the NIH due to the lack of adequate research safeguards.[185] There is no evidence that the NIH took any action to meaningfully address the concerns.

A regulatory shift would arise after the public became aware of the cancer study at the Jewish Chronic Disease Hospital. In 1966, the Public Health Service (PHS) instituted a new policy that encompassed extramural research—studies funded by PHS but conducted externally at hospitals, universities, prisons, and institutions. The new guidelines encompassed all NIH-funded research, and contained four key elements: (1) peer review of projects to examine whether research protocols appropriately balance the rights and welfare of research subjects with anticipated benefits of the study; (2) informed consent; (3) continuing review by an institutional peer review committee to ensure a project maintains an acceptable balance of risks and benefits; and (4) documentation of adherence to the guidelines.[186] Within the rules, however, key terms such as "informed consent" and "rights and welfare of potential subjects" were not adequately defined, and these omissions resulted in confusion and arbitrary implementation of the new policy.[187]

Within months of the issuance of the new PHS guidelines, Beecher published an article, *Ethics and Clinical Research*, wherein he identified twenty-two studies of published research as unethical, involving hundreds of research subjects who were not fully informed of research risks or who were completely unaware that they were research subjects.[188] Beecher noted that many of the studies were funded and conducted by the military, VA, NIH, medical schools, and private hospitals.[189] Although Beecher did not name names, over the years, scholars pieced together that the majority of the studies involved scientists at elite institutions such as Harvard, Emory, Duke, NYU, and Georgetown, and that the work was funded by the Atomic Energy Commission (AEC), PHS, Parke-Davis, and Merck.[190] Given his stature as a leading physician and researcher, Beecher's article helped promote acceptance of the new PHS requirements.[191] As he concluded in the article: "An experiment is ethical or not at its inception; it does not become ethical *post hoc*—ends do not justify means."[192]

Also significant was Maurice Pappworth's 1967 book, *Human Guinea Pigs*, which detailed unethical research conducted across the United States and Britain. As Pappworth explained: "I have frequently been attacked by doctors who contend that by such publication I am doing a great disservice to my profession, and, more especially, that I am undermining the faith and trust that lay people have in doctors."[193] As he was working on the book, Pappworth said he often would receive "telephone calls, almost entirely from strangers, in an attempt to persuade me to abandon the project."[194] Interestingly, some physicians urged Pappworth to limit his publications to medical journals, to "avoid completely any discussion outside professional circles."[195] Yet, when Pappworth inquired with medical journals or spoke at medical conferences on the topic of human experimentation, he often was dismissed, and the problems that he raised were ignored.[196] As Pappworth contends: "Science is not the ultimate good, and the pursuit of new scientific knowledge should not be allowed to take precedence over moral values where the two are in conflict."[197]

Thereafter, in 1971, the Department of Health, Education, and Welfare (DHEW) published a guidebook that explained the agency's policy for human subjects research and outlined the basic elements of informed consent. Notwithstanding the additional details, the guidebook underscored that "a flexible policy is essential" and that the agency permits interpretation of the rules based on "sound professional judgment of reasonable men."[198]

The ethical and legal implications of lax research guidelines were meticulously detailed in a 1972 treatise published by Jay Katz, a renowned physician and law professor who fled Nazi Germany in the late 1930s and served in the US military during WWII. Katz's treatise, which spans over eleven hundred pages, includes a plethora of articles, court opinions, governmental reports, and personal letters published over the previous quarter century from physicians, attorneys, scientists, and philosophers.[199] The materials reveal that, for decades, experts had been debating the contours of informed consent, fair selection of research subjects, researcher obligations, and the societal implications of human experimentation.

Despite new rules governing PHS-funded research, the DHEW guidebook, the FDA's 1962 guidelines for pharmaceutical research, and rising calls for adherence to ethical principles, untoward research continued. Orchestrating a culture shift in the research enterprise was challenging. Generations of physicians and scientists enjoyed broad research freedoms and deeply believed their work was essential to the progress of science and society. Paternalism enveloped the practice of medicine, and patients did not feel empowered to ask questions or inquire into a physician's judgment. Eugenics policies peppered throughout colleges and medical schools from decades earlier undoubtedly hung in the minds of some researchers, who viewed the illiterate and disabled as inferior beings who owed debts to society. The privileges of the research elite contrasted mightily with the tribulations of research subjects, who typically were poor and marginalized. Jim Crow laws, mass incarceration under "law and order" policies, structural and institutional racism, and widespread discrimination meant that African Americans suffered disproportional research harm. For example, within prisons, there were suspicions that African American prisoners were recruited into the most dangerous experiments.[200]

The insider environment of medicine and research left little room for whistleblowers, outside criticism, or civil rights advocates. Lawsuits against physicians, researchers, or institutions were rare, despite the fact that individuals were legally protected from improper bodily intrusions by state common law principles such as battery. According to a medico-legal treatise published in 1971, a line of court decisions required that physicians "make a simple, quiet, but honest disclosure commensurate with the risk in all cases and let the patient choose what risks he wishes to run with his body."[201] As the treatise further noted, however, there were few decisions that guided potentially therapeutic or nontherapeutic experimentation.[202] Taken together, although courts generally had upheld a patient's right to be duly informed prior to consenting to medical treatment, the chance of a research-related lawsuit was very small, and fear of lawsuits did not deter questionable research practices.

Meaningful legal protections for research subjects would arise following public disclosure of the Tuskegee Syphilis Study in 1972. The study, which began in 1932 and was funded and conducted by the PHS, tracked the development of syphilis in hundreds of poor African American men in Alabama for forty years. Researchers withheld treatment from the men despite the fact that, for over twenty years, physicians knew that penicillin cured the devastating disease. Throughout the Tuskegee study, the men were told that they were being treated for "bad blood," and were never informed that they were participants in an experiment.[203] Many men fought in WWII, at which point the PHS convinced the military to continue to withhold treatment so the study would not be compromised.[204]

In 1965, student protests organized by Students for a Democratic Society called for an end to the Tuskegee study; the group was portrayed as radical leftists, and their call to end the experiment was simply disregarded.[205] The study was not halted until 1972—six years after the implementation of the PHS guidelines and one year after the publication of the DHEW guidebook—after a journalist raised ethical concerns in a publication that reached mainstream American society.[206] In a 1972 news article that discussed the study, one of the lead PHS physician-researchers, John R. Heller, remarked: "There was nothing in the experiment that was unethical or unscientific."[207]

Thereafter, the federal government acknowledged that the United States did not have adequate laws to protect research subjects.[208] In 1974, DHEW issued new rules governing research with human subjects that applied to projects funded by the department: the new protections included informed consent, institutional review of research protocols, and institutional assurances that federal law would be adhered to.[209] That same year, Congress passed the National Research Act. Among its provisions, the law established the National Commission for the Protection of Human Subjects of Biomedical and Behavioral Research, a group charged with issuing recommendations on ethical principles to frame federal protections for research subjects.[210]

The commission issued its findings—commonly referred to as the Belmont Report—in 1979. The Belmont Report recommended that federal law adhere to three ethical principles: respect for persons (treating people as autonomous agents and affording additional protections to individuals with diminished autonomy), beneficence (an obligation to do no harm, maximize possible benefits, and minimize potential harms), and justice (treating individuals fairly, and fairly distributing benefits and risks across the population). These principles, the authors of the Belmont Report explained, can be achieved by requiring informed consent, independent assessment of a research project's risks and benefits, and protocols to guarantee fair selection of research subjects.[211]

In the same year that the Belmont Report was issued, DHEW was separated into distinct departments for education (Department of Education) and health (Department of Health and Human Services [HHS]). In light of the recommendations in the Belmont Report, during the 1980s, HHS revised its guidelines governing research with human subjects.[212] The guidelines were codified into federal law in 1991 and adopted by sixteen federal agencies, including the DoD.[213]

Summary

Exploitation of the German and Japanese war science enterprises did not stem from haphazard decisions. American leaders carefully considered the risks of recruiting and employing recent enemies in sensitive areas of military science, including the chance for sabotage or traitorous endeavors. Moral qualms of collaborating with Nazis and war criminals were debated and judged against the desire to enrich military science. Leaders at the highest levels of government and the military echoed the value of capitalizing on information from experiments that would be difficult or unethical to replicate. These judgments were not uniquely American, but rather represented a calculus mirrored by nations throughout the world in a global quest for scientific and military superiority.

The impact of Operation Paperclip and the biowarfare pact with Ishii and his colleagues reverberated for decades. The knowledge gained from the programs helped propel the development of military science in several areas. As the United States built upon the war machines of Germany and Japan, however, the justifications that motivated the Germans and Japanese into conducting abhorrent experiments infiltrated America's Cold War scientific culture. A moral stance that favored national security over research ethics and pursued science at any price became firmly embedded in both civilian and military research establishments. Although this driving force had been present at various times throughout US history, it escalated exponentially during the Cold War alongside America's global efforts to safeguard democratic capitalism. As detailed in chapters 10 and 11, in the three decades between WWII and the mid-1970s, military research in atomic, chemical, and biological warfare careened out of control amid a regulatory abyss.

Notes

1. World War II Casualties by Country, *World Population Review* (accessed February 21, 2024).
2. "The Holocaust," *National WWII Museum* (accessed February 21, 2024); "How Many People Did the Nazis Murder?," *Holocaust Encyclopedia* (accessed February 21, 2024); "Children During the Holocaust," *Holocaust Encyclopedia* (accessed February 21, 2024).
3. Hugh J. Knerr, Memorandum for Commanding General, US Strategic Air Forces in Europe (June 1, 1945).
4. Stephen Wertheim, "The Price of Primacy: Why America Shouldn't Dominate the World," *Foreign Affairs* 99, no. 2 (March/April 2020): 19–29; Daniel Immerwahr, *How to Hide an Empire* (2019).
5. Hunt, *Secret Agenda*, 1–5; John Gimbel, "German Scientists, United States Denazification Policy, and the 'Paperclip Conspiracy,'" *International History Review* 12, no. 3 (1990): 441–65. In many documents and reports, the program is referred to as Project Paperclip, though

in recent years convention has been to call the program Operation Paperclip. For the sake of simplicity, throughout the book I will use the title Operation Paperclip.

6. Moreno, *Undue Risk*, 92.
7. Gimbel, "German Scientists," 442–43.
8. Hunt, *Secret Agenda*, 1–5.
9. Gimbel, "German Scientists," 447–51.
10. Hunt, *Secret Agenda*, 6–21.
11. Annie Jacobsen, *Operation Paperclip* (2014), 219.
12. Quoted in Hunt, *Secret Agenda*, 33.
13. Alexander Mitscherlich and Fred Mielke, "Seven Were Hanged," in *The Nazi Doctors*, edited by Annas and Grodin, 105–7.
14. Jacobsen, *Operation Paperclip*, 286.
15. Mitscherlich and Mielke, "Seven Were Hanged," 106–7.
16. Guillemin, *Hidden Atrocities*, 297.
17. Telford Taylor, "Opening Statement of the Prosecution, December 9, 1946," in *The Nazi Doctors*, edited by Annas and Grodin, 91.
18. Jacobsen, *Operation Paperclip*, 247–49.
19. Schmidt, *Justice at Nuremberg*, 108–9.
20. U.S. Naval Technical Mission, *German Aviation Medical Research at the Dachau Concentration Camp*, Report No. 331–45 (October 1945), quoted in Schmidt, *Justice at Nuremberg*, 109.
21. "U.S., Britain Hold German Experts, Berlin Communist Papers Charge," *The New York Times*, October 27, 1946; "Nazi Brains Help U.S.," *Life*, December 9, 1946; Herbert Shaw, "Wright Field Reveals 'Project Paperclip,'" *Dayton Daily News*, December 4, 1946.
22. Hunt, *Secret Agenda*, 58–59.
23. Quoted in Hunt, *Secret Agenda*, 58–59.
24. Robert Patterson, Memorandum to William D. Leahy, "German Scientists" (May 28, 1945), quoted in Jacobsen, *Operation Paperclip*, 106.
25. Moreno, *Undue Risk*, 98–101; Hunt, *Secret Agenda*, 48–61.
26. John Gimbel, "U.S. Policy and German Scientists: The Early Cold War," *Political Science Quarterly* 101, no. 3 (1986): 433–51.
27. Gimbel, "U.S. Policy," 439.
28. Gimbel, "U.S. Policy," 439–40.
29. Moreno, *Undue Risk*, 92–94; Hunt, *Secret Agenda*, 125–42.
30. Tucker, *War of Nerves*, 116–17.
31. Hunt, *Secret Agenda*, 120.
32. Tucker, *War of Nerve*, 116–17; Hunt *Secret Agenda*, 35, 97, 143–44.
33. Harris and Paxman, *A Higher Form of Killing*, 140–43; Jacobsen, *Operation Paperclip*, 165.
34. Jacobsen, *Operation Paperclip*, 66–69.
35. Moreno, *Undue Risk*, 92–94.
36. Moreno, *Undue Risk*, 94.
37. Gimbel, "U.S. Policy," 462–63; Moreno, *Undue Risk*, 92–94; Hunt, *Secret Agenda*, 78–93.
38. Jacobsen, *Operation Paperclip*, 250–64.
39. Quoted in Jacobsen, *Operation Paperclip*, 341.
40. Schmidt, *Justice at Nuremberg*, 111.
41. Schmidt, *Justice at Nuremberg*, 108–11.
42. Hunt, *Secret Agenda*, 132–35.
43. Jacobsen, *Operation Paperclip*, xii; Hunt, *Secret Agenda*, 44.
44. Moreno, *Undue Risk*, 93–96.

45. Harris, *Factories of Death*, 54–59, 86–87; Harris and Paxman, A Higher Form of Killing, 76–83; Guillemin, Hidden Atrocities, xi–xix.
46. Craughwell, *The War Scientists*, 223.
47. Harris and Paxman, *A Higher Form of Killing*, 76–83; Guillemin, *Hidden Atrocities*, xii–xiii.
48. Hal Gold, *Unit 731 Testimony* (1997), 40–41.
49. Quoted in Gold, *Unit 731*, 154.
50. Harris and Paxman, *A Higher Form of Killing*, 80.
51. Yuki Tanaka, *Hidden Horrors* (1996), 136–37.
52. Arthur Kleinman, Jing-Bao Nie, and Mark Selden, "Introduction: Medical Atrocities, History and Ethics," in *Japan's Wartime Medical Atrocities*, edited by Jing-Bao Nie et al. (2011), 4–5.
53. "Japanese Biological Warfare Crimes Documented," *Xinhua Press*, January 11, 2014.
54. Hersh, *Chemical and Biological Warfare*, 16–17; Guillemin, *Hidden Atrocities*, 227.
55. Harris and Paxman, *A Higher Form of Killing*, 81.
56. Craughwell, *The War Scientists*, 223–24; Harris, *Factories of Death*, 93–94, 113.
57. Harris, *Factories of Death*, 5.
58. John W. Powell, "A Hidden Chapter in History," *The Bulletin of the Atomic Scientists* (October 1981): 44–52; Harris and Paxman, *A Higher Form of Killing*, 81.
59. Harris, *Factories of Death*, 63, 91.
60. Harris, *Factories of Death*, 78–79; Harris and Paxman, *A Higher Form of Killing*, 80–82; Guillemin, *Hidden Atrocities*, xviii, 120–31.
61. Franklin D. Roosevelt, Press Conference (June 5, 1942), quoted in Brown, *Chemical Warfare*, 201.
62. Harris and Paxman, A Higher Form of Killing, 78; Guillemin, *Hidden Atrocities*, xviii–xix; Barnaby, *The Plague Makers*, 117.
63. Norbert H. Fell, Report to Assistant Chief of Staff (June 24, 1947), quoted in Harris, *Factories of Death*, 198.
64. Guillemin, *Hidden Atrocities*, 265–70.
65. Military Intelligence Division, War Department, *Biological Warfare: Activities and Capabilities of Foreign Nations* (March 30, 1946), quoted in Guillemin, *Hidden Atrocities*, 192–93.
66. General MacArthur to War Department General Intelligence Division, "Operational Priority" (May 6, 1947), quoted in Guillemin, *Hidden Atrocities*, 262.
67. Edward Wetter and H. I. Stubblefield, "Interrogation of Certain Japanese by Russian Prosecutor" (July 1, 1947), quoted in Powell, "A Hidden Chapter in History," 47–48.
68. Robert McQuaill, Army Intelligence, "Summary of Information: Subject Ishii" (January 10, 1947), quoted in Harris, *Factories of Death*, 189.
69. Guillemin, *Hidden Atrocities*, 265-70.
70. Harris, *Factories of Death*, 202.
71. Guillemin, *Hidden Atrocities*, 288–91.
72. Harris and Paxman, *A Higher Form of Killing*, 80–81.
73. Guillemin, *Hidden Atrocities*, 194–203.
74. Sandra Wilson et al., *Japanese War Criminals* (2017), 1–11, 74, 270.
75. Harris, *Factories of Death*, 192, 216–18; Wilson et al., *Japanese War Criminals*, 271–72.
76. Wetter and Stubblefield, "Interrogation of Certain Japanese," quoted in Powell, "A Hidden Chapter in History," 47–48.
77. Boris G. Yudin, "Research on Humans at the Khabarovsk War Crimes Trial, in *Japan's Wartime Medical Atrocities*, edited by Nie et al., 61–62.
78. *Materials on the Trial of Former Servicemen of the Japanese Army Charged with Manufacturing and Employing Bacteriological Weapons* (1950).

79. *Materials on the Trial*, 268.
80. Harris and Paxman, *A Higher Form of Killing*, 143.
81. "No Knowledge, MacArthur Says," *The New York Times*, December 27, 1949.
82. *Materials on the Trial.*
83. *Treatment of American Prisoners of War in Manchuria: Hearings before the Subcommittee on Compensation, Pension and Insurance of the House Committee on Veterans' Affairs*, 99th Cong. 16–19 (September 17, 1986) (statement of Frank James).
84. *Veterans Administration Programs in Montana: Hearing before the Subcommittee on Oversight and Investigations of the House Committee on Veterans' Affairs*, 97th Cong. 19–21 (June 19, 1982) (statement of Warren W. Whelchel).
85. *Treatment of American Prisoners of War*, 99th Cong. 24–27 (statement of Gregory Rodriquez).
86. *Veterans Administration Programs*, 97th Cong. 18 (statement of Gregory Rodriquez).
87. Harris, *Factories of Death*, 201–3, 218; Alexander Kelle, Kathryn Nixdorff, and Malcolm Dando, *Preventing a Biochemical Arms Race* (2012), 21.
88. Edwin V. Hill, "Summary Report on B.W. Investigations" (December 12, 1947), quoted in Powell, "A Hidden Chapter in History," 47.
89. Quoted in Harris, *Factories of Death*, 219.
90. Harris and Paxman, *A Higher Form of Killing*, 157.
91. Guillemin, *Hidden Atrocities*, 332–39.
92. Harris, *Factories of Death*, 222.
93. Franklin D. Roosevelt, Letter to Vannevar Bush (November 17, 1944).
94. Bush, *Science*, 5–9, 33–34.
95. Bush, *Science*, 17, 43–45.
96. Joint Letter to National Academy of Sciences, quoted in Bush, *Science*, 17.
97. Bush, *Science*, 21–27.
98. National Science Foundation Act of 1947, S. 526, 80th Cong. (1947).
99. Bush, *Science*, xix–xx.
100. National Science Foundation Act of 1950, Pub. L. No. 507, 81st Cong. (1950).
101. An Act to Amend the National Science Foundation Act of 1950, Pub. L. No. 86–232, 73 Stat. 467 (September 8, 1959).
102. Bush, *Science*, xix–xx.
103. National Security Act of 1947, Pub. L. No. 235, 61 Stat. 496 (July 26, 1947).
104. John Prados, *Presidents' Secret Wars* (1996), 18–29.
105. National Security Council Directive on Office of Special Projects, NSC 10/2 (June 18, 1948); Central Intelligence Agency, "Office of Policy Coordination, 1948–1952," *Studies in Intelligence* (Summer 1973): 3.
106. CIA, "Office of Policy Coordination," 4.
107. Dwight D. Eisenhower, Letter to Lt. Gen. James H. Doolittle, "Panel of Consultants on Covert Activities of the Central Intelligence Agency" (July 26, 1954).
108. James H. Doolittle et al., *Report on the Covert Activities of the Central Intelligence Agency* (1954), 2.
109. Doolittle et al., *Report*, 2.
110. Doolittle et al., *Report*, 2.
111. Proclamation 2914, Proclaiming the Existence of a National Emergency (December 16, 1950); Executive Order 10,193, Providing for the Conduct of the Mobilization Effort of the Government (December 16, 1950).
112. Defense Production Act of 1950, Pub. L. No. 81–774, 64 Stat. 798 (1950).
113. National Institutes of Health, *Assumptions Underlying NIH Defense Planning* (August 2, 1952), quoted in Final Report, *The Human Radiation Experiments*, 13.

114. Philip Morrison, "The Laboratory Demobilizes . . .," *Bulletin of the Atomic Scientists* 2, no. 9–10 (November 1, 1946): 6.
115. Ann Finkbeiner, *The JASONS* (2006), 20–41.
116. Sharon Weinberger, *The Imagineers of War* (2017), 5, 50–67; Annie Jacobsen, *The Pentagon's Brain* (2015), 51–59.
117. Audra J. Wolfe, *Freedom's Laboratory* (2018), 91–99.
118. Weinberger, *The Imagineers of War*, 93–96.
119. Alex Abella, *Soldiers of Reason* (2008).
120. Bush, *Science*, xii; Allen M. Hornblum, *Acres of Skin* (1998), 86.
121. Bush, *Science*, xxiv; *The Twentieth Annual Report of the National Science Foundation* (1970).
122. National Science Foundation, *Federal Funds for Research, Development, and Other Scientific Activities* (Washington DC: Government Printing Office, 1964), 5.
123. Bradford H. Gray, *Human Subjects in Medical Experimentation* (1975), 10.
124. William J. Curran, "Governmental Regulation of the Use of Human Subjects in Medical Research: The Approach of Two Federal Agencies," in *Experimentation with Human Subjects*, edited by Paul A. Freund (1970), 402–54.
125. NIH, "Group Consideration of Clinical Research Procedures Deviating from Accepted Medical Practice of Involving Unusual Hazard," *Memorandum Approved by the NIH Director* (1953), cited in "Federal Protection for Human Research Subjects," *Congressional Research Service Report* (June 2, 2005): 70.
126. NIH, "Group Consideration."
127. NIH, "Group Consideration."
128. Curran, "Governmental Regulation."
129. Army Regulation 70-25, *Use of Volunteers as Subjects of Research* (1962), reprinted in Henry K. Beecher, *Research and the Individual* (1970), 252–55.
130. Army Regulation 70–25, reprinted in Beecher, *Research*, 252.
131. Army Regulation 70–25, reprinted in Beecher, *Research*, 252.
132. Final Report, *The Human Radiation Experiments*, 98–99.
133. James H. Kim and Anthony R. Scialli, "Thalidomide: The Tragedy of Birth Defects and the Effective Treatment of Disease," *Toxicological Sciences* 122, no. 1 (July 2011): 1–6.
134. "Federal Protection for Human Research Subjects," *Congressional Research Service Report*, 70–71.
135. Drug Amendments of 1962, Pub. L. No. 87–781, 76 Stat. 780 (October 10, 1962).
136. Federal Food, Drug, and Cosmetic Act, Pub. L. No. 75–717, 52 Stat. 1040 (1938).
137. Irving Ladimer and Roger W. Newman, eds., *Clinical Investigation in Medicine* (1963), 311.
138. 21 C.F.R. § 130 (1963).
139. Final Report, *The Human Radiation Experiments*, 98–99; Gray, *Human Subjects*, 10–11.
140. Hornblum et al., *Against Their Will*, 59–60.
141. Jonas Salk, Letter to Harry Weaver (June 16, 1950), quoted in David M. Oshinsky, *Polio: An American Story* (New York: Oxford University Press, 2005), 151.
142. Hornblum et al., *Against Their Will*, 81–109.
143. Howard Howe, "Antibody Response of Chimpanzees and Human Beings to Formal-Inactivated Trivalent Poliomyelitis Vaccine," *American Journal of Hygiene* 56 (November 1952): 265–86.
144. David J. Rothman and Sheila M. Rothman, *The Willowbrook Wars* (1984), 260–67; Hornblum et al., *Against Their Will*, 98–102.
145. Rothman and Rothman, *The Willowbrook Wars*, 260–67; Hornblum et al., *Against Their Will*, 98–102.

146. Editorial, "Is Serum Hepatitis Only a Special Type of Infectious Hepatitis?," *Journal of the American Medical Association* 200, no. 5 (May 1, 1967): 136–37; Robert Ward et al., "Infectious Hepatitis: Studies of Its Natural History and Prevention," *The New England Journal of Medicine* 258, no. 9 (February 27, 1958): 407–16.
147. Hornblum et al., *Against Their Will*, 111–23.
148. Hornblum et al., *Against Their Will*, 151–76.
149. Hornblum, *Acres of Skin*, 191–95.
150. Moreno, *Undue Risk*, 230.
151. Hornblum, *Acres of Skin*, 3–28.
152. Adolf Katz, "Prisoners Volunteer to Save Lives," *Philadelphia Bulletin*, February 27, 1966, quoted in Hornblum, *Acres of Skin*, 37.
153. Quoted in Hornblum, *Acres of Skin*, 37.
154. Hornblum, *Acres of Skin*, 4–8.
155. Quoted in Hornblum, *Acres of Skin*, 11.
156. Hornblum, *Acres of Skin*, 20–21, 120, 195.
157. Hornblum, *Acres of Skin*, 39–71.
158. Hornblum, *Acres of Skin*, 233.
159. W. B. Rankin, FDA, Letter to Drug Companies (July 19, 1966), quoted in Hornblum, *Acres of Skin*, 54.
160. W. B. Rankin, FDA, Letter to Drug Companies (August 19, 1966), quoted in Hornblum, *Acres of Skin*, 55.
161. Hornblum, *Acres of Skin*, 54–61.
162. Quoted in Hornblum, *Acres of Skin*, 70.
163. Katz, "Prisoners Volunteer to Save Lives," quoted in Hornblum, *Acres of Skin*, 38.
164. Quoted in Hornblum, *Acres of Skin*, 70.
165. Hornblum, *Acres of Skin*, 39.
166. Hornblum, *Acres of Skin*, 163–83.
167. Adele S. Allen, Secretary to Albert M. Kligman, Letter to Verald Keith Rowe (March 10, 1965), quoted in Hornblum, *Acres of Skin*, 166.
168. "Inmates in 60's Test of a Poison Sought," *The New York Times*, January 18, 1981.
169. Hornblum, *Acres of Skin*, 175–83.
170. John W. Melone, Environmental Protection Agency, "Status of Action Plan for Holmesburg Prison" (August 14, 1981), quoted in Hornblum, *Acres of Skin*, 177.
171. Hornblum, *Acres of Skin*, 175–83.
172. *Roach v. Kligman*, 412 F. Supp. 521 (E.D. Pa. 1976) (indicating that plaintiffs settled claims with respect to Kligman, the drug company, and other researchers); *Bradford v. City of Philadelphia*, Civ. Action No. 06-cv-5121, 2007 WL 2345278 (E.D. Pa. August 16, 2007) (dismissing case as time-barred under applicable statute of limitations).
173. *Roach*, 412 F. Supp. at 523 (dismissing claims against prison officials and the City of Philadelphia).
174. Philadelphia Mayor's Office Press Release, "City Issues Formal Apology for Experiments Conducted at Holmesburg Prison Decades Ago" (October 6, 2022).
175. Hornblum, *Acres of Skin*, 89–91; Moreno, *Undue Risk*, 230.
176. Hornblum, *Acres of Skin*, 93–95.
177. "New Approach to Cancer," *The New York Times*, May 24, 1956.
178. Jay Katz, *Experimentation with Human Beings* (1972), 9–65.
179. Hornblum et al., *Against Their Will*, 9, 20.
180. Henry K, Beecher, "Experimentation in Man," *Journal of the American Medical Association* 169, no. 5 (January 31, 1959): 470.
181. Beecher, "Experimentation in Man," 468.

182. "Federal Protection for Human Research Subjects," *Congressional Research Service Report*, 72.
183. Ladimer and Newman, eds., *Clinical Investigation.*
184. Ladimer and Newman, eds., *Clinical Investigation*, 169.
185. Robert B. Livingston, Associate Chief for Program Development, Memorandum to the NIH Director, *Progress Report on Survey of Moral and Ethical Aspects of Clinical Investigation* (November 4, 1964), quoted in "Federal Protection for Human Research Subjects," *Congressional Research Service Report*, 72.
186. Gray, *Human Subjects*, 11–16.
187. "Federal Protection for Human Research Subjects," *Congressional Research Service Report*, 72–73.
188. Henry K. Beecher, "Ethics and Clinical Research," *The New England Journal of Medicine* 274, no. 24 (June 16, 1966): 1354–60.
189. Beecher, "Ethics and Clinical Research," 1354.
190. Hornblum et al., *Against Their Will*, 78–79.
191. Gray, *Human Subjects*, 11–16.
192. Beecher, "Ethics and Clinical Research," 1360.
193. Pappworth, *Human Guinea Pigs*, ix–x.
194. Pappworth, *Human Guinea Pigs*, x.
195. Pappworth, *Human Guinea Pigs*, x.
196. Pappworth, *Human Guinea Pigs*, x.
197. Pappworth, *Human Guinea Pigs*, 27.
198. *Institutional Guide to DHEW Policy on Protection of Human Subjects* (1971), iii.
199. Katz, *Experimentation with Human Beings.*
200. Hornblum, *Acres of Skin*, 198–99.
201. R. Crawford Morris and Alan R. Moritz, *Doctor and Patient and the Law* (1971), 155.
202. Morris and Moritz, *Doctor and Patient*, 346–50.
203. James H. Jones, *Bad Blood* (1993), 1–15.
204. Jones, *Bad Blood*, 177–78.
205. Washington, *Medical Apartheid*, 168.
206. Final Report, *The Human Radiation Experiments*, 101–3.
207. "Ex-Chief Defends Syphilis Project," *The New York Times*, July 28, 1972.
208. "Federal Protection for Human Research Subjects," *Congressional Research Service Report*, 73.
209. 45 C.F.R. 46 (1974); "Research Projects Involving Human Subjects," *NIH Guide for Grants and Contracts*, vol. 3, no. 12 (August 26, 1974).
210. National Research Act, Pub. L. No. 93–348, 88 Stat. 342 (July 12, 1974).
211. *The Belmont Report: Ethical Principles and Guidelines for the Protection of Human Subjects of Research* (1979).
212. "Federal Protection for Human Research Subjects," *Congressional Research Service Report*, 14–15, 73–76.
213. Federal Policy for the Protection of Human Subjects, 56 Fed. Reg. 28,003 (June 18, 1991); "Federal Protection for Human Research Subjects," *Congressional Research Service Report*, 5–6.

10
Radiation Experiments and Atomic Weapons Research

Introduction

The bombings of Hiroshima and Nagasaki catapulted atomic science onto the world stage. Writing in *Harper's Magazine* shortly after the war, Secretary of War Henry Stimson characterized "the recent unlocking of atomic energy" as "a new control by man over the primal forces of nature."[1] Stimson underscored atomic energy's "great, but as yet unexploited, promise for the well-being of civilization" but warned that "man's ability to destroy himself is very nearly complete."[2] The fear and awe aspects that Stimson identified guided post-World War II (WWII) atomic research, which was led by the cadre of scientists, physicians, government officials, and military leaders assembled for the wartime Manhattan Project.[3]

The Atomic Energy Commission (AEC), a newly established civilian agency, commanded America's atomic energy program. The AEC was charged with overseeing the field of atomic science, coordinating military and civilian uses of atomic energy, and protecting the public from the harmful effects of radiation. By placing control in a civilian agency rather than the military, the government sought to alleviate public fears of an atomic apocalypse. The commission inherited wartime facilities, including Los Alamos, and was the central entity responsible for America's nuclear weapons program. The AEC expanded atomic research beyond weapons development and into medical and industrial technologies, areas envisioned but not thoroughly explored during the war.[4] Congress granted the AEC generous funding and exemptions from standard oversight procedures, benefits that afforded the commission great latitude and shielded its activities from public review.

The Cold War motivated much of the AEC's work. The scope of activities was extraordinary: the AEC collaborated with hospitals, universities, public health agencies, industry, and the military. Service members, veterans, prisoners, pregnant women, children, and others took part in AEC studies, though many were coerced into participation or were unsuspecting research subjects. With AEC oversight, the military administered field tests with nuclear weapons to prepare service members for nuclear war, physicians conducted biomedical radiation research, and scientists secretly inundated large swaths of America with intentional radiation releases. The projects were developed to promote military preparedness and examine the impact of radiation on humans, animals, crops, and the environment.

America's Military Biomedical Complex. Efthimios Parasidis, Oxford University Press.
 DOI: 10.1093/9780199351473.003.0011

Preparing for Nuclear War

The military dedicated substantial resources to prepare service members for nuclear battle and to examine how much radiation a person could be exposed to and still fight effectively. Projects included field tests with nuclear weapons, flights near nuclear detonations, and studies to analyze the health impact of radiation exposure and the psychological effects of witnessing a nuclear explosion.

Between 1945 and 1962, more than 200,000 service members participated in one or more of 200+ atmospheric tests involving the detonation of a nuclear weapon.[5] In a typical test, as depicted in Figure 10.1, a nuclear bomb was loaded atop a tower hundreds of feet tall. Surrounding the bomb were buildings, vehicles, radiation-recording instruments, and live animals such as pigs and sheep.[6] Before detonation, and often in the presence of armed guards, service members were ordered to sign a confidentiality agreement that outlined disclosure penalties that included a $10,000 fine and ten years in a military prison.[7]

Moments before a test blast, an officer informed the service members: "You are here to become acquainted with the effects of nuclear weapons."[8] To provide context to the simulation, participants were asked to imagine that the United States had been invaded and that a nuclear bomb would be dropped on enemy positions, after which the troops would storm the bombed area.[9] After detonation—as AEC personnel observed from miles away donning protective clothing and hooded respirators—service members were ordered to march toward the blast wearing dungarees and a steel helmet.[10] After a test explosion, soldiers were bused back to base, and their uniforms were brushed off with household brooms.[11] In addition to training service members to fight and maneuver after a nuclear detonation, investigators analyzed the health effects of radiation exposure.

The first nuclear tests conducted after the conclusion of WWII took place on the Marshall Islands, a series of atolls in the South Pacific that had been under Japanese control. After WWII, the United States constructed military bases on the islands. Between 1946 and 1958, America administered sixty-seven nuclear tests across the atolls, equating to an energy yield of seven thousand Hiroshima bombs.[12] Among the sixty-seven explosions was the 1954 Castle Bravo test at Bikini Atoll, the most powerful nuclear bomb ever detonated by the United States. The blast spread radioactive material across the entire globe and caused acute radiation sickness in individuals living on neighboring atolls. For decades, the Marshallese have suffered significant health ailments and economic devastation due to widespread environmental contamination caused by the blasts. The Marshallese have been forced to migrate to safer lands, and to this day, many atolls cannot be inhabited.[13]

In 1951, after years of detonations in the South Pacific, the military began nuclear testing within the continental United States in remote desert areas of Nevada that were sparsely populated.[14] Notwithstanding the harmful effects of fallout and radiation, the Department of Defense (DoD) underscored the importance of conducting atmospheric nuclear blasts to build field experience for service members. According to Dr. Richard Meiling, chair of the top medical advisory board for the Secretary of Defense: "Fear of radiation is almost universal among the uninitiated and unless it

Figure 10.1 Service members watch an atomic bomb test blast in Nevada.
Photograph from the Library of Congress.

is overcome in the military forces it could present a most serious problem if atomic weapons are used."[15]

The responsibility for establishing and monitoring safety protocols for test explosions was split between the DoD and AEC: the DoD created guidelines for military personnel, while the AEC set safety measures at test sites in Nevada and elsewhere. The jurisdiction of the agencies overlapped, and often there was disagreement on which measures should be adopted. For example, military leaders objected to AEC recommendations that service members be kept at least seven miles from the blast, contending that the distance would not provide an adequate battlefield simulation.[16] Differences in appropriate safety protocols were irreconcilable—in turn, the DoD petitioned for "full responsibility for the physical and radiological safety of troops and all observers accompanying troops."[17] In January 1953, the AEC granted the DoD's request, stating that "the tactical use of atomic weapons, as well as the hazards which military personnel are required to undergo during their training, must be evaluated and determined by the Department of Defense."[18] The transfer afforded the DoD substantial authority over nuclear test conditions.

Shortly thereafter, the DoD doubled the AEC-recommended radiation exposure limits, stationed service members four miles from the blast site, and ordered the men to march toward ground zero immediately after a detonation. In subsequent tests, service members were placed closer and closer to the explosion, and at times, men were ordered to within a few hundred meters of the detonation.[19] Some blasts were coordinated between the United States, Canada, and Britain. During one tripartite meeting, Britain and Canada refused to endorse American protocols, contending that they posed unnecessarily high risks to military personnel.[20] Decades later, the US military disclosed that radiation from the tests was comparable to the bombings at Hiroshima and Nagasaki.

Scientists measured the psychological impact of witnessing an explosion and marching toward it. Complete terror was the typical response.[21] Although researchers noted that their "results can be expected to appear superficially trivial," they nonetheless concluded that the inquiry is "of such extreme importance" and, thus, "As many men as possible ought to be exposed to this experience under safe conditions."[22] Service members often participated in several detonations on the premise that additional tests would help them overcome their atomic fears. As a commanding officer stationed in the South Pacific explained: "That men were placed in such a remote environment, exposed to repeated nuclear blasts, and then expected to perform like soldiers was ludicrous. I had seen soldiers turn into animals as they lost touch with reality."[23]

Coupled with the psychological studies, researchers examined the physiological impact of experiencing a nuclear explosion.[24] One atomic veteran recalled that, with his "eyes tightly closed, I could see the bones in my forearm as though I were examining a red x-ray."[25] In their book, *Countdown Zero*, atomic veterans Thomas Saffer and Orville Kelly recount that "our superiors assured us that our nuclear test experience would be comparable to observing the explosion of a large conventional weapon. There was nothing to worry about, they promised, but we were cautioned again and again not to discuss our activities with anyone."[26]

The military did not inform service members of potential health risks or provide the men with an opportunity to opt out of participation. Instead, the field tests were viewed as a component of service and an occupational risk of military training.[27] As one official candidly remarked during a DoD medical sciences committee meeting in 1950, "When you start thinking militarily of this, if men are going out on these missions anyway, a high percentage is not coming back, the fact that you may get cancer 20 years later is just of no significance to us."[28]

The massive mushroom clouds from the atmospheric detonations were impossible to hide, and several newspapers questioned the potential risks to the public. The AEC downplayed the risks by asserting that "the levels of radiation produced outside the test control area were in no way harmful to humans, animals or crops."[29] Yet, shortly after the first detonations on American soil, one-quarter of sheep herds in surrounding communities died—the AEC attributed the deaths to unprecedented cold weather. Although the AEC publicly denied any potential harms, classified reports reveal that the AEC was acutely aware of significant risks to human health and the environment.[30] In addition, the AEC privately warned film manufacturers that fallout could damage their products.[31]

During the 1950s, a group of scientists from St. Louis initiated a campaign to collect baby teeth to measure whether atomic tests caused an increase in radioactive elements absorbed by infants and children. More than 320,000 baby teeth were donated, and a series of reports published in the early 1960s found a significant increase in strontium-90, a radioactive by-product of atomic testing.[32] The studies received substantial public attention and helped lead to the 1963 Partial Nuclear Test Ban Treaty, where America, Britain, and the Soviet Union agreed to halt atmospheric tests.[33] In 1997, scientists at the National Cancer Institute estimated that the Nevada atmospheric tests alone may have caused up to seventy-five thousand thyroid cancers, 75 percent of which developed in people aged five or younger at the time of the blasts in the 1950s.[34] Studies of the atomic-age baby teeth have continued into the 2020s.[35]

In addition to nuclear warfare exercises in the field, military pilots were instructed to fly near atomic clouds to gather data on radiation and practice maneuvering around an atomic explosion. Some pilots were ordered to fly through an atomic cloud in a cabin without an air filter so researchers could examine the hazards of inhaling radioactive particles. In other studies, researchers measured radiation exposure by forcing pilots to swallow film placed in a watertight capsule attached to a string; after a test flight, researchers pulled up the film to compare radiation exposure inside and outside the pilot's body. The tests occurred despite previous research—including flights with mice and monkeys—which found harmful radiation levels within a two-and-a-half-mile radius of an atomic cloud.[36]

Pilots were also subjected to flash blindness studies, which involved watching a nuclear explosion to analyze retinal exposure and test new goggles that might protect against the temporary loss of vision that occurs from witnessing a nuclear flash. Although flash blindness was a concern across the armed forces, it was particularly salient for pilots, who could not afford to lose even a few seconds of vision during flight. The military characterized the pilot studies as occupational hazards of service and training.[37]

Decontamination efforts after a test blast were another source of field training and radiation exposure, impacting thousands of service members. Various cleanup methods were studied as scientists experimented with best practices to limit radiation exposure during decontamination.[38] Men were directed to wash away radioactive material after "black rain" would fall on Navy ships following a detonation; others were ordered to use their bare hands to wipe down aircraft following an atomic cloud flythrough.[39] After test explosions in the Marshall Islands, soldiers performed backbreaking work to remove six inches of contaminated topsoil from the land and dump it into massive pits sealed with cement eighteen inches thick.[40] The military did not place protective liners around the pits. Thus, over time, the cement cracked, raising concerns that radioactive material was leaking into surrounding land and waterways, and prompting the United Nations to issue a contamination warning in 2019.[41] In various parts of the United States, radioactive waste from atomic experiments, some dating back to the Manhattan Project, still poses health and environmental hazards.[42]

A 1952 Army publication explained that a "soldier is not a casualty until he requires treatment. Even though he has been exposed to a lethal dose of radiation, he can perform his combat mission until symptoms appear."[43] As atomic tests increased, however, Veterans Administration (VA) officials were worried about widespread health consequences that would overwhelm their department.[44] Despite the risks, AEC Commissioner Thomas Murray asserted during a private committee meeting that "we must not let anything interfere with this series of tests—nothing."[45]

Atomic veterans have suffered cancer and other health ailments at alarming rates and have endured reproductive health complications, including infertility, children born with congenital disabilities, and premature death of their children.[46] Atomic veterans also have suffered from significant adverse psychological effects stemming from their ongoing fears of developing radiation-related diseases, difficulties in obtaining medical care for their ailments, and anger at their government for placing them in an unreasonably dangerous condition and then refusing to take responsibility for the affiliated harms.[47] Data further suggest that radiation exposure has caused reproductive health impairments in children and grandchildren of atomic veterans. However, causal studies have been elusive due to a confluence of environmental factors and challenges in obtaining accurate health records for atomic veterans.[48]

For years, the DoD opposed compensation for atomic veterans, contending that radiation exposure experienced by service members did not cause significant adverse health conditions. During the early 1980s, the DoD lobbied against a proposed compensation fund, stating that it would have a disastrous impact on the military's nuclear weapons program and civilian nuclear facilities because it would suggest that exposure to low-level radiation poses health hazards.[49] Contemporaneously, relying on principles of sovereign immunity, courts routinely dismissed claims brought by service members and veterans who sought remedies for radiation injuries caused by the atomic experiments.[50]

Some atomic veterans sued private contractors, such as Lawrence Livermore Laboratory and Sandia Laboratory, which had worked with the military on atomic experiments. These entities did not have the protection of sovereign immunity since they were private companies, but the government had agreed to indemnify the companies against all legal claims stemming from the experiments.[51] Under the terms of the contract, if the companies were sued, the government was responsible for paying the claims and the company's attorney fees, which easily could have reached into the millions of dollars.

Due to the substantial financial risks to the government, in 1984, Congress enacted a law allowing the government to intervene in a lawsuit against a private contractor, to remove the private company from the lawsuit, and to place the government as the sole defendant.[52] The government then used this new law in several ongoing lawsuits brought by atomic veterans. Once the government was the sole defendant, it asserted sovereign immunity and was able to dismiss the claims. In addition to saving the government considerable money, the dismissal of the claims under sovereign immunity precluded the atomic veterans from seeking discovery from the government and private contractors, which might have revealed the then-secret information regarding known health and environmental hazards from the nuclear tests.[53]

Atomic veterans continued to organize and protest what they viewed as unfair treatment and abandonment by the military and government. Thereafter, legislation was enacted that afforded some compensation for a subset of atomic veterans.[54] Many veterans fell outside the scope of the laws because compensation was available only if a veteran took part in a designated atomic test event and manifested a designated disease. Veterans frequently lamented the expense and difficulty in obtaining information, and many faced insurmountable obstacles due to missing or inaccurate records.[55]

Records gaps and the atomic test secrecy mandate have hindered the ability of veterans to receive appropriate healthcare and service-related disability benefits.[56] The secrecy law was repealed in 1996 but was not widely publicized. A 2019 study found that many surviving atomic veterans—most of whom were in their eighties and nineties—were reluctant to speak about their test experiences and health ailments because they believed that the confidentiality agreement still applied, along with the potential for imprisonment.[57]

Preparations for nuclear war have had a lasting impact on service members and veterans—not just in terms of adverse health outcomes but also in the form of loss of trust in military leaders and the government. Many veterans expressed the view that it was unnecessary to second-guess the military's decision to conduct field tests with nuclear weapons, but that they nonetheless felt betrayed by the military and the government for their failure to treat the men fairly during the studies and after they manifested service-related illnesses.[58]

Biomedical Radiation Research

The AEC's budget was devoted primarily to weapons development, but the commission also maintained an expansive program dedicated to biomedical radiation research. In June 1947, just months after the AEC was established, the agency's Medical Board of Review determined that there was an urgent need for radiation research due to "the extraordinary danger of exposing living creatures to radioactivity" and the lack of "defensive measures (in the military sense) against radiant energy."[59] Housed within the AEC's Division of Biology and Medicine, the cutting-edge biomedical radiation research program conducted in-house studies and collaborated with hospitals, universities, and research institutes. A public relations campaign heralded the potential health benefits of the AEC's ambitious biomedical research agenda. Yet, behind the scenes the commission secretly conducted dangerous experiments on prisoners, veterans, children, pregnant women, hospital patients, and others.[60]

Scientists intentionally selected people from poor, disenfranchised, and relatively powerless segments of society. Most of the research subjects had no idea they were being experimented upon. Across the studies, African Americans suffered disproportionally—whereas African Americans comprised nearly 10 percent of the US population, they represented an estimated 60 percent of biomedical radiation research subjects.[61] Over three decades, between the 1940s and 1970s, the AEC's biomedical division sponsored thousands of studies to examine radioisotopes, the effects of radiation on the human body, radiation countermeasures, and other biomedical techniques.[62]

The public was not aware of many research details until the 1990s, decades after the work was conducted. In 1993, a series of articles published by Eileen Welsome in the *Albuquerque Tribune* garnered national attention by describing government-sponsored studies whereby unsuspecting civilians were injected with plutonium.[63] Within weeks of the revelations, President Bill Clinton commissioned a committee to investigate the human radiation experiments.[64] The Final Report of the Advisory Committee on Human Radiation Experiments, published in 1995, provides an overview of the expansive biomedical radiation research program. Although the report and its three supplemental volumes exceed twenty-five hundred pages, the committee highlighted that, for most experiments, few details could be found in archival records.[65] Nevertheless, reviewing several case studies highlights the various research-related goals, risks, and protocols.

The AEC devoted significant resources to understand occupational hazards from working with radioactive elements. To simulate exposure in occupational settings, researchers would inject test subjects with polonium, plutonium, uranium, or other radioactive elements, and then observe how the elements impacted the body. The research was conducted on various individuals, including service members, veterans, civilians, and children. Some research subjects were victims of circumstance. One man entered a hospital emergency room after a car accident and was injected with plutonium without his knowledge or consent; another unsuspecting patient received a plutonium injection as he was being treated for Addison's disease. The injections typically were not recorded in patient medical records.[66]

At the time, there were no known short-term risks, and individuals typically did not exhibit any acute reactions following an injection. However, there were concerns about long-term harms, including the development of cancer. Researchers monitored the impact of the injections on the unsuspecting research subjects and, at times, secretly placed the individuals in long-term studies. During the follow-up tests, some continuing into the 1970s, physicians lied to the research subjects, informing them that the myriad blood and diagnostic tests were routine components of their health examinations. In some cases, scientists exhumed bodies of research subjects to measure plutonium levels in skeletal remains; families were not informed of the real reason for exhumation.[67] According to the Presidential Advisory Committee on Human Radiation Experiments, "the ethical questions raised by these experiments would be revisited in debates that themselves were long kept secret."[68]

In addition to analyzing occupational health hazards, the AEC coordinated experiments to examine potential medical applications of radioisotopes. Scientists viewed radioisotopes as promising clinical tools because they trace metabolic functions as they travel through the body, help distinguish normal from abnormal functioning, and can carry a lethal dose of radiation to target and destroy unhealthy cells. To facilitate radioisotope research, the AEC funded cancer research centers, provided equipment and radioisotopes free of charge, and sponsored research at hospitals and universities throughout the country. In addition, the VA conducted thousands of experiments at VA facilities, often in collaboration with local universities. VA officials encouraged the creation of a "publicized" program and a "confidential" one, candidly, albeit internally, indicating that the confidential program would help limit legal liability in the event veterans manifested research-related harms from radiation exposure.[69]

At Vanderbilt University Hospital, AEC-sponsored researchers provided more than eight hundred pregnant women and breast-feeding mothers with radioactive cocktails: the experiments began in the 1940s, with follow-up studies conducted into the 1960s. The research was conducted to observe the effects of ingested radioactive elements to analyze iron absorption and help determine nutritional requirements for iron during pregnancy. The study was also motivated by concerns raised by radioactive iron exposure in the nearby Oak Ridge, Tennessee nuclear facility.[70]

Oak Ridge scientists knew that ingestion of radioactive iron was dangerous and that it exposed a person to considerable radiation. Nevertheless, Vanderbilt physicians instructed the women to drink the radioactive cocktails, which they described as a vitamin mix that would benefit them and their babies. Most of the women were poor, and none were told they were research subjects. Among the research cohort, women and children suffered severe adverse reactions, including losing their hair and teeth, bizarre rashes, obscure blood disorders, and cancer. Several children died from cancer at a rate higher than that of the general population.[71] The health ailments were similar to those experienced by service members subjected to field tests with nuclear weapons.

During the 1940s and 1950s, MIT scientists fed breakfast cereal laced with radioactive elements to students at the Walter E. Fernald School, a Massachusetts institution initially created for children with cognitive disabilities and behavioral disorders. Over time, the school grew to include orphans and children from poor or dysfunctional homes. The research was funded by the AEC, the National Institutes of Health, and the Quaker Oats Company, and was designed to study how a child's body absorbs iron, calcium, and other minerals.[72] Some parents were told that the school was forming a group "of our brighter patients . . . to receive a special diet."[73] Other parents were asked if they objected to their child joining a "Science Club" wherein they would receive a special diet created in collaboration with MIT scientists. Membership in the club, the parents were told, brought "many additional privileges," such as being "taken to a baseball game, to the beach, and to some outside dinners." A letter to parents stated, "If you have not expressed any objections we will assume that your son may participate."[74] Families were not informed of the true purpose of the experiments or the risks to their children. Details were not publicly revealed until the 1990s, and shortly thereafter MIT and Quaker Oats agreed to a civil settlement with the families for $1.85 million.[75]

An analogous experiment was performed during the 1960s at the Wrentham State School, a Massachusetts facility for children with cognitive disabilities and behavioral disorders. In conjunction with physicians from Harvard Medical School, Massachusetts General Hospital, and Boston University School of Medicine, scientists administered radioactive iodine to Wrentham students to test radiation uptake in the body and examine potential radiation countermeasures. Seventy children—the majority aged one to eleven—were fed radioactive iodine daily for three months. The Wrentham experiment built upon clandestine AEC-sponsored studies conducted during the 1950s at Johns Hopkins University (where thirty-four children, aged two months to fifteen years, were injected with radioactive iodine), the University of Tennessee (where seven newborn infants, each less than three days old, received radioactive injections), and the University of Michigan (where sixty-five infants, most less than two weeks old, were orally administered radioactive iodine).[76]

Conditions at institutions like Fernald and Wrentham were dreadful; sexual assaults were rampant, and children as young as nine were forced to perform grueling work in the fields and throughout the institution from morning to night.[77] As an investigation into the Fernald and Wrentham studies explained, "the buildings were dirty and in disrepair, staff shortages were constant, brutality was often accepted, and programs were inadequate or nonexistent."[78] According to one surviving research subject from Fernald: "I won't tell you now about the severe physical and mental abuse, but I can assure you, it was no Boys' Town. The idea of getting consent for experiments under these conditions was not only cruel but hypocritical. They bribed us by offering us special privileges, knowing that we had so little that we would do practically anything for attention."[79] When children refused, researchers worked to induce participation, going as far as describing the studies as a debt the children owed to society due to the benefits they gained from the institution.[80]

According to Gordon Shattuck, a Fernald survivor, when he tried to terminate his role in the experiments due to the pain caused by the daily needles, blood draws, and physical examinations, researchers told him that he could not leave. When Shattuck persisted in his efforts to resist the blood draws, which occurred six times per day, the research team locked him in solitary confinement. After eight days in confinement, being provided only bread and water, Shattuck acquiesced and was placed back into the study.[81] The experiments at Fernald, Hopkins, and Wrentham were not aberrations. In 1995, the Presidential Advisory Committee on Human Radiation Experiments disclosed evidence for eighty-one pediatric radiation research projects between 1944 and 1974 and noted that the projects "by no means constitute all the pediatric radiation research conducted during this time."[82]

Full-body irradiation was another area explored by the AEC. In collaboration with the military and hospitals throughout the country—such as MD Anderson Hospital, Baylor University College of Medicine, and the Memorial Sloan-Kettering Cancer Research Institute—the AEC placed thousands of patients in experiments examining radiation's impact on disease and human health.[83] Some of the experiments were led by Nazi physicians brought to the United States under Operation Paperclip.[84] The primary purpose of the research was not to find cancer treatments, but rather to analyze occupational risks faced by military personnel in the development of atomic weapons and nuclear-powered aircraft. Specifically, the experiments sought to unravel the acute effects of radiation exposure, the psychological effects of varying doses of radiation, and whether there is a biological dosimeter that can measure how much radiation a person has received.[85]

Some government officials recommended that the research be limited to animal studies due to known risks of radiation, a lack of data indicating potential benefits for humans from radiation exposure, and the concern that such a study "would have a little of the Buchenwald touch."[86] The latter point was an unmistakable reference to Nazi experiments conducted at the German concentration camp. After a series of meetings on the ethics of the research, the AEC refused to sanction human studies with healthy volunteers due to "serious repercussions from a public relations standpoint," but permitted the experiments on unsuspecting cancer patients.[87]

In one well-documented experiment, which took place at the University of Cincinnati, a team led by radiologist Eugene Saenger conducted irradiation studies on eighty-eight patients between 1960 and 1972. The patients were not informed of potential risks, including nausea or bone marrow suppression, which could lead to death.[88] As one of Saenger's collaborators unabashedly explained, the researchers purposely experimented on "slum" patients because "these persons don't have any money and they're black and they're poorly washed."[89] In other experiments conducted across the country between 1953 and 1954, the AEC authorized irradiation studies on 235 African American newborns without obtaining parental consent or describing the risks or purposes of the study.[90] Once the Cincinnati study was publicly disclosed, a series of investigations and lawsuits followed, and in 1999, the university settled the case for $5.4 million.[91]

AEC-sponsored research in prisons was no less disturbing. Between 1963 and 1973, more than 130 prisoners in Washington and Oregon were subjected to testicular irradiations that had no potential therapeutic benefit. The studies—conducted by leading endocrinologists—analyzed the effects of radiation on the male reproductive system to help determine occupational risks and safety protocols for individuals working with atomic energy. In addition to having their testicles bombarded with radiation, the prisoners underwent testicular biopsies. They were vasectomized at the conclusion of the research to eliminate the chance that they could father children with abnormalities.[92]

The inmates suffered research-related injuries that included severe rashes and blisters on the scrotum, pain during sexual intercourse, difficulty maintaining erections, and testicles shrinking in size. Some of the symptoms persisted for years after the experiments. The prisoners received $5 per month of participation, $10 for each biopsy, and $100 for the vasectomy.[93] As detailed in chapter 9, research in American prisons was not uncommon. However, some contemporary ethicists expressed concerns with prison research due to inmates' lack of autonomy and their vulnerability to coercion or undue influence. The prison studies were halted following the 1972 disclosure of the Tuskegee Syphilis Study.

The AEC's expansive biomedical research program did not occur in an ethical vacuum, and for years, AEC officials debated legal, moral, and ethical aspects of research protocols. AEC attorneys initially proposed a "written release" to be signed by each research participant, but the proposal was rejected by Stafford Warren, the leader of the AEC's medical advisory board and the former medical director of the Manhattan Project. Warren countered that, instead of a signed release, two physicians would certify that an individual provided oral consent after being informed of the purpose and risks of the study. AEC lawyers concurred but remained concerned about backlash and lawsuits should word of the studies fall into the public domain. To mitigate the risks, AEC officials decided to keep the experiments classified.[94]

Another ethical concern contemplated by AEC officials was whether nontherapeutic research should be conducted. In the fall of 1947, Carroll Wilson—an MIT scientist who became the AEC's first general manager—supported a policy that prohibited the administration of any "substances known to be, or suspected of being, poisonous or harmful" unless there was "reasonable hope" that the substance would provide therapeutic benefit to the research subject.[95] Wilson also advocated for a policy that required a signed consent form for each research participant that allowed the participant to opt out of the research at any time.[96]

There is no evidence that Wilson's policies ever were applied. Instead, AEC officials consistently informed scientists that no written policy governing research involving human subjects existed.[97] While the evidentiary record illustrates that some AEC leaders recognized the need for protocols to govern atomic research, there was no consensus on the parameters of the rules, and researchers working on AEC-sponsored studies did not receive notice of binding research protocols.[98] As bioethicist Jonathan Moreno explains, Wilson and others who shared his concerns "were no match for a powerful medical establishment that, at the end of the day, called the tune."[99]

Intentional Radiation Releases

Between the 1940s and 1960s, the AEC oversaw hundreds of intentional radiation releases whereby scientists deliberately emitted radioactive elements into densely populated cities and other locations to test radiation detection devices and study the impact of radiation on humans, animals, and the environment. The experiments doubled as field tests to analyze the feasibility of radiological warfare—which involves the disbursement of radioactive elements with the intent to harm people, animals, crops, or water—a tactic explored by Manhattan Project scientists but not utilized during WWII.[100] Although massive nuclear blasts and mushroom clouds made atmospheric atomic tests impossible to hide, intentional radiation releases could be concealed easily since there was no detonation, and detection required specialized equipment that civilians did not typically own. The studies were conducted throughout the United States and kept secret for nearly half a century.

When the radiation releases were first orchestrated in the 1940s, officials debated publicly disclosing the program. Those who advocated for disclosure contended that "better work could be done from the scientific and medical standpoint" with a declassified program.[101] With disclosure, physicians could track their patients and better understand adverse health conditions, while public health officials could compare rates of disease in areas exposed to radiation to rates in those areas not exposed. Public disclosure also would allow citizens to comment on the legitimacy of the experiments and change their daily habits to mitigate the potential harm from radiation exposure. Ultimately, disclosure was not the consensus, as officials recognized that public fears of widespread radiation contamination would jeopardize their work and derail the studies.[102]

Notwithstanding the decision to keep the radiation releases secret, officials discussed how best to structure the experiments to minimize the potential harms to humans and the environment.[103] Despite these efforts, as the Presidential Advisory

Committee on Human Radiation Experiments explained in its 1995 report, the risks to society "were often subordinated to concerns for national security, which were sometimes joined or melded with concerns for public relations."[104]

Following each release, scientists measured radioactivity levels in humans, homes, vegetation, animals, water, and the air. Some measurements found levels four hundred times greater than AEC-permissible limits, though the public was unaware of the risks. Children were particularly susceptible to radiation-related health impairments from the releases, risks that were compounded by drinking milk derived from cows that grazed on contaminated pastures.[105]

In Hanford, Washington—which maintained a massive nuclear production facility and the world's first plutonium factory, both constructed during WWII—people grew accustomed to providing scientists with urine samples and allowing researchers with Geiger counters to examine their homes. Government researchers told Hanford residents that the measurements were to monitor their safety given their proximity to the nuclear plant. Many were proud to live in America's Atomic City.[106] Unbeknownst to the residents, not only were they exposed to radiation accidentally released from the facility, but under a project code-named Green Run, the government intentionally bombarded their community with radiation. Decades later, when the public learned of the intentional releases at Hanford, Cold War patriotism turned to disbelief and a lasting sense of distrust.

The Hanford experiments built off studies conducted during WWII. For example, Manhattan Project researchers examined the impact of radioactive contamination by mixing radioactive elements into a barrel of water and secretly spraying the radioactive mix across crops and animal farms near the University of Rochester. Via Green Run, Hanford residents were deliberately exposed to radiation equivalent to thousands of X-rays, a dose high enough to have an adverse impact on health and to cause cancer. Thousands have suffered ill effects, including benign and malignant tumors, miscarriages, and children born with abnormalities.[107] To this day, Hanford remains the most contaminated area in America due to the intentional releases, accidental releases from the nuclear facility, and leaks from underground storage tanks filled with radioactive waste. As to the latter, the Hanford site maintains 177 underground storage tanks containing fifty-six million gallons of "highly radioactive and chemically hazardous waste," more than one million gallons of which have leaked into the surrounding community.[108] Due to the widespread contamination, Hanford has been dubbed America's Chernobyl.

The public first learned of the intentional radiation releases in 1986 when, in response to a Freedom of Information Act request to disclose information related to the Hanford facility, individuals reviewing over 19,000 pages of documents realized that government researchers deliberately released radioactive material into the community to measure the effects on humans, animals, plants, and the environment. Following the shocking revelation, Senator John Glenn asked the General Accounting Office (now known as the Government Accountability Office) to investigate if other releases were conducted. By 1993, the government disclosed twelve more intentional releases, and in 1995, the government admitted that it had performed hundreds of intentional radiation releases throughout the United States.[109]

Summary

In the decades following WWII, atomic science expanded significantly to become a prominent component of military and civilian research agendas. Projects were created to prepare service members for nuclear war, test protective clothing and equipment, understand how radiation impacts humans and the environment, set occupational safety measures, treat radiation injuries, create effective decontamination protocols, and examine industrial and medical uses of radiation and atomic elements. Many of the studies facilitated military, technological, or medical breakthroughs. As noted in the first treatise on Atomic Medicine, published in 1949: "Never before in history have the interests of the weaponeers and those who practice the healing arts been so closely related."[110]

The promise and power of atomic energy accelerated a global pursuit for atomic equilibrium. As the nuclear weapons program of the Soviet Union evolved and fear of nuclear attack resonated across America, the United States structured research programs and field tests to account for the unique circumstances presented by nuclear weapons. Scientists, physicians, government officials, and military leaders worked vigorously to further military preparedness, analyze the effects of radiation, and craft potential treatments for radiation injuries. A toxic mix of fear, power, and hubris blanketed the nation, contributing to a Cold War hysteria that served as a justification for elaborate, risky, and clandestine programs.

The research protections outlined in the 1953 DoD policy and 1962 Army Regulation were largely ignored, primarily because the military viewed the atomic studies as military training, not research. Although, at times, some officials recognized and debated ethical and legal considerations, the concerns did not lead to the implementation of adequate safety protocols due to the high-stakes race for scientific and military superiority. While Cold War atomic and radiation projects served to further pressing national security interests and to advance medical knowledge—and may not have been crafted with malicious intent—they often disproportionally impacted marginalized populations. The programs also instilled a lasting sense of mistrust and resentment among veterans and the public, and called into question the credibility of the government's risk-related information disclosures. As the 1995 report of the Presidential Advisory Committee on Human Radiation Experiments succinctly concluded, "It was a time of arrogance and paternalism on the part of government officials and the biomedical community."[111]

Notes

1. Henry L. Stimson, "The Bomb and the Opportunity," *Harper's Magazine*, March 1946. While the terms "atomic" and "nuclear" often are used interchangeably, there is a slight nuance to highlight. Atomic energy refers to energy stored within an entire atom, while nuclear energy refers only to the energy contained in the nucleus of an atom: thus, nuclear energy is a subset of atomic energy. Atomic bombs, however, are one form of nuclear weapons. Nuclear weapons generate their explosive power from fission, or a combination of

fission and fusion. The former are referred to as atomic bombs, whereas the latter are called hydrogen bombs or thermonuclear bombs.

2. Henry L. Stimson, "The Bomb and the Opportunity"; Henry L. Stimson, "The Decision to Use the Atomic Bomb," *Harper's Magazine*, February 1947.
3. Final Report, *The Human Radiation Experiments*, 1.
4. Stimson, "The Decision to Use the Atomic Bomb."
5. Final Report, *The Human Radiation Experiments*, 284–306.
6. Claudia Grisales, "Conspiracy of Silence: Veterans Exposed to Atomic Weapons Wage Final Fight," *Stars and Stripes*, June 16, 2019.
7. Thomas H. Saffer and Orville E. Kelly, *Countdown Zero* (1983), 28–29.
8. "Atomic Testing Ends in Sickness and Death," *Enlisted Times*, May 1979, 6, quoted in Saffer and Kelly, *Countdown Zero*, 19.
9. Final Report, *The Human Radiation Experiments*, 287–88.
10. Saffer and Kelly, *Countdown Zero*, 32, 50.
11. *Gaspard v. United States*, 713 F.2d 1097 (5th Cir. 1983).
12. Hart Rapaport and Ivana Nikolic Huges, "The U.S. Must Take Responsibility for Nuclear Fallout in the Marshall Islands," *Scientific American*, April 4, 2022.
13. Walter Pincus, *Blown to Hell* (2021).
14. Saffer and Kelly, *Countdown Zero*, 30–31; Final Report, *The Human Radiation Experiments*, 284–90.
15. Richard L. Meiling, Chairman, Armed Forces Medical Policy Council, Memorandum to Deputy Secretary of Defense, "Military Medical Problems Associated with Military Participation in Atomic Energy Commission Tests" (June 27, 1951), quoted in Final Report, *The Human Radiation Experiments*, 286.
16. Final Report, *The Human Radiation Experiments*, 297–98.
17. Quoted in Final Report, *The Human Radiation Experiments*, 298.
18. M. W. Boyer, AEC General Manager, Letter to Major General H. B. Loper (January 8, 1953), quoted in Final Report, *The Human Radiation Experiments*, 298.
19. Saffer and Kelly, *Countdown Zero*, 20, 32–45.
20. Howard L. Rosenberg, *Atomic Soldiers* (1980), 63.
21. Final Report, *The Human Radiation Experiments*, 287–90.
22. Joint Panel on the Medical Aspects of Atomic Warfare (September 8, 1952), quoted in Final Report, *The Human Radiation Experiments*, 289–90.
23. Quoted in Saffer and Kelly, *Countdown Zero*, 125.
24. Final Report, *The Human Radiation Experiments*, 286–91.
25. Quoted in Saffer and Kelly, *Countdown Zero*, 43.
26. Saffer and Kelly, *Countdown Zero*, 28.
27. Final Report, *The Human Radiation Experiments*, 285.
28. Department of Defense, Research and Development Board, Committee on Medical Sciences, Proceedings of 23 May 1950, quoted in Final Report, *The Human Radiation Experiments*, 301.
29. Quoted in Howard Ball, "Downwind from the Bomb," *The New York Times*, February 9, 1986.
30. Ball, "Downwind from the Bomb."
31. Moreno, *Undue Risk*, 9.
32. L. Z. Reiss, "Strontium-90 Absorption by Deciduous Teeth," *Science* 134 (November 24, 1961): 1669–73.
33. Ellen Gerl, "Scientist-Citizen Advocacy in the Atomic Age: A Case Study of the Baby Tooth Survey, 1958-1963," *PRism Journal* 11, no. 1 (2014): 1–14.

34. Matthew L. Wald, "Thousands Have Thyroid Cancer from Atomic Tests," *The New York Times*, August 2, 1997.
35. Michele Munz, "Harvard University Research Revives Use of Decades-Old Famous St. Louis Baby Tooth Survey," *St. Louis Post-Dispatch*, March 26, 2021.
36. Final Report, *The Human Radiation Experiments*, 285–96.
37. Final Report, *The Human Radiation Experiments*, 285, 291–96.
38. Final Report, *The Human Radiation Experiments*, 296–97.
39. Claudia Grisales, "Conspiracy of Silence: Veterans Exposed to Atomic Weapons Wage Final Fight," *Stars and Stripes*, June 16, 2019; Final Report, *The Human Radiation Experiments*, 296–97.
40. Claudia Grisales, "Conspiracy of Silence: After Atomic Blasts, a Dangerous Cleanup Scarred Troops for Life," *Stars and Stripes*, June 18, 2019.
41. Grisales, "Conspiracy of Silence: After Atomic Blasts"; "Nuclear 'Coffin' May be Leaking Radioactive Material into Pacific Ocean, U.N. Chief Says," *CBS News*, May 16, 2019.
42. Sophie Pinkham, "The Chernobyl Syndrome," *The New York Review of Books*, April 4, 2019, 22–24.
43. *U.S. Army Infantry School Quarterly*, quoted in Saffer and Kelly, *Countdown Zero*, 32–33.
44. Final Report, *The Human Radiation Experiments*, 299–300.
45. Quoted in Ball, "Downwind from the Bomb."
46. Institute of Medicine, *Adverse Reproductive Outcomes in Families of Atomic Veterans* (1995), 1–8; Final Report, *The Human Radiation Experiments*, 299–303.
47. Henry M. Vyner, "The Psychological Effects of Ionizing Radiation," *Culture, Medicine and Psychiatry* 7, no. 3 (September 1983): 241–61.
48. IOM, *Adverse Reproductive Outcomes*, 1–8.
49. Welsome, *The Plutonium Files*, 269.
50. *Jaffee v. United States*, 663 F.2d 1226 (3d Cir. 1981) (citing several cases).
51. William A. Fletcher, "Atomic Bomb Testing and the Warner Amendment: A Violation of the Separation of Powers," *Washington Law Review* 65, no. 2 (1990): 286, 304–9. Sandia Laboratories changed its name to Sandia National Laboratories in 1979, when Congress made Sandia a Department of Energy national laboratory.
52. Department of Defense Authorization Act of 1985, Pub. L. No. 98-525, § 1631, 98 Stat. 2492, 2646–47 (1984).
53. Fletcher, "Atomic Bomb Testing," 286, 304–9.
54. Veterans Dioxin and Radiation Compensation Standards Act of 1984, Pub. L. No. 98-542, 98 Stat. 2725 (October 24, 1984); Radiation-Exposed Veterans Compensation Act of 1988, Pub. L. No. 100-321, 102 Stat. 485 (May 20, 1988).
55. Final Report, *The Human Radiation Experiments*, 303.
56. Final Report, *The Human Radiation Experiments*, 303.
57. Grisales, "Conspiracy of Silence: After Atomic Blasts."
58. Final Report, *The Human Radiation Experiments*, 303–6.
59. *Report of the Medical Board of Review* (June 20, 1947), quoted in Final Report, *The Human Radiation Experiments*, 8.
60. Final Report, *The Human Radiation Experiments*, 8–10; Schmidt, *Justice at Nuremberg*, 275–77.
61. Washington, *Medical Apartheid*, 216–43.
62. Final Report, *The Human Radiation Experiments*, 8–10, 135.
63. Welsome, *The Plutonium Files*, 1–11.
64. Final Report, *The Human Radiation Experiments*, xxi–xxiii.
65. Final Report, *The Human Radiation Experiments*, xxv–xxvii.
66. Final Report, *The Human Radiation Experiments*, 139–63; Moreno, *Undue Risk*, 128.

67. Final Report, *The Human Radiation Experiments*, 143, 156–58.
68. Final Report, *The Human Radiation Experiments*, 7.
69. Final Report, *The Human Radiation Experiments*, 7–10, 172–89.
70. Welsome, *The Plutonium Files*, 219–28; Final Report, *The Human Radiation Experiments*, 213–16.
71. Welsome, *The Plutonium Files*, 219–28; Final Report, *The Human Radiation Experiments*, 213–16.
72. Hornblum, Newman, and Dober, *Against Their Will*, 125–50; Final Report, *The Human Radiation Experiments*, 196, 210–13.
73. Task Force on Human Subject Research, *A Report on the Use of Radioactive Materials in Human Subject Research that Involved Residents of State-Operated Facilities Within the Commonwealth of Massachusetts from 1943 through 1973* (April 1973), quoted in Moreno, *Undue Risk*, 213–14.
74. Clemens E. Benda, Fernand Clinical Director, Letter to Parents (May 1953), reprinted in Final Report, *The Human Radiation Experiments*, 210–11.
75. Moreno, *Undue Risk*, 10.
76. Final Report, *The Human Radiation Experiments*, 196–205; Hornblum et al., *Against Their Will*, 145–50.
77. Hornblum et al., *Against Their Will*, 125–50.
78. Task Force on Human Subject Research, *A Report on the Use of Radioactive Materials*, 33, quoted in Final Report, *The Human Radiation Experiments*, 211.
79. Quoted in Final Report, *The Human Radiation Experiments*, 212.
80. Final Report, *The Human Radiation Experiments*, 211–12.
81. Hornblum et al., *Against Their Will*, 7–8, 125–50.
82. Final Report, *The Human Radiation Experiments*, 197.
83. Final Report, *The Human Radiation Experiments*, 227–53.
84. Washington, *Medical Apartheid*, 229–40.
85. Final Report, *The Human Radiation Experiments*, 227–79.
86. Joseph Hamilton, Letter to Shields Warren (November 28, 1950), quoted in Final Report, *The Human Radiation Experiments*, 235.
87. Quoted in Final Report, *The Human Radiation Experiments*, 235.
88. Final Report, *The Human Radiation Experiments*, 239–48.
89. Quoted in Washington, *Medical Apartheid*, 10–11.
90. Washington, *Medical Apartheid*, 238–39.
91. "Radiation Victims Get $5.4 Million: Cold War Cancer Patients Weren't Told of Hazards," *Associated Press*, May 6, 1999.
92. Final Report, *The Human Radiation Experiments*, 263–71; Welsome, *The Plutonium Files*, 362–82.
93. Welsome, *The Plutonium Files*, 367–72.
94. Final Report, *The Human Radiation Experiments*, 45–67.
95. Carroll Wilson, General Manager, AEC, Letter to Robert Stone, University of California (November 5, 1947), quoted in Final Report, *The Human Radiation Experiments*, 49.
96. Final Report, *The Human Radiation Experiments*, 48–50.
97. Moreno, *Undue Risk*, 140–43.
98. Final Report, *The Human Radiation Experiments*, 45–67.
99. Moreno, *Undue Risk*, 143.
100. Final Report, *The Human Radiation Experiments*, 11–13, 317–44.
101. Atomic Energy Commission, Advisory Committee for Biology and Medicine, Proceedings of June 11–12, 1948, quoted in Final Report, *The Human Radiation Experiments*, 327.

102. Final Report, *The Human Radiation Experiments*, 317–29.
103. Final Report, *The Human Radiation Experiments*, 338–39.
104. Final Report, *The Human Radiation Experiments*, 339.
105. Final Report, *The Human Radiation Experiments*, 319–22.
106. Michael D'Antonio, *Atomic Harvest* (1993), 19–21.
107. Final Report, *The Human Radiation Experiments*, 7–8; D'Antonio, *Atomic Harvest*, 1–7, 271.
108. State of Washington, Department of Ecology, *Hanford Overview: Underground Nuclear Waste Storage Tanks* (January 18, 2024).
109. Final Report, *The Human Radiation Experiments*, 317.
110. C. F. Behrens ed., *Atomic Medicine* (New York: Thomas Nelson and Sons, 1949), 3.
111. Final Report, *The Human Radiation Experiments*, xxxi.

11
Expanding America's Biological and Chemical Warfare Programs

Introduction

America's recipe for radiation experiments and atomic weapons research served as the blueprint for its biological and chemical warfare programs—hire top scientists, provide abundant resources, encourage innovative solutions, and shield deliberations and activities with top-secret classifications and legal immunities. The military viewed biological and chemical weapons as humane and cost-effective methods of war that would instill fear and destruction in the enemy but cause fewer casualties and structural damage than nuclear weapons. Although international treaties limited the situations under which a nation could deploy biological or chemical weapons, the agreements did not restrict research, development, or stockpiling. The Cold War was the primary impetus guiding the programs, and the military and the Central Intelligence Agency (CIA) collaborated with dozens of universities, hospitals, and research institutes. These endeavors were bolstered by Operation Paperclip and the Ishii biowarfare pact: Nazi scientists worked in American chemical weapons labs, while handbooks prepared by Japanese researchers advanced US biological warfare efforts.

As with atomic and radiation studies, haphazard adherence to research subject protections were hallmarks of America's biological and chemical warfare programs. In many instances, studies involving service members were deemed training exercises; in others, clandestine research was conducted on unsuspecting civilians. Projects sought to enhance military preparedness and garner knowledge on offensive and defensive measures: this included weaponization of pathogens, field tests with biological weapons, development of mind-altering drugs for use in military and intelligence operations, and formulation of noxious chemicals that could devastate enemy forces, land, and food supplies.

Weaponizing Pathogens

America's World War II (WWII) biological weapons program was publicly revealed in a 1946 report prepared by George Merck, a pharmaceutical executive and wartime biological weapons adviser. The Merck Report, discussed in chapter 7, advocated for a comprehensive postwar biological weapons program. In addition to bolstering military preparedness, the report stated that a biowarfare program would lead to societal benefits "of great value to public health, agriculture, industry, and the fundamental sciences."[1] Replicating the post-WWI public relations campaign orchestrated by the chemical industry in support of a broad chemical weapons program, Merck and his

America's Military Biomedical Complex. Efthimios Parasidis, Oxford University Press.
 DOI: 10.1093/9780199351473.003.0012

pharmaceutical industry colleagues emphasized the great military potential of biological warfare. They underscored the need for a robust postwar program to keep pace with other nations. As they lobbied for an expansive program and government contracts, the industrialists portrayed biological weapons as merely another despicable aspect of war, no worse than nuclear or chemical weapons.[2]

The military value of biological weapons was succinctly articulated by William Creasy, director of the Research and Engineering Command of the Chemical Corps: "Man's dread of disease is universal. The mysteriousness and invisibility of bacteriological warfare agents, the knowledge that they strike via the simplest and most basic sources of man's security—food, drink, and the air he breathes—and a feeling of helplessness in dealing with the unknown, all add to the psychological potential."[3] On the morality of weaponizing disease, one influential committee report explained: "A fallacious concept has developed that weapons can be divided into moral and immoral types."[4]

By the late 1940s, leaders from the military, academia, and industry supported a broad biological weapons program that included offensive and defensive measures. By 1951, the military had entered into more than two hundred biowarfare-related contracts, most cloaked in secrecy mandates. A stark increase in government funding followed: from $3.5 million in 1950 to $345 million in 1951–1953. Between 1950 and 1952, personnel dedicated to biological warfare research more than doubled, and facilities were updated at Dugway Proving Ground, Utah, and Pine Bluff, Arkansas. The increase in funding also supported a massive expansion at Camp Detrick, Maryland. The site became a vast complex with modern laboratories, production plants, greenhouses, an animal breeding farm, and specialized chambers for testing toxic substances. The expansion contributed to a renaming of the base as Fort Detrick.[5]

The biowarfare research enterprise focused on highly infectious agents that were easy to reproduce in a lab and stable enough to survive on the battlefield. Scientists screened dozens of pathogens, including anthrax, tularemia, plague, glanders, and typhus. In addition to weaponizing the diseases, researchers worked on developing vaccines, broad-spectrum antibiotics, and other medical treatments and prophylaxis. Anti-crop agents were manufactured, including pathogens that could destroy soybeans, sweet potatoes, cotton, sugar beets, and tobacco.[6] As detailed in a secret presentation, the CIA and the military also investigated the ability to infect water supplies with cholera, dysentery, or other agents "by pumping the agent into a faucet located near a principal water main."[7]

As a 1951 Joint Chiefs of Staff memorandum reported, "the destruction of the enemy's food supply by the use of anti-animal Biological Warfare agents would be strategic in its effect."[8] Erich Traub—a Paperclip scientist who developed a weaponized foot-and-mouth virus that was dispersed during WWII over livestock in Russia—worked on dozens of lethal viruses with American researchers. One of the primary sites for the research was Plum Island, a remote location off the northeastern coast of Long Island, New York. At Plum Island, researchers conducted top-secret studies on animal diseases such as Rift Valley fever, African swine fever, foot-and-mouth disease, and rinderpest. Plum Island scientists, many of whom were employed by the US Department of Agriculture (USDA), worked closely with biological warfare experts at Fort Detrick. Some reports allege that diseases examined in Plum Island labs, most notably Lyme disease, inadvertently leaked into surrounding communities and impacted animals and people in New York and beyond.[9]

Biowarfare delivery devices, including munitions, missiles, aerosols, dart guns, and live insects, were also studied extensively. Officials reasoned that the effectiveness of biological agents would increase if used in conjunction with conventional bombs since key infrastructure would be disrupted and access to healthcare would be hindered. Germs were particularly useful as they could reach enemy forces in secluded areas such as tunnels, caves, and secret installations. Surprise was a key element in catching off guard enemy health systems and maximizing the psychological effects of hopelessness and despair. Officials likewise calculated that the scale of a biological attack should be large to inundate enemy forces and medical facilities, thus permitting aggressive follow-up military incursions on enemy positions.[10]

Shiro Ishii and his team—whom the United States shielded from war crimes prosecution in exchange for sharing biological warfare secrets with the US military—coached the United States to use bioagents that cause illnesses which plausibly could be linked to a natural outbreak.[11] In a secret report dated December 21, 1951, President Harry Truman's Secretary of Defense, Robert Lovett, ordered the Joint Chiefs of Staff to achieve "actual readiness" in biological warfare "in the earliest practicable time."[12] Tactical plans were devised to disperse biological agents that harm humans, agriculture, livestock, and water supplies. Some plans proposed the simultaneous deployment of biological and chemical weapons, including nerve gases, to ensure maximum destruction.[13]

From the late 1940s into the mid-1950s, the United States maintained biowarfare laboratories and research teams in Japan, some of which collaborated with Japanese public health organizations. Army reports referred to the Japanese experts as "Ishii scientists" because many had worked in Japan's biological warfare program. In 1951, the commanding officer of an American research unit in Japan required all enlisted personnel to complete training in biological warfare, including the use of insects as disease-carrying agents.[14]

During the Korean War (1950–1953), North Korea and China claimed that American planes dropped bombs and packages filled with feathers, fleas, ticks, spiders, mosquitos, bedbugs, and ants, many of which tested positive for cholera, meningitis, plague, or other diseases. Following the flyovers, there were reports of concentrated pockets of insects laying on top of snow and localized outbreaks of plague, anthrax, and cholera. Dozens of American soldiers captured by North Korea admitted to engaging in biological warfare. American officials denied the accusations, contending that the warfighters were brainwashed into confessions. Upon their return to the United States, the men retracted their statements, indicating that they were taken under duress.[15]

Following the alleged attacks, China assembled a committee comprised of biological warfare experts from Brazil, Britain, France, Italy, Sweden, and the Soviet Union. The committee unanimously concluded in a comprehensive report that "the peoples of Korea and China have indeed been the objective of bacteriological weapons. These have been employed by units of the U.S.A. armed forces, using a great variety of different methods for the purpose, some of which seem to be developments of those applied by the Japanese army during the second world war."[16] The report identified various tactics that were utilized, including anthrax-laden feathers and fleas, lice, and mosquitoes carrying plague or yellow fever.[17] As detailed in Figure 11.1, the report also provided photographs of bacteriological warfare bombs that the committee alleged were dropped by the United States.

The United States denied the allegations and called for an independent review by the United Nations, a move rebuffed by China and North Korea. Years later, the United States confirmed that it maintained the ability to conduct biological attacks, though it continued to deny that it engaged in biological warfare.[18] Decades later, a British representative from the international committee, Joseph Needham of Cambridge University, reaffirmed his belief that biological weapons were deployed. However, he noted that "mostly it was experimental work, as far as we could see."[19]

After the Korean War, the United States accelerated biowarfare research and development. By the mid-1950s, America was manufacturing tularemia and the bacteria that causes brucellosis. As government researchers were honing new biological agents, they also contracted with universities to develop vaccines to protect against the diseases. The United States then amended its military doctrine to allow a first strike with chemical or biological weapons as long as the president provides explicit approval.[20] As the 1956 Army field manual explained, "The United States is not a party to any treaty, now in force, that prohibits or restricts the use in warfare of toxic or nontoxic gases, of smoke or incendiary materials, or of bacteriological warfare."[21] The policy shift was not publicly acknowledged until 1977.[22]

Biowarfare was an integral component of the Cold War scientific arms race. Soviet researchers examined anthrax, Ebola, smallpox, and plague, among other diseases. The Soviets also investigated anti-crop and anti-livestock pathogens such as foot-and-mouth disease, rinderpest, and African swine fever. The Soviet program, which employed thousands at locations across the country, also involved research into genetically engineered viruses and bacteria. As with other nations, the Soviets justified their work on principles of national security, noting that international treaties did not prohibit research and development for defensive purposes.[23]

Under the administration of President John F. Kennedy, the US government embarked on a public relations campaign that used press conferences, private meetings with journalists, and congressional hearings to convince the public of the humane aspects of chemical and biological warfare. During the 1960s and 1970s, there were several reports of biowarfare attacks against Cuba that targeted crops and animals. In one instance, the United States allegedly infected Cuban turkeys with a fatal virus that killed eight thousand birds; the United States claimed the deaths were due to natural causes or neglect by Cuban farmers. In another, an outbreak of African swine flu—the first ever in the Western hemisphere—led to the slaughter of 500,000 Cuban pigs; the disease was one of many studied by American biowarfare experts.[24]

A further expansion of biological weapons research and development occurred during the Vietnam War. The Army Medical Service and the Chemical Corps embarked on a joint venture known as Project Whitecoat. This two-decade-long research program tested biological agents at government labs, including a massive chamber at Fort Detrick known as the Eight Ball.[25] The Eight Ball was a state-of-the-art facility—an enclosed, airtight, bombproof circular structure, four stories tall with

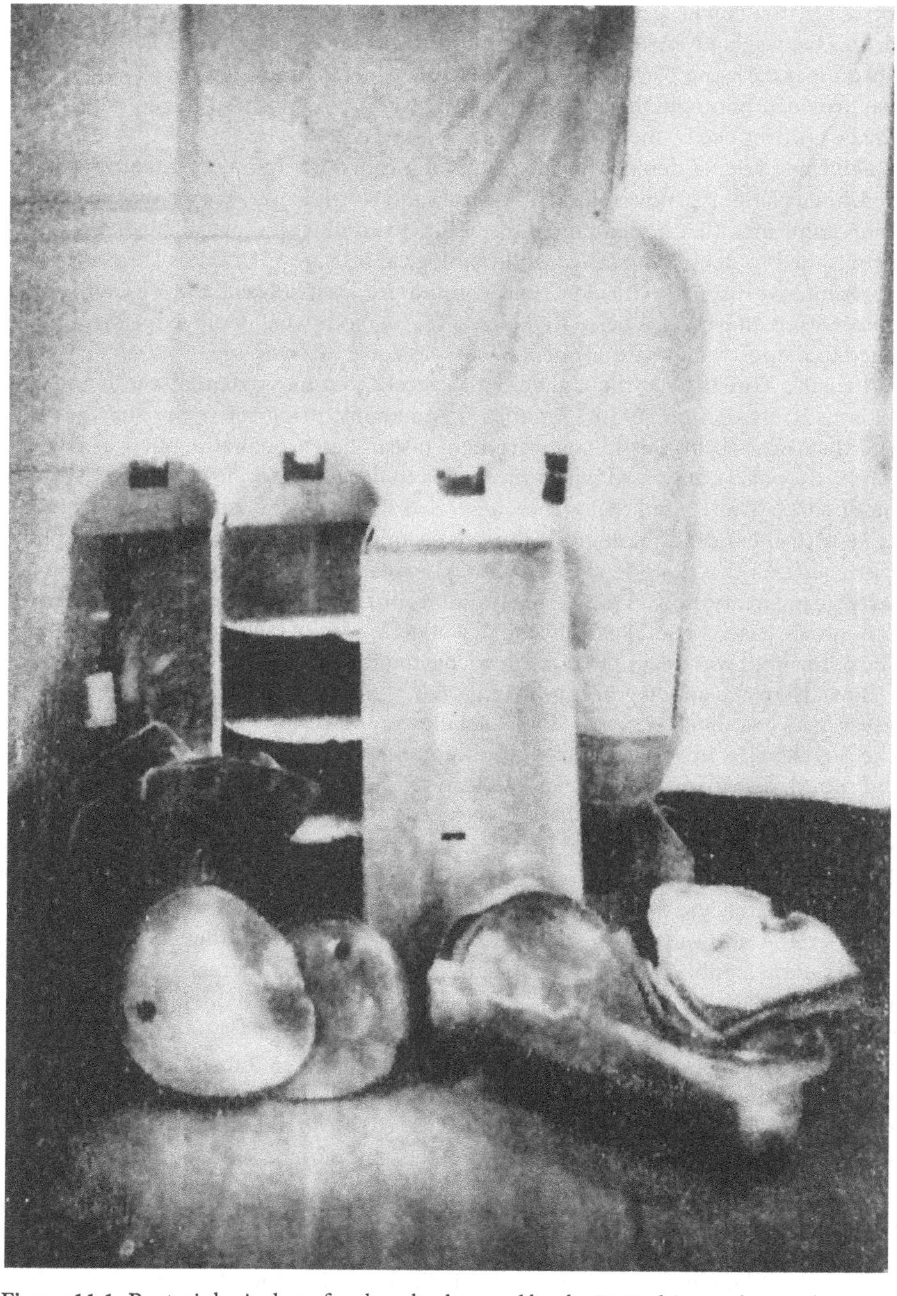

Figure 11.1 Bacteriological warfare bombs dropped by the United States during the Korean War, according to the 1952 Report of the International Scientific Commission for the Investigation of the Facts Concerning Bacterial Warfare in Korea and China.
Image from Appendix Y of the Report.

thick steel walls. The facility was strikingly similar to a massive octagonal-shaped chamber utilized by Japanese biological warfare researchers, the contours of which were revealed to the Americans via the Ishii biowarfare pact.[26]

When Project Whitecoat began, Colonel William Tigertt negotiated with the Seventh-Day Adventist Church to recruit research volunteers. Since church members were conscientious objectors and could be relieved from battle duties if drafted, Colonel Tigertt hoped the draftees would agree to participate in biological warfare studies. His efforts were successful. Although the conscientious objectors protested armed conflict, they were willing to be research subjects and support the progress of military science. More than two thousand individuals agreed to participate in the research, where they were exposed to Q fever, tularemia, Venezuelan equine encephalitis, and other agents to study the lethality of the diseases and create prophylaxis.[27] During the studies, the men were closely monitored and were provided medical care to address health conditions caused by pathogen exposure. Project Whitecoat ended with the draft in 1973, though it helped inspire a biowarfare research model based on volunteers, transparency, and post-research medical care.[28]

Much of America's offensive biological weapons research came to an abrupt halt in 1969 when President Richard Nixon stated: "The United States shall renounce the use of lethal biological agents and weapons, and all other methods of biological warfare."[29] Several factors influenced Nixon's pronouncement, including backlash to the widespread use of Agent Orange and other toxic chemicals in Vietnam, reports of chemical weapons leaks during cross-country transport, and a growing environmental movement (motivated in large part by Rachel Carson's groundbreaking book, *Silent Spring*).[30] In addition, by 1969, there were no less than five hundred occupational infections documented at Fort Detrick and other biological warfare facilities, resulting in no less than three deaths of biowarfare researchers.[31]

Nixon announced that the United States would only conduct research and development for defensive purposes. He created a new institute at Fort Detrick, the US Army Medical Research Institute of Infectious Diseases (USAMRIID), to engage in this work. In 1972, the United Nations General Assembly adopted the Biological Weapons Convention, whereby state parties agreed "never in any circumstances to develop, produce, stockpile or otherwise acquire or retain" biological weapons.[32] Signatories included the United States, the Soviet Union, and scores of other nations. The Biological Weapons Convention came into effect in 1975, the same year that the United States ratified the 1925 Geneva Protocol.[33]

Despite Nixon's policy and US accession to international treaties, the military and the CIA continued with biological weapons research, though publicly the government maintained that it was engaged solely in defensive work. Notwithstanding the assurances, during the 1975 Church Committee hearings that investigated improper conduct by American intelligence agencies, CIA Director William Colby acknowledged that the agency still maintained a stockpile of incapacitating and lethal agents that were ready for operational use, as well as dart devices for clandestine attacks that could imperceptibly inoculate a person with a biological or chemical agent.[34] Internally, the CIA group dedicated to these efforts was colloquially referred to as the Health Alteration Committee, and CIA officials referred to the secret vault that stored the devices and toxic agents as the Cave of Bugs.[35] Colby admitted that the

CIA maintained biological warfare records dating back to 1952, but indicated that "the records are very incomplete" since many were intentionally destroyed during the early 1970s in contravention of record retention policies.[36] When pressed by senators on whether the CIA held toxins for offensive use, in contradiction of US policy that only permitted research, Colby responded that the stockpiled toxins were "in a quantity which certainly is excessive for research purposes."[37] As the CIA's Chief of Counterterrorism, James Angleton, testified days after Colby's admissions: "It is inconceivable that a secret arm of the government has to comply with all the overt orders of the government."[38]

The CIA stockpiles included tuberculosis, anthrax, encephalitis, Rift Valley fever, salmonella, botulinum toxin, smallpox, and other agents.[39] Smallpox—which had ravaged humanity for centuries and nearly destabilized the Continental Army during the American Revolutionary War—was one of the most lethal pathogens studied by US scientists. The work created an incongruous scenario: as researchers attempted to make a weaponized version of smallpox that was more lethal than the natural one, public health officials were working to eradicate the disease from the globe. Eradication was achieved in 1980, but thereafter the Soviet Union and United States maintained smallpox stockpiles, rejecting calls from the World Health Organization to destroy them.[40]

Biological warfare research continued, with a focus on developing prophylactic vaccines and evaluating new biothreats, such as novel toxins and genetically engineered pathogens that are more contagious and lethal than naturally occurring diseases. The military also examined new detection devices and no less than thirty-one pathogens as potential biological weapons, including agents characterized as opioid-like "hibernation triggers" that cause hibernation-like states to temporarily incapacitate a person.[41] In 1990, biowarfare experts characterized the hibernation drugs as a "bizarre BW [biological warfare] scenario,"[42] yet in 2002 Russian special forces used such substances to temporarily incapacitate terrorists who were holding hundreds of people hostage inside a Moscow theater.[43] To this day, biowarfare stockpiles are housed at Fort Detrick—headquarters for defensive biological weapons work under the auspices of USAMRIID, America's leading biosecurity institute.

Biological Warfare Field Tests

Seeking information beyond the confines of a controlled laboratory environment, from the 1940s through the 1970s, the United States conducted hundreds of clandestine biological warfare field tests at military and civilian locations. The tests sought to understand American biowarfare vulnerabilities and produce information to bolster offensive strategies. As one secret Army report explained, "Open air testing of infectious biological agents was considered essential to an ultimate understanding of BW potentialities because of the many unknown factors affecting the degradation of microorganisms in the atmosphere."[44] Studies included using insects as pathogen-delivery agents, spraying diseases over humans standing in fields, and conducting mock germ warfare exercises in buildings, transportation hubs, cities, and the countryside.

Weaponizing insects was a key research area. During the work, the military gained access to a Trinidadian man who had been infected with yellow fever. Scientists took serum from the man, injected the serum into a monkey, removed infected plasma from the monkey, and combined the plasma with mosquito larvae. The infected mosquitos then bit laboratory mice, which contracted yellow fever.[45] The project amounted to public health research in reverse. It was modeled on Japanese experiments that Ishii and his colleagues had revealed to the United States in the post-WWII biowarfare immunity pact. The research was conducted a half-century after Walter Reed's momentous yellow fever experiments, where scientists uncovered the role of mosquitos in transmitting the disease.

Shortly after the mid-century yellow fever studies, scientists at Fort Detrick were engineering colonies of infected mosquitos and had devised a plan to produce 130 million infected mosquitos a month.[46] In 1956, several field tests were conducted in Florida and Georgia whereby uninfected mosquitos were released from aircraft to examine whether they would eventually bite humans.[47] According to a report from the Chemical Corps, "Within a day the mosquitoes had spread a distance of between one and two miles and had bitten many people." The report further indicated, "These tests showed that mosquitoes could be spread over areas of several square miles by means of devices dropped from planes or set up on the ground. And while these tests were made with uninfected mosquitoes, it is a fairly safe assumption that infected mosquitoes could be spread equally well."[48] Fleas, ticks, bedbugs, spiders, and flies were also viewed as effective vectors for biological warfare, and by the end of the 1950s the military was reported to have maintained various insects infected with yellow fever, malaria, dengue, plague, tularemia, cholera, anthrax, and dysentery.[49]

Coupled with the use of insects to transmit pathogens, military researchers examined the feasibility of dispersing bacteria in building ventilation systems, subways, and across vast expanses of land. The work included secret mock attacks on the Pentagon, the Capitol, and the White House—Army researchers walked into the buildings, dropped pints of bacteria into the ventilation system, and measured how far and fast the bacteria spread.[50] In a study conducted in June 1966 in New York City, military researchers nonchalantly dropped light bulbs filled with bacteria on ventilation grills at street level and onto the tracks as a subway car entered a station; the bacteria quickly spread throughout the metro system.[51] The New York subway tests—which were uncovered in 1980 by members of the Church of Scientology, which has long advocated for a moratorium on biological warfare research—were similar to clandestine tests conducted by German agents in the Paris metro in 1933 and British scientists in the London metro in 1963.[52] Following the American studies, no medical monitoring was undertaken to determine if the bacteria caused adverse health effects.[53]

The aforementioned experiments represent a small sample of more than two hundred secret biological warfare field tests conducted nationwide between 1949 and 1969.[54] In Minneapolis, scientists sprayed bacteria from rooftops and trucks, and then measured contamination inside and outside homes and schools.[55] In other experiments conducted throughout 1957 and 1958, military cargo planes dispersed tons of zinc cadmium sulfide across America to simulate a biological attack and examine the feasibility of covering large areas with microscopic agents. Zinc cadmium sulfide is a tracer compound that creates a fluorescent mist when placed under ultraviolet light.

In some studies, the chemical mix was detected up to twelve hundred miles from the point of release.[56]

Thirty-five zinc cadmium sulfide releases took place over St. Louis, where officials selected poor neighborhoods and sought the assistance of local police to help control any public resistance.[57] One secret report characterized the St. Louis test site area as consisting "principally of a densely populated slum district."[58] In some instances, when locals inquired as to the purpose of the tests, researchers stated that the Army was experimenting with "smoke screens" to protect the city from radar detection.[59] In 2023, survivors of the studies petitioned Congress to allocate funding for health studies and compensation for injuries and premature deaths they claim were caused by the secret tests. As one survivor, Ben Phillips, stated, "We were experimented on. That was a plan. And it wasn't an accident."[60]

In the early 1990s, hundreds of former students from a Minneapolis school who were unwittingly subjected to a simulated biowarfare attack came forward and revealed high rates of obscure health conditions.[61] Thereafter, Congress funded a study to examine the impact of the 1950s experiments. A 1997 National Academy of Sciences report acknowledged that "people were outraged by being exposed to chemicals by the government without their knowledge or consent" and noted that the health impact of the zinc cadmium sulfide tests is unknown.[62] Nevertheless, the report concluded, "it is extremely unlikely that anyone in the test areas developed adverse health effects."[63] A separate report—prepared by the Army—similarly determined that the releases "should not have posed any adverse health effects for residents in the test areas."[64] These reports contradicted peer-reviewed literature which found that, although causal evidence of harm was elusive, "it is more likely that such poisoning has been of a low-level chronic nature" due to the toxic properties of the elements used in the releases.[65]

While some biological warfare tests may have been innocuous, others were likely the cause of significant health ailments. In an Alabama county where bacteria were dispersed in an open-air field test, pneumonia cases tripled shortly after the test, and over the next two years the number of cases reverted back to the mean number of cases that was recorded in the year prior to the test.[66] These figures suggest a link between the field test and the increase in pneumonia cases. During a nine-day period in September 1950, the Army conducted six "biological warfare trials" over San Francisco, exposing all 800,000 residents to a bacterial spray.[67] Immediately thereafter, local hospitals treated no less than eleven cases of rare bacterial infections. The Army claimed the outbreak was coincidental, even though the bacterial infections were identical to one of the two bacterial agents used during the field tests.[68]

In 1955, the CIA disbursed *Hemophilus pertussis*, the bacteria that causes whooping cough, along Florida's Gulf coast. Whooping cough is a nasty disease that is particularly harmful to infants and young children; the incidence of whooping cough increased from 339 cases and 1 death in 1954 to 1,080 cases and 12 deaths in 1955.[69] That same year at Dugway Proving Ground, thirty service members stood in a line more than a half-mile long, joined by caged monkeys and guinea pigs. Scientists released a mist containing the bacteria that causes Q-fever—a disease with a mortality rate of about 4 percent—and measured the impact on the men and animals.[70]

Coupled with the biowarfare studies within US borders, American researchers collaborated with Canadian and British scientists to conduct biowarfare field tests in Canada and on Caribbean islands under British control.[71] One of the most hazardous field tests was conducted in the 1960s near Johnston Atoll in the South Pacific, where military scientists sprayed anthrax over a thirty-two-mile span and measured the infectiousness of the bacteria until it dissipated.[72] When the public learned of the experiments during the 1970s, officials justified the studies by pointing to "our fear of world dominion by the Communist countries, primarily the Soviet Union."[73]

During 1977 congressional hearings into biological warfare experiments, a Maryland public health official testified that the simulated attacks "constituted an unjustifiable health hazard for a particular segment of the population."[74] George Connell, Assistant Director at the Centers for Disease Control and Prevention (CDC), added: "There is no such thing as a microorganism that cannot cause trouble."[75] The military acknowledged that it did not conduct medical monitoring after the field tests, and further added that similar experiments might resume if an "area of vulnerability" were discovered.[76] By the mid-1980s, the Department of Defense (DoD) resumed testing and acknowledged using *Bacillus subtilis* in open-air tests.[77] Though harmless to most people, the bacteria can be harmful to individuals with open wounds or a compromised immune system.

Edward Nevin died from a bacterial infection that his family claims was caused by the 1950 San Francisco biowarfare field test. His family sued, but in 1983 a federal court dismissed the lawsuit, finding that the government maintained full immunity for "discretionary functions" related to "planning-level decisions."[78] The court held that it did not have the authority to review the decision to engage in mock germ warfare because "judicial evaluation would impair the effective administration of the government."[79]

Biowarfare field tests continue to the present day. While the military and government maintain that the tests are harmless, citizens continue to express concerns about adverse health effects, particularly for tests near schools.[80]

Mind Control

In addition to the expansive biowarfare program, the CIA and military devoted significant resources to mind control. Through a series of classified programs—many of which involved collaborations with universities and hospitals—researchers experimented on unsuspecting civilians and service members to ascertain whether drugs or psychological tactics could be used to modify behavior or facilitate interrogations. Parallel CIA research examined potential military uses of telepathy, psychokinesis, remote viewing, and other extrasensory powers.[81] The CIA studies built off WWII Office of Strategic Services (OSS) studies and concentration camp experiments conducted by Nazi scientists, some of whom were recruited to the United States under Operation Paperclip.[82] Interest in behavioral modification expanded significantly as reports from the Korean War indicated that captured American warfighters were brainwashed into confessions.[83]

Military and intelligence officials heralded mind control as a humane and effective method of procuring information. As a 1961 CIA report detailed: "Any technique that promises an increment of success in extracting information from an uncompliant source is *ipso facto* of interest in intelligence operations."[84] The potential benefits of mind control were documented in a series of secret reports, but officials also acknowledged ethical and legal concerns. The sentiment was captured in an internal CIA memo written in 1963 by Deputy Director Richard Helms, who would go on to become Director of the CIA: "While I share your uneasiness and distaste for any program which tends to intrude upon an individual's private and legal prerogatives, I believe it is necessary that the Agency maintain a central role in this activity, keep current on enemy capabilities the [sic] manipulation of human behavior, and maintain an offensive capability."[85] As a CIA Inspector General report candidly explained: "Precautions must be taken not only to protect operations from exposure to enemy forces but also to conceal these activities from the American public in general. The knowledge that the Agency is engaging in unethical and illicit activities would have serious repercussions in political and diplomatic circles and would be detrimental to the accomplishment of its mission."[86]

Countless studies were conducted via projects identified by cryptonyms, some of which wryly captured aspects of underlying work. Project CHATTER, a Navy program administered from 1947 to 1953, tested drugs to aid interrogations.[87] Experiments under Project CHATTER were based on studies documented in concentration camp experiments that were conducted by German physicians at Dachau and Auschwitz using mescaline and other hallucinogenic substances.[88] Project BLUEBIRD and Project ARTICHOKE, both operational during the 1950s, examined drugs for interrogations, mind control, hypnosis, memory enhancement, and inducing amnesia.[89]

MKNAOMI, an expansive program that was operational from the early 1950s to 1970, involved stockpiling incapacitating and lethal substances to harm humans, crops, and animals; creating devices for disseminating biological and chemical weapons; and field tests to study the substances and devices.[90] The CIA and the Army Special Operations Division at Fort Detrick collaborated on MKNAOMI projects, including the development of biological agents for use in covert operations.[91] One internal memo from 1952 advocated for collaborations with scientists outside the United States since other countries might permit "certain activities which were not permitted by the United States government."[92]

Via QKHILLTOP, which began in 1954, CIA and military researchers investigated brainwashing and interrogation techniques.[93] To facilitate partnerships with universities and research institutes, one QKHILLTOP project established a foundation at Cornell University—the Society for the Investigation of Human Ecology, later renamed the Human Ecology Fund—which funded projects of interest to the CIA and military.[94] QKHILLTOP projects eventually were folded into MKULTRA, a principal CIA "umbrella project" created by CIA Director Allan Dulles in 1953. MKULTRA was dedicated to the research and development of chemical, biological, and radiological materials for use in clandestine operations.[95]

MKULTRA funding was categorized into three realms: searching for substances that are suitable for mind control and behavioral modification, testing the substances on human subjects in the lab, and testing the substances in real-life situations.[96] MKDELTA oversaw MKULTRA projects undertaken outside the United States, including the administration of drugs during interrogations conducted by US agents on foreign soil.[97] In one MKDELTA project, the CIA administered mind-altering substances in an attempt to facilitate interrogations of suspected double agents. The work was conducted at the Panama Canal Zone, in prison cells that originally were constructed to temporarily house drunk or disorderly sailors.[98]

Between 1966 and 1973, MKULTRA projects continued under a program named MKSEARCH, which sought to "develop, test, and evaluate capabilities in the covert use of biological, chemical, and radioactive material systems and techniques" for altering human behavior.[99] In addition, the CIA and DoD collaborated on Project OFTEN and Project CHICKWIT, programs that experimented with various drugs that might influence human behavior.[100]

In 1973, CIA Director Richard Helms bypassed a CIA records retention policy and ordered the destruction of all MKULTRA and MKSEARCH files.[101] Two years later, during the Church Committee hearings investigating improper conduct by US intelligence agencies, CIA officials testified that no documents survived the destruction order and that no person could recall any details of the projects. In 1977, in response to a Freedom of Information Act (FOIA) request for MKULTRA documents, a CIA records officer located several boxes that had inadvertently survived the destruction order because they were stored in archives for the CIA's Budget and Finance materials rather than the CIA's central Retired Records Center. Although the surviving files totaled nearly eight thousand pages, they were merely a subset of the entire MKULTRA record. They primarily included financial records and budget requests, with limited project details and no post-study reports. Nevertheless, the documents revealed a far more extensive mind control program than that admitted by the agency during the 1975 congressional hearings.[102]

The files detailed 149 MKULTRA subprojects conducted over a decade, including collaborations with 185 non-governmental researchers and more than eighty institutions, including Princeton, Harvard, Columbia, Stanford, Johns Hopkins, Cornell, Ohio State, Penn State, Emory, and the universities at Minnesota, Denver, Illinois, Oklahoma, Richmond, Maryland, Indiana, Wisconsin, Houston, Florida, and Texas.[103] Many researchers were witting CIA partners, while others were kept in the dark as to the funding source or received grants through shell foundations that laundered CIA money.[104] As Sidney Gottlieb—a leading spymaster and director of the CIA Technical Services Staff—explained during a 1977 Senate investigation, the agency sought "to harness the academic and research community of the United States to provide badly needed answers to some pressing national security problems."[105]

The breadth of MKULTRA experiments is stunning. In addition to research into LSD and other drugs, projects examined psychotherapy, electro-shock, extrasensory perception, hypnosis, aspects of the magician's art that may be useful in covert operations, and the development of toxic substances to harm humans, animals, and crops.[106] In a project funded by the CIA and conducted at McGill University, more than one hundred patients suffering from severe mental illness were administered LSD, confined in an isolation chamber, and bombarded with a taped message that repeated continuously for one week, sixteen hours per day. The experiment—dubbed psychic driving—was led by Ewen Cameron, a world-renowned psychiatrist, and sought to erase personality traits and redirect a person's thoughts and actions. There is no evidence that any patient volunteered for the experiment.[107]

In a separate project, a decade-long study was conducted at the Lexington Rehabilitation Center, a treatment facility in Kentucky that was maintained by the Public Health Service and housed individuals convicted of drug offenses. Working with the National Institute of Mental Health, researchers tested various drugs on recovering addicts. Many were surreptitiously administered LSD, barbiturates, and other drugs, at times for several months. Others were provided an illicit drug of their choice if they agreed to also ingest a substance selected by researchers. Many research subjects became addicted to the drugs provided during the study.[108]

The MKULTRA LSD experiments were conducted despite research findings, some dating to the mid-1950s, which warned that "almost all patients" administered LSD "become distressed," and that many hallucinate, exhibit "neurotic symptoms," suffer an "anxiety attack," or "pass into an acute state of violent trembling."[109] In light of the concerns, one peer-reviewed publication warned that, since an individual provided LSD "may potentially be a danger to himself and to others" and "too little as yet is known about individual differences in susceptibility," researchers should "refrain from using the drug in out-patients or in day-hospital patients, and restrict its use to inpatients—and then only when constant supervision by trained personnel is available."[110] Due to the MKULTRA records destruction, it is unclear how many LSD studies adhered to the precautions. As CIA Director Stansfield Turner testified, it was agency policy that MKULTRA projects "maintain no records of the planning and approval of test programs."[111]

CIA and Army drug experiments also were conducted on prisoners and individuals housed in mental institutions.[112] During Senate hearings in 1977, Turner admitted: "There are cases in penal institutions where it is not clear whether the prisoner was given a choice or not."[113] At the New York State Psychiatric Institute, a state institution staffed by Columbia University psychiatrists, physicians injected patients with mescaline to examine whether the drug could be used as a psycho-chemical agent.[114] The Psychiatric Institute study began in 1951, under a contract with the Army, and sought to derive "new technical data" that would "provide a firmer basis for the utilization of psychochemical agents both for offensive use as sabotage weapons and for protection against them."[115]

The Army provided the drugs, and Psychiatric Institute physicians working on the study had to undergo security clearances to perform the experiments.[116] One casualty of the study was Harold Blauer, a professional tennis player receiving treatment at the institute, who suffered a heart attack and died after a series of mescaline injections.[117] Blauer, aged forty-two, was voluntarily admitted to the institute due to depression he suffered following his divorce. During his treatment, Blauer agreed to be injected with experimental drugs that, physicians told him, were potentially therapeutic. There was no mention of the Army's involvement in the experiment or the true goal of the study, which was to evaluate chemical warfare agents.[118]

Blauer suffered headaches and tremors after receiving the first three doses, and he informed the researchers that he no longer wanted to continue with the experiment. Blauer's physician told him that if he did not continue, he would be transferred to

Bellevue or Roosevelt Hospital, two institutions where Blauer had had horrible experiences as a patient. Blauer reluctantly agreed to continue with the study. After the fifth injection, Blauer began sweating profusely and was wildly flailing his arms: his body stiffened, his teeth clenched, and he started frothing at the mouth. Within two hours of the injection, Blauer fell into a coma; he was pronounced dead about an hour later.[119]

In Army records, the Chemical Corps indicated that it wanted to avoid "embarrassment and adverse publicity." It was concerned that public disclosure of Blauer's death would jeopardize similar contracts with other institutions.[120] The State of New York and the Psychiatric Institute also sought to conceal their involvement in the Army's chemical warfare studies. Thereafter, the State of New York and the US government split an $18,000 settlement for Blauer's former wife in exchange for her agreeing to abandon all legal claims for Blauer's wrongful death. At no time was she provided any evidence regarding the chemical warfare studies; instead, government lawyers simply informed her that Blauer had died from a drug overdose.[121] The government kept Blauer's cause of death secret for more than twenty years.

In June 1975, the Rockefeller Commission—which was created by President Gerald Ford and led by Vice President Nelson Rockefeller, and charged with investigating untoward activities by the CIA and other US intelligence agencies—issued a report that revealed that the military had experimented with mind-altering drugs on unsuspecting subjects.[122] The revelations alarmed the public, and a series of congressional hearings followed. Although initially the military refused to comment, it gradually disclosed that it conducted a small number of drug experiments and revealed the names of some individuals who died during the studies, including Blauer.[123] Blauer's eldest daughter sued the United States, the Army, the State of New York, the Psychiatric Institute, Army officials, physicians at the Psychiatric Institute, and the former New York State Commissioner of Mental Hygiene.[124]

After more than ten years of litigation, a federal district court in New York explained that, under the doctrine of sovereign immunity, it is not authorized to second-guess "the planning and institution of Government programs."[125] Analogizing Blauer's claim to that of Nevin's following the military's San Francisco mock biowarfare experiments, the court held that "the decision to test potential chemical warfare agents on human subjects at the Psychiatric Institute, while perhaps disturbing, involved weighing these same policy factors."[126] Unlike with Nevin, however, where the court dismissed the entire case, the court granted Blauer's family an obscure legal avenue for recovery of money damages from the government.

The court held that the government was responsible for the wrongful death of Blauer because Army Chemical Corps scientists were negligent in their performance of mescaline toxicity studies that they shared with physician-researchers at the Psychiatric Institute. Specifically, the court reasoned that the toxicity study data that the Army provided to the Psychiatric Institute were based on negligently performed animal studies, which caused the physicians to administer mescaline doses that were far higher than that which was safe for human experimentation. The court awarded Blauer's estate $702,044 as compensation for the pain and suffering he endured prior to his death, for funeral expenses, and for lost parental support, nurture, and guidance that Blauer's daughters experienced.[127]

Blauer's experience at the Psychiatric Institute represents one example of government-civilian collaborations involving experimentation on unsuspecting patients in hospitals. The CIA also funneled hundreds of thousands of dollars via a foundation to support the construction of a six-story wing at Georgetown University Medical Center.[128] Via an agreement with Georgetown, one floor of the new wing was dedicated space for CIA studies, and three full-time hospital employees would be planted by the CIA's Chemical Division, thus providing "full professional cover" for the spies and a scenario whereby "Agency sponsorship of sensitive research projects will be completely deniable."[129] The top floor of the new wing included a radioisotope lab, which was provided by the Atomic Energy Commission and utilized for secret biomedical radiation research.[130] As detailed in CIA records, between 1959 and 1967 the arrangement ensured a steady flow of patients for experiments, some of whom were terminally ill.[131] One CIA memo characterized the Georgetown facility as "the equivalent of a hospital safehouse" and explained that patients "for experimental use will be available under controlled clinical conditions."[132] As a CIA physician told a room of recruits, "our guiding light is not the Hippocratic Oath, but the victory of freedom."[133]

Apart from studies conducted in institutional settings, MKULTRA field tests included projects whereby CIA agents would frequent bars and observe the behavior of individuals whose drinks they had secretly spiked with LSD. From 1952 to 1965, the CIA rented apartments in San Francisco and New York's Greenwich Village for use as research safehouses. Via projects Operation MIDNIGHT and Operation CLIMAX, the agency paid prostitutes to bring men back to the apartments, which were outfitted with elaborate drapes, luxurious accessories, microphones, and recording devices. The unsuspecting men were provided drinks laced with LSD or other drugs, as CIA agents watched behind two-way mirrors and documented how much information a man would reveal at various stages of a sexual encounter.[134] Stephen Lukasik, ARPA's former director, who worked with Sidney Gottlieb, a chemist who was director of the MKULTRA program, observed: "They didn't so much care about the laws, but in terms of creative people who are just told they don't have to worry about anything, and just create. They were good people."[135]

The CIA also surreptitiously administered LSD to its own personnel. In November 1953, during a CIA retreat to Deep Creek Lake, Maryland, Gottlieb and CIA officer Robert Lashbrook slipped LSD into the drinks of agents sharing a bottle of Cointreau. Within days, the mental state of one agent, Frank Olson, deteriorated significantly.[136] Among his responsibilities, Olson collaborated with Army researchers at Fort Detrick on biological weapons research and field tests.[137] CIA officials took Olson to Manhattan, where he met with a physician and a magician, both of whom had CIA clearance.[138] The magician, John Mulholland, had written a manual on deception for the CIA and reportedly tried to hypnotize Olson to gain insights into his inner mind.[139] The hypnosis was unsuccessful, and the CIA determined that Olson should be institutionalized.[140]

The next evening, Olson and Lashbrook retired to their room at the Statler Hotel, across the street from New York's Penn Station. In the middle of the night, Olson crashed through the window of Room 1018A, falling over one hundred feet to his death.[141] When the police arrived shortly after 2:30 a.m. that winter morning, they found Olson on the pavement, wearing only underwear and a t-shirt, and Lashbrook sitting on the toilet in the hotel room.[142] According to the hotel operator, shortly after Olson's fall, a man from Room 1018A called and said, "He's gone"; the voice on the other line replied, "That's too bad."[143] During congressional hearings, evidence revealed that Gottlieb was the person who was called from the room.[144] The New York City Police Department commenced an investigation into whether Olson's death was a homicide, but within two days, the case was closed, and his death was deemed a suicide.[145]

Thereafter, the CIA concealed Olson's association with the agency.[146] During an internal investigation, CIA staff remarked that Olson's death was "just one of the risks running with scientific experimentation."[147] Twenty-two years later, the 1975 Rockefeller Commission Report noted that an Army employee jumped from a hotel window after being given LSD during a CIA study—this was the first time that the Olson family linked Frank's death to the agency.[148] President Ford apologized to the family, and the government provided them with a $750,000 settlement in exchange for abandoning all legal claims.[149]

Olson's son, Eric, was tormented by his father's suspicious death, and for decades, he worked to uncover the true story. In 1994, the Olson family had Frank's body exhumed and autopsied. Although in 1953, the CIA told the family that there needed to be a closed casket because Frank's body was mangled, the exhumation revealed that Frank's body was embalmed and in a respectable condition. Moreover, the 1994 autopsy report stated that Frank had suffered a substantial blow to his head before hitting the ground. Thereafter, the Manhattan District Attorney's Office reopened the case to investigate whether it was a homicide. Though prosecutors were unable to make a conclusive determination, they changed Olson's cause of death from suicide to unknown.[150]

Eric Olson and others have posited that the CIA killed Frank because he was a dissident who questioned the morality and legality of the CIA's biological warfare program and enhanced interrogation techniques. The spiked Cointreau was not an agency experiment, they contend, but an attempt to get Frank to reveal his true thoughts about the CIA's work.[151] In 1997, the CIA inadvertently declassified a 1953 memo on clandestine assassinations that indicated that "the most efficient accident, in simple assassination, is a fall of 75 feet or more onto a hard surface." The assassination manual further noted that one useful method was to "stun or drug the subject before dropping him."[152] Eric Olson claims to have spoken with retired CIA agents who told him that the agency hired mafia hitmen to throw his father out the window.[153]

By the early 1960s, the CIA recognized that LSD had limited use in covert operations. One report explained that "information obtained from a person in a psychotic drug state would be unrealistic, bizarre, and extremely difficult to assess" and that "no such magic brew as the popular notion of truth serum exists."[154] The report further noted that LSD could be used "to produce anxiety or terror in medically unsophisticated subjects unable to distinguish drug-induced psychosis from actual insanity" but that "an enlightened operative could not be thus frightened, however, knowing that the effect of these hallucinogenic agents is transient in normal individuals."[155]

The CIA's authority to engage in mind control and related areas is grounded in the broad mandate outlined in the National Security Act of 1947, which created the CIA and authorized the agency "to perform such other functions and duties related to intelligence affecting the national security as the National Security Council may from time to time direct."[156] Allen Dulles, an attorney and wartime OSS officer who became CIA director, helped formulate the expansive clause.[157] Two years later, the Central Intelligence Act of 1949 explicitly exempted the CIA from standard congressional reporting requirements to protect intelligence sources and methods.[158]

Permissive laws, secrecy authorizations, and reduced reporting obligations frame the CIA's far-reaching intelligence enterprise. A complex web of intelligence sources bolsters the CIA's work: between 1948 and 1959, more than forty thousand individuals and companies—from academia, science, hospitals, and industry—provided intelligence to the agency.[159] In an article published in *The Washington Post* in 1963, former President Truman stated: "I never had any thought that when I set up the CIA that it would be injected into peacetime cloak-and-dagger operations." Truman lamented that "the CIA has been diverted from its original assignment" as an intelligence agency and transformed into "an operational and at times a policy-making arm of the government." He called for the CIA's operational aspects to be terminated or transferred to another agency.[160] Neither occurred.

Although the CIA and military collaborated on several mind control programs, at times, the organizations were plagued by jurisdictional conflict. For example, the Army attempted to conceal from the CIA its overseas testing of LSD and other drugs, at which point the CIA engaged in covert measures to uncover the details.[161] CIA leadership also prepared a memo stating that the agency should "restrain others in the intelligence community (such as the Department of Defense) from pursuing operations."[162]

Despite the CIA's attempts to stymie the DoD, the military vigorously pursued mind control research throughout the 1950s and 1960s.[163] Via Project THIRD CHANCE and Project DERBY HAT, the Army used LSD and other drugs to facilitate interrogations of foreign intelligence agents and others located outside the United States. This included, for example, US Army Private James Thornwell, stationed in France, who was suspected of stealing classified documents.[164] In addition to using the drugs to help elicit information, Army researchers examined whether LSD had a different effect on "Orientals."[165]

The military also experimented with BZ, a psychotropic drug that could be distributed via an airborne spray. In 1962, the DoD invested $2 million to build a BZ factory at Pine Bluff Arsenal in Arkansas, and more than 100,000 pounds of BZ were produced there over the next two years. Field tests with BZ gas clouds were conducted at Dugway Proving Ground in Utah and Hawaii, though in the end, the military concluded that BZ was not an effective weapon.[166]

Like the CIA, the military contemplated the legal and ethical implications of its work. A memo prepared by the Army Intelligence Center, dated October 15, 1959, reads: "It was always a tenet of Army Intelligence that the basic American principle of the dignity and welfare of the individual will not be violated." The memo explains: "In intelligence, the stakes involved and the interests of national security may permit a more tolerant interpretation of moral-ethical values, but not legal limits." The report concluded that secret experiments should be conducted and suggested that legal liability could be mitigated by implementing "proper security and appropriate operational techniques" to cover up the experiments.[167]

James Stanley's experience is illustrative. In 1958, aged twenty-four and stationed at Fort Knox, Kentucky, Stanley volunteered for a study ostensibly designed to test new gas masks. Stanley reported to the Edgewood Arsenal at the Aberdeen Proving Ground in Maryland each Wednesday for four consecutive weeks. While trying on various masks, Stanley was given a glass of clear liquid that appeared to be water. The water was laced with LSD, and the actual objective of the gas mask research was to study human response to psychotropic drugs to determine if they could be used to develop chemical weapons or facilitate interrogations.[168]

During and after the tests, Stanley experienced intense hallucinations and periods of incoherence. As he later explained, "I held onto the bed. I felt I was floating through space and through time. . . . The drinking water . . . caused total chaos. I didn't know where I was at or what I was doing."[169] Realizing his bizarre state of mind, Stanley questioned the purity of the water. Officials assured him that it had been tested and was clean.[170]

The adverse reactions caused by the clandestine research intensified. During one episode, Stanley threw a piece of iron weighing several hundred pounds against the wall. In another, he awoke in the middle of the night and, without reason, violently beat his wife and six children. He had no memory of either incident. In addition to uncontrollable fits, hallucinations, and flashbacks, Stanley would leave his military post without authorization and would be unable to account for his whereabouts. Two years after the Edgewood experiments, the Army demoted him for "inefficiency." For years following the secret experiments, Stanley doubted his sanity; his children remained fearful of their unpredictable father, and later, his wife divorced him. Simply stated, Stanley's personal and professional life were ruined.[171]

Nearly two decades after the research, Stanley received a letter from the Army. The letter stated that the Army was conducting a follow-up study of "participants" from the LSD experiments and wanted to learn what long-term effects Stanley had experienced. As Stanley recalled, "When I opened the letter, I stood there and cried."[172] This was the first time Stanley learned that he had been secretly administered LSD. Though his erratic behavior now could be traced to a tangible source, Stanley was infuriated by the acts of the researchers and the callous disregard for his health and well-being. He later filed a lawsuit against the Army and researchers.

A federal district court in Florida dismissed Stanley's case against the military on the grounds that the government maintains sovereign immunity, but allowed Stanley to bring a claim against the individual researchers, including Sidney Gottlieb and several University of Maryland physicians who worked on the experiments.[173] The court described the clandestine LSD experiments as "an egregious intrusion on the most precious right protected by the Constitution—the right not to be deprived of life, liberty, or property without due process of law."[174] The court further reasoned that "by donning a uniform, a member of the Army does not volunteer himself to be duped and deceived into becoming a guinea pig for his superiors, and consequently to be deprived of his most fundamental constitutional and human rights."[175]

A federal appellate court affirmed the lower court decision that allowed Stanley's case to proceed against the individual researchers; the appellate court also reversed the lower court decision that barred the claim against the military. As to the latter, the court ruled that sovereign immunity does not apply since the "inescapable demands of military discipline and obedience to orders are not implicated by the facts of this case," and there is no concern "that the peculiar and special relationship of the soldier to his superiors might be disrupted."[176] The appellate court remanded the case for further proceedings.[177]

In 1987, after a prolonged legal battle that lasted nearly a decade, Stanley's case eventually reached the US Supreme Court. In a 5–4 decision, the high court ruled against Stanley on all counts. Writing for the majority, Justice Antonin Scalia held that courts should refrain from questioning military decisions, and that an investigation of the LSD experiments would "disrupt the military regime."[178] Scalia further explained that the LSD experiments represent an exercise of military discretion that civilian courts are precluded from questioning.[179] Thus, as was the case with Nevin and the San Francisco mock biowarfare attack, the military and researchers were completely immune from legal responsibility.

Stanley's case was not an aberration—the military conducted secret drug experiments on more than one thousand service members throughout the 1950s and 1960s.[180] Even in instances where the Army utilized volunteers, the consent forms did not identify the drugs that would be administered or notify the research subjects of potential adverse reactions or health risks. The forms did, however, threaten soldiers with court-martial should they disclose any information related to the experiments, including disclosures to their physicians should health ailments materialize.[181]

Decades after the experiments, veterans continued to express their frustration and sense of abandonment. As Army Private Tim Josephs explained during an interview conducted in 2012, officials told him that he would "test new Army field jackets, clothes, weapons, and things of that nature, but no mention of drugs or chemicals."[182] When Josephs arrived at Edgewood in 1968, he saw men in white lab coats and expressed second thoughts about participating in the studies. An officer then told him: "You volunteered for this. You're going to do it. If you don't, you're going to jail. You're going to Vietnam either way—before or after."[183]

Following the experiment, Josephs was hospitalized for Parkinson's-like tremors. The symptoms continued for decades, and eventually, he was diagnosed with Parkinson's disease. In addition to suffering near-constant tremors, Josephs was personally responsible for nearly $2,000 a month in out-of-pocket costs for medications. In 2011, the VA granted Josephs partial disability benefits for his Parkinson's, based on his exposure to Agent Orange during the Vietnam War, but the VA benefits decision made no mention of the LSD experiments.[184]

During the late 1970s, after public disclosure of the LSD experiments, the Army conducted a follow-up study to assess the long-term effects of the LSD research. Of the 1,000+ veterans who took part in the LSD experiments, only 220 were given a physical examination.[185] Many who the Army contacted refused to participate in the follow-up health examinations, fearing that the medical care was merely a front for additional secret experiments. Those who underwent a physical examination reported being treated horribly: for example, Stanley indicated that he was told that he would be put up in a hotel and given a physical at Walter Reed Medical Center.[186]

"Instead," Stanley stated, "I was taken to a psychiatric hospital and treated like a crazy person. I was furious."[187]

In a report published in 1980, the Army Medical Department outlined a long list of health conditions experienced by veterans of the LSD experiments, including flashbacks, somatic complaints, depression, personality changes, anxiety, nightmares, paranoia, neurosis, memory loss, and psychosis.[188] Yet, the Army report denied that these conditions were caused by the experiments, concluding that "the majority of subjects evaluated did not appear to have sustained any significant damage from their participation in the LSD experiments; and in those cases where there were abnormalities either by history or on examination, LSD could not generally be identified conclusively as the causative agent."[189] Decades later, in a 2012 report, a lead scientist from the Edgewood experiments, James S. Ketchum, defended the LSD studies, stating that "it needed to be done." As Ketchum elaborated: "I have always had the feeling that I am doing more the right thing than the wrong thing."[190]

One notable physician who consulted for the CIA and the military on LSD research was Harvard's Henry Beecher, discussed in chapter 9, who for decades has been widely regarded as a pioneer in the field of bioethics. Beecher's LSD work is less well known than his ethics contributions. Also less publicized is the role that Beecher played in preventing the implementation of the 1953 DoD policy that adopted the Nuremberg Code for military research. During the early 1960s, when the Army was negotiating a contract with Harvard, Beecher protested the Army's desire to include protections for research subjects that were modeled on the Nuremberg Code, contending that the Army simply should trust the virtues of Harvard investigators.[191]

Proliferation of America's Chemical Weapons Program

America's chemical weapons program also grew substantially after WWII. US leaders characterized chemical weapons as more humane than atomic bombs and reasoned that an expansive chemical weapons stockpile would help deter a chemical attack. Over the next four decades, the military worked closely with universities and the chemical industry to create toxic substances that could ravage people, animals, crops, and land. MIT assisted with research on phosgene, mustard gas, and other chemicals. At the same time, Harvard and Columbia helped develop incendiary bombs and napalm, which were used extensively during the wars in Korea and Vietnam.[192] By 1969, the United States maintained a fearsome forty-two-thousand-ton chemical weapons arsenal that included mustard gas, phosgene, V agents, Sarin, Soman, Tabun, and other agents such as tear gas.[193]

Service members and civilians participated in studies to examine the lethality of various chemicals and to develop protective clothing and prophylaxis. Experiments were conducted at several locations, including Dugway Proving Ground and the Eight Ball at Fort Detrick. At Holmesburg prison in Philadelphia, the Army conducted various experiments, some of which were collaborations with physicians from the University of Pennsylvania. These included "skin-hardening" tests that required significant inflammation and crusting of a prisoner's skin to see if scientists could create a natural body armor to protect against chemical irritants that soldiers might face on the battlefield. Over a dozen toxic substances were applied to the naked bodies of inmates, including hazardous chemicals such as hydrochloric acid and benzene. The men often were not informed of research-related risks, and many suffered long-term health ailments.[194]

Of the multitude of chemical weapons developed during the Cold War, V agents were the most lethal. The military was particularly interested in VX, which slows the heart, inhibits muscle movement, and leads to death within minutes, usually from asphyxia. VX is highly viscous, attaches well to surfaces, evaporates slowly, and is more toxic than other nerve agents.[195] A VX production plant was constructed in Alabama, and the military performed more than twelve hundred field tests to examine the toxicity of the agent.[196]

Between 1962 and 1973, through Project Shipboard Hazard, the Navy dispersed Sarin and VX over ships, exposing more than fifty-eight hundred sailors to the chemicals. The tests were not publicly revealed until 2002, and a 2007 Institute of Medicine (IOM) report noted an increased risk of heart disease among some sailors from the studies.[197] The IOM also identified "poorer overall physical and mental health among" the test participants.[198] In addition to the Shipboard Hazard experiments, a 1968 open air VX field test at Dugway Proving Ground went awry when a chemical tank malfunctioned and the pilot was unable to stop the airplane from spraying VX. The plane dispersed the toxic agent far outside the planned test area at Dugway Proving Ground, contaminating acres of land and killing more than six thousand sheep.[199] Figure 11.2 is an example of an Army field test during which a plane sprayed a chemical agent over soldiers.

In many instances, chemical weapons field tests and laboratory studies were coordinated with Canada and Britain. Experiments were conducted at Britain's Porton Down and over a one-thousand-square-mile testing ground in Suffield, Canada. Similar to their American counterparts, British military officials coopted service members into experiments, often withholding research-related risks and forcing men to sign confidentiality agreements. In Britain, service members were promised time off if they volunteered for an experiment to find a cure for the common cold. When service members reported for the studies, British scientists exposed the men to Sarin and other noxious agents, sometimes by locking them in gas chambers for extended periods of time.[200]

Coupled with research into chemicals that were lethal to humans, US military scientists investigated chemicals that could destroy crops and farmland. From the 1940s through the 1960s, American researchers examined more than twenty-six thousand chemicals, some of which were later marketed to the public as weed killers.[201] ARPA played an instrumental role in the development of herbicides for military purposes, drawing on Britain's wartime experiences using similar chemicals during colonial battles in Malaya.[202] The US military's first large-scale defoliation field test was conducted in 1959 in upstate New York.[203] By the early 1960s, the Chemical Corps selected six chemicals for operational use, code-named Agents Blue, Green, Orange, Pink, Purple, and White, after the color of the fifty-gallon drums in which they were stored.[204]

Against the advice of some of his advisers, President John F. Kennedy authorized the use of the chemicals in Vietnam.[205] One adviser, Roger Hilsman, wrote that the

Figure 11.2 An airplane spraying a chemical mist over US service members during a chemical warfare field test.
Photograph from the Library of Congress.

"political repercussions would outweigh any possible gains. Defoliation was just too reminiscent of gas warfare."[206] A senior Pentagon engineer, Seymour Deitchman, raised concerns about the potential adverse health effects. During a meeting to discuss operational use, he asked: "Well, do you know what effect that is going to have on the people caught up by it?"[207] Notwithstanding the concerns, American leaders justified the deployment of the chemicals under principles of national security and military necessity, assuring the public that the agents "are not harmful to people, animals, soil or water."[208] Secretary of State Dean Rusk advised President Kennedy that "the use of defoliant does not violate any rule of international law concerning the conduct of chemical warfare and is an accepted tactic of war."[209]

Between 1961 and 1971, the United States sprayed more than 11 million gallons of Agent Orange throughout Vietnam, destroying more than 5 million acres of land and 370,000 acres of crops.[210] The chemical was sprayed over scores of villages that had a combined estimated population of two to four million people.[211] Agent Orange, manufactured primarily by Monsanto and Dow Chemical, is a defoliant that causes vegetation to rapidly grow and self-destruct. One of its key components is dioxin, a known toxic substance with devastating properties: a few ounces are sufficient to poison the entire water supply of a city the size of New York.[212] By the end of the war, American forces had showered more than 366 kilograms of dioxin over Vietnam.[213]

Among American warfighters and the Vietnamese, Agent Orange caused various health conditions such as cancer, skin ailments, and reproductive complications, including tens of thousands of stillborn children and children born with missing limbs, cranial deformities, and other congenital disorders.[214]

Apart from the deployment of Agent Orange, the United States dispersed more than 6 million gallons of other toxic substances in Vietnam and conducted classified field tests with anti-crop agents in Thailand.[215] American forces also deployed more than 13.7 million pounds of tear gas during the war.[216] Media outlets from France, Germany, Japan, and elsewhere denounced America's use of tear gas—which causes burns and respiratory problems—analogizing the conduct to the use of poison gases during WWI.[217] As a Canadian physician who treated Vietnamese tear gas victims explained:

> Those patients that have come to my attention were very ill with signs and symptoms of gas poisoning similar to those that I have seen in veterans from the First World War treated at Queen Mary Veteran's Hospital in Montreal. The only difference between the cases was that these Vietnamese patients were more acutely ill.... Patients are feverish, semi-comatose, severely short of breath, vomit, are restless and irritable. Most of the physical signs are in the respiratory and circulatory systems.... The mortality rate in adults is about 10 per cent while the mortality rate in children is about 90 per cent.[218]

An editorial in *The New York Times* condemned the use of tear gas, stating that "in Vietnam, gas was supplied and sanctioned by white men against Asians. This is something that no Asian, Communist or not, will forget. No other country has employed such a weapon in recent warfare."[219]

Throughout the Vietnam War, American officials maintained that Agent Orange, defoliants, and tear gases were not chemical weapons and that their use did not violate any international treaties.[220] Indeed, in 1975, when the United States ratified the Geneva Protocol of 1925, it retained the right to utilize defoliants or riot control agents under certain circumstances, and also retained the right to retaliate in kind with biological or chemical weapons.[221] During the war, the United States focused on a legalistic interpretation of its chemical weapons obligations under international law, rather than engaging in an assessment of whether the military's actions were morally appropriate.

In the years following the Vietnam War, political backlash to the deployment of chemical weapons led to a modest decline in funding for America's chemical warfare program. The funding cuts were short-lived, and during the 1980s, appropriations for chemical weapons increased significantly. In part, the increased funding was used to expand America's arsenal of binary chemical weapons. With conventional chemical weapons, chemical agents are filled into a weapon and stored, ready to be deployed. Since most agents are corrosive, however, long-term storage is challenging. To address this concern, the United States began stockpiling binary chemical weapons, which use a safer and more efficient method whereby chemicals are stored separately within the weapon and combine to form a toxic agent once the weapon is deployed.

Following America's upgrade, an arms race ensued whereby France, Britain, and the Soviet Union began stockpiling binary chemical weapons.[222]

Although several aspects of the history of America's biological and chemical weapons programs are now in the public domain, evidence suggests that much more information remains classified or buried in government archives. During the mid-1990s investigation regarding human radiation experiments, the presidential committee uncovered documents suggesting untoward conduct in chemical and biological weapons research. Since the scope of the committee was limited to radiation experiments, the committee did not pursue many of the leads.[223] According to one committee member, bioethicist Jonathan Moreno, a senior Army physician interviewed during the radiation experiments investigation remarked: "If you ask me, you won't stop here. I think you go on to biological and chemical (weapons). That's where the real action is."[224] To this day, no such investigation has been conducted.

Summary

New scientists, new enemies, and post-WWII confidence in the American way of life coalesced to cultivate a massive expansion of America's biological and chemical warfare programs. Universities dedicated faculty and laboratories to further research and development, while the engine of industry hummed to the tune of military preparedness. Military and intelligence agencies sometimes paused to consider legal and ethical considerations, though the concerns were typically sidestepped in a race to generate innovative solutions to Cold War problems. Theodor Rosebury, a biological warfare scientist who worked at Fort Detrick, aptly captured the predicament: "We were fighting a fire, and it seemed necessary to risk getting dirty as well as burnt."[225]

To propel advancements in military science, the national security establishment coopted thousands of service members and civilians into secret experiments and honed offensive and defensive methods of biological and chemical warfare by subjecting millions of Americans to clandestine field tests. The government worked diligently to hide the scope of its actions and deter legal and moral accountability, casting individual liberties aside in the quest for military and geopolitical superiority. In instances of research-related harms, secrecy classifications and legal doctrines shielded wrongful actions and created insurmountable obstacles for those injured by the experiments. Contemporary observers may find it difficult to comprehend the ethical and legal calculus of America's scientific and military elite, yet many of the research protocols were typical for the era and were not considered morally troubling. It was an era of unbridled proliferation for biological and chemical weapons.

Notes

1. George W. Merck, "Biological Warfare: Report to the Secretary of War" (January 3, 1946): 6.
2. Endicott and Hagerman, *The United States and Biological Warfare*, 43–50.

3. William M. Creasy, "Presentation to the Secretary of Defense's Ad Hoc Committee on CEBAR" (February 24, 1950): 21, quoted in Endicott and Hagerman, *The United States and Biological Warfare*, 67.
4. *Report of the Secretary of Defense's Ad Hoc Committee on Chemical, Biological and Radiological Warfare* (June 30, 1950), 9.
5. Endicott and Hagerman, *The United States and Biological Warfare*, 43–50, 64–65.
6. Rose C. Engelman, ed., *A Decade of Progress* (1971), 161–62; Harris and Paxman, *A Higher Form of Killing*, 101–2, 163–65.
7. Creasy, "Presentation to the Secretary of Defense's Ad Hoc Committee on CEBAR," quoted in Harris and Paxman, *A Higher Form of Killing*, 157–58.
8. Quoted in Michael C. Carroll, *Lab 257* (2005), 46.
9. Carroll, *Lab 257*, 3–11, 18–27, 46–52, 199–206.
10. Endicott and Hagerman, *The United States and Biological Warfare*, 64–72.
11. Jeanne Guillemin, *Biological Weapons* (2005), 85.
12. Quoted in Endicott and Hagerman, *The United States and Biological Warfare*, 104.
13. Endicott and Hagerman, *The United States and Biological Warfare*, 80–87.
14. Endicott and Hagerman, *The United States and Biological Warfare*, 143–54.
15. Harris and Paxman, *A Higher Form of Killing*, 165; Endicott and Hagerman, *The United States and Biological Warfare*, 1–20, 106, 156, 163.
16. *Report of the International Scientific Commission for the Investigation of the Facts Concerning Bacterial Warfare in Korea and China* (1952), 62.
17. *Report of the International Scientific Commission*, 26–46.
18. Harris and Paxman, *A Higher Form of Killing*, 165–66.
19. Quoted in Harris and Paxman, *A Higher Form of Killing*, 165–66.
20. *Armed Forces Doctrine for Chemical and Biological Weapons Employment and Defense* (April 1964), 3.
21. Department of the Army, *The Law of Land Warfare*, Field Manual 27-10 (July 1956), 18.
22. Wright, *Preventing a Biological*, 32.
23. Ken Alibek and Stephen Handelman, *Biohazard* (1999), ix, 8–9, 18–19, 38–44, 153–67.
24. Wright, *Preventing a Biological*, 33; William Blum, *Killing Hope* (2004), 184–93.
25. Moreno, *Undue Risk*, 258–62.
26. Guillemin, *Hidden Atrocities*, 289.
27. Smith, *American Biodefense*, 49–50.
28. Moreno, *Undue Risk*, 258–62.
29. Richard Nixon, "Presidential Statement" (November 25, 1969).
30. Guillemin, *Hidden Atrocities*, 325–27.
31. George W. Christopher et al., "Biological Warfare: A Historical Perspective," *Journal of the American Medical Association* 278, no. 5 (August 6, 1997): 412–17.
32. Convention on the Prohibition of the Development, Production, and Stockpiling of Bacteriological (Biological) and Toxin Weapons and on Their Destruction, art. 1 (1972).
33. Gerald R. Ford, Remarks Upon Signing Instruments of Ratification of the Geneva Protocol of 1925 and the Biological Weapons Convention (January 22, 1975).
34. *Intelligence Activities: Unauthorized Storage of Toxic Agents, Hearings before the Senate Select Committee to Study Governmental Operations with Respect to Intelligence Activities*, 94th Cong. 9–13 (September 16–18, 1975) (statement of William E. Colby).
35. Loch K. Johnson, *A Season of Inquiry Revisited* (2015), 45, 70–74.
36. *Intelligence Activities: Unauthorized Storage of Toxic Agents*, 94th Cong. 9–13 (statement of William E. Colby), 18–26 (testimony of William E. Colby).
37. *Intelligence Activities: Unauthorized Storage of Toxic Agents*, 94th Cong. 14 (testimony of William E. Colby).

38. After further questioning, Angleton agreed to withdraw the statement but refused to admit that he did not mean it. *Intelligence Activities: Huston Plan, Hearings before the Senate Select Committee to Study Governmental Operations with Respect to Intelligence Activities*, 94th Cong. 73 (September 23–25, 1975) (testimony of James Angleton).
39. *Intelligence Activities: Unauthorized Storage of Toxic Agents*, 94th Cong. 189–190 (Exhibit 1, Memorandum from Thomas Karamessines to Director of Central Intelligence, "Contingency Plan for Stockpile of Biological Warfare Agents" (February 16, 1970)).
40. Kathleen Sebelius, "Why We Still Need Smallpox," *The New York Times*, April 25, 2011.
41. Susan Wright, "Evolution of Biological Warfare Policy, 1945-1990," in Wright, *Preventing a Biological*, 48–55; Susan Wright and Stuart Ketcham, "The Problem of Interpreting the U.S. Biological Defense Research Program," in Wright, *Preventing a Biological*, 178–85.
42. Wright and Ketcham, "The Problem of Interpreting," 183.
43. Mike Allen, "Bush Defends Putin in Handling of Siege," *The Washington Post*, November 19, 2002.
44. Department of the Army, *US Army Activity in the U.S. Biological Warfare Programs*, vol. 1 (February 24, 1977), 2–4
45. United States Army Chemical Corps, *Summary of Major Events and Problems: Fiscal Year 1959* (January 1960), 101–4.
46. U.S. Army Chemical Corps, *Summary of Major Events*, 104.
47. U.S. Army Chemical Corps, *Summary of Major Events*, 103.
48. U.S. Army Chemical Corps, *Summary of Major Events*, 103.
49. U.S. Army Chemical Corps, *Summary of Major Events*; Harris and Paxman, *A Higher Form of Killing*, 168–70.
50. *Biological Testing Involving Human Subjects by the Department of Defense, 1977: Hearings before the Subcommittee on Health and Scientific Research of the Senate Committee on Human Resources*, 95th Cong. 13 (March 8 and May 23, 1977) (testimony of Colonel George A. Carruth); "Ex-Pentagon Researcher Says the Army Waged Mock Attack on Nixon," *The New York Times*, March 11, 1977; Ken Ringle, "Army Sprayed Bacteria on Unsuspecting Travelers," *The Washington Post*, December 5, 1984; Harris and Paxman, *A Higher Form of Killing*, 157.
51. Department of the Army, Special Operations Division, *A Study of the Vulnerability of Subway Passengers in New York City to Covert Attack with Biological Agents* (January 1968).
52. Moreno, *Undue Risk*, 235; Barnaby, *The Plague Makers*, 94–95; Leonard A. Cole, *Clouds of Secrecy* (1988), 65–71.
53. Cole, *Clouds of Secrecy*, 6.
54. George C. Wilson, "Army Conducted 239 Secret, Open-Air Germ Warfare Tests," *The Washington Post*, March 9, 1977.
55. U.S. Army Chemical Corps, "Behavior of Aerosol Clouds Within Cities," *Joint Quarterly Report*, no. 3 (January–March 1953); U.S. Army Chemical Corps, "Behavior of Aerosol Clouds Within Cities," *Joint Quarterly Report*, no. 4 (April–June 1953).
56. National Research Council, *Toxicologic Assessment of the Army's Zinc Cadmium Sulfide Dispersion Tests* (1997), 1–2; Leonard A. Cole, *The Eleventh Plague* (1997), 18–28.
57. U.S. Army Chemical Corps, "Behavior of Aerosol Clouds," *Joint Quarterly Report*, no. 3; U.S. Army Chemical Corps, "Behavior of Aerosol Clouds," *Joint Quarterly Report*, no. 4.
58. U.S. Army Chemical Corps, "Behavior of Aerosol Clouds," *Joint Quarterly Report*, no. 4, 27.
59. U.S. Army Chemical Corps, "Behavior of Aerosol Clouds," *Joint Quarterly Report*, no. 3, 28–31.
60. Quoted in "'We Were Experimented On': Victims of Secret Cold War Testing in St. Louis Demand Compensation," *NBC News*, September 25, 2023.
61. NRC, *Toxicologic Assessment*, 1–2.

62. NRC, *Toxicologic Assessment*, 12.
63. National Research Council, *Toxicologic Assessment of the Army's Zinc Cadmium Sulfide Dispersion Tests: Answers to Commonly Asked Questions* (Washington, DC: National Academies Press, 2017), 10–11.
64. U.S. Army Environmental Hygiene Agency, "Assessment of Health Risk, Minneapolis, Minnesota," *Health Risk Assessment Study* (July 10, 1994): 1, quoted in Cole, *The Eleventh Plague*, 28.
65. L. Arthur Spomer, "Fluorescent Particle Atmospheric Tracer: Toxicity Hazard," *Atmospheric Environment* 7 (1973): 353.
66. *Biological Testing Involving Human Subjects*, 95th Cong. 18.
67. U.S. Chemical Corps Biological Laboratory, "Biological Warfare Trials at San Francisco, California, 20–27 September 1950" (January 22, 1951), quoted in Cole, *Clouds of Secrecy*, 78–81.
68. Department of the Army, *US Army Activity in the U.S. Biological Warfare Programs*, vol. 2 (February 24, 1977), appendix 2, annex E, p. 4.
69. Bill Richards, "Report Suggests CIA Involvement in Fla. Illnesses," *The Washington Post*, December 17, 1979; Cole, *Clouds of Secrecy*, 18.
70. Ed Regis, *The Biology of Doom* (1999), 3–6.
71. Harris and Paxman, *A Higher Form of Killing*, 158.
72. Thomas V. Inglesby et al., "Anthrax as a Biological Weapon: 2002 Updated Recommendations for Management," *Journal of the American Medical Association* 287, no. 17 (May 1, 2002): 2236–52.
73. U.S. Army, *Information Sheet* (January 12, 1977), quoted in Harris and Paxman, *A Higher Form of Killing*, 162.
74. *Biological Testing Involving Human Subjects*, 95th Cong. 270 (testimony of J. M. Joseph, Maryland State Department of Health and Mental Hygiene).
75. *Biological Testing Involving Human Subjects*, 95th Cong. 270 (testimony of George Connell).
76. *Biological Testing Involving Human Subjects*, 95th Cong. 18 (testimony of Colonel George A. Carruth).
77. U.S. Department of Defense, "Biological Defense Program," *Report to the House Committee on Appropriations* (May 1986), cited in Cole, *Clouds of Secrecy*, 3.
78. *Nevin v. United States*, 696 F.2d 1229, 1230 (9th Cir. 1983).
79. *Nevin*, 696 F.2d at 1230.
80. Oliver O'Connell, "New York to Test Readiness for Biological Attack by Deploying 'Safe Gas' in Subways and Parks," *The Independent*, October 12, 2021; Rhiannon Poolaw, "Residents are Concerned Over Biological Weapons Testing in OK," *KSWO-TV7 News Southwest Oklahoma*, November 10, 2017; "Harmless Gas Released into New York Subway to Prep for Biological Attack," *The Guardian*, May 9, 2016; "Scientists Simulate Boston Subway Terror Attack," *CBS News*, August 20, 2010.
81. Annie Jacobsen, *Phenomena* (2017).
82. Hunt, *Secret Agenda*, 1–5; Waller, *Wild Bill Donovan*, 101–3.
83. Marks, *The Search*, 125–46.
84. Central Intelligence Agency, " 'Truth' Drugs in Interrogation," *Report* (Spring 1961).
85. *Project MKULTRA, The CIA's Program of Research in Behavioral Modification: Joint Hearing before the Select Committee on Intelligence and the Subcommittee on Health and Scientific Research of the Committee on Human Resources*, 95th Cong. 74 (August 3, 1977) (Memorandum for the Deputy Director of Central Intelligence from the Deputy Director for Plans, December 17, 1963, 2–3).
86. *Project MKULTRA*, 95th Cong. 74 (Inspector General Survey of TSD, 1957, 217).

87. *Project MKULTRA*, 95th Cong. 67.
88. Ohler, *Blitzed*, 209–11.
89. *Project MKULTRA*, 95th Cong. 67–68; Jeffrey T. Richelson, *The Wizards of Langley* (2001), 9–11.
90. *Project MKULTRA*, 95th Cong. 68–69.
91. *Project MKULTRA*, 95th Cong. 68–69.
92. *Project MKULTRA*, 95th Cong. 101 (ARTICHOKE Memorandum, June 13, 1952).
93. *Project MKULTRA*, 95th Cong. 171.
94. *Project MKULTRA*, 95th Cong. 171
95. *Project MKULTRA*, 95th Cong. 4 (statement of Admiral Stansfield Turner, Director of Central Intelligence), 69–72, 171.
96. *Project MKULTRA*, 95th Cong. 4 (statement of Admiral Stansfield Turner, Director of Central Intelligence), 69–72.
97. *Project MKULTRA*, 95th Cong. 69–72.
98. Weiner, *Legacy of Ashes*, 64–66.
99. *Project MKULTRA*, 95th Cong. 169.
100. *Project MKULTRA*, 95th Cong. 169.
101. *Project MKULTRA*, 95th Cong. 2–4, 83–89.
102. *Project MKULTRA*, 95th Cong. 2–4, 4–8 (statement of Admiral Stansfield Turner, Director of Central Intelligence), 14.
103. *Project MKULTRA*, 95th Cong. 2–4, 4–8 (statement of Admiral Stansfield Turner, Director of Central Intelligence); *MKULTRA Briefing Book* (January 1, 1976), in Stephen Foster, ed., *The Project MKULTRA Compendium* (2009), 369, 372, 379, 403, 425, 430–31, 433–34, 459, 461, 464, 469–70, 478, 481, 484, 504, 507–8, 524–27, 529, 533, 536–46, 553, 558–59, 571, 575, 584–85, 592.
104. *Project MKULTRA*, 95th Cong. 4–8 (statement of Admiral Stansfield Turner, Director of Central Intelligence), 35–38.
105. *Human Drug Testing by the CIA, 1977: Hearings before the Subcommittee on Health and Scientific Research of the Committee on Human Resources*, 95th Cong. 172 (September 20–21, 1977) (statement of Sidney Gottlieb).
106. *Project MKULTRA*, 95th Cong. 4–8 (statement of Admiral Stansfield Turner, Director of Central Intelligence).
107. *MKULTRA Briefing Book*, 505–6; David Remnick, "25 Years of Nightmares," *The Washington Post*, July 27, 1985.
108. *Project MKULTRA*, 95th Cong. 23, 32 (reprint of Central Intelligence Agency, "'Truth' Drugs in Interrogation"), 69–72.
109. R. A. Sandison, A. M. Spencer, and J. D. A. Whitelaw, "The Therapeutic Value of Lysergic Acid Diethylamide in Mental Illness," *Journal of Mental Science* 100 (April 1954): 491–507.
110. C. Elkes, J. Elkes, and W. Mayer-Gross, "Hallucinogenic Drugs," *The Lancet* 265 (April 2, 1955): 719.
111. *Project MKULTRA*, 95th Cong. 4–8 (statement of Admiral Stansfield Turner, Director of Central Intelligence).
112. Marks, *The Search*, 201–2.
113. *Project MKULTRA*, 95th Cong. 22.
114. *Barrett v. United States*, 660 F. Supp. 1291 (S.D.N.Y. 1987); Moreno, *Undue Risk*, 194–98.
115. *Barrett*, 660 F. Supp. at 1295 (quoting the contract between the Army and the Psychiatric Institute).
116. *Barrett*, 660 F. Supp. at 1295–96.
117. *Project MKULTRA*, 95th Cong. 65.

118. *Barrett*, 660 F. Supp. at 1295–1300.
119. *Barrett*, 660 F. Supp. at 1298–1300.
120. *Barrett*, 660 F. Supp. at 1302–3 (quoting Army documents).
121. *Barrett*, 660 F. Supp. at 1300–1306.
122. *Report to the President by the Commission on CIA Activities Within the United States* (1975), 226–29.
123. Joseph B. Treaster, "Army Discloses Man Died in Drug Test It Sponsored," *The New York Times*, August 13, 1975.
124. *Barrett*, 660 F. Supp. at 1291–95; Joseph B. Treaster, "$8.5 Million Sought from Army in 1953 Drug Death," *The New York Times*, September 4, 1975.
125. *Barrett*, 660 F. Supp. at 1313–34.
126. *Barrett*, 660 F. Supp. at 1314.
127. *Barrett*, 660 F. Supp. at 1315–23.
128. *Project MKULTRA*, 95th Cong. 38–41, 128–34; *Human Drug Testing by the CIA*, 95th Cong. 84–86.
129. *Project MKULTRA*, 95th Cong. 38–41, 128–34; *Human Drug Testing by the CIA*, 95th Cong. 84–86.
130. *Human Drug Testing by the CIA*, 95th Cong. 94–99.
131. *Project MKULTRA*, 95th Cong. 38–41, 128–34; *Human Drug Testing by the CIA*, 95th Cong. 84–86.
132. *Project MKULTRA*, 95th Cong. 38–41, 128–34; *Human Drug Testing by the CIA*, 95th Cong. 84–86.
133. Quoted in Hornblum et al., *Against Their Will*, 74.
134. *Project MKULTRA*, 95th Cong. 45–48; *Human Drug Testing by the CIA*, 95th Cong. 181–85.
135. Quoted in Weinberger, *The Imagineers of War*, 223.
136. *Project MKULTRA*, 95th Cong. 74–83.
137. Michael Ignatieff, "What Did the CIA Do to His Father?" *The New York Times Magazine*, April 1, 2001; Jacobsen, *Operation Paperclip*, 366–72.
138. *Project MKULTRA*, 95th Cong. 74–83.
139. Ignatieff, "What Did the CIA Do."
140. *Project MKULTRA*, 95th Cong. 74–83.
141. *Project MKULTRA*, 95th Cong. 74–83.
142. Ignatieff, "What Did the CIA Do"; Ted Gup, "The Coldest," *The Washington Post*, December 16, 2001.
143. Quoted in Ignatieff, "What Did the CIA Do."
144. *Project MKULTRA*, 95th Cong. 74–83.
145. Jacobsen, *Operation Paperclip*, 366–72.
146. *Project MKULTRA*, 95th Cong. 74–83.
147. *Project MKULTRA*, 95th Cong. 74–83.
148. *Report to the President*, 226–29; Ignatieff, "What Did the CIA Do."
149. Marks, *The Search*, 85.
150. *Wormwood*, Documentary Mini-Series (Netflix, 2017); Ignatieff, "What Did the CIA Do."
151. Ignatieff, "What Did the CIA Do."
152. Quoted in Ignatieff, "What Did the CIA Do."
153. *Wormwood*, Documentary.
154. CIA, "'Truth' Drugs in Interrogation."
155. CIA, "'Truth' Drugs in Interrogation."
156. National Security Act of 1947, Pub. L. No. 253, §102(d)(5), 61 Stat. 496, 498 (1947). In 2004, pursuant to the intelligence reform measures enacted after 9/11, the term "National

Security Council" in the 1947 National Security Act was replaced with "President or the Director of National Intelligence." Intelligence Reform and Terrorism Prevention Act of 2004, Pub. L. No. 108-458, §104A(d)(4), 118 Stat. 3638, 3661 (2004).

157. Victor Marchetti and John D. Marks, *The CIA and the Cult of Intelligence* (1974), 22.
158. Central Intelligence Act of 1949, Pub. L. No. 110, § 7, 63 Stat. 208, 211–12 (1949).
159. Anthony F. Czajkowski, "Techniques of Domestic Intelligence Collection," *Studies in Intelligence* 3, no. 1 (Winter 1959), 69–83, reprinted in H. Bradford Westerfield, ed., *Inside CIA's Private World* (1995), 51–62.
160. Harry S. Truman, "Limit CIA Role to Intelligence," *The Washington Post*, December 22, 1963.
161. *Project MKULTRA*, 95th Cong. 66–67.
162. *Project MKULTRA*, 95th Cong. 100 (Memorandum from the DDP to the DCI, November 9, 1964, 2).
163. *Project MKULTRA*, 95th Cong. 91–92.
164. *Project MKULTRA*, 95th Cong. 91–92; *Thornwell v. United States*, 471 F. Supp. 344 (D.D.C. 1979).
165. *Project MKULTRA*, 95th Cong. 91–92.
166. Harris and Paxman, *A Higher Form of Killing*, 192–93.
167. *Project MKULTRA*, 95th Cong. 96–97 (USAINTC Staff Study, "Material Testing Program EA1729," October 15, 1959).
168. *Stanley v. United States*, 549 F. Supp. 327 (S.D. Fla. 1982).
169. Quoted in "Ruling Reopens Wound for Bitter Ex-Soldier," *The New York Times*, June 30, 1987.
170. *United States v. Stanley*, 483 U.S. 669 (1987).
171. *Stanley*, 483 U.S. at 671–72.
172. Quoted in William E. Gibson, "LSD Case Provides Acid Test on Whether GIs Can Sue Army," *South Florida Sun Sentinel*, April 22, 1987.
173. *Stanley v. United States*, 549 F. Supp. at 332; *Stanley v. United States*, 786 F.2d 1490, 1492 n.1 (11th Cir. 1986).
174. *Stanley*, 549 F. Supp. at 331.
175. *Stanley v. United States*, 549 F. Supp. 474, 482 (S.D. Fla. 1983).
176. *Stanley*, 786 F.2d at 1496 (internal citations omitted).
177. *Stanley*, 786 F.2d at 1499.
178. *Stanley*, 483 U.S. at 683.
179. *Stanley*, 483 U.S. at 683.
180. *Stanley*, 483 U.S. at 688–90 (Brennan, J., concurring in part and dissenting in part); U.S. Army Medical Department, *LSD Follow-Up Study* (October 1980), 1.
181. *Project MKULTRA*, 95th Cong. 96–98.
182. Quoted in David S. Martin, "Vets Feel Abandoned After Secret Drug Experiments," *CNN*, March 1, 2012.
183. Quoted in Martin, "Vets Feel Abandoned."
184. Martin, "Vets Feel Abandoned."
185. U.S. Army, *LSD Follow-Up Study*, 11.
186. Hunt, *Secret Agenda*, 236.
187. Quoted in Hunt, *Secret Agenda*, 236.
188. U.S. Army, *LSD Follow-Up Study*, 41–61.
189. U.S. Army, *LSD Follow-Up Study*, Executive Summary, 2.
190. Quoted in Raffi Khatchadourian, "Operation Delirium," *The New Yorker*, December 17, 2012, 46–64.

191. Moreno, *Undue Risk*, 242–43; Jonathan D. Moreno, "Reassessing the Influence of the Nuremberg Code on American Medical Ethics," *Journal of Contemporary Health Law & Policy* 13, no. 2 (1997): 347–60.
192. Schmidt, *Secret Science*, 138–39; Guillemin, *Hidden Atrocities*, 71.
193. Harris and Paxman, *A Higher Form of Killing*, 198–99; Wright, "Evolution of Biological Warfare Policy," 48.
194. Hornblum, *Acres of Skin*, 138–45.
195. McCamley, *Secret History*, 135; Hersh, *Chemical and Biological Warfare*, 45.
196. Tucker, *War of Nerves*, 203–12; Harris and Paxman, *A Higher Form of Killing*, 219–21.
197. Thom Shanker and William J. Broad, "Sailors Sprayed with Nerve Gas in Cold War Test, Pentagon Says," *The New York Times*, May 24, 2002; Institute of Medicine, *Long-Term Health Effects of Participation in Project SHAD (Shipboard Hazard and Defense)* (2007), 1–2; Institute of Medicine, *Assessing Health Outcomes Among Veterans of Project SHAD (Shipboard Hazard and Defense)* (2016), S-1.
198. IOM, *Assessing Health Outcomes*, S-1.
199. *Environmental Dangers of Open-Air Testing of Lethal Chemicals: Hearings before a Subcommittee of the House Committee on Government Operations*, 91st Cong. 2–5, 225–28 (May 20–21, 1969).
200. Tucker, *War of Nerves*, 108–11, 147–51, 164–65.
201. Harris and Paxman, *A Higher Form of Killing*, 101, 194.
202. *In re Agent Orange Product Liability Litigation*, 373 F. Supp. 2d 7, 20–21 (E.D.N.Y. 2005); Weinberger, *The Imagineers of War*, 125–28.
203. *In re Agent Orange*, 373 F. Supp. 2d at 20–21.
204. *In re Agent Orange*, 373 F. Supp. 2d at 30; Harris and Paxman, *A Higher Form of Killing*, 194.
205. Wright, "Evolution of Biological Warfare Policy," 35.
206. Quoted in Hersh, *Chemical and Biological Warfare*, 148.
207. Quoted in Weinberger, *The Imagineers of War*, 132.
208. Letter from Dixon Donelly, Assistant Secretary, Department of Defense (September 1966), quoted in Harris and Paxman, *A Higher Form of Killing*, 195.
209. Dean Rusk, Secretary of State, Memorandum to President Kennedy, "Defoliant Operations in VietNam" (November 24, 1961).
210. Henry Eschwege, Director U.S. General Accounting Office, Letter to The Honorable Ralph H. Metcalfe, House of Representatives (August 16, 1978); *Agent Orange: What Efforts are Being Made to Address the Continuing Impact of Dioxin in Vietnam?: Hearing before the Subcommittee on Asia, the Pacific and the Global Environment of the House Committee on Foreign Affairs*, 111th Cong. 36–37 (June 4, 2009) (statement of Professor Vo Quy, Vietnam National University, Member US-Vietnam Group on Agent Orange/Dioxin).
211. *In re Agent Orange*, 373 F. Supp. 2d at 19 (citing academic studies).
212. Harris and Paxman, *A Higher Form of Killing*, 195–99.
213. *Agent Orange*, 111th Cong. 36–37 (statement of Professor Vo Quy).
214. *In re Agent Orange Product Liability Litigation*, 373 F. Supp. 2d at 32–35; Harris and Paxman, *A Higher Form of Killing*, 195–99.
215. Eschwege, Letter to Metcalfe; Department of the Air Force, *Project CHECO Report: Base Defense in Thailand* (February 18, 1973); Hersh, *Chemical and Biological Warfare*, 148–49.
216. Gordon Burck, "Biological, Chemical and Toxin Warfare Agents," in Wright, *Preventing a Biological*, 360.

217. Harris and Paxman, *A Higher Form of Killing*, 198–99; Hersh, *Chemical and Biological Warfare*, 170.
218. Alje Vennema, Letter to Dr. E. W. Pfeiffer, University of Montana (November 23, 1967), quoted in Hersh, *Chemical and Biological Warfare*, 183–84.
219. "Gas (Nonlethal) in Vietnam," *The New York Times*, March 24, 1965.
220. Secretary of State, Memorandum for the President, "The Geneva Protocol" (February 11, 1971); John W. Finney, "2 Experts Back Ban on Tear Gas and Sprays in War," *The New York Times*, March 27, 1971; *The Chemical Weapons Convention: Senate Report*, Exec. Report 104–33, 104th Cong. 12 (September 11, 1996).
221. U.S. Department of State, Bureau of International Security and Nonproliferation, *Narrative: Geneva Protocol* (September 25, 2002).
222. Kelle, Nixdorff, and Dando, *Preventing a Biochemical*, 18–19.
223. Moreno, *Undue Risk*, x–xvii.
224. Quoted in Moreno, *Undue Risk*, xiii.
225. Theodor Rosebury, "Medical Ethics and Biological Warfare," *Perspectives in Biology and Medicine* 6 (Summer 1963): 514–15, quoted in Endicott and Hagerman, *The United States and Biological Warfare*, 34.

12
A Global Military Biomedical Establishment

Introduction

America's expansive post-World War II (WWII) programs in atomic, biological, and chemical weapons comprised just a portion of the military biomedical complex: healthcare for military personnel remained the core mission. Resources dedicated to force health protection broadened during the Cold War, particularly as military operations extended to encompass the Korean War, personnel stationed at NATO bases in Europe, and the Vietnam War. A doctor draft to help fill the ranks of military physicians was created in 1950 and remained active for more than two decades. In addition to assisting with wartime health needs, thousands of physicians fulfilled their draft duties in research-related posts, some of which were with the National Institutes of Health (NIH). By the late 1980s, the military and the Veterans Administration (VA) each maintained robust healthcare enterprises that conducted cutting-edge research and provided world-class medical care in facilities around the globe.

Coupled with its overt military engagements during the Cold War, the United States increasingly relied on covert operations to pursue political, economic, and military goals. The Department of Defense (DoD) created a special forces command for elite warfighters and collaborated with the Central Intelligence Agency (CIA) on a plethora of clandestine ventures. The covert operations are now known to have included secret missions in more than forty countries, some of which involved psychological warfare and assassination attempts with poisonous substances. Contemporaneously, weapons of mass destruction proliferated across the world, and dozens of countries maintained or were developing atomic, biological, or chemical weapons. To combat the threats, the military accelerated its efforts to create medical countermeasures that could be integrated into health preparedness plans.

Modernization of Military Healthcare

As with postwar periods of the past, in the years immediately following WWII, demobilization and the massive downsizing of the armed forces gutted the military's healthcare infrastructure. Funding for military health diminished, medical divisions shrank or were eliminated, hospitals were decommissioned, and medical personnel transitioned from military to civilian jobs.[1] Coupled with the rollback of the healthcare infrastructure for active-duty military personnel, addressing the healthcare needs of more than 15 million WWII veterans was immensely challenging due to personnel and resource shortages. Countless veterans languished without adequate care.[2] To help WWII veterans heal their war wounds and assimilate back into civilian life, the government built new hospitals, expanded access to healthcare for veterans,

America's Military Biomedical Complex. Efthimios Parasidis, Oxford University Press.
 DOI: 10.1093/9780199351473.003.0013

and enacted the GI Bill, which afforded veterans unemployment compensation, low-cost mortgages, college tuition subsidies, and other benefits.

The military downsizing was short-lived. Cold War proxy battles demanded more service members and an American military presence worldwide, and the DoD scrambled to recruit medical personnel and reassemble adequate healthcare facilities. This was particularly challenging during the Korean War (1950–1953), during which 1.7 million Americans served. To compensate for the dearth of military physicians, Congress created a doctor draft in 1950, and by 1952, nearly 90 percent of American physicians in Korea were draftees. Few had training in combat medicine. As military historians John Greenwood and Clifton Berry explain: "More often than not, pediatricians, gynecologists, and even dermatologists became surgeons once they reported to their units . . . debriding frostbitten tissue, amputating shattered limbs, suturing lacerated kidneys and perforated intestines, and extracting shrapnel and bullets from every part of the human body."[3] The Doctor Draft remained active throughout the Korean War and during the Vietnam War, until an all-volunteer force was established in 1973.[4]

Guerilla warfare and the rugged Korean terrain complicated the provision of medical care in the field. To help address the challenges, the military relied heavily on medics and mobile army surgical hospitals (MASHs). A typical MASH, as seen in Figure 12.1, was a sixty-bed, tented structure, though some were two-hundred-bed semipermanent facilities. Modeled after portable medical clinics used during WWII, MASHs handled seriously wounded service members, performed life-saving surgeries, and offered care to patients who could not be transported. A MASH routinely handled two hundred to four hundred patients per day, and some units moved up to fifty times. In addition to MASHs, Navy hospital ships and air evacuations were instrumental elements of the wartime medical infrastructure. Diseases such as smallpox, typhus, cholera, malaria, tuberculosis, and Japanese B encephalitis were endemic, and poor sanitation and unclean water led to dysentery and other gastrointestinal ailments. Notwithstanding the challenges, due to improved medical care and a robust healthcare infrastructure, the number of casualties due to disease was far less than in previous conflicts.[5]

Coupled with America's substantial military commitment in Korea, the creation of the North Atlantic Treaty Organization (NATO) in 1949 brought substantial US military obligations in Europe. NATO was formed as a political and military alliance to deter Soviet expansion. General Dwight Eisenhower was named NATO's first Supreme Allied Commander, and the United States constructed and staffed military bases in several NATO countries. By 1961—the year the Berlin Wall was built—more than 228,000 American service members were stationed in Europe.[6] For decades, America's foreign policy was guided by the twin goals of rolling back Soviet territorial gains and containing Soviet expansion and influence.[7]

During the 1950s and 1960s, Congress expanded military health benefits to include dependents of military personnel. Previously, dependents were provided care ad hoc and only if resources were available.[8] The patient population of the military health care system ballooned; in the European efforts alone, the Army Medical Department was responsible for the healthcare needs of hundreds of thousands of soldiers and their dependents.[9]

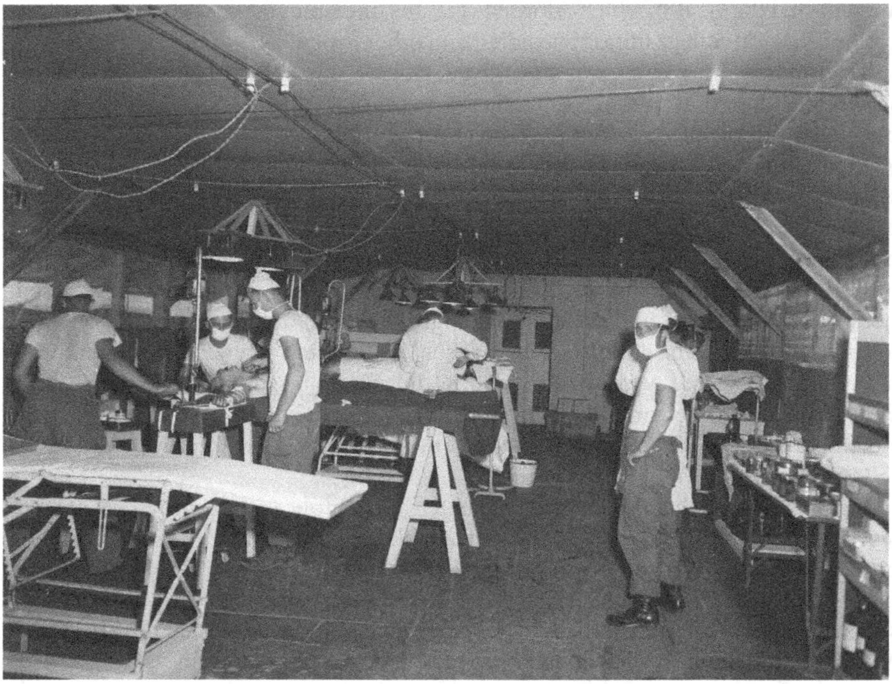

Figure 12.1 A MASH operating room during the Korean War.
Photograph from the Otis Historical Archives, National Museum of Health and Medicine.

The military health system was further strained by the Vietnam War (1955–1975); more than 9 million Americans served during the war, 2.7 million in the war theater, with a peak field strength of 540,000 in 1968. Medical facilities in Vietnam included hospital ships, MASHs, and medical unit, self-contained transportables (MUSTs). MUSTs were modular, were packaged in shipping containers, and could be quickly converted into functioning field hospitals.[10] Although MUSTs were more difficult to assemble than MASHs, they provided a cleaner and more comfortable environment that included air conditioning and a reliable source of electricity for operating rooms, X-rays, and other healthcare needs.[11]

During the 1960s, mandatory inoculations for service members included vaccines to protect against smallpox, polio, yellow fever, typhus, cholera, tetanus, and diphtheria, though some of the mandates applied only to personnel stationed in certain regions. Although mosquito control and other preventive health measures were utilized, service members commonly suffered from malaria and other insect-borne diseases. As many battles occurred in dense jungles and away from field hospitals, medics typically administered front-line emergency care, and medevac helicopters were integral to transporting injured warfighters. Hospital mortality was cut in half from WWII to the Vietnam War, from 4.5 percent to 2.2 percent. Of wounded warriors requiring hospitalization, a record 87 percent returned to duty.[12]

Notwithstanding the wartime health achievements, the signature ailments of the Vietnam War would not be widely recognized until after the war: combat-related posttraumatic stress disorder (PTSD) and toxic side effects from Agent Orange and other chemical defoliants.[13] The understanding that combat causes significant mental stress dates back at least to the American Civil War, but PTSD was not officially recognized in the *Diagnostic and Statistical Manual of Mental Disorders* until 1980, five years after the Vietnam War ended.[14] Nonetheless, studies estimate that at least 30 percent of Vietnam War veterans experienced PTSD and that more than 10 percent were suffering from PTSD forty years after the war.[15] Some researchers dispute the percentages, but practically all concur on PTSD-related health, familial, and societal harms: these include substance abuse, addiction, chronic pain, anxiety, depression, and heart disease, among others.[16] Approximately one in five Vietnam War veterans became addicted to heroin, most of whom received little medical care upon their return to the United States. Thousands were dishonorably discharged for failing a drug test. In 1971, in New York City alone, an estimated thirty thousand to forty-five thousand veterans were addicted to heroin; many were homeless.[17]

Widespread deployment of Agent Orange and other chemical agents caused innumerable health concerns such as cancer, skin and respiratory ailments, and congenital disabilities in children born to individuals exposed to the chemicals. Countless veterans were denied disability benefits because they could not prove—and the government and chemical companies would not admit—that the chemicals caused adverse health conditions. Over time, however, a series of lawsuits and congressional investigations revealed that both industry and the government were aware of the dangers of the deployed chemicals. Epidemiological research of Vietnam War veterans confirmed a link.[18]

Thereafter, the government created a program to provide benefits and compensation for injured veterans.[19] In addition, Monsanto and other chemical companies settled a class action lawsuit brought by veterans and their families: a 100 percent disabled Vietnam War veteran was eligible for approximately $100 per month for ten years, whereas the spouse of a veteran who died from exposure to Agent Orange was eligible for up to $3,400.[20] Of eligible veterans, the average compensation was $5,700.[21] The settlement amounts provided little relief to ailing veterans and their families, and if a veteran accepted the payout, they risked hindering their ability to collect other benefits or public assistance.[22] It was not until 2010 that the VA acknowledged that Agent Orange was a cause for several ailments; the total cost to process 230,000 related claims exceeded $3.6 billion.[23]

By the late 1960s, a majority of Americans opposed the Vietnam War and supported an end to the draft and a withdrawal of military personnel from Vietnam. Public opinion of the US government and military was largely negative, and many Americans abhorred the military's destructive tactics, such as the widespread use of tear gas and chemical defoliants. Within this social turmoil, it was difficult for the armed forces to acquire, train, and retain adequate numbers of medical professionals. To encourage accession, the military offered tuition grants to prospective medical students if they agreed to a reserve commission after completing their studies. In addition, several military facilities became teaching hospitals for residents and specialized medical centers. The military also provided opportunities for postgraduate education and continuing education in medicine, dentistry, and nursing.[24] These elements helped facilitate the training and retention of health professionals, and placed military medical centers on par with leading civilian centers.

Furthermore, under the mid-century doctor draft, approximately three thousand physician-draftees fulfilled their draft duties by working in NIH research posts under the Commissioned Corps of the Public Health Service. Within military medical circles, resentment grew toward the so-called Yellow Berets, a moniker that implied cowardice since the men escaped field service in Vietnam and worked at the NIH. After their service, many Yellow Berets obtained prestigious appointments at medical schools and research institutes, and ten won Nobel Prizes.[25] One physician-scientist who fulfilled his draft duties in the program was Anthony Fauci, who became a household name due to his leadership role at the NIH during the coronavirus pandemic that began in 2020.

Another significant step was taken in 1972 when Congress chartered the Uniformed Services University (USU), a medical school dedicated to uniformed physicians. Students receive a free education in exchange for a service commitment. Over the years, USU has ensured a steady stream of health professionals trained with an understanding of specialized medical needs facing service members. A core curricular philosophy of USU is eliminating compartmentalized training for Army, Navy, and Air Force physicians. While historically each was treated separately, USU faculty combined the fields, ran joint exercises, and created an ethos of military medicine as a uniform discipline. The first class of USU physicians graduated in 1980.[26] Today, USU maintains separate colleges for medicine, nursing, dentistry, and allied health sciences, as well as twenty specialized research centers that focus on various areas such as biotechnology, traumatic stress, and military precision health.

Coupled with research centers established under the auspices of USU, the military created an expansive network of medical research facilities encompassing no less than fifteen hospitals and fourteen research institutes. Two notable biomedical research institutes established during this era were the Walter Reed Army Institute of Research and the US Army Medical Research Institute for Infectious Diseases (USAMRIID), located at Silver Spring and Fort Detrick, Maryland, respectively. For decades, these cutting-edge centers have engaged in a wide array of biomedical research, including trauma, surgery, psychiatry, preventive medicine, and infectious diseases. As detailed in chapter 11, by the early 1970s, USAMRIID had become the centerpiece of America's biosecurity program.

A component of USAMRIID's infectious disease research has overlapped with biological warfare studies, including work to weaponize pathogens and develop medical countermeasures. The work has occasionally involved field testing of experimental vaccines. For example, USAMRIID scientists were developing a vaccine to protect against Rift Valley Fever. This virus primarily impacts animals, but also can be transmitted to humans via an infected animal or blood-feeding insect. Rift Valley Fever can cause temporary or permanent blindness, hallucinations, convulsions, hemorrhagic fever, and death. The disease is prevalent in sub-Saharan Africa, and several nations have attempted to weaponize the virus. In 1977, when a natural outbreak of Rift Valley Fever arose in Egypt, the Army decided to field test an experimental vaccine on hundreds of American service members stationed in Cairo to examine the extent to which the vaccine prevented infection and whether there were significant side effects. The vaccine had undergone human testing, mainly on Project Whitecoat volunteers, but it was not FDA-approved.[27] Although no significant adverse effects were reported, the data were inconclusive on whether the vaccine was effective.[28] As of early 2024, there is still no vaccine licensed for human use to protect against the virus.

The 1977 field testing of the Rift Valley Fever vaccine exemplifies the challenging calculus faced by military officials when there is no FDA-approved countermeasure to combat a health risk and clinical trials are difficult to structure because it would be unethical to expose individuals to dangerous pathogens to test countermeasure efficacy. Field testing under such circumstances might produce a health benefit if the countermeasure ends up working as anticipated. Even if the countermeasure is not effective, the field test can inform future research because data on safety and effectiveness can be collected during the test. In these circumstances, decision makers must balance uncertainties regarding the safety of an experimental vaccine against potential benefits for force health protection and the military mission. The military's benefit-risk calculus to administer an experimental vaccine may be different in combat situations, where the vaccine might provide an operational benefit and vaccine adverse events are viewed as one of many risks faced by warfighters during battle.

To be sure, the administration of experimental vaccines also raises ethical concerns that go beyond operational considerations. When a person joins the military, they expect to be exposed to combat risks and are informed that they must follow the chain of command, including commands to adhere to health-readiness protocols. At the same time, a reasonable person could presume that military health protocols will follow medical best practices and will only include FDA-approved medical products.

In other words, the risks one assumes by joining the military do not necessarily include exposure to health risks from compelled administration of unlicensed medical products. Compelled administration of experimental medical products blurs the line between medical research and the practice of medicine, each of which is guided by separate laws and norms. A divergence in protections between military and civilian populations—whereby actions that are permissible if conducted on military personnel but would be illegal if conducted on civilians—raises significant ethical questions regarding military exceptionalism and the scope of bodily autonomy, individual liberties, and informed consent. As will be discussed in chapters 13 and 14, these concerns accelerated during the 1991 Persian Gulf War and after the 9/11 attacks.

The Rise of Covert Operations and Special Forces

The role of covert action in American foreign policy broadened during the Cold War. Many of the thousands of Office of Special Services (OSS) agents from WWII went on to work for the CIA. Four OSS veterans—Allen Dulles, Richard Helms, William Colby, and William Casey—would become CIA directors, and the Wild West spirit of the OSS heavily influenced the CIA's activities, particularly covert action. Covert action involves activities that seek to influence political, economic, or military conditions, where it is intended that the role of a nation will not be apparent or publicly acknowledged, and will be plausibly deniable. Covert action represents an alternative to overt war and diplomacy, though sometimes covert actions are conducted in conjunction with diplomacy or overt war. Between 1961 and 1974, the CIA engaged in thousands of covert actions. Examples include paramilitary activities, election interference, assassination attempts, clandestine assistance to politicians and warfighters, propaganda, sabotage, and other activities to influence a nation's politics, society, or economy.[29] At a tactical level, the Cold War secrecy rationale supporting covert action paralleled that guiding military science: national security and the success of a military mission was the ultimate goal, even if it entailed violations of law or questionable ethical conduct.

As with several areas of military science that were once secret but later revealed to the public, dozens of covert operations eventually entered the public domain. It is now known that, during the Cold War, the United States engaged in covert operations in Afghanistan, Albania, Algeria, Angola, Argentina, Australia, Bolivia, Brazil, British Guiana, Bulgaria, Cambodia, Chad, Chile, China, Congo, Costa Rica, Cuba, the Dominican Republic, Ecuador, El Salvador, France, Germany (East and West), Ghana, Greece, Grenada, Guatemala, Haiti, Indonesia, Iran, Iraq, Italy, Jamaica, Laos, Libya, Morocco, Nicaragua, Panama, Peru, the Philippines, Seychelles, Suriname, Syria, Uruguay, the USSR, Vietnam, and Zaire. In many cases, the United States sought to influence elections to suppress public support for candidates who maintained socialist policies or rejected continuing colonial profiteering; in others, the United States provided assistance to right-wing military coups to overthrow democratically elected leaders. Some biological and chemical weapons were developed to assist in covert operations, including assassinations.[30]

One top CIA target was Cuban leader Fidel Castro, whom the CIA tried to kill several times, sometimes by spiking his milkshakes with various poisons, including botulinum toxin, other times by infecting him with tuberculosis or placing a toxic fungus inside his diving suit.[31] Another target was Patrice Lumumba, the first elected prime minister of Congo. Lumumba was an African nationalist who supported Congolese independence from Belgium and sought to create a self-sustaining nation that focused on state-led economic development and neutrality in foreign affairs.[32] Notwithstanding the extensive use of force by the Belgians in their efforts to enrich themselves while subjugating native Congolese, Lumumba remarked: "We have chosen just one weapon for our struggle, and that weapon is non-violence."[33]

In his drive to free Congo from colonial rule and exploitation of its natural resources, Lumumba petitioned for assistance from the United States and the United Nations, but both refused. The United States supported Belgium's attempts to undermine Lumumba's reforms, and Lumumba turned to the Soviet Union for help. Thereafter, President Eisenhower approved a plot to assassinate Lumumba: the CIA sent a diplomatic pouch to Congo that contained rubber gloves, masks, a hypodermic syringe, and a lethal biological agent concealed within a tube of toothpaste. Before the substance could be utilized, Belgium-supported secessionists staged a military coup, captured Lumumba, and executed him. His body was dissolved in sulfuric acid and Belgian police confiscated his skeletal remains. Some reports indicate that the CIA assisted in apprehending Lumumba.[34] In 2002, Belgium formally apologized for its role in Lumumba's assassination.[35] Twenty years later, Belgian authorities returned Lumumba's sole remains—a gold-capped tooth—to his family in Congo.[36]

Although an analysis of all of America's publicly known Cold War covert actions is beyond the scope of this book, examining the motivations behind covert operations helps inform and contextualize the Cold War military biomedical complex. The key goals were to stifle Soviet influence and promote American political and economic interests—a philosophy where the ends justified the means enveloped the American political and military establishments. For military science, the rationale for far-reaching research programs was in line with America's Cold War mindset that prioritized secrecy over transparency and national security over human rights.

The number and extent of covert actions increased during the 1960s, an expansion that paralleled the proliferation of clandestine research related to atomic, biological, and chemical warfare. President John F. Kennedy embraced the idea of paramilitary units engaged in counterinsurgency and clandestine missions, calling for "a whole new kind of strategy, a wholly different kind of force, and therefore a new and wholly different kind of military training."[37] While covert operations furthered America's Cold War goal of Soviet containment, the work also sowed the seeds of anti-Americanism due to relentless interference in the domestic affairs of dozens of nations. America's neo-colonial policies aided corrupt autocrats and the suppression of civil rights. These actions, particularly in Central America and Africa, have had long-lasting consequences that have contributed to widespread wealth inequality, disenfranchisement, poverty, and migration.[38]

Covert operations evolved alongside Cold War priorities and were structured to help the United States achieve its economic and foreign policy goals. As a CIA official remarked in the context of covert operations in Albania in the late 1940s that attempted to stall Soviet expansion, US covert actions there were "a clinical experiment to see whether larger rollback operations would be feasible elsewhere."[39] In other words, similar to chemical and biological weapons, covert operations sometimes were field tested in confined environments to evaluate their potential for further deployment.

Throughout the Cold War, the CIA was granted broad discretion to direct its covert operations with little oversight—a fact that several CIA directors candidly acknowledged after secret operations were publicly revealed. Allen Dulles stated that he would "fudge the truth a little" to a congressional oversight committee but would tell the committee chairman the truth if he "wants to know."[40] As William Colby further explained: "The old tradition was that you don't ask. It was a consensus that intelligence was apart from the rules . . . that was the reason we did step over the line in a few cases, largely because no one was watching. No one was there to say don't do that."[41] And, as Richard Helms detailed, "the nation must to a degree, take it on faith that we too are honorable men devoted to her service."[42]

In addition to withholding critical information from congressional oversight committees, the CIA has cultivated close relationships with executives at *The New York Times*, *The Washington Post*, and other national publications. In order to keep receiving inside scoops on intelligence operations, these media titans often acquiesced to the CIA's requests to edit or plant stories regarding the agency's work. The CIA also sought to suppress news of its untoward actions and maintain a positive image in the public.[43]

It was not until the mid-1970s that mainstream media outlets began publishing investigative reports on clandestine operations conducted by the CIA and the military, including assassination attempts, experiments with mind-altering drugs, propaganda campaigns, and spying on American citizens. These disclosures led to congressional hearings and new oversight laws.[44] Although the laws included new reporting requirements and new congressional committees—primarily, the Senate Select Committee on Intelligence and the House Intelligence Committee—the laws did not shift the consensus view that covert operations were essential foreign policy tools.

During the 1980s, under the administration of President Ronald Reagan, the CIA expanded its covert operations—particularly in the Middle East and Central America—and the military created its own intelligence units that conducted covert operations and clandestine missions throughout the world, including in El Salvador, Honduras, Iran, Iraq, Israel, Italy, Laos, Lebanon, Nicaragua, Panama, Saudi Arabia, and West Germany. The missions included operations in counterterrorism, hostage rescue, psychological warfare, and guerrilla warfare. By conducting covert operations through the military rather than through the CIA, the Reagan administration recognized that it could largely conceal the missions from congressional oversight since the intelligence oversight law only required that CIA covert actions be reported to Congress. Furthermore, in 1985, Reagan issued a secret order that gave retroactive approval for all previous covert actions. Reagan also instructed the CIA to withhold information on covert actions from congressional oversight committees unless he authorized disclosure. Each of these measures was a stark departure from customary CIA practices and oversight.[45]

Some CIA endeavors caused significant public backlash once they became public. This included the Yellow Fruit program, a secret intelligence unit within the DoD that worked closely with the CIA to provide military, intelligence, and financial support to rebels in Honduras, Guatemala, Nicaragua, El Salvador, Panama, and elsewhere. Also controversial was the Iran-Contra affair, whereby the US military and the CIA sold weapons to Iran (which was subject to a congressional arms embargo) and transferred the proceeds to the Contras (right-wing militants seeking to overthrow the Sandinista government in Nicaragua). The secret scheme—which was conducted with assistance from Israel and involved a release of hostages held in Lebanon—violated the Boland Amendment, which explicitly prohibited US government funding of the Contras. Dozens of Reagan officials were indicted due to their involvement in the Iran-Contra affair, though many were pardoned by President George H. W. Bush, who was Vice President at the time of the scandal and previously had been director of the CIA. Bush's pardons included Reagan's Secretary of Defense, Casper Weinberger, and several high-ranking CIA officials.[46]

Concurrently with the rise in CIA covert actions, the military expanded its special forces and increasingly deployed these men into special operations around the globe. During the 1980s—after a failed attempt to rescue American hostages in Iran—the DoD created the Joint Special Operations Command (JSOC), an elite group of warfighters that includes special forces from the Army, Navy Air Force, and Marines.[47] Due to infighting between the services on the role, scope, and command of the special forces, in 1987 Congress created the US Special Operations Command (USSOCOM).[48] At the time, some military leaders objected to the rise in special forces due to concerns that funding for traditional units would be in jeopardy if specialized units were particularly successful.[49]

USSOCOM, headquartered in Tampa, Florida, oversees more than seventy thousand military personnel, including warfighters from the Green Berets, Army Rangers, Navy SEALs, SEAL Team 6, air commandos, and other units. These elite warriors are meticulous, brilliant, brave, and physically fit. USSOCOM maintains a high level of autonomy within the DoD and has separate divisions for psychological warfare, counterterrorism, special reconnaissance, research and development, direct military action, unconventional warfare, and other areas. Special operations have become an integral component of America's national security apparatus, and at times USSOCOM special forces are loaned to the CIA's paramilitary unit.[50] As will be discussed in chapter 15, reliance on special forces and covert operations expanded significantly after 9/11.

Global Proliferation of Atomic, Biological, and Chemical Weapons

Although the Cold War centered on weapons proliferation in the United States and the Soviet Union, countries around the world worked diligently to enhance their military arsenals. During the 1960s, Egypt reportedly deployed chemical weapons to help defeat royalist forces during the Yemeni Civil War. Egypt's chemical weapons program was developed with assistance from Nazi scientists who were recruited to Egypt after WWII.[51] By the 1970s, intelligence reports indicated that several countries were pursuing the development or acquisition of biological or chemical weapons, including Algeria, Britain, Canada, China, Cuba, Egypt, France, Iraq, Israel, North Korea, Poland, South Africa, Spain, Sweden, Syria, West Germany, and Yugoslavia.[52]

A 1979 outbreak of anthrax in Sverdlovsk, Russia—where the Soviet Union maintained a military microbiology facility—raised fears that the Soviets were continuing with offensive biological weapons research notwithstanding prohibitions outlined in international treaties such as the 1925 Geneva Protocol and the 1972 Biological Weapons Convention.[53] During the 1980s, a classified intelligence memo warned that "Third World proliferation is getting worse, and the possible consequences are extremely serious."[54] By the 1990s, the list of countries developing or seeking to acquire biological or chemical weapons expanded to include Bulgaria, Czechoslovakia, East Germany, Hungary, India, Iran, Laos, Libya, Romania, Taiwan, and Vietnam.[55]

In many instances, the US government supplied pathogens and research materials to assist the budding programs. Pathogens also were exported by US companies, such as the American Type Culture Collection, a global supplier of biological materials. The private sales were approved by the Department of Commerce, which maintains an advisory committee on biological exports.[56] In 1990, when media outlets reported the government-approved transactions, Centers for Disease Control and Prevention (CDC) officials expressed little fear that the pathogens would be used to further the development of biological weapons since, the officials noted, "We know these people." At the same time, the CDC admitted that it did not maintain complete records of the sales because the transactions were handled "informally."[57] Notwithstanding the

CDC's assurances, behind the scenes, the CIA warned of the spread of chemical and biological weapons, indicating that "at least 10 countries are working to produce both previously known and futuristic biological weapons."[58]

Some nations conducted disturbing research. For example, Israel reportedly worked on a genetically engineered biological weapon that could harm Arabs but not Jews. Although many scientists questioned whether such a weapon could be produced, one Israeli source indicated that they have "succeeded in pinpointing a particular characteristic in the genetic profile of certain Arab communities, particularly the Iraqi people."[59] South Africa maintained a sophisticated program that examined cholera, Ebola, anthrax, botulinum toxin, and other toxic agents. South Africans developed umbrellas and walking sticks that could inject harmful toxins into an unsuspecting person, deodorant laced with typhoid, and whisky contaminated with poisonous snake venom. Via the apartheid-era Project Coast, South African researchers investigated whether they could create bacteria that would only affect specific races and a vaccine to block human fertility that would be administered to Black South Africans, who would be told it was a vaccine to protect against yellow fever. South African academics working on behalf of the military also prepared massive quantities of LSD and ecstasy in an effort to destabilize Black communities.[60] Archbishop Desmond Tutu characterized Project Coast as "the most diabolical aspect of apartheid."[61]

According to Wouter Basson, a South African physician who directed Project Coast, officials from Britain, Canada, Germany, Japan, and the United States provided "an incredible amount" of information regarding their respective chemical and biological weapons programs in exchange for intelligence on Soviet-backed African nations.[62] Basson and other Project Coast scientists further stated that British and American officials freely shared information with apartheid leaders but did not want the incoming, democratically elected administration of Nelson Mandela to have access to the information.[63] Years earlier, the CIA assisted the apartheid regime when it arrested and imprisoned Mandela, who was receiving support from the Soviet Union.[64]

The Iran-Iraq War (1980–1988) marked a pivotal moment for chemical and biological warfare. Prior to the war, Iraqi leader Saddam Hussein viewed chemical and biological weapons as strategic tools that could help Iraq fight against a larger and better-equipped foe. During the 1970s, Iraq began purchasing foundational elements for its biological weapons program from the United States, including spores of anthrax and botulinum toxin. Biological weapons experts at Fort Detrick developed one of the anthrax strains that Iraq purchased.[65]

The Americans sent the materials to an Iraqi agency that the CIA had identified as a front for Iraq's biological weapons program.[66] Iraq also purchased substantial amounts of growth media (which is used to promote the growth of bacteria and viruses) from Oxoid, a British company. Iraqi scientists later admitted to UN inspectors that they had grown nineteen thousand liters of botulinum toxin and eight thousand liters of anthrax, and had researched camel pox virus (which could be engineered to create weaponized smallpox).[67] Inspectors also found evidence of gas chamber experiments with anthrax and field tests with biological weapons: reports indicated that human subjects included Iraqi criminals, political prisoners, Kurds, and Iranian prisoners of war.[68]

America supported Iraq during the Iran-Iraq War, in large part due to US antipathy toward the 1979 Iranian revolution and the Iran hostage crisis from 1979–1981. The Iranian revolution ousted the American-backed Iranian leader, Mohammed Reza Pahlavi (commonly referred to as Mohammed Reza Shah, or simply the Shah), who gained power following an American-supported coup in 1953 that overthrew Mohammed Mosaddegh, the elected prime minister of Iran and a champion of secular democracy. Mosaddegh took measures to build the Iranian nation after years of foreign exploitation: he instituted fair labor standards, sought to eliminate corrupt businesses and political practices, took steps to build a national infrastructure and a social safety net, and sought to reclaim a fair share of Iran's oil fields. As to the latter, Britain vehemently opposed Mosaddegh's actions due to its extensive and longstanding control over Iranian oil.[69]

The British sought US support to topple Mosaddegh and stall his reforms. President Harry Truman refused to condone what Secretary of State Dean Acheson characterized as Britain's destructive rule or ruin policy in Iran, but American foreign policy shifted dramatically when President Eisenhower took office in January 1953. Despite Mosaddegh's opposition to communism, Winston Churchill convinced Eisenhower that Mosaddegh's reforms would have negative economic implications for American interests and that, over time, Mosaddegh would turn to the Soviet sphere.[70]

Secretary of State John Foster Dulles instructed the CIA, led by his younger brother Allen Dulles, to craft a secret operation to oust Mosaddegh. Prior to serving in government, the Dulles brothers were partners at Sullivan & Cromwell, a prestigious law firm that represented a company that had controlled a substantial portion of Iranian oil.[71] Led by CIA agent Kermit Roosevelt, who was an OSS agent during WWII, the CIA station in Tehran crafted a propaganda campaign to oust Mosaddegh. CIA operatives posed as Mosaddegh agents and threatened local religious leaders; these actions created great resentment against Mosaddegh. The CIA also collaborated with the Shah—Iran's ruler prior to Mosaddegh, who had briefly fled to Italy in exile when Mosaddegh came to power—to create a coup to overthrow Mosaddegh. Violent, CIA-orchestrated protests followed, and pro-Shah military forces stormed the capital and the prime minister's official residence. Mosaddegh surrendered and the Shah returned from Italy, appointing the CIA-backed Fazlollah Zahedi as Iran's new prime minister.[72]

For twenty-five years, the Shah was a very close American ally: he allowed the United States to create military bases in Iran and launch surveillance flights from Iran into the Soviet Union, and he permitted espionage agents to flow across the border from Iran into the USSR. The Shah also purchased US military equipment and allowed American companies to exploit Iranian oil and other resources.[73] So close was the American-Iranian alliance that, during the mid-1970s, the US military devised a plan to launch two hundred nuclear weapons at the Soviet Union should it invade Iran.[74] For Iranians, life under the Shah was rife with poverty, police abuse, and suppression of civil rights. The Iranian secret police, which was formed with assistance from America and Israel, sought out and tortured Iranian dissidents throughout the world. The CIA reportedly paid the Shah and other Iranian leaders up to $400 million annually.[75]

The arrangement continued until 1979, when the Iranian Revolution—a largely nonviolent movement—ousted the Shah and replaced him with Ayatollah Khomeini, who instituted a theocratic-republican political system whereby he became the supreme leader of the newly formed Islamic Republic of Iran. Khomeini criticized America's undue influence in Iranian politics and sought to eliminate the power channels through which the United States benefited.[76] Following the 1979 revolution and ensuing hostage crisis, America turned against Iran and its people, a geopolitical position that continues into the 2020s.

During the Iran-Iraq War, the United States supplied Iraq with military support, intelligence, and financial assistance, and encouraged other countries to sell military equipment to Iraq.[77] Two West German companies built chemical plants in Iraq. At the same time, more than thirty suppliers—including companies from Australia, Austria, Belgium, Egypt, France, Holland, India, Italy, Poland, Switzerland, the United Kingdom, the United States, and West Germany—entered into agreements with Iraq to support weapons development.[78]

UN inspectors verified that Iraq used Tabun, mustard gas, and other toxic substances against Iranian soldiers and civilians, a conclusion that had been confirmed years earlier by American intelligence officials in classified reports.[79] Iran's calls for the international community to condemn Iraq's actions were ignored, a silence that motivated Iraq to expand its chemical attacks and Iran to commence its own chemical weapons program.[80] As a 1987 UN report concluded, "the continued use of chemical weapons in the present conflict increases the risk of their use in future conflicts."[81]

Meanwhile, the United States continued its support for Iraq, notwithstanding the fact that Iraq's chemical attacks caused thousands of casualties.[82] By 1988, chemical attacks had become a standard component of Hussein's military strategy.[83] Iran did not possess chemical weapons at the start of the war—Khomeini opposed chemical warfare on religious grounds because the Koran forbids the use of poisonous weapons. As Iraq's chemical attacks continued, however, Khomeini issued a secret order that permitted use of chemical weapons.[84] Even after Iran and Iraq agreed to a ceasefire, Hussein deployed chemical weapons against Iraq's Kurdish citizens (which included the Peshmerga, Iraqi-Kurds who had allied with Iran).[85] By the end of the war, Iraq's chemical attacks had caused more than fifty thousand injuries and about five thousand deaths,[86] numbers that approximate America's WWI casualties from gas warfare.

In 1989—due to heightened concerns regarding an attack against American interests with biological and chemical weapons—the United States banned the sale of anthrax and other pathogens to Iran, Iraq, Libya, and Syria.[87] By then, however, worldwide proliferation was on an unstoppable trajectory. The UN's failure to punish or condemn Iraq's actions highlighted the limits of international treaties and had a corrosive impact on the norms governing the use of chemical weapons. Just months after the end of the Iran-Iraq war, Iran's Parliamentary leader stated that "chemical and biological weapons are . . . the poor man's atomic bomb."[88] Thereafter, Iran signed agreements with Western companies to build chemical weapons production facilities. Other nations—including Egypt and Syria—followed suit, creating new programs or expanding existing ones.[89]

International doctrines—such as the 1925 Geneva Protocol, the 1972 Biological Weapons Convention, and the 1993 Chemical Weapons Convention—have set a global norm against the deployment of chemical and biological weapons and have been ratified by the vast majority of nations around the world. However, research continues, stockpiles remain, and verification mechanisms are inadequate to assure compliance. In addition, some nations (such as Egypt, Israel, and North Korea) have not ratified one or more of the treaties.[90]

During the 1990s, the United States engaged in several secret anthrax projects, including recreating an anthrax strain that was resistant to the anthrax vaccine utilized by Russia, reconstructing a Soviet-era anthrax bomb model, and building an anthrax spore production factory. This work, publicly revealed in 2001, underscored the amorphous line between offensive and defensive biological warfare research. One renowned scientist who worked on the secret projects, Joshua Lederberg, reportedly asked for governmental legal assurances that the work did not violate any law or treaty.[91]

In 2013, the Syrian government deployed chemical weapons—including chlorine gas, Sarin, and other toxic agents—against rebel forces during the country's civil war.[92] In the months following the attack, following international pressure, Syria agreed to voluntarily destroy its chemical weapons stockpiles, though questions remain whether the nation has complied with its representations. In 2023, Israel deployed white phosphorus—an incendiary chemical similar to napalm that causes excruciating burns, respiratory damage, and severe long-term health conditions—in attacks in Gaza and Lebanon.[93] This was even though the nation pledged a decade earlier to phase out its use.[94] The chemical was provided to the Israeli army by the United States. Although American officials expressed concern that Israel was using the chemical in violation of the law of armed conflict (also referred to as international humanitarian law), the United States continued to provide substantial funding and military assistance to Israel following the white phosphorus attacks.[95]

Coupled with the global proliferation of biological and chemical weapons, by the late 1980s, eight countries were known to possess nuclear weapons: America, Britain, China, France, India, Israel, South Africa, and the Soviet Union.[96] The United States maintained thousands of nuclear weapons throughout Europe and created nuclear-sharing agreements whereby NATO allies could utilize American weapons deployed to Europe during a war against the Soviet Union.[97] Reports surfaced that the Soviet Union had created the "Dead Hand," a semiautomatic machine that would launch nuclear weapons in the event Soviet leadership had been incapacitated by a nuclear attack.[98] The former Soviet states of Belarus, Kazakhstan, and Ukraine maintained nuclear weapons after the breakup of the USSR, but later each reportedly transferred its weapons to Russia. During the transition, as Soviet scientists scrambled to find new employers, intelligence analysts surmised that core elements of nuclear capabilities were secretly sold or misappropriated.[99] South Africa reportedly dismantled its nuclear arsenal in the early 1990s, though Pakistan and North Korea later joined the nuclear club.[100] Intelligence experts suspect that additional nations and non-state actors are pursuing the development or acquisition of nuclear weapons.[101] As French President Charles de Gaulle observed decades ago, "No country without an atom bomb could properly consider itself independent."[102]

Summary

America's post-WWII rise as a global superpower brought paradigm-shifting changes to the military biomedical complex. The medical and research establishment grew exponentially, while America's Cold War policy of Soviet containment led to a significant increase in service members stationed throughout the world. Military engagements in Korea and Vietnam stretched the resources of the military healthcare system, while covert operations became a strategic tool to achieve America's political, economic, and military goals. Meanwhile, the global proliferation of atomic, biological, and chemical weapons contributed to a perilous feedback loop that accelerated US research and development in these fields.

Iraq's chemical attacks during the Iran-Iraq War represented a disregard for international treaties and norms, but also underscored a tactical calculus that nations face during an existential threat. The international community did little to punish Iraq for its actions or deter other countries from engaging in similar conduct in the future. Less than three years after the conclusion of the Iran-Iraq War, when the United States came to the aid of Kuwait following an Iraqi invasion, America was confronted with the possibility that Iraq would deploy chemical or biological weapons against US military personnel.

Notes

1. Greenwood and Berry, *Medics*, 117.
2. Rostker, *Providing*, 174, 215, 262.
3. Greenwood and Berry, *Medics*, 128.
4. Greenwood and Berry, *Medics*, 123–28.
5. Congressional Research Service, "American War and Military Operations Casualties: Lists and Statistics" (July 29, 2020); Greenwood and Berry, *Medics*, 119–28.
6. Department of the Army, Office of the Chief of Military History, *U.S. Army Expansion, 1961-62* (1963), 100–101.
7. X, "The Sources of Soviet Conduct," *Foreign Affairs* 25, no. 4 (July 1947): 566–82 (it was later acknowledged that George Kennan wrote the article under the pseudonym X); John Lewis Gaddis, *Strategies of Containment* (1982); Ernest May, *American Cold War Strategy* (1993).
8. Michelle Dolfini-Reed and Jennifer Jebo, *The Evolution of the Military Health Care System: Changes in Public Law and DOD Regulations* (Alexandria, VA: Center for Naval Analyses, 2000), 4–6, 15–18.
9. Greenwood and Berry, *Medics*, 133–35
10. Donald L. Custis, "Military Medicine from World War II to Vietnam," *Journal of the American Medical Association* 264, no. 17 (November 7, 1990): 2259–62.
11. Custis, "Military Medicine"; Ginn, *The History*, 313.
12. Engelman, *A Decade of Progress*, 34, 171–77; Custis, "Military Medicine."
13. Eric T. Dean, *Shook Over Hell* (1997), 7–25.
14. American Psychiatric Association, *Diagnostic and Statistical Manual of Mental Disorders*, 3rd ed. (1980); G. J. Turnbull, "A Review of Post-Traumatic Stress Disorder," *Injury* 29, no. 2 (March 1998): 87–91.

15. Charles R. Marmar et al., "Course of Posttraumatic Stress Disorder 40 Years After the Vietnam War," *JAMA Psychiatry* 72, no. 9 (September 2015): 875–81.
16. Marmar et al., "Course of Posttraumatic Stress"; Stefan G. Hofmann, Brett T. Litz, and Frank W. Weathers, "Social Anxiety, Depression, and PTSD in Vietnam Veterans," *Journal of Anxiety Disorders* 17, no. 5 (2003): 573–82; Grace Macdonald-Gagnon et al., "Generalized Anxiety and Mild Anxiety Symptoms in U.S. Military Veterans: Prevalence, Characteristics, and Functioning," *Journal of Psychiatric Research* 171 (March 2024): 263–70
17. Kamienski, *Shooting Up*, 191–217.
18. David Burnham, "Dow Says U.S. Knew Dioxin Peril of Agent Orange," *The New York Times*, May 5, 1983; Ralph Blumenthal, "Files Show Dioxin Makers Knew of Hazards," *The New York Times*, July 6, 1983; Philip Shabecoff, "Hazard of Dioxins Assailed in Study," *The New York Times*, May 30, 1986.
19. U.S. Department of Veterans Affairs, "Agent Orange Exposure and VA Disability Compensation" (accessed April 23, 2024).
20. Milena Jovanovitch, "Agent Orange Suit: A Veteran's Legacy," *The New York Times*, August 21, 1988.
21. Leonard Buder, "Judge Schedules Start of Agent Orange Payments," *The New York Times*, July 6, 1988.
22. Fred A. Wilcox, *Scorched Earth* (2011), 69.
23. Coffey, *American Arsenal*, 242.
24. Greenwood and Berry, *Medics*, 134–35.
25. Sandeep Khot, Buhm Soon Park, and W. T. Longstreth, "The Vietnam War and Medical Research: Untold Legacy of the U.S. Doctor Draft and the NIH 'Yellow Berets,'" *Academic Medicine* 86, no. 4 (February 2011): 502–8.
26. Greenwood and Berry, *Medics*, 151–53.
27. Carroll, *Lab 257*, 119–20.
28. G. A. Eddy and C. J. Peters, "The Extended Horizons of Rift Valley Fever," in *New Developments with Human and Veterinary Vaccines*, edited by A. Mizrahi, I. Hertman, M. A. Klingberg, and A. Kohn (New York: A.R. Liss, 1980), 179–91.
29. Amy B. Zegart, *Spies, Lies, and Algorithms* (2022), 60–65; Loch K. Johnson, *Spy Watching* (2018), 327–52.
30. Blum, *Killing Hope*; Stansfield Turner, *Burn Before Reading* (2005); Prados, *Presidents' Secret Wars*; Weiner, *Legacy of Ashes*; Michael McClintock, *Instruments of Statecraft* (1992).
31. *Alleged Assassination Plots Involving Foreign Leaders*, Interim Report of the Senate Select Committee to Study Governmental Operations with Respect to Intelligence Activities, Senate Report No. 94-465 (1975), 71–180; David Belin, Executive Director of CIA Commission, *Summary of Facts: Investigation of CIA Involvement in Plans to Assassinate Foreign Leaders* (May 30, 1975).
32. Gordon Corera, *The Art of Betrayal* (2012), 94–134.
33. Quoted in Daniel Immerwahr, "Make It Hurt: Amid the Ebbing of Empire, Frantz Fanon and Ian Fleming Agreed on One Thing," *The New Yorker*, January 15, 2024, 54.
34. *Alleged Assassination Plots*, 13–70; *Foreign Relations of the United States: Congo, 1960-1968* (Washington, DC: US Government Printing Office, 2013); Corera, *The Art of Betrayal*, 94–134; Johnson *A Season of Inquiry*, 46; Ted Gup, "The Coldest," *The Washington Post*, December 16, 2001.
35. "Belgium: Apology for Lumumba Killing," *The New York Times*, February 6, 2002.
36. Jean-Pierre Stroobants, "Belgium Prime Minister Officially Apologizes for the Death of Patrice Lumumba," *Le Monde*, June 21, 2022.

37. John F. Kennedy, Remarks at West Point to the Graduating Class of the U.S. Military Academy (June 6, 1962).
38. Hal Brands, *Latin America's Cold War* (2012); Greg Grandin, *The Last Colonial Massacre* (2011); Elizabeth Schmidt, *Foreign Intervention in Africa* (2013).
39. Quoted in W. Thomas Smith, *Encyclopedia of the Central Intelligence Agency* (New York: Facts on File, 2003), 8.
40. Quoted in Tom Braden, "What's Wrong with the CIA?," *Saturday Review*, April 5, 1975, 14.
41. Quoted in Bob Wiedrich, "Watching the CIA with Fewer Eyes," *Chicago Tribune*, February 3, 1976.
42. Quoted in "Excerpts from Speech by Helms to Society of Newspaper Editors," *The New York Times*, April 15, 1971.
43. Weiner, *Legacy of Ashes*, 77–79.
44. Steven Emerson, *Secret Warriors* (1988), 32–34.
45. Bob Woodward, *Veil* (1987); Stephen Schlesinger and Stephen Kinzer, *Bitter Fruit* (1982); Emerson, *Secret Warriors*, 7–11; John Rizzo, *Company Man* (2014), 107–8.
46. Emerson, *Secret Warriors*; Weiner, *Legacy of Ashes*, 401–12.
47. Douglas C. Waller, *The Commandos* (1994), 32–36; Jeremy Scahill, *Dirty Wars* (2013), 48–60.
48. "U.S. Special Operations Forces (SOF): Background and Issues for Congress," *Congressional Research Service* (May 6, 2021).
49. Emerson, *Secret Warriors*, 31–32.
50. Waller, *The Commandos*, 32–36.
51. Corey J. Hilmas, Jeffery K. Smart, and Benjamin A. Hill, "History of Chemical Warfare," in *Medical Aspects of Chemical Warfare*, edited by Shirley D. Tuorinsky (Washington, DC: Borden Institute, 2008), 57–58; Tucker, *War of Nerves*, 190–94.
52. Tucker, *War of Nerves*, 153, 187–89, 196, 236; Hersh, *Chemical and Biological Warfare*, 281–99.
53. George W. Christopher et al., "Biological Warfare: A Historical Perspective," *Journal of the American Medical Association* 278, no. 5 (August 6, 1997): 412–17.
54. U.S. Army Science Board, *Final Report of the Ad Hoc Subgroup on Army Biological Defense Research Program* (July 1987): 6, quoted in Susan Wright, "Evolution of Biological Warfare Policy, 1945-1990," in Wright, *Preventing a Biological*, 56.
55. Tucker, *War of Nerves*, 254; Regis, *The Biology of Doom*, 220.
56. Barnaby, *The Plague Makers*, 68–69.
57. *Foreign Operations, Export Financing, and Related Programs Appropriations for 1992: Hearings before a Subcommittee of the House Committee on Appropriations*, 102nd Cong. 77 (February 26, 1991) (reprint of Eric Nadler and Robert Windrem, "Deadly Contagion").
58. *Global Spread of Chemical and Biological Weapons: Hearings before the Senate Committee on Governmental Affairs and Its Permanent Subcommittee on Investigations* 101st Cong. 10 (February 9, 1989) (testimony of William H. Webster, CIA Director).
59. Quoted in Uzi Mahnaimi and Marie Colvin, "Israel Developing an Ethno-Bomb," *Sunday Times, London*, November 15, 1998.
60. Chandré Gould and Peter Folb, *Project Coast* (2002); Harris and Paxman, *A Higher Form of Killing*, 248–49; Barnaby, *The Plague Makers*, 119–22.
61. Desmond Tutu, *No Future Without Forgiveness* (1999), 183.
62. Quoted in "Africa 'Doctor Death' Implicates West," *BBC News*, July 31, 1998.
63. Barnaby, *The Plague Makers*, 119–22.
64. Weiner, *Legacy of Ashes*, 361–63.

65. Tucker, *War of Nerves*, 249–51; Judith Miller, Stephen Engelberg, and William Broad, *Germs* (2002), 88–89; Barnaby, *The Plague Makers*, 68–69.
66. Miller et al., *Germs*, 88–89.
67. Paul Aebersold, "FDA Experience with Medical Countermeasures Under the Animal Rule," *Advances in Preventive Medicine* 2012 (2012): 1–11.
68. Moreno, *Undue Risk*, 2–3.
69. Prados, *Presidents' Secret Wars*, 91–98; Weiner, *Legacy of Ashes*, 81–92.
70. Prados, *Presidents' Secret Wars*, 91–98; Weiner, *Legacy of Ashes*, 81–92.
71. Blum, *Killing Hope*, 71; Prados, *Presidents' Secret Wars*, 94.
72. Prados, *Presidents' Secret Wars*, 91–98; Weiner, *Legacy of Ashes*, 81–92.
73. Blum, *Killing Hope*, 64–72.
74. Niall Ferguson, "Kissinger and the True Meaning of Détente," *Foreign Affairs* 103, no. 2 (March/April 2024): 120–33.
75. Blum, *Killing Hope*, 64–72.
76. Nikki R. Keddie, "The Iranian Revolution and U.S. Policy," *SAIS Review of International Affairs* 2, no. 1 (1981): 13–26.
77. Tucker, *War of Nerves*, 262, 270.
78. Cole, *The Eleventh Plague*, 81–102; Tucker, *War of Nerves*, 249–51, 270–72.
79. Central Intelligence Agency, *The Iraqi Chemical Weapons Program in Perspective: An Intelligence Assessment* (January 1985); Central Intelligence Agency, *Impact and Implications of Chemical Weapons Use in the Iran-Iraq War: Interagency Intelligence Memorandum* (April 1988); Central Intelligence Agency, *Iraq's Chemical Warfare Program: More Self-Reliant, More Deadly* (August 1990); Julian Perry Robinson, and Jozef Goldblat, "Chemical Warfare in the Iran-Iraq War 1980-1988: Fact Sheet," *Stockholm International Peace Research Institute* (May 1984); U.N. SC Res. 620 (August 26, 1988).
80. Tucker, *War of Nerves*, 250, 259, 272–73; Patrick G. Eddington, *Gassed in the Gulf* (1997), 9–11.
81. P. Dunn, *Chemical Aspects of the Gulf War, 1984-1987: Investigations by the United Nations* (Ascot Vale, Australia: Materials Research Laboratories, 1987), quoted in Hilmas et al., "History of Chemical Warfare," 63.
82. James R. Clapper, *Facts and Fears* (2018), 57.
83. Trends and Developments, Biweekly Report, December 15–31, 1990, quoted in Eddington, *Gassed in the Gulf*, 10; Tucker, *War of Nerves*, 278–88.
84. Tucker, *War of Nerves*, 272–73.
85. Robert Draper, *To Start a War* (2020), 56–59.
86. Tucker, *War of Nerves*, 285.
87. Department of Commerce, *Removal of Unilateral National Security Controls; Additional Controls on Chemicals and Biological Agents and Precursors*, 54 Fed. Reg. 8281 (February 28, 1989).
88. Ali Akbar Hashemi Rafsanjani, *Islamic Republic News Agency*, Tehran, English broadcast, transcribed in *FBIS Daily Report: Near East and South Asia*, October 19, 1988, 55–56, quoted in Tucker, *War of Nerves*, 286–87.
89. Tucker, *War of Nerves*, 259, 272–73, 286–88.
90. State Parties and Signatories, International Humanitarian Law Databases, *International Committee of the Red Cross* (accessed January 26, 2024).
91. Jeanne Guillemin, *American Anthrax* (2011), 124.
92. Human Rights Watch, *Death by Chemicals: The Syrian Government's Widespread and Systemic Use of Chemical Weapons* (May 1, 2017).

93. William Christou, Alex Horton, and Meg Kelly, "Israel Used U.S.-Supplied White Phosphorus in Lebanon Attack," *The Washington Post*, December 11, 2023; "Evidence of Israel's Unlawful Use of White Phosphorus in Southern Lebanon as Cross-Border Hostilities Escalate," *Amnesty International* (October 31, 2023); "Israel: White Phosphorous Used in Gaza, Lebanon," *Human Rights Watch* (October 12, 2023).
94. Isabel Kershner, "Israel: Military to Stop Using Shells Containing Phosphorus," *The New York Times*, April 26, 2013.
95. Zolan Kanno-Youngs and Ephrat Livni, "U.S. Raises Concern Over Israel's Possible Use of U.S.-Supplied White Phosphorus," *The New York Times*, December 11, 2023; John Hudson, "U.S. Signs Off on More Bombs, Warplanes for Israel," *The Washington Post*, March 29, 2024; Patsy Widakuswara, "Why Biden Won't Put Conditions on Military Aid to Israel," *Voice of America*, March 29, 2024.
96. Hans M. Kristensen and Robert S. Norris, "Global Nuclear Weapons Inventories," *Bulletin of the Atomic Scientists* 69, no. 5 (September/October 2013): 75–81.
97. Keir A. Lieber and Daryl G. Press, "The Return of Nuclear Escalation," *Foreign Affairs* 102, no. 6 (November/December 2023): 45–55.
98. David E. Hoffman, *The Dead Hand* (2009), 23–24, 365–66, 422–23.
99. Hoffman, *The Dead Hand*, 24, 384–411.
100. J. W. de Villiers, Roger Jardine, and Mitchell Reiss, "Why South Africa Gave Up the Bomb," *Foreign Affairs* 72, no. 5 (November/December 1993): 98–109; Kristensen and Norris, "Global Nuclear Weapons Inventories."
101. *The Commission on the Intelligence Capabilities of the United States Regarding Weapons of Mass Destruction* (2005), 516–20.
102. Quoted in "The Thoughts of Charles de Gaulle," *The New York Times Magazine*, May 12, 1968, 102.

PART IV
1991–2024
From Protecting to Enhancing the Fighting Force

13
Military Medicine and the Persian Gulf War

Introduction

Fears of a chemical or biological attack came to the fore during the Persian Gulf War, when America faced a cognizable threat from Iraq, a nation that, just years earlier, it helped arm with chemical and biological weapons. To counter Iraq's invasion of Kuwait and its provocations against other countries in the Middle East, the United States deployed 694,550 service members. It mobilized eighty-seven thousand medical personnel, the largest medical force since World War II (WWII). The Navy staffed several hospital ships, the Army prepared more than thirteen thousand hospital beds in Saudi Arabia, and the Air Force equipped hospitals in Germany and England.[1] The Department of Defense (DoD) instituted several preventive health policies, including sanitary measures, immunizations, insect control measures, and food and water inspections. Military health officials made an extensive effort to address endemic diseases such as malaria, sandfly fever, and Rift Valley fever.[2] General Norman Schwarzkopf, commander-in-chief of Central Command, heralded the medical build-up as "nothing short of spectacular."[3]

The Persian Gulf War (1990–1991) occurred at a pivotal moment for the US government, the military, and the biomedical establishment. The end of the Cold War and the dissolution of the Soviet Union altered America's geopolitical priorities. These momentous events also raised complex questions regarding the scope of American investments in military matters and the allocation of US military resources across the globe. Some US officials argued for a substantial decrease in military spending, while others supported an increase in military funding and an eastward expansion of NATO's borders.[4]

Concurrently, the Federal Policy for the Protection of Human Subjects—also referred to as the Common Rule—was codified in 1991 and adopted by fifteen federal agencies (including the DoD).[5] The Central Intelligence Agency (CIA) did not adopt the Common Rule,[6] but a 1981 Executive Order requires that the intelligence community adhere to Department of Health and Human Services (HHS) policies regarding human experimentation. Via the Executive Order, the CIA is required to comply with the Common Rule because HHS has adopted the rule.[7] As detailed in chapter 9, codification of the Common Rule was the culmination of a research ethics endeavor that lasted over a decade, galvanized in large part by public backlash to the 1970s revelations regarding the Tuskegee Syphilis Study and other untoward research conducted by the military, the CIA, and civilian scientists. The Common Rule imbued ethical considerations into research protocols and motivated a shift in the ethos of medicine and research away from paternalistic policies and toward principles of justice and fairness.

As the Common Rule was being finalized, however, the DoD sought to carve out exemptions to informed consent laws and other research protections to maintain the

America's Military Biomedical Complex. Efthimios Parasidis, Oxford University Press.
 DOI: 10.1093/9780199351473.003.0014

discretion to mandate investigational medical products during military missions. In preparation for the Persian Gulf War, this process centered on pyridostigmine bromide tablets (PB) and the botulinum toxoid (BT) vaccine.[8] Neither was approved by the Food and Drug Administration (FDA) as prophylaxis for chemical or biological warfare. However, the DoD sought to use both as mandatory biodefense countermeasures due to fears that Iraq would deploy chemical or biological weapons.

After a series of closed-door meetings with the DoD, the FDA altered its regulations and created a new rule that allowed for a waiver of informed consent requirements in instances of military exigency. The FDA then issued informed consent waivers for PB and the BT vaccine. Service members sued the FDA and the DoD, but a federal court ruled in favor of the government and dismissed the case. After the war, speculation arose that the countermeasures were contributing factors to Gulf War Illness, a debilitating ailment that has impacted hundreds of thousands of Gulf War veterans.

Questions then were raised regarding the FDA's role in permitting the administration of the countermeasures: specifically, whether the agency maintained adequate independence from the military, whether it was sufficiently fulfilling its mandate to protect the public health, and whether the agency maintained appropriate safeguards for medical products administered to service members. In light of continuing public concerns regarding Gulf War Illness, the FDA revoked the rule that granted the military special regulatory exemptions from informed consent requirements, but only after Congress enacted a new law granting the president the authority to waive informed consent policies for military missions. The new law firmly embedded a divergent policy whereby mandates for experimental medical products were permissible in the military even though they would be illegal in a civilian context.

FDA Review of PB and the BT Vaccine Before the Persian Gulf War

During the early 1980s, the DoD considered whether PB could serve as prophylaxis to protect against Soman, a nerve agent that causes convulsions, respiratory failure, paralysis, and death. PB was FDA-approved in 1955 as a treatment for myasthenia gravis, a rare disease that leads to muscle weakness. The effectiveness of PB as a Soman countermeasure for humans was unclear because no human clinical trials had been conducted for that use. Animal studies revealed some risks of using PB as a chemical weapon countermeasure, but also showed some potential benefits.

The animal studies suggested that, for PB to be effective, it must be ingested several hours before Soman exposure and, immediately after exposure, two other products, Atropine and 2-PAM, must be administered rapidly.[9] Animal studies further revealed that PB may amplify the ill effects of Soman if the drug is used less than several hours before Soman exposure, during exposure, or after exposure. Data also indicated that PB could not protect against low-level Soman exposure, and that the drug may exacerbate harm if a person who ingests PB is exposed to other nerve agents such as Sarin, VX, or Tabun.[10] Following the animal studies, in 1984, the Army applied for FDA approval of PB as prophylaxis for Soman.[11] As the FDA review was ongoing, the Army stockpiled PB for potential future use, though the FDA application was still pending during military preparations for the Persian Gulf War.[12]

The BT vaccine was developed in the 1950s by Fort Detrick scientists, and as of 1990, the most recent batch had been manufactured in 1970. Although the vaccine was not FDA-approved, the FDA granted the Centers for Disease Control and Prevention (CDC) preapproval authority to administer the vaccine to individuals at risk of exposure to botulinum toxin. This group primarily included biosecurity researchers. Studies had found that, for the vaccine to provide protection, administration must consist of three doses over a twelve-week period and an annual booster thereafter. The vaccine protected against five of the seven known toxin types of botulinum. As of October 1990, approximately three thousand individuals had been vaccinated.[13]

By 1990, there was no evidence that PB or the BT vaccine caused serious or long-term adverse effects. Yet, both products were categorized as investigational medical products since they were not FDA-approved as prophylaxis for chemical or biological weapons. One reason for the investigational status centered on the ethical implications of conducting human studies. To evaluate whether the BT vaccine was effective, human subjects would receive the vaccine and then be exposed to botulinum toxin. A similar study would need to be conducted with PB and Soman. Although scientists had conducted analogous experiments with chemical and biological agents during the Cold War, new guidelines for human subjects research made it difficult to obtain approval from an institutional review board (IRB) for biodefense clinical studies.

To understand the legal, medical, and ethical concerns raised by the DoD's decision to administer PB and the BT vaccine, one must first have a general understanding of FDA regulations governing the approval process for medical products and how the use of medical products is regulated after FDA approval. Pursuant to the Federal Food, Drug, and Cosmetic Act (FDCA), medical products undergoing FDA review are deemed investigational, and the FDA can approve a new medical product only if data provide substantial evidence that the product is safe and effective. Standard FDA protocols require that investigational medical products be offered primarily to participants enrolled in clinical trials and only upon obtaining written informed consent from the research subjects.[14]

An exception to this general rule is where obtaining consent is "not feasible." As of August 1990, this was limited to circumstances of emergency use where (1) an individual was "confronted by a life-threatening situation necessitating the use of the test article," (2) informed consent could not obtained "because of an inability to communicate with or obtain legally effective consent from the subject," (3) "there is insufficient time to obtain consent from the subject's legal representative," and (4) "there is no available approved or generally recognized therapy that provides an equal or greater likelihood of saving the life of the subject."[15] To qualify as an emergency use, all four requirements must be met.

If the FDA determines that safety and efficacy data are sufficient, it will approve a medical product for a specific indication, which is termed an on-label use. Physicians are free to prescribe FDA-approved medications for other indications—which are called off-label uses—but only if there is adequate evidence that the product is safe and effective for off-label use. The FDA does not evaluate the safety and effectiveness of off-label uses, nor does the agency regulate physician prescribing for off-label indications.[16] Rather, medical practice guidelines govern physician prescribing, and a claim that a physician was negligent in prescribing a medicine for off-label use may be brought as a disciplinary hearing before a state medical board, as a court case alleging medical malpractice, or both.[17] This is a hybrid legal framework whereby the FDA regulates the approval of medical products while physician use of medical products (which constitutes medical treatment) is governed by state medical practice guidelines and state tort law.

For civilians, a physician must obtain informed consent from a patient prior to initiating medical treatment. The exception is when there is a medical emergency and the patient cannot give consent, in which case a physician can perform emergency care without consent as long as the care is provided to address the medical emergency. In the military, however, as has been detailed throughout this book, a commanding officer can order that a subordinate submit to medical treatment deemed necessary for military missions. This includes the administration of FDA-approved medical products for on-label indications, such as vaccines or other prophylactic countermeasures. At the time the military was preparing for the Persian Gulf War, forced administration was not permitted for investigational medical products or for off-label uses of approved medical products. Rather, in these two scenarios, administration was permissible only if military personnel provided consent after being informed of the risks and benefits of the medical product.

DoD and FDA Negotiate Informed Consent Waivers

Shortly after Iraq's invasion of Kuwait, which began on August 2, 1990, the DoD and the FDA commenced discussions on medical management for anticipated biological and chemical warfare. DoD and FDA officials considered several issues, including (1) whether federal law prohibited administration of investigational products without informed consent; (2) whether use of the investigational products constituted medical treatment or medical research; and (3) the proper process that the FDA must employ in considering whether informed consent is "not feasible" in military contingencies.[18] The DoD requested that it be excused from complying with FDA labeling and informed consent requirements. Under standard FDA protocols, investigational products must be labeled *Caution: New Drug—Limited by Federal Law to Investigational Use*. The DoD indicated that this disclaimer would undermine confidence in the medical product and would discourage use of the product. In the alternative, the DoD suggested that the label read: *For Military Use and Evaluation Only*. The DoD further contended that "the chaotic nature of armed conflict" required that it be excused from recordkeeping requirements and that it should be provided additional time to report countermeasure side effects to the FDA.[19]

Early in the discussions, the FDA explained to the DoD that if PB and the BT vaccine were exported from the United States and administered overseas, or if the products were produced and administered outside the United States, the FDA's informed consent and labeling requirements would not apply because the FDA's jurisdiction extends only to within US borders. In making this assertion, the FDA provided the DoD with a method of bypassing the FDA's requirements. The FDA further stated that if the medical products were to be administered in the United States, amendments to the FDA regulations would be necessary to permit the administration of the countermeasures without the informed consent of military personnel. The DoD affirmed that it needed the flexibility to administer the medical products both in the United States and overseas, and thus advocated for regulatory amendments.[20]

A significant component of the DoD-FDA discussions centered on whether administration of investigational medical products, such as PB and the BT vaccine, constitutes medical practice or medical research. Under existing federal law, the DoD was required to obtain informed consent from research participants prior to conducting medical research. This rule was first established as a component of the 1972 DoD appropriations bill after the public became aware of several research-related abuses, most notably the Tuskegee Syphilis Study.[21] For years, it remained a component of defense appropriations, and in 1984, the rule was codified into federal law as 10 U.S.C. § 980.[22] In 1990, DoD lawyers opined that Section 980 did not apply to PB and the BT vaccine because the proposed uses were medical treatment, not medical research.[23] As such, the DoD contended, informed consent was not required, since military law permits commanding officers to order medical treatment deemed necessary for military missions.

The FDA was hesitant to interpret its existing regulations to allow the administration of investigational products without informed consent and suggested to the DoD that the FDA create a new regulation—specifically tailored to the military—that would grant the FDA the ability to issue an informed consent waiver on a case-by-case basis. Codifying a new regulation would require approval from HHS, the National Institutes of Health Office for the Protection of Research Risks, and the White House Office of Management and Budget. Under the proposed framework, the DoD would need to request an informed consent waiver and explain why informed consent was not feasible.[24]

In addition to its discussions with the FDA, the DoD held several internal meetings regarding PB and the BT vaccine. During a US Army Medical Research Institute of Infectious Diseases (USAMRIID) Human Use Committee meeting on October 4, 1990, which was held to discuss the proposed BT vaccine mandate, Medical Division Lieutenant Colonel Kelly McKee indicated that the "intent of this protocol is benefit rather than scientific inquiry."[25] McKee further stated that "thorough documentation of informed consent . . . would be a logistical nightmare" and that the supply of the vaccine was low and, therefore, "not enough to vaccinate all the U.S. troops in the Middle East."[26] Accordingly, McKee concluded, a "risk analysis of the theater of operations will determine which troops get vaccinated and which do not."[27] The chairman of the committee, Colonel Arthur Anderson, added: "The limited availability of vaccine could have adverse effects on morale and extensive informed consent documentation would hasten spread of this knowledge."[28] Nonetheless, Anderson stated, "I feel oral informed consent would be expedient and would show a desire to adhere to the high principles of the Declaration of Helsinki."[29] The committee unanimously approved the protocol "with the recommendation that requirement of full compliance with documented informed consent . . . be waived and in its stead an abbreviated oral informed consent statement be substituted and administered."[30]

By the beginning of December 1990, intelligence experts posited that Iraq probably had brought chemical weapons into Kuwait and was prepared to use them.[31] On December 21, 1990, the FDA announced that "because of the urgency created by current military operations in Operation Desert Shield," the agency was "issuing an interim regulation to amend its current informed consent regulations."[32] The interim rule took effect immediately, without the standard period of public review and comment that federal law typically requires of new regulations.[33]

Under the interim rule, the FDA Commissioner maintained the authority to determine "that obtaining informed consent from military personnel for the use of an investigational drug or biologic is not feasible in certain battlefield or combat-related situations."[34] The new regulation expanded the definition of when informed consent is "not feasible" to include "combat (actual or threatened) circumstances in which the health of the individual or the safety of other military personnel, may require that a particular drug or biologic for prevention or treatment be provided to a specified group of military personnel, without regard to any individual's personal preference for no treatment or for some alternative treatment."[35]

This was a drastic expansion of an existing regulation, 21 C.F.R. § 50.23, which defines the term "not feasible." For decades, the FDA interpreted the term "not feasible" in that provision strictly in a medical sense—to encompass instances where a patient was in a medical emergency and obtaining informed consent for life-saving medical treatment was not practicable because the patient was incapacitated due to their injury or illness.[36] Under the new rule, the "not feasible" exception to informed consent now would also include combat situations and war preparations.

One week after the FDA adopted the new regulation, the DoD requested informed consent waivers for PB and the BT vaccine.[37] The DoD stated that "obtaining informed consent is not feasible . . . because of military combat exigencies."[38] The DoD further indicated that administration of the vaccine "without informed consent has been concurred in by a duly constituted institutional review board."[39] The DoD concluded that it would not be feasible to obtain informed consent since a soldier's "personal preference" does not take precedence over the military's view that the drug and vaccine would contribute to the "safety of other personnel in a soldier's unit and the accomplishment of the combat mission."[40] The DoD assured the FDA that "we have nothing exotic in the works," and highlighted that "In all peacetime applications, we believe strongly in informed consent and its ethical foundations."[41]

The FDA granted the informed consent waivers within seventy-two hours of the DoD's request, finding that "informed consent is not feasible and that withholding treatment would be contrary to the best interests of military personnel."[42] In its waiver request to the FDA, however, the DoD did not disclose the findings of the USAMRIID ethics committee (which had advocated for oral informed consent), but rather commissioned a second ethics committee that did not recommend that informed consent, oral or written, be obtained. This was despite the fact that, as Anderson explained, the USAMRIID committee had "primary jurisdiction over protocols prepared by USAMRIID investigators regardless whether the subject population is in house or located anywhere in the world."[43] The recommendation of the first USAMRIID ethics committee was not publicly revealed until 1997.[44]

Lawsuits and Unanticipated Twists

On January 11, 1991—days after the FDA issued the informed consent waivers—American service members filed a lawsuit contending that (1) the FDA exceeded its regulatory authority in issuing the interim rule; (2) the DoD's decision to administer PB and the BT vaccine without obtaining informed consent violated the DoD Authorization Act, which prohibits research without informed consent; and (3) use of the countermeasures without informed consent violated the constitutional rights of service members.[45] The court issued its decision on January 31, 1991, two weeks after US military personnel entered Kuwait.

The court held that, pursuant to the political question doctrine, "the DoD's decision to administer the unapproved drugs is a military decision that is not subject to judicial review."[46] This finding was sufficient to dismiss the entire case. The court explained that even if the DoD's decision were subject to judicial review, the service members still would lose since the court must defer to the FDA's interpretation of the phrase "not feasible."[47] Since the FDA's definition was not arbitrary or capricious, the FDA did not exceed its regulatory authority, and principles of administrative law dictate that the court could not overrule the agency's decision.[48]

The court further held that the countermeasure mandates do not constitute research because the "primary purpose of administering the drugs is military, not scientific."[49] The court noted that, in other instances, the FDA "has interpreted the FDCA to permit using unapproved drugs in a treatment-investigational setting," and thus, the agency "does not view every use of unapproved drugs as research."[50] In making this analogy, the court cited a regulation that permits treatment-related access to non-FDA-approved products in life-threatening situations where no satisfactory treatment exists.[51] This provision is often referred to as "compassionate use," whereby a patient may gain access to an experimental treatment if the patient, their doctor, and the drug company each provide consent. Finally, the court held that the countermeasure mandates and informed consent waivers are rationally related to a legitimate government purpose, and thus, there is no violation of the service members' constitutional rights.[52]

Thus, the FDA's informed consent waivers were validated by a federal judge. Given the limited supply of the BT vaccine, military leaders ordered that vaccination be based on a classified framework whereby troops deployed to risky areas would first be immunized. The antitoxin used to create the BT vaccine is derived from the blood of animals exposed to botulinum. As of 1990—despite over five decades of defensive research into biological warfare and years of intelligence that detailed a buildup of biological weapons across the globe—the United States had only one horse that could produce the antitoxin.[53] As General Robert Belihar testified in 1996, the decision on how to allocate the vaccine was made by General Schwarzkopf and kept classified "because the patterns of vaccine administration would indicate, perhaps, how those troops were going to be employed."[54]

In the end, despite months of negotiations and repeated affirmations that military discipline would suffer without an informed consent waiver, in some instances, informed consent was not waived, and service members could refuse the BT vaccine.[55] Notwithstanding the DoD's legal victory in the informed consent waiver lawsuit,

General Schwarzkopf ordered that all service members be given a release form to read and sign, and that any service member who wanted to refuse the vaccine could. Adherence to Schwarzkopf's order was inconsistent, and some service members were administered the vaccine without being given a choice to refuse.[56] Approximately eight thousand service members received the BT vaccine; most received two doses, even though immunity was likely to confer only after three doses.[57]

About 150,000 service members also received an anthrax vaccine,[58] prophylaxis that was not included in the informed consent waivers issued by the FDA despite the fact that the DoD sought to use the vaccine for an off-label indication as a countermeasure to protect against weaponized inhalation anthrax. The vaccine earned regulatory approval in 1970 as prophylaxis for cutaneous anthrax, which occurs when anthrax comes into contact with human skin. Cutaneous anthrax had been reported in individuals who handled anthrax spores in a lab or worked with grazing animals such as sheep and cattle, where anthrax can grow naturally. By the late 1980s, the vaccine had been administered to a very limited population—primarily wool workers, veterinarians, and biosecurity researchers.[59]

The DoD was concerned that Iraq had weaponized anthrax as an aerosol. Since the FDA had not evaluated the vaccine as prophylaxis for inhalation anthrax, the FDA did not have any guidelines on what dosage was appropriate in that context; accordingly, the DoD adopted the cutaneous anthrax dose schedule, whereby six doses over an eighteen-month period were needed to confer immunity. Even the six-dose series was speculative—decades earlier the dosage was doubled from three to six after some vaccinated wool workers contracted cutaneous anthrax.[60] Given time pressures and limited vaccine supply, few service members deployed to the Gulf received more than one dose of the anthrax vaccine.[61]

Warfighters also were issued a Mark I Nerve Agent Antidote Kit, which contained PB, Atropine, and 2-PAM. The hope was that preexposure treatment with PB, when combined with postexposure treatment with Atropine and 2-PAM, would enable a service member to survive Soman exposure.[62] However, not all service members were provided PB, and commanders had the responsibility for deciding when to begin or discontinue use of the drug.[63] As a point of comparison, coalition forces from Canada and Britain received anthrax and plague vaccines as prophylaxis against anticipated biological warfare, but did not receive PB or the BT vaccine.[64]

After the war, the DoD informed the FDA that PB "tablets were used without prior informed consent."[65] The DoD estimated that more than 250,000 service members used PB, but that there was great variation in use.[66] The FDA instructions for PB required one tablet every eight hours over a seven-day period. However, service members self-administered PB; many reported that they did not take the drug, took less than the required twenty-one pills, or used an alternative schedule and did not take one pill every eight hours.[67] Some service members took more than the recommended amount due to the mistaken belief that more pills would provide better protection.[68] Others simply tossed their pills into the desert.

Overall, administration of the countermeasures differed significantly from the FDA requirements. As a condition of FDA permission to use the medical products without informed consent, the DoD agreed to (1) provide a PB information sheet to all service members, including information on expected side effects and the reason for use of the tablets; (2) collect, review, and make reports of adverse events; (3) label PB "For military use and evaluation"; (4) ensure that each dose of the BT vaccine was recorded in each service member's medical record; and (5) maintain adequate records related to the receipt, shipment, and disposition of the BT vaccine.[69] The DoD failed to comply with each of the requirements.[70] Moreover, in surveys conducted after the war, many service members indicated that they were told PB was FDA-approved.[71] This was a true but misleading statement, since PB was FDA-approved as a treatment for myasthenia gravis but not as prophylaxis for Soman.

Medical recordkeeping was woefully inadequate, and failed to adhere to the FDA's requirements and the DoD's own internal policies. As one report found:

> The secrecy of the vaccination program complicated recordkeeping and created some confusion and fear among service members. Medical personnel in the field received instructions that receiving the shots was classified "Secret" and that the shots were not to be discussed with anyone. DOD asserts the secrecy protected troops since it limited Iraq's knowledge of U.S. defensive capabilities. When the vaccinations were recorded in medical records retained by individual service members, they were encoded to eliminate document classification problems. . . . According to testimony presented to the Committee, in the flurry of personnel anxious to come home at the end of the Gulf War, much of the documentation about vaccinations was lost or destroyed.[72]

After the war, officials concluded that intelligence reports were incorrect and that Iraq did not possess Soman. As such, PB ended up being unnecessary since it could only potentially mitigate the effects of Soman. In fact, there is a chance that PB was harmful since the drug is contraindicated for individuals exposed to low-level nerve agents other than Soman. And, there is evidence that US troops were exposed to low-level Sarin. Moreover, PB is contradicted for individuals with underlying health conditions such as asthma, peptic ulcers, or diseases of the liver, kidney, or heart—yet, a 1994 Senate report found that service members were not screened for, or informed about, these contraindications.[73]

Operation Desert Storm involved forty-one days of an intensive air campaign and one hundred hours of a ground attack, though the legal battle over informed consent waivers continued long after the war. On July 16, 1991, a three-judge appellate court panel reversed part of the district court's decision. Specifically, the appellate court held that, although "deference is owed to the political branches in military matters," the court can review the issue here because it centers on FDA's issuance of the Interim Rule, not "military action."[74]

Notwithstanding this important component of the decision, which rejected the military's stance that the political question doctrine precludes court review, the appellate court reasoned that the FDA did not exceed its authority since the agency's interpretation of "not feasible" was not arbitrary and capricious, and thus was within the agency's discretion.[75] Despite this finding, the court noted that the "not feasible" standard set forth in the interim rule "might at some future time be applied in violation of" a service member's constitutional rights.[76]

The Elusive Causes of Gulf War Illness

Of the 694,550 American service members deployed to the Gulf region, 148 suffered combat deaths and 145 died due to accident or disease; 467 were wounded.[77] These figures represent the lowest casualty rates experienced by the United States in any major conflict during the twentieth century.[78] Following the war, however, veterans began suffering from debilitating chronic illnesses with symptoms that included fatigue, joint pain, headache, dermatitis, and memory loss.[79] Other health conditions included gastrointestinal disorders, musculoskeletal problems, respiratory issues, sensitivity to chemicals, mental health issues, cancer, reproductive health issues, and children born with congenital disorders at alarming rates.[80] A November 1995 article in *Life* magazine, *The Tiny Victims of Desert Storm*, highlighted the tragic plight of Gulf War veterans and their families, with a focus on children born without limbs or with other abnormalities.[81]

Researchers conducted scores of studies, and several committees investigated what eventually would be named Gulf War Illness, a multisymptomatic disorder that has affected approximately 200,000 Gulf War veterans.[82] Nearly three decades after the war, causal links remain elusive, though suspected risk factors include PB, the BT vaccine, the anthrax vaccine, occupational exposure to paint and petroleum products, psychological and physical stress, insecticides and pesticides, depleted uranium, sand, smoke from oil fires, exposure to chemical or biological weapons, endemic infectious diseases, or some combination thereof.[83] During the war, the United States deployed advanced depleted uranium shells, rockets, and missiles: these weapons were utilized despite studies that warned the weapons would produce radioactive clouds and toxic rubble that could lead to cancer or other long-term health concerns.[84] US forces also bombed two nuclear reactors, which may have caused radioactive contamination and health hazards.[85]

Some studies have found that PB may cause neurological disorders if taken under high-stress conditions, such as those experienced on a battlefield.[86] Other studies observed a synergistic negative effect between PB and nerve agents, pesticides, or insect repellants utilized during the war.[87] The BT vaccine is known to cause fatigue and muscle problems, two common ailments of Gulf War veterans. Moreover, the BT vaccine doses were twenty years old at the time they were administered, and some experts feared that the vaccines became toxic over time.[88]

In a comprehensive study published in 2008, PB and pesticides were identified as the two "significant risk factors" among the long list of potential causes of Gulf War Illness.[89] The report also uncovered dose-response effects, "indicating that veterans who took PB for longer periods of time have higher illness rates than veterans who took less PB."[90] The 2008 finding echoed the findings of a 1993 animal study, which found that exposure to PB and DEET—a common insect repellant that was used during the war—increased the toxicity of both products significantly. The 1993 study concluded that dual exposure "could explain the serious neurological symptoms experienced by so many Gulf War veterans."[91] Although the causes remain elusive, research into Gulf War Illness continues; for example, a 2022 study found that veterans who had a gene that helped metabolize Sarin were less likely to develop symptoms.[92]

While the US military and government maintain that Iraq did not utilize biological or chemical weapons during the war, service members were exposed to chemical agents. Allied bombings damaged or destroyed Iraqi chemical munitions filled with Sarin and mustard gas, and on several occasions, coalition forces detected Sarin and mustard agents in the air.[93] Although soldiers from coalition forces were ordered to use their gas masks, US commanders refused to issue similar orders because they did not want to create a panic among military personnel.[94] This decision was made despite the fact that PB was contraindicated for anyone exposed to Sarin. Figure 13.1 depicts a cache of Iraqi chemical weapons that date back to the 1980s, which was later uncovered by the US Army.[95]

According to Patrick Eddington, a former CIA analyst who worked on intelligence matters related to the Persian Gulf War, "the United States was not prepared to face a chemically armed opponent," let alone one "that we ourselves had helped to arm."[96] In his book, *Gassed in the Gulf*, Eddington documents over fifty instances where monitors detected chemical weapons exposure in Kuwait, Iraq, and Saudi Arabia.[97] Eddington also details a series of meetings where he and other CIA officials discussed the DoD's reluctance to publicly state that Gulf War Illness may be linked to repeated low-level exposure to chemical weapons. He posits that Iraq may have intentionally used "limited CW [chemical weapon] payloads" that would cause low-level exposure "with the purpose of causing long-term casualties against Coalition forces."[98] As

Figure 13.1 Cache of Iraqi chemical weapons, dating back to the 1980s, uncovered by the US Army in 2005. During the post-9/11 war in Iraq, hundreds of American service members suffered adverse health effects from exposure to degrading chemicals from the old weapons.
U.S. Army photograph.

Eddington elaborates, such a tactic is "consistent with Iraq's military strategy during the Iran-Iraq war, where Iraqi forces sought to wear down their more numerous Iranian enemies with both conventional and unconventional chemical attacks."[99]

Indeed, the Presidential Advisory Committee on Gulf War Veterans' Illnesses (Gulf War PAC), established by President Bill Clinton in May 1995, "found substantial evidence of site-specific, low-level exposures to chemical warfare agents."[100] Perhaps most importantly, the Gulf War PAC "found DOD's investigations to date superficial and unlikely to provide credible answers to veterans' and the public's questions."[101] The committee candidly indicated that the DoD "did not act in good faith in this regard."[102] As the Gulf War PAC further observed: "The most striking feature of our evaluation of the government's response to Gulf War veterans' health issues has been the parallels between the experiences of these veterans and veterans of previous conflicts."[103] The committee highlighted difficulties in access to care that many veterans faced, including "Inadequate information, delays in scheduling appointments, insensitive personnel, and inadequate follow-up."[104]

The committee explained that one key hurdle to identifying the causes of Gulf War Illness was the lack of baseline data on the health of service members prior to deployment.[105] Going forward, the Gulf War PAC recommended that the DoD conduct standardized predeployment physical examinations that are structured to facilitate postconflict medical surveillance and epidemiological studies.[106] In response to the Gulf War PAC's findings, the DoD acknowledged that "more needs to be and will be done," indicating that its goal was to ensure "that every service member is fully informed during orientation and training of the health risks, benefits, and proper use of all medical countermeasures, and, that when used, such countermeasures are documented and maintained as part of the individual's health record."[107]

The Gulf War PAC highlighted the public's mistrust of the DoD and the government, observing that the "DOD's slow and erratic efforts to release information to the public have further served to erode the public's trust."[108] One salient example occurred in 1996 when approximately four hundred declassified documents were removed from GulfLINK, a website created specifically to provide information on Gulf War Illness. The documents were not reclassified, but it took nearly nine months for them to be restored to the website.[109] The Gulf War PAC noted that this lack of trust is not limited to veterans, but rather represents a nationwide perspective and "increasingly strongly held view that DOD is still withholding relevant information from concerned veterans and the public."[110]

It was later revealed that, during discussions regarding the FDA's issuance of the informed consent waivers, the DoD withheld important information from the FDA. For example, the DoD did not disclose to the FDA that PB could lessen the protection that service members would otherwise receive from Atropine and 2-PAM against nerve agents such as Sarin or VX.[111] The Gulf War PAC also criticized the DoD's implementation of the informed consent waivers for PB and the BT vaccine, and recommended that the DoD "establish a quality assurance program to ensure compliance with pre-, during, and postdeployment medical assessment policies."[112] The committee further indicated that it was "concerned that FDA had failed, in the five years since the Gulf War, to devise better long-term methods governing military use of drugs and vaccines for CBW defense."[113]

FDA Revokes the Interim Rule

In 1996, three organizations—the National Veterans Legal Services, the National Gulf War Resource Center, and Public Citizen (a nonprofit public interest group dedicated to public health and consumer welfare)—filed a petition with the FDA that requested that the agency repeal the interim rule.[114] The petition questioned the ethical foundations of the rule, highlighted various implementation deficiencies, and concluded that "not only did the Interim Rule fail to operate in the manner the FDA intended, but it also allowed the military to circumvent the safeguards the FDA offered to rationalize this departure from its ordinary rules on informed consent."[115] The DoD urged the FDA to deny the petition, underscoring the important national security interests at play and stating that the "current rule is fully consistent with law and ethics."[116]

On July 31, 1997, more than six years after issuance of the interim rule, the FDA published a request for public comment on whether the agency should adopt, modify, or revoke the rule.[117] The FDA also requested comment regarding "the evidence needed to demonstrate safety and effectiveness for such investigational drugs that cannot ethically be tested on humans."[118] The FDA acknowledged that "the interim final rule did not work the way that the agency anticipated"[119] and highlighted that, "at the time, FDA gave considerable deference to the DOD's judgment and expertise regarding the feasibility of obtaining informed consent under battlefield conditions."[120]

According to a 1999 RAND report, behind the scenes, "it was known that FDA had proposed to revoke the authority the Interim Rule established, that DoD had objected to revocation, and that the matter was under discussion at the Office of Management and Budget."[121] Thereafter, language was inserted into the Defense Authorization Act for Fiscal Year 1999, which created a new law that vested the authority to issue informed consent waivers with the President of the United States.[122] The law was enacted in October 1998 and codified at 10 U.S.C. § 1107(f), shifting decision-making power from the FDA to the president.

The FDA's interim rule became moot, and the agency officially revoked it in October 1999.[123] In doing so, the FDA explained:

> Experience with the use of the waiver provision of the 1990 interim rule suggests two conclusions: (1) To the extent possible, military personnel should receive treatments whose safety and effectiveness have been fully evaluated; (2) where it is necessary to utilize investigational agents and to waive informed consent, new standards and criteria for doing so should be developed that will better ensure protection of the troops receiving the investigational product.[124]

Coupled with its revocation of the 1990 interim rule, the FDA highlighted that 10 U.S.C. § 1107(f) "is silent about the standards and criteria that the President is to apply in making a determination that obtaining consent is not in the interests of national security."[125]

Coinciding with the FDA's revocation of the interim rule, President Clinton issued Executive Order 13139, titled *Improving Health Protection of Military Personnel Participating in Particular Military Operations*. Under the Executive Order, the DoD can request that the president waive informed consent requirements for investigational drugs. Such a waiver can be granted only if the president provides a written determination "that obtaining consent: (1) is not feasible; (2) is contrary to the best interests of the member; or (3) is not in the interests of national security."[126] Executive Order 13139 outlines the procedure for seeking a presidential waiver of informed consent but does not provide substantive standards that govern the consideration of a request. Rather, the Executive Order indicates that the president must apply the standards set forth by the FDA in 21 C.F.R. § 50.23(d).[127]

Following issuance of the Executive Order, the FDA amended 21 C.F.R. § 50.23(d) to identify criteria that the president should assess when considering an informed consent waiver under 10 U.S.C. § 1107(f). Similar to the FDA's guidelines under the 1990 interim rule, the new FDA regulation required that the president balance military needs with available data regarding safety and efficacy and consider the viability of alternative treatments or preventive measures.[128] The guidelines further mandate that, in the petition for an informed consent waiver, the DoD must include the meeting minutes from IRB deliberations.[129] The latter requirement likely was included because of the DoD's suppression of the USAMRIID IRB recommendations from the informed consent waiver applications for PB and the BT vaccine.

The updated 21 C.F.R. § 50.23(d) indicates that informed consent waivers may need to be classified, and thus permits public notice of a waiver "as soon as practicable and consistent with classification requirements."[130] The FDA further stated that the agency is developing new guidelines for countermeasures in instances where studies in humans cannot ethically be conducted.[131] As will be discussed in chapter 14, 9/11 and the 2001 anthrax letter attacks accelerated the adoption of these new regulatory mechanisms.

Summary

The DoD's decision to administer investigational drugs and vaccines to hundreds of thousands of service members was not conducted with malice, but rather represented an exercise of military judgment that sought to balance the success of a military mission with warfighter protection. To be sure, the DoD did not act honorably in withholding information regarding countermeasure risks and the USAMRIID ethics committee recommendation regarding informed consent, nor did the DoD act properly when it failed to abide by the informed consent waiver conditions imposed by the FDA. The lack of trust created by these shortcomings was exacerbated by the DoD's shoddy efforts to unpack the causes of Gulf War Illness.

As the principal agency responsible for ensuring the safety and efficacy of medical products, the FDA could have served as a check on the DoD's desire to use PB and the BT vaccine as countermeasures despite scant evidence of safety and effectiveness. Rather than doing so, the FDA facilitated the DoD's goals by bypassing standard regulatory requirements and creating military-specific protocols that increased the risk that service members would be exposed to unsafe or ineffective medical products. The medical experience of the Persian Gulf War reveals that a conflict with excellent medical preparations and very low casualty rates can evolve into a postwar health tragedy. In addition, the regulatory exceptions and court rulings set a precedent that military exigencies can justify mandatory administration of unlicensed medical products. Legal and ethical considerations regarding mandatory countermeasures became more complex after 9/11 and growing fears of attacks with biological or chemical agents.

Notes

1. Greenwood and Berry, *Medics*, 158–66.
2. Final Report, *Presidential Advisory Committee on Gulf War Veterans' Illnesses* (1996), 117–18.
3. General H. Norman Schwarzkopf, Letter to Members of the U.S. Army Medical Department, reprinted in *AMEDD Journal* (May/June 1993), quoted in Ginn, *The History*, 428.
4. M. Wade Markel, Alexandra Evans, Miranda Priebe, Adam Givens, Jameson Karns, and Gian Gentile, *The Evolution of U.S. Military Policy from the Constitution to the Present*, vol. 4 (2019), 85–123.
5. 45 C.F.R. Pt. 46 (1991); 32 C.F.R. Pt. 219 (1991).
6. HHS Office for Human Research Protections, "Federal Policy for the Protection of Human Subjects ('Common Rule')" (accessed April 24, 2024).
7. Executive Order 12333, § 2.10 (December 4, 1981).
8. *Doe v. Sullivan*, 938 F.2d 1370, 1372 n.1 (D.C. Cir. 1991).
9. Richard A. Rettig, *Military Use of Drugs Not Yet Approved by the FDA for CW/BW Defense* (1999), 3–7.
10. Michael A. Dunn and Frederick R. Sidell, "Progress in Medical Defense Against Nerve Agents," *Journal of the American Medical Association* 262, no. 5 (August 4, 1989): 649–52; Stanley L. Hartgraves and Michael R. Murphy, "Behavioral Effects of Low-Dose Nerve Agents," in *Chemical Warfare Agents*, edited by Satu M. Somani (New York: Academic Press, 1992), 125–54, cited in Eddington, *Gassed in the Gulf*, 1819.
11. IND No. 23509 (March 1984), cited in Rettig, *Military Use*, 6.
12. Dunn and Sidell, "Progress in Medical Defense."
13. "Minutes of the Ninety-Third Meeting of the USAMRIID Human Use Committee (4 Oct. 1990): Memorandum for Record" (October 5, 1990): 2–4.
14. 21 U.S.C. § 355 (2024); 21 C.F.R. § 50.25 (2024); FDA, *Understanding Investigational Drugs* (accessed April 24, 2024).
15. Food and Drug Administration Interim Rule: Informed Consent for Human Drugs and Biologics; Determination That Informed Consent Is Not Feasible, 55 Fed. Reg. 52,814 (December 21, 1990).
16. Food and Drug Administration, *About FDA: Patient Q&A* (accessed April 4, 2024); Shariful Syed, Brigham A. Dixon, Eduardo Constantino, and Judith Regan, "The Law and Practice of Off-Label Prescribing and Physician Promotion," *Journal of the American Academy of Psychiatry and the Law* 49, no. 1 (March 2021): 53–59.
17. Syed et al., "The Law and Practice of Off-Label Prescribing"; Judith G. Edersheim and Theodore A. Stern, "Liability Associated with Prescribing Medications," *Primary Care Companion to the Journal of Clinical Psychiatry* 11, no. 3 (2009): 115–19.
18. George H. Sisson, "Memorandum for Record: Meeting with FDA, Friday, 14 Sept. 1990" (September 17, 1990), cited in Rettig, *Military Use*, 19.
19. Craig R. Lehman, "Memorandum for Record: Proceedings of Meeting Between FDA and DOD Regarding Operation Desert Shield" (August 30, 1990), cited in Rettig, *Military Use*, 16–18.

20. Rettig, *Military Use*, 16–17.
21. Pub. L. 92-570, § 745, 86 Stat. 1203 (October 26, 1972).
22. Pub. L. 98-525, 98 Stat. 2615 (October 19, 1984).
23. Robert L. Gilliat, "Memorandum for the Assistant Secretary of Defense (Health Affairs): Applicability of Human Subject Research Restrictions to Potential Medical Treatments in Connection with Operation Desert Shield" (September 14, 1990), cited in Rettig, *Military Use*, 19–21.
24. Rettig, *Military Use*, 22–23.
25. "Minutes of the Ninety-Third Meeting," 3.
26. "Minutes of the Ninety-Third Meeting," 3.
27. "Minutes of the Ninety-Third Meeting," 3.
28. "Minutes of the Ninety-Third Meeting," 3.
29. "Minutes of the Ninety-Third Meeting," 3.
30. "Minutes of the Ninety-Third Meeting," 5.
31. Eddington, *Gassed in the Gulf*, xiii.
32. FDA Interim Rule, 55 Fed. Reg. 52,814.
33. FDA Interim Rule, 55 Fed. Reg. 52,814.
34. FDA Interim Rule, 55 Fed. Reg. 52,814.
35. FDA Interim Rule, 55 Fed. Reg. 52,814.
36. William J. Curran, "Governmental Regulation of the Use of Human Subjects in Medical Research: The Approach of Two Federal Agencies," in Freund, *Experimentation with Human Subjects*, 422.
37. Enrique Mendez, Assistant Secretary of Defense, Letter to David Kessler, Commissioner of Food and Drugs (December 28, 1990).
38. Enrique Mendez, Assistant Secretary of Defense, Letter to David Kessler, Commissioner of Food and Drugs (December 28, 1990).
39. Enrique Mendez, Assistant Secretary of Defense, Letter to David Kessler, Commissioner of Food and Drugs (December 28, 1990).
40. *Doe v. Sullivan*, 938 F.2d at 1373.
41. Letter from Assistant Secretary of Defense (Health Affairs) to Assistant Secretary for Health, Department of Health and Human Services (October 30, 1990), reprinted in FDA Interim Rule, 55 Fed. Reg. 52,814–815.
42. David Kessler, Commissioner of Food and Drugs, Letter to Enrique Mendez, Assistant Secretary of Defense (December 31, 1990).
43. Arthur Anderson, Email to David Franz and Theresa Haupt (September 29, 1994).
44. Keith Epstein and Bill Sloat, "Objection to Gulf War Vaccine Was Overridden," *Cleveland Plain Dealer*, December 21, 1997.
45. *Doe v. Sullivan*, 756 F. Supp. 12, 14 (D.D.C. 1991).
46. *Doe v. Sullivan*, 756 F. Supp. at 14.
47. *Doe v. Sullivan*, 756 F. Supp. at 15–17.
48. *Doe v. Sullivan*, 756 F. Supp. at 16–17. In 2024, the US Supreme Court altered the legal standard of deference that courts should apply in determining whether an administrative agency has acted within its statutory authority or has properly interpreted an ambiguous statute. *Loper Bright Enterprises v. Raimondo*, 144 S. Ct. 2244 (2024). *Loper Bright* grants courts more discretion to review and overrule an agency action. However, all of the court decisions discussed in this book were issued prior to the *Loper Bright* ruling.
49. *Doe v. Sullivan*, 756 F. Supp. at 15–16.
50. *Doe v. Sullivan*, 756 F. Supp. at 16.
51. *Doe v. Sullivan*, 756 F. Supp. at 17–18 (citing 21 C.F.R. § 312.34).
52. *Doe v. Sullivan*, 756 F. Supp. at 17–18.

53. Miller et al., *Germs*, 106.
54. Presidential Advisory Committee on Gulf War Veterans' Illnesses, Transcript of January 12, 1996, quoted in Rettig, *Military Use*, 34–35.
55. Rettig, *Military Use*, 335.
56. Miller et al., *Germs*, 119.
57. Food and Drug Administration: Accessibility to New Drugs for Use in Military and Civilian Exigencies When Traditional Human Efficacy Studies Are Not Feasible; Determination Under the Interim Rule That Informed Consent Is Not Feasible for Military Exigencies; Request for Comments, 62 Fed. Reg. 40,996, 40,999 (July 31, 1997).
58. Interim Report, *Presidential Advisory Committee on Gulf War Veterans' Illnesses* (1996), 21.
59. Institute of Medicine, *The Anthrax Vaccine* (2002), 40–49.
60. Miller et al., *Germs*, 188–90.
61. Smith, *American Biodefense*, 97.
62. Tucker, *War of Nerves*, 304–5.
63. Rettig, *Military Use*, 32.
64. Mark Peakman, Ania Skowera, and Matthew Hotopf, "Immunological Dysfunction, Vaccination and Gulf War Illness," *Philosophical Transactions of the Royal Society of Britain* 361 (2006): 681–87.
65. Enrique Mendez, Assistant Secretary of Defense, Letter to David Kessler, Commissioner of Food and Drugs (March 15, 1991).
66. Rettig, *Military Use*, 32.
67. Jill R. Keeler, Charles G. Hurst, and Michael A. Dunn, "Pyridostigmine Used as a Nerve Agent Pretreatment Under Wartime Conditions," *Journal of the American Medical Association* 266, no. 5 (August 7, 1991): 693–95.
68. Anthony Swofford, *Jarhead* (2003), 183–84.
69. Food and Drug Administration: Human Drugs and Biologics; Determination that Informed Consent is NOT Feasible or Is Contrary to the Best Interests of Recipients; Revocation of 1990 Interim Final Rule; Establishment of New Interim Final Rule, 64 Fed. Reg. 54,180, 54,184–185 (October 5, 1999).
70. FDA: Human Drugs and Biologics, 64 Fed. Reg. 54,184–185.
71. FDA: Accessibility to New Drugs, 62 Fed. Reg. 40,999.
72. Interim Report, *Presidential Advisory Committee*, 22.
73. Staff of Senate Committee on Veterans' Affairs, "Is Military Research Hazardous to Veterans' Health?: Lessons Spanning Half a Century," 103rd Cong. (December 8, 1994), 28–31.
74. *Doe v. Sullivan*, 938 F.2d at 1379–81.
75. *Doe v. Sullivan*, 938 F.2d at 1381–82.
76. *Doe v. Sullivan*, 938 F.2d at 1383.
77. Final Report, *Presidential Advisory Committee*, 1.
78. Final Report, *Presidential Advisory Committee*, 66.
79. Interim Report, *Presidential Advisory Committee*, 1.
80. Final Report, *Presidential Advisory Committee*, 34, 84–85.
81. Derek Hudson, Kenneth Miller, and Jimmie Briggs, "The Tiny Victims of Desert Storm," *Life*, November 1995.
82. Research Advisory Committee on Gulf War Veterans' Illnesses, *Gulf War Illness and the Health of Gulf War Veterans* (2008), 4.
83. Interim Report, *Presidential Advisory Committee*, 25–26; Final Report, *Presidential Advisory Committee*, 125; Peakman et al., "Immunological Dysfunction."
84. Blum, *Killing Hope*, 334.
85. Blum, *Killing Hope*, 334.

86. Robert W. Haley et al., "Evaluation of Neurologic Function in Gulf War Veterans," *Journal of the American Medical Association* 277, no. 3 (January 15, 1997): 223–30.
87. Research Advisory Committee, *Gulf War Illness*, 6–15.
88. Moreno, *Undue Risk*, 272.
89. Research Advisory Committee, *Gulf War Illness*, 8.
90. Research Advisory Committee, *Gulf War Illness*, 8.
91. Staff of Senate Committee on Veterans' Affairs, "Is Military Research Hazardous," 32.
92. Robert W. Haley et al., "Evaluation of a Gene–Environment Interaction of *PON1* and Low-Level Nerve Agent Exposure with Gulf War Illness: A Prevalence Case-Control Study Drawn from the U.S. Military Health Survey's National Population Sample," *Environmental Health Perspectives* 130, no. 5 (May 2022): 1–16.
93. Final Report, *Presidential Advisory Committee*, 38–42.
94. Philip Shenon, "Czechs Say They Warned U.S. of Chemical Weapons in Gulf," *The New York Times*, October 19, 1996.
95. Patrick Dickson, "Troops in Iraq: We Were Exposed to Chemical Warfare Agents," *Stars and Stripes*, December 23, 2014.
96. Eddington, *Gassed in the Gulf*, xxi.
97. Eddington, *Gassed in the Gulf*, 116–17, 126–27, 291–99.
98. Eddington, *Gassed in the Gulf*, 94.
99. Eddington, *Gassed in the Gulf*, 261.
100. Final Report, *Presidential Advisory Committee*, ix.
101. Final Report, *Presidential Advisory Committee*, ix.
102. Final Report, *Presidential Advisory Committee*, 7.
103. Final Report, *Presidential Advisory Committee*, 7.
104. Final Report, *Presidential Advisory Committee*, 20.
105. Final Report, *Presidential Advisory Committee*, 19.
106. Final Report, *Presidential Advisory Committee*, 18–19, 50–56.
107. Persian Gulf Veterans Coordinating Board, "Response to the Presidential Advisory Committee on Gulf War Veterans' Illnesses" (March 7, 1997): 13, quoted in Rettig, *Military Use*, 38.
108. Final Report, *Presidential Advisory Committee*, 45.
109. Final Report, *Presidential Advisory Committee*, 45.
110. Final Report, *Presidential Advisory Committee*, 45–46.
111. Sidney M. Wolfe, "Statement on Petition to Repeal DOD/FDA Regulation," *Public Citizen* (May 7, 1996).
112. Final Report, *Presidential Advisory Committee*, 18.
113. Final Report, *Presidential Advisory Committee*, 18.
114. FDA: Accessibility to New Drugs, 62 Fed. Reg. 41,000.
115. FDA: Accessibility to New Drugs, 62 Fed. Reg. 41,000.
116. FDA: Accessibility to New Drugs, 62 Fed. Reg. 41,000.
117. FDA: Accessibility to New Drugs, 62 Fed. Reg. 40,996.
118. FDA: Accessibility to New Drugs, 62 Fed. Reg. 40,996.
119. FDA: Accessibility to New Drugs, 62 Fed. Reg. 40,996.
120. FDA: Accessibility to New Drugs, 62 Fed. Reg. 40,997.
121. Rettig, *Military Use*, xiv.
122. Rettig, *Military Use*, xiv.
123. FDA: Human Drugs and Biologics, 64 Fed. Reg. 54,180.
124. FDA: Human Drugs and Biologics, 64 Fed. Reg. 54,184.
125. FDA: Human Drugs and Biologics, 64 Fed. Reg. 54,185.

126. Executive Order No. 13139 (September 30, 1999), Improving Health Protection of Military Personnel Participating in Particular Military Operations, 64 Fed. Reg. 54,175 (October 5, 1999).
127. Executive Order No. 13139, 64 Fed. Reg. 54,176.
128. FDA: Human Drugs and Biologics, 64 Fed. Reg. 54,185–186.
129. FDA: Human Drugs and Biologics, 64 Fed. Reg. 54,185–186.
130. FDA: Human Drugs and Biologics, 64 Fed. Reg. 54,186.
131. FDA: Human Drugs and Biologics, 64 Fed. Reg. 54,185.

14
New Laws to Facilitate the Development and Administration of Medical Countermeasures

Introduction

The controversy surrounding the Food and Drug Administration's 1990 interim rule and the devastating impact of Gulf War Illness did not diminish the ongoing need to develop countermeasures to combat chemical, biological, radiological, and nuclear (CBRN) agents. The rising national security threats and related public health concerns were underscored by a series of terrorist acts, most notably 9/11 and the 2001 anthrax letter attacks. As detailed in chapter 15, these events catapulted the United States into a "war on terror" that brought substantial military commitments around the world and served as a justification for far-reaching programs such as extraordinary rendition, enhanced interrogation, and dragnet surveillance.

For the military biomedical complex, a component of the response to CBRN vulnerabilities was the creation of anthrax and smallpox immunization programs. Both were controversial. Political turmoil, vaccine development challenges, and repeated violations of vaccine manufacturing requirements plagued the anthrax vaccine program. Congress publicly rebuked the program, military personnel filed lawsuits to protest the administration of the vaccine, and a federal court halted the program twice due to the failure of the Department of Defense (DoD) to follow informed consent requirements and the failure of the Food and Drug Administration (FDA) to follow its own vaccine approval rules. Concurrently, public health leaders widely criticized the smallpox vaccine program as a medically unnecessary political tactic that sought to inflate public fears of a biological attack to gain public support for President George W. Bush's geopolitical and military agenda. As the smallpox vaccine program was implemented among service members and civilian first responders, the vaccine caused significant side effects, including more than one hundred heart attacks and several deaths.

Meanwhile, during litigation that challenged mandatory inoculations under the anthrax vaccine program—and following extensive lobbying from the DoD and pharmaceutical industry—Congress created new laws regarding CBRN countermeasures. The laws provided the FDA with new regulatory pathways that bypassed standard FDA protocols for safety and efficacy review. The laws also earmarked significant funding for biosecurity, and granted broad legal shields for countermeasure-induced injuries that protected the government and industry from lawsuits filed by injured patients. Although the laws have facilitated the development and administration of CBRN countermeasures, they also have increased the risk that individuals will be exposed to unsafe or ineffective medical products. As of mid-2024, the laws are still in force.

America's Military Biomedical Complex. Efthimios Parasidis, Oxford University Press.
 DOI: 10.1093/9780199351473.003.0015

Rising Threats

Terrorist attacks throughout the 1990s and early 2000s reshaped America's national security landscape. In 1993, al Qaeda operatives detonated a bomb in the subterranean parking garage of the World Trade Center in New York, killing six people, injuring more than a thousand, and causing panic throughout the United States. The bomb carved a hundred-foot crater and caused massive damage across seven floors, six of which were below ground.[1] That same year, a congressional report estimated that a hundred-kilogram release of anthrax outside an American city would match or exceed the damage of a nuclear bomb.[2] In March 1995, members of the Japanese religious group Aum Shinrikyo released Sarin gas in the Tokyo metro by using their umbrellas to puncture holes in packages placed under their seats. Within fifteen seconds, Sarin's toxic effects kicked in: twelve people died and more than five thousand were injured. Less than a year earlier, the group killed seven people in a Sarin attack at a mountain resort outside Tokyo, and between 1990 and 1993, the group sprayed botulinum toxin and anthrax from buildings and moving cars in various neighborhoods near Tokyo. Further investigations revealed that Aum Shinrikyo was a worldwide movement with over $1 billion in assets and more than sixty thousand members in Australia, Germany, Russia, Taiwan, Ukraine, and the United States.[3] At one point, the group attempted to acquire the Ebola virus from a source in Zaire.[4]

In April 1995, Americans Timothy McVeigh and Terry Nichols discharged a massive truck bomb that ripped apart a federal building in Oklahoma City. The bomb injured more than 680 and killed 168, including 19 children. Within weeks of the bombing, Larry Harris of the American white supremacist group Aryan Nations ordered three vials of plague bacteria from the American Type Culture Collection, a Maryland supplier that sells biological agents to researchers throughout the world. As the vials were in transit, Harris called to inquire why it was taking so long for the bacteria to arrive. This raised suspicions, and the supplier reported the incident to federal authorities, who arrested Harris.[5]

The attempt by Aryan Nations to obtain plague bacteria followed a March 1995 conviction of an American group called the Patriots Council, which possessed ricin and planned on poisoning government officials by spreading the toxic substance on their doorknobs.[6] The actions of Aryan Nations and the Patriots Council mirrored a terrorist blueprint set by fringe American groups from the 1970s and 1980s. During the early 1970s, members of a right-wing group called the Order of the Rising Sun were arrested for possession of typhoid bacteria: they had planned to poison water supplies in Chicago, St. Louis, and other cities.[7] In 1984, members of the Rajneesh movement used typhoid bacteria to poison several salad bars in an attempt to sicken locals and sway an election in rural Oregon—more than 750 people became ill and dozens were hospitalized.[8]

By the mid-1990s, reports indicated that more than a dozen countries and terrorist groups maintained biological or chemical weapons programs, and warned of an attack with a weapon of mass destruction.[9] In August 1996 and February 1998, al Qaeda leader Osama bin Laden issued fatwas, calling on all Muslims to fight against nations that support Israel or maintain military forces in Islamic countries.[10] In 1997, Saddam Hussein expelled US members from an international team that was inspecting Iraq's biological weapons program. Previously, UN inspectors had determined that Iraq had placed more than 180 biological bombs and warheads at military installations throughout the country. The United States suspected that Iraq still maintained biological weapons and was expanding its chemical warfare program.[11]

On August 7, 1998—eight years to the day when US military personnel were ordered to Saudi Arabia in preparation for the Persian Gulf War—al Qaeda exploded bombs at US embassies in Kenya and Tanzania, killing 224 (including twelve Americans) and injuring more than 5,000.[12] Less than two weeks later, the United States conducted air strikes on al Qaeda-affiliated sites in Afghanistan and Sudan, one of which was a Sudanese pharmaceutical factory suspected of producing chemical weapons.[13] Leaders from government, industry, and academia held a 1999 symposium that underscored bioterrorism risks and how to structure a medical and public health response to an attack.[14] Thereafter, in October 2000, al Qaeda suicide bombers attacked the *USS Cole* as it was docked in Yemen, tearing a massive hole in the side of the ship and killing seventeen sailors.

The public health and national security concerns amplified exponentially following the 9/11 attacks and a series of anthrax-laden letters sent to lawmakers and media outlets during the autumn of 2001. By the end of the anthrax letter scare, twenty-two people were diagnosed with anthrax, five of whom died. Several buildings were contaminated and underwent extensive decontamination procedures, and more than 1.8 million letters and parcels were quarantined.[15]

Congressional hearings conducted after the anthrax letter attacks revealed that the Centers for Disease Control and Prevention (CDC) did not maintain accurate records regarding the identification of all US labs which handled pathogens that could be weaponized. The hearings also brought to light the fact that approximately 550 American labs and 1,000 labs in other nations were equipped to handle such biological agents.[16] These disclosures—coupled with the series of terrorist attacks throughout the 1990s—underscored the potential for future attacks with chemical or biological agents.

In light of the rising threats, CBRN countermeasures were viewed as an integral component of military preparedness. Congress increased funding for biosecurity, and established new laws to facilitate the development of CBRN countermeasures. In addition, the military instituted an anthrax vaccine immunization program, and the government created a smallpox vaccination program for military personnel and civilian first responders. Each of these endeavors was mired in controversy.

The Anthrax Vaccine Immunization Program

The anthrax vaccine was one of several countermeasures administered to service members during the Persian Gulf War. Due to the ongoing threat that a nation would deploy weaponized anthrax against US interests, after the Persian Gulf War the DoD considered whether it should create a force-wide anthrax vaccine immunization program (AVIP). To institute the AVIP, several regulatory and logistical concerns had to be addressed. One central issue was FDA approval. Although the vaccine was FDA-approved as prophylaxis to protect against cutaneous anthrax—which is anthrax that comes into contact with skin—the DoD sought to expand the vaccine's on-label indication to include protection against weaponized aerosol anthrax.

On October 5, 1995, the US Army Medical Research and Material Command wrote to the Michigan Department of Public Health, the manufacturer of the anthrax vaccine. The Army detailed a plan to file an application with the FDA to expand the vaccine's on-label indication to include protection against aerosol exposure to anthrax spores.[17] A decade earlier, the FDA had stated that the efficacy of the vaccine "against inhalation anthrax is not well documented" and concluded that the "anthrax vaccine is safe and effective under the limited circumstances for which this vaccine is employed," specifically, to protect against cutaneous anthrax.[18] Weeks after the October 1995 letter, an internal DoD memo noted that use of the vaccine to protect against aerosol exposure to anthrax "can only legally be done if the FDA licenses the vaccine for that specific schedule and indication."[19] The application to expand the on-label indication to include inhalation anthrax was filed with the FDA in September 1996.[20]

In January 1997, while the FDA application was still pending, Deputy Secretary of Defense John White indicated that the AVIP was ill-timed due to the ongoing controversy regarding the possible role of medical countermeasures in Gulf War Illness. White suggested that the AVIP be deferred for a few years and ordered a new review of the safety and efficacy of the anthrax vaccine. White also asked his subordinates to create a plan for how the DoD could attempt to persuade service members to accept administration of the vaccine.[21]

Notwithstanding White's recommendations and previous DoD discussions regarding the need for FDA approval to expand the vaccine label to include protection against inhalation anthrax, on March 4, 1997, the Pentagon's top civilian health official, Assistant Secretary of Defense for Health Affairs, Stephen Joseph, wrote to FDA Deputy Commissioner Michael Friedman, stating that the "DoD has long interpreted the scope of the license to include inhalation exposure, including that which would occur in a biological warfare context." Joseph inquired "whether the FDA has any objection to our interpretation of the scope of the licensure for the anthrax vaccine."[22]

Nine days later, and while the 1996 vaccine label change application was still pending, Friedman responded: "While there is a paucity of data regarding the effectiveness of Anthrax Vaccine for prevention of inhalation anthrax, the current package insert does not preclude this use."[23] FDA regulations deem such response letters as nonbinding informal communications that do not necessarily represent the formal positions of the agency; thus, Friedman's letter did not equate to an official FDA decision that the evidence supports a label change to include inhalation anthrax as an on-label indication for the vaccine.[24] Nevertheless, based on the representations made in Friedman's letter, the DoD proceeded to implement the AVIP, even though the label expansion application had not been granted.[25]

The Michigan Department of Public Health operated the only lab that produced the anthrax vaccine, and two days prior to Friedman's letter the FDA had issued a warning to the lab.[26] FDA inspectors found rusty lab equipment, live and dead insects throughout the facility, and equipment that was not properly sterilized.[27] The warning was the third in four years whereby the FDA found "significant deviations" from FDA manufacturing requirements at the lab, and the FDA indicated that it would revoke the lab's license to produce vaccines absent immediate action to address the concerns.[28]

In light of the FDA's findings regarding the lab deficiencies, the DoD decided to test the lots of its stockpiled vaccines. Some officials questioned the approach. For example, Michael Gilbreath, of the Joint Program Office for Biological Defense, wrote: "Are you sure you folks really want to go down this path?"[29] Gilbreath and others were concerned that the tests would reveal vaccine doses that did not meet sterility, potency, or safety standards—something that could stall immunizations or sidetrack the AVIP.[30]

Coupled with vaccine manufacturing shortcomings, behind the scenes, there was a fundamental disagreement on whether the AVIP was necessary. Several high-ranking military officers doubted the need for anthrax vaccinations, contending that the risk of an anthrax attack was small and could be deterred by the threat of retaliation.[31] Military leaders also raised concerns about vaccine-induced adverse events, balked at the high cost of the program, and noted that other nations might interpret the AVIP as a sign that the United States was preparing an anthrax attack.[32] In addition, scientists from Russia's State Research Centre for Applied Microbiology had recently published work on a genetically engineered strain of anthrax that was antibiotic-resistant and for which there was no known vaccine.[33] This development, coupled with Soviet defector Ken Alibek's disclosure that the USSR had worked on genetically modified biological weapons—including anthrax—raised a new question: was the anthrax vaccine, which was developed decades ago, obsolete?[34] Notwithstanding reservations among scientists and uniformed officers at the Pentagon, senior civilian leaders strongly supported mandatory vaccinations.[35]

The AVIP was launched in December 1997, with immunizations scheduled to begin within three months. The vaccine was mandatory for all 2.5 million active duty and reserve service members, as well as members of the Coast Guard and certain civilian employees, regardless of where an individual was stationed or set to deploy. In a press release that accompanied the launch, the DoD set forth several AVIP protocols: (1) supplemental testing, consistent with FDA standards, to assure sterility, potency, and purity of the vaccine stockpile; (2) implementation of a system for fully tracking personnel who receive the anthrax vaccinations; (3) approval of appropriate operational plans to administer the immunizations and communications plans to inform military personnel of the overall program; and (4) review of medical aspects of the program by an independent expert.[36] In addition, Congress directed the CDC to create a research program to evaluate the safety and efficacy of the vaccine.[37]

In February 1998, FDA inspectors found continuing problems with the Michigan lab. Although the lab addressed some of the previous issues, the inspection revealed additional concerns with anthrax production, including a contaminated vaccine lot (which contains about 200,000 vaccine doses).[38] Thereafter, the DoD provided $1.8 million for lab renovations and $15 million to expand the facilities. The lab was closed from March 1998 to May 1999 for the work to be completed.[39] Meanwhile, in March 1998, Secretary of Defense William Cohen was publicly vaccinated, and mandatory inoculations under the AVIP began.[40]

During the DoD-funded renovations, Michigan sold the lab for $25 million to BioPort, a private company, with $21.75 million of the purchase price coming from loans provided by the state.[41] BioPort assembled a team of military and government elites to serve as advisers and board members, including Admiral William Crowe, Jr.—former chairman of the Joint Chiefs of Staff under Presidents Ronald Reagan and George H. W. Bush, and the former chair of President Bill Clinton's Intelligence Advisory Board—who received a 13 percent stake in BioPort but did not invest any money in the venture.[42] Other BioPort board members included Louis Sullivan, Secretary of the Department of Health and Human Services (HHS) from 1989 to 1993, Joseph Allbaugh, former director of the Federal Emergency Management Agency, and Jerome Hauer, a former HHS official specializing in public health emergency preparedness.[43] BioPort followed the conventional approach of private companies seeking government funding: assemble a board of individuals with insider knowledge and connections that could help garner government contracts.

In September 1998—just days after BioPort finalized the purchase of the Michigan lab—the DoD awarded the company a $29 million contract to produce and supply the anthrax vaccine.[44] The government agreed to pay $4.36 per dose during the contract's first year and $2.26 per dose during year two.[45] Less than one year later, BioPort indicated that it could not produce the vaccine without additional government funding, ostensibly because of increased manufacturing costs. BioPort requested a $10 million advance to pay its creditors, a steep increase in price per dose, and the right to sell up to 20 percent of the vaccine on the open market.[46] As one member of Congress observed, "In order to maintain any production capability for its own needs, DoD must pay more money for less vaccine."[47]

Initially, DoD auditors examined BioPort's request and concluded that it did not meet the legal requirements for additional funding.[48] Of the DoD money already provided to BioPort, about $1 million was used to renovate and furnish executive offices, including $23,000 for the CEO's office furniture. More than $1.25 million was spent on executive bonuses, and millions more were used for consulting fees and other purposes.[49] Senior DoD officials overruled the auditors' determination, stating that national security interests called for additional funding and "extraordinary contractual relief."[50] Though it could do so, Congress did not nullify the contract that provided additional funds to BioPort. As a congressional investigation later revealed, the DoD was "captive to the demands of the sole-source provider," BioPort, for a "critical force protection" vaccine.[51]

BioPort invested heavily in well-connected lobbyists and public relations experts, sponsoring public education seminars to build support for stockpiling the anthrax vaccine and other countermeasures. This practice was analogous to that of the chemical industry following World War I, which had lobbied extensively for funding to combat the threat of chemical warfare, and the pharmaceutical industry following World War II (WWII), which had lobbied for funding to counter emerging biological warfare threats. According to an investigative report published in 2005, BioPort paid former military officials to spread the company's message and had officials sign ghost-written articles; these officials included Marine Major General Randy West and former Army Surgeon General Ronald Blanck.[52] The report also found that BioPort funded a faculty member at George Washington University School of Public Health, who created a group that petitioned the FDA to approve the inhalation indication for the anthrax vaccine; the group's website did not identify BioPort as a funding source.[53]

The restructured DoD contract provided BioPort an additional $24.1 million, including an $18.7 million advance payment.[54] The United States more than doubled the amount it agreed to pay per dose to $10.64, and lowered the number of doses BioPort was required to deliver from 7.9 million to 4.6 million.[55] Additionally, the government indemnified BioPort against legal liability for claims related to vaccine adverse reactions or the failure of the vaccine to confer immunity.[56] In other words, in addition to paying a handsome fee per dose, the government would be financially responsible for vaccine-induced injuries if the vaccine turned out to be unsafe or ineffective. Furthermore, notwithstanding vaccine supply shortages that the DoD characterized as causing serious national security concerns, it granted BioPort permission to sell up to 300,000 doses of the anthrax vaccine to foreign parties, even though BioPort had failed to meet the dose production it promised to the DoD.[57]

In November 1999, after the new contract and despite the influx of tens of millions of dollars, the lab once again failed an FDA inspection. FDA inspectors found more than thirty deficiencies. One of the most significant was that the lab did not have adequate measures to ensure that each vaccine lot met the appropriate specifications for purity and potency.[58] In addition, supplemental testing of the vaccine lots, as Gilbreath and others feared, revealed that eighteen of thirty-one lots were "unavailable" because of "unresolved purity, potency or sterility issues."[59] By February 2000, the FDA had yet to approve the new facility or any new vaccine lots. In turn, BioPort quarantined eleven lots and held back additional lots after FDA inspections identified further issues with vaccine potency.[60]

Notwithstanding the deficiencies, the DoD proceeded with the AVIP. Within weeks of the first vaccinations, many service members began to refuse the vaccine due to concerns about adverse events. Local commanders had the discretion to implement disciplinary proceedings against those who refused the vaccine, including court-martial.[61] By July 2000, more than 440 service members had been disciplined, and some served time in a military prison.[62] Hundreds of reserve pilots and aircrew quit the Air Force rather than submit to the administration of the vaccine.[63] In response to these developments—and because the anthrax vaccine was still being evaluated as a possible causal agent of Gulf War Illness—Congress held several hearings and investigations.

A congressional report, *The Department of Defense Anthrax Vaccine Immunization Program: Unproven Force Protection*, published in 2000, described the AVIP as "a well-intentioned but overwrought response to the threat of anthrax as a biological weapon."[64] The report characterized the DoD process that led to the AVIP as "more predetermined than deliberative" and labeled the AVIP an "unrealistic program" that was based on an "unstable supply" of a "dated" vaccine of "uncertain safety" and "uncertain efficacy."[65] The report concluded that "the AVIP compromises the practice of medicine to achieve military objectives," and highlighted that many service members do not "trust DoD has learned the lessons of past military medical mistakes: atomic testing, Agent Orange, Persian Gulf War drugs and vaccines."[66]

Congress also found "an institutional culture resistant, even hostile, to reports attributing ill health to the anthrax vaccine."[67] The DoD monitored adverse reactions primarily by relying on voluntary reporting by service members, but some military personnel testified that DoD "medical providers saw the issue of identifying vaccine reactions as 'politically sensitive'" and sought to avoid it.[68] Several service members testified that they were discouraged from reporting adverse events; others withheld reports for fear that admitting adverse health conditions would have a negative impact on their career. Notwithstanding widespread underreporting of adverse reactions, studies of reported reactions found vaccine-related adverse events to be significantly higher than anticipated. Moreover, no study had evaluated long-term health effects or whether the anthrax vaccine might harm a fetus, cause cancer, or impair fertility.[69]

The congressional report recommended that the AVIP be suspended until an improved anthrax vaccine is FDA-approved. The congressional report further indicated that, should the DoD decide to continue with the AVIP using the current vaccine, the DoD must either obtain the informed consent of each service member before vaccination or request that the president waive the informed consent requirement pursuant to 10 U.S.C. § 1107(f), a law enacted following the informed consent controversy during the Persian Gulf War.[70] Within days of the publication of the congressional report, the Institute of Medicine (IOM) issued a report that aligned with many of the congressional findings. The IOM report also raised concerns about the lack of data on the long-term effects of the anthrax vaccine, and stated that the DoD should publicly disclose its internal but unpublished studies.[71]

Apart from the reports, legislation introduced on July 16, 1999, sought to override the AVIP and make anthrax immunizations optional for service members.[72] A separate bill, introduced three days later, proposed to suspend the AVIP pending additional scientific review of the vaccine.[73] Neither bill progressed, but the unusual level of congressional action against a single DoD immunization requirement was extraordinary. Notwithstanding the reports and proposed legislation, the DoD forged ahead with the AVIP, did not seek informed consent from service members, and did not obtain a presidential waiver of informed consent. Moreover, the DoD continued to discipline service members who refused the vaccine.

Due to ongoing manufacturing issues and doubts about whether BioPort could produce enough vaccines that could meet FDA standards, during several congressional hearings, lawmakers and DoD officials indicated that they were considering abandoning their investment in BioPort. Remarkably, notwithstanding millions in funding from the DoD and a floundering facility, BioPort's President and CEO, Fuad El-Hibri, stated: "As a commercial entity, BioPort cannot continue to subsidize the DoD."[74]

Abandoning BioPort—the sole American facility producing an anthrax vaccine—became untenable after 9/11 and the anthrax letter attacks. The anthrax letters underscored the health risks of biowarfare and the need to maintain anthrax countermeasures. In an abrupt shift, lawmakers began questioning why the anthrax vaccine was in such short supply. Although support for the AVIP increased, substantial concerns remained with respect to BioPort's ability to manufacture safe and effective anthrax vaccines that would meet the DoD's needs.

For example, amid the anthrax letter attacks during the fall of 2001, testimony during a congressional hearing revealed that, in the early 1990s, the Michigan lab that produced the anthrax vaccine failed to report to the FDA that it altered key aspects of the manufacturing process. FDA regulations require such reporting because the process of vaccine manufacture impacts the quality of the vaccine. The manufacturing changes included the installation of different filters and fermenters that may have led to levels of protective antigen—the vaccine's main immunizing component—one hundred times greater than the FDA-approved levels.[75] After the manufacturing changes, the lab did not conduct studies to examine whether vaccine quality was affected.[76] And, for an extended period, the Michigan lab refused to allow FDA inspectors to enter rooms with manufacturing and production equipment, ostensibly because the inspectors were not vaccinated against anthrax.[77]

When FDA inspectors finally gained access to the facilities, they found serious deficiencies that could have impacted the safety, purity, and potency of the vaccines. These deficiencies motivated the FDA's decision to halt anthrax vaccine production in 1998.[78] Months prior to the inspection, when the FDA requested additional information on whether the filters altered the quality of the vaccine, the data provided by BioPort was inadequate because it failed to include vaccine quality tests from before and after the filter change.[79] Thus, it was impossible to measure the impact of the manufacturing changes on the safety and effectiveness of the vaccine.

Based on the available information, the data indicated that the rate and severity of adverse reactions for doses produced after the manufacturing changes were far more than those identified prior to the changes. While the FDA-approved vaccine insert indicated that 30 percent of recipients should experience mild adverse reactions and 0.2 percent more serious, systemic reactions, data on the anthrax vaccine used during the Persian Gulf War and thereafter found that 76.2 percent experienced mild reactions and 23.8 percent experienced systemic reactions.[80] Based on this series of events, Nancy Kingsbury, the Managing Director of Applied Research and Methods in the General Accounting Office (which, in 2004, was renamed the Government Accountability Office) (GAO), testified that if the FDA were to reinstate BioPort's license to produce the anthrax vaccine, the FDA and DoD should institute "an aggressive active surveillance program to ensure the early identification and analysis of adverse reactions."[81] Kingsbury's recommendation was not adopted.

In January 2002, the FDA announced that the BioPort facility could reopen—it still was the sole anthrax vaccine manufacturing facility in the United States. The FDA also released three vaccine lots (approximately 600,000 doses) that had been quarantined due to previous concerns regarding safety and potency. Some commentators stated that the approval was hastily granted due to political pressure, though government officials denied the allegations.[82]

Nearly a year before the FDA announcement, the CDC's Advisory Committee on Immunization Practices concluded that, due to safety and efficacy concerns, administration of the anthrax vaccine to "emergency first responders, federal responders, medical practitioners, and private citizens . . . is not recommended," but that the vaccine is acceptable for military personnel.[83] On the day the FDA released the quarantined lots and approved the BioPort facility, HHS leader Tommy Thompson stated during a press conference that the vaccine still is not recommended for the general public.[84] This double standard represented a continuing governmental policy of exposing service members to medical risks deemed inappropriate for civilians.

Months after the FDA announcement, a report revealed that the CDC manipulated data to make the anthrax vaccine seem safer and more effective than it actually was. For example, the CDC relied on studies that omitted negative information and overstated the reliability of adverse event reports collected by the FDA.[85] As of December 2003, at least six deaths were linked to the vaccine, and the pregnancy risk increased from Category C (risk cannot be ruled out) to Category D (positive evidence of risk).[86] BioPort was renamed Emergent BioSolutions in 2004, and the company would go on to receive several lucrative biosecurity government contracts.[87] In the spring of 2021, during the coronavirus pandemic, Emergent BioSolutions made headlines when it was forced to destroy millions of doses of coronavirus vaccines because of quality control problems.[88] According to a congressional report, for weeks, FDA regulators warned the company of the manufacturing deficiencies, but the company did not take adequate steps to address them and, at times, hid evidence of their mistakes.[89] The doses were wasted during the pandemic when vaccines were in short supply, and the company received hundreds of millions of dollars in government funding to produce the unusable vaccines.[90]

Meanwhile, in 2010, a seven-year Department of Justice (DOJ) investigation into the anthrax letter attacks revealed startling details. Dubbed the Amerithrax Investigation, the DOJ's efforts were extensive, involving more than 600,000 investigator hours, 10,000 witness interviews on six continents, 5,730 environmental samples from sixty locations, and 6,000 pieces of evidence. Dozens of government, university, and commercial laboratories assisted in the investigation. Steven Hatfill and Bruce Ivins—USAMRIID scientists who were two of the country's foremost experts on weaponized anthrax, and who had spent decades researching and developing the anthrax vaccine—were key suspects. During the investigation, the DOJ eliminated Hatfill as a suspect, and Ivins died by suicide. The DOJ later concluded that Ivins—acting alone—mailed the anthrax letters. According to the DOJ, Ivins had the requisite motive to commit the attack because the AVIP was failing, and anthrax vaccine safety and effectiveness were being questioned widely. These events, the DOJ asserted, jeopardized biosecurity work to which Ivins had dedicated much of his career.[91]

A 2011 National Academy of Sciences report found serious deficiencies in the DOJ investigation and questioned the DOJ's conclusion that Ivins alone perpetrated the attack.[92] In 2014, a comprehensive two-year GAO audit of the Amerithrax Investigation found that the FBI used poorly designed sampling and statistical methods to validate the bureau's conclusion that Ivins was the sole perpetrator. The GAO concluded that others might have been involved.[93] Notwithstanding who was responsible for mailing the letters, the anthrax attacks dampened the likelihood of rolling back the AVIP, led to expedited procurement of the anthrax vaccine, and influenced the creation of new laws governing the development and administration of medical countermeasures.

Anthrax Vaccine: From Lawsuits to New Laws

By the summer of 2000, approximately 447,000 service members had started the anthrax vaccine series, and 442 had been disciplined for refusing the vaccine. Dozens were court-martialed, and some were dishonorably discharged.[94] Captain John Buck was one of several military physicians court-martialed for failing to submit to vaccination and refusing to administer the anthrax vaccine to service members without their informed consent. Personally, Buck was concerned about vaccine-induced adverse events, and professionally, he sought to uphold the medical autonomy of his patients to determine whether to accept off-label administration of a vaccine with uncertain risks and benefits.[95] On the day he was sentenced, Buck stated: "I was at the cross roads between the oath of an officer and the oath of a physician. The only way I could have peace about the apparent conflict was to do what I knew to be right as a physician and to stare down the barrel of the gun with the courage of an officer."[96] A military court fined Buck $21,000, reversed his pending promotion, and confined him to a base for two months.[97] Buck's case reflects conflicts that may arise due to dual loyalties faced by military health professionals: they are ethically responsible for the well-being of their patients but also have responsibilities to promote military missions and obey lawful orders from the chain of command.

There are few publicly available military court decisions, but those that are accessible reveal that military judges consistently granted requests from military prosecutors to exclude evidence concerning vaccine safety and efficacy.[98] Relying on DoD directives that characterized the anthrax vaccine as "an FDA-licensed product and not an IND requiring informed consent for its administration," military courts repeatedly ruled that informed consent was unnecessary.[99] Despite a long line of losing legal efforts, many service members continued to refuse the vaccine and challenge the imposition of sanctions.

In 2003, six plaintiffs went outside the military court system and filed a lawsuit in a federal court seeking to stop the DoD from continuing the AVIP.[100] Three of the six had submitted to at least one dose, though none had completed the series.[101] During the case, the FDA supported the DoD, contending that since the vaccine's license does not specify the route of exposure, the license encompassed cutaneous and inhalation anthrax, and informed consent was not required.[102]

The court, however, was aware of the FDA's 1985 review of the anthrax vaccine, wherein the agency unambiguously indicated that there was insufficient evidence to evaluate the effectiveness of the vaccine against inhalation anthrax.[103] The court also referenced the 1996 application to the FDA, prepared by the DoD, which specified that "the ultimate purpose of this IND is to obtain a specific indication for inhalation anthrax and a reduced vaccination schedule."[104]

On December 22, 2003, Judge Emmet Sullivan ruled in favor of the plaintiffs. Judge Sullivan was mindful of the deference civilian courts must grant to military affairs and recognized that civilian courts may be ill-equipped "to intervene between soldiers and their military superiors."[105] Nevertheless, the judge noted that the law grants civilian courts the ability to review "internal military affairs if there is an allegation that a constitutional right has been deprived or an allegation that the military has acted in violation of applicable statutes or regulations."[106] Judge Sullivan ruled that the AVIP constituted off-label use of the anthrax vaccine because the vaccine was not FDA-approved as prophylaxis for inhalation anthrax. Accordingly, the law mandated that the DoD either obtain informed consent from each service member or have the president issue an informed consent waiver. Because the DoD failed to comply with either predicate, the court issued an injunction that halted the mandatory aspect of the AVIP. The court further indicated that inoculations could continue only if consent were obtained.[107]

In rendering its decision, the court considered and rejected the DoD's arguments that obtaining informed consent would negatively impact military preparedness and unduly burden the military mission and command structure. These were significant findings. As it did in the lawsuit challenging countermeasure mandates from the Persian Gulf War, for the AVIP, the DoD argued that complying with consent laws would disrupt the smooth functioning of the military, hinder military readiness, and reduce the military's ability to protect service members. The DoD further argued that, should individuals who refused the anthrax vaccine be injured by anthrax exposure, these injuries would have a detrimental effect on the military mission.[108]

In response, the court held that "the right to bodily integrity and the importance of complying with legal requirements, even in the face of requirements that may potentially be inconvenient or burdensome, are among the highest public policy concerns one could articulate." The court further reasoned that, if obtaining informed consent would have a detrimental impact, the DoD could seek a presidential waiver. As the court noted, if the President "determines that this is truly an exigent situation, then obtaining a presidential waiver would be an expeditious end to this controversy." Judge Sullivan's concluding remarks were particularly striking: "The women and men of our armed forces put their lives on the line every day to preserve and safeguard the freedoms that all Americans cherish and enjoy. Absent an informed consent or presidential waiver, the United States cannot demand that members of the armed forces also serve as guinea pigs for experimental drugs."[109]

Judge Sullivan's decision was not the end of the AVIP. Within days of the court order, the FDA issued a final rule that permitted the use of the anthrax vaccine "independent of the route of exposure," a move that captured the indication of inhalation anthrax.[110] In a press release that accompanied the issuance of the final rule, the FDA indicated that it disagreed with the court's ruling and "does not regard the approved anthrax vaccine as 'investigational' for protection against inhalation anthrax."[111] Since inhalation anthrax was now an on-label indication, the DoD could mandate the vaccine without obtaining consent from service members or seeking a presidential waiver.

A lawsuit against the FDA ensued. In the previous case, Judge Sullivan considered whether the DoD abided by informed consent laws; in the second case, he examined whether the FDA followed the appropriate procedures when it expanded the vaccine label to encompass inhalation anthrax. In his analysis of the FDA's actions, Judge Sullivan highlighted that courts must provide deference to administrative agencies, particularly in cases where the underlying substantive decision involves scientific information. As the judge further noted, "it is well within this Court's scope of authority to ensure that the agency adheres to its own procedural requirements."[112] In other words, the court was not going to conduct an independent analysis as to whether the data were sufficient to find that the anthrax vaccine was a safe and effective prophylaxis against inhalation anthrax. Instead, the court would analyze whether the FDA followed the relevant procedures in rendering its finding.

The facts were not on the FDA's side. By issuing a hasty approval just days after the court's order that halted the AVIP, legal experts opined that it seemed clear that the FDA did not follow its own guidelines in approving the new indication. A court decision that found the FDA's approval to be unlawful would further complicate the DoD's ability to continue with mandatory anthrax immunizations.

During court deliberations, BioPort and the DoD sought to bypass a second court-mandated halt to the AVIP. They lobbied Congress for the enactment of a new law to permit emergency use of non-FDA-approved medical products. Their efforts were successful, and on July 21, 2004, Congress enacted the Project BioShield Act. Among its provisions, the act created the Emergency Use Authorization (EUA) pathway, which allows the FDA the authority to grant healthcare providers with emergency authorization to treat patients with an unapproved medical product or to administer an approved medical product for an off-label use.[113]

Since the EUA process is a separate pathway to market that is distinct from the FDA's standard approval mechanism and requires less safety and efficacy data, an EUA may be issued only in emergency circumstances and only if there are no adequate, approved, and available alternatives; this includes instances where there are insufficient supplies of approved products. An emergency can be declared if there is an actual attack or a heightened risk of attack against US interests, anywhere in the world, or a public health emergency. Issuance of an EUA does not equate to FDA approval. The FDA's standard of review for issuing an EUA is whether, based on available data, the product "may be effective"; in contrast, to earn FDA approval, a company must provide "substantial evidence" of effectiveness.[114] This is a legal standard with a meaningful clinical difference—products authorized for use via the EUA mechanism are subject to a lower bar.

On October 27, 2004—three months after the enactment of the BioShield Act—Judge Sullivan ruled that the FDA breached its own protocols and ordered that the agency reconsider the inhalation anthrax indication.[115] It is the obligation of courts, Judge Sullivan wrote, "to ensure that FDA follow the law in order to carry out its vital role in protecting the public's health and safety."[116] The court was unwilling to allow the FDA and the DoD to circumvent their respective legal requirements. The judge concluded: "The men and women of our armed forces deserve the assurance that the vaccines our government compels them to take into their bodies have been tested by the greatest scrutiny of all—public scrutiny. This is the process the FDA in its expert judgment has outlined, and this is the course this Court shall compel FDA to follow."[117] Judge Sullivan's decision halted the AVIP for a second time.

Weeks later, the DoD filed an EUA application to be able to use the anthrax vaccine as prophylaxis against inhalation anthrax.[118] The FDA granted the request, and on January 27, 2005, the anthrax vaccine became the first medical product authorized for emergency use via the EUA pathway.[119] In issuing the EUA, the FDA indicated that military personnel could refuse the vaccine and that one who refuses could not be punished or subject to disciplinary action.[120] Products authorized for use via the EUA mechanism can be administered only with consent, though a presidential order can waive consent for military personnel.[121] President Bush did not issue an informed consent waiver, and thus, anthrax vaccines were optional for service members.[122]

On December 19, 2005, while the anthrax vaccine was still optional under an EUA, the FDA issued a final rule that added inhalation anthrax as an on-label indication to the vaccine.[123] Because inhalation anthrax was now an approved indication, the DoD could mandate immunizations and did not need to obtain informed consent from service members or a presidential waiver of consent. In turn, on February 9, 2006, a federal appeals court overturned Judge Sullivan's injunction that halted the AVIP.[124]

Thereafter, the DoD announced that the anthrax vaccine would be mandatory for most military units and civilian contractors deployed to certain regions of the world, including Afghanistan, Iraq, and South Korea.[125] According to Assistant Secretary of Defense for Health Affairs, Dr. William Winkenwerder, the FDA "came to the very unambiguous and clear conclusion that the vaccine was safe and it was effective against all forms of exposure."[126] During the period when vaccination was voluntary, about 50 percent of service members opted to receive the vaccine.[127]

Litigation continued against the FDA. Service members sued to invalidate the FDA's December 2005 decision, arguing that there was insufficient data for the FDA to conclude that the anthrax vaccine was effective prophylaxis against weaponized aerosol anthrax.[128] This case challenged the substantive decision of the FDA, unlike the previous case, which alleged that the FDA failed to comply with procedural requirements. The lawsuit was unsuccessful. In an opinion dated February 29, 2008, the court stated that it must afford "a high level of deference" to the FDA's decision, and that the agency's decision could be set aside only if the court found that the FDA made "a clear error of judgment."[129] While the court noted that reasonable scientific minds could differ in analyzing the safety and efficacy of the anthrax vaccine, the court held that the FDA acted within its discretion. On September 29, 2009, an appellate court affirmed the lower court's ruling.[130] That decision was the last stage of over ten years of litigation, spanning military and civilian courts, regarding anthrax immunizations.

During the decade of litigation, more than 2.1 million service members received the anthrax vaccine.[131] Between 1998 and 2008, the FDA received 575 reports of vaccine-related hospitalizations or permanent disabilities, and more than 5,000 reports of other adverse events. Twenty-five people died from vaccine-induced causes, though the CDC and FDA noted that these numbers likely undercount the number of vaccine-related injuries.[132]

A 2008 study found that most surveyed service members believed that the vaccine was unsafe and ineffective, and only 30 percent felt that the DoD was "protecting my health by instructing that I receive the vaccine."[133] Less than 25 percent thought that military officials are knowledgeable about adverse events or "seem to care when I describe my concerns related to the anthrax vaccine."[134] One in twelve respondents indicated leaving the military due to the AVIP.[135] Although the AVIP may have been created to address national security concerns, the program's various shortcomings have contributed to lasting mistrust. The government's smallpox vaccination program exacerbated this mistrust.

The Smallpox Vaccination Program

Amid the AVIP controversy, the United States commenced a smallpox vaccination program. The program was publicly revealed on December 13, 2002, as the Bush administration was contemplating a war against Iraq, grounded on the position that Iraq maintained weaponized smallpox and might use biological weapons against Americans. As national security expert Frank Smith surmised, "In the spirit of threat inflation, [Vice President] Cheney favored a nationwide—even compulsory—smallpox vaccination program."[136] The goal was to vaccinate 500,000 service members and 10 million civilian first responders.[137]

Smallpox was eradicated from the world in 1980, and the last documented case in the United States was in 1949. In 1972, the United States halted routine immunizations for civilians. Still, the military continued inoculations because the disease was present in various parts of the world, and there was a slight chance that a nation had weaponized the virus. To limit the military threat, in 1984, the World Health Organization (WHO) adopted a resolution that allowed only the United States and the Soviet Union to maintain their smallpox stockpiles. Six years later, the US military stopped routine smallpox immunizations. Although biological warfare concerns remained, the vaccine was riskier than most other vaccines and was known to cause serious side effects and death.[138]

During the early 1990s, Russia and the United States announced that they would map the DNA sequence of the smallpox virus and destroy their stockpiles. Once sequencing was completed, many questioned the wisdom of the plan. If the virus reemerged, the stockpiles could help scientists create a new vaccine. Moreover, studying smallpox could further research into related viruses that were endemic in parts of the world, such as vaccinia, cowpox, and mpox. While destroying the stockpiles might demonstrate moral leadership on the part of Russia and the United States, intelligence reports warned that other nations or terrorist groups might have access to smallpox and could weaponize the virus.[139] As Soviet defector and biological warfare expert Ken Alibek explained: "A world no longer protected from smallpox was a world newly vulnerable to the disease."[140] As of mid-2024, the stockpiles remain.

The 2002 smallpox vaccination program was widely criticized. The CDC director stated that "we have no imminent threat" of an attack.[141] At the same time, HHS officials supported the creation of a voluntary immunization program if a safer vaccine earned FDA approval.[142] Public health officials also expressed concerns about secondary infections since the smallpox vaccine contains a live vaccinia virus that can be transmitted to others.[143] In addition, about 1:1000 vaccinees were predicted to endure a severe adverse event, including life-threatening adverse reactions.[144] If the government's plan were successful and 10.5 million individuals were inoculated, thirty-two vaccine-induced deaths were predicted.[145]

Several additional factors cautioned against the vaccination program. Unlike anthrax—which can survive outside a host for decades—smallpox can survive only for minutes.[146] Smallpox is less infectious than measles or influenza, and transmission requires considerable exposure in close quarters.[147] Of the 945 reported smallpox cases between WWII and 1980, the majority occurred at the bedside of an infected individual.[148] Public health experts explained that, in the event of an outbreak, transmission could be contained by surveillance and containment of cases.[149] These evidence-based positions had little impact on the Bush administration's decision to implement the program.

The United States invaded Iraq in March 2003, and on May 1, 2003, President Bush declared victory and the end of major combat operations. The victory declaration quickly dissolved as Iraq fell into civil war and the United States committed significant personnel and resources to the region. Between December 2002 and June 2003, more than 450,000 service members were administered the smallpox vaccine.[150] In

September 2003, an American-led team indicated that it did not find any evidence of smallpox—or any weapons of mass destruction—in Iraq.[151] About one year later, a CIA report concluded that there is "no direct evidence that Iraq either retained or acquired smallpox virus isolates or proceeded with any follow-up smallpox related research."[152] Notwithstanding the findings, mandatory immunizations continued, and by January 2005, the number of vaccinated military personnel had risen to 730,580.[153]

In terms of adverse events, more than seven hundred service members were hospitalized due to vaccine-induced adverse events, and no less than fifty-two secondary vaccinia infections were reported in spouses, children, friends, and intimate contacts.[154] One infant contracted vaccinia from breastfeeding.[155] At least 214 neurologic complications were reported, one of which was fatal.[156] More than one hundred vaccinees suffered cardiac adverse events, leading to at least five deaths.[157] It was previously unknown that the smallpox vaccine causes heart inflammation and heart attacks, and the link alarmed public health officials and served to caution against further administration of the vaccine.[158] Although the vaccination program sought to exclude pregnant women due to known vaccine-induced reproductive harms, at least 236 pregnant women received the vaccine; many suffered preterm births and spontaneous pregnancy loss, though researchers could not conclusively determine whether the vaccine was a contributing factor.[159] As a 2005 IOM report recounted, the smallpox vaccination program was "an extraordinary policy decision: to vaccinate people against a disease that does not exist with a vaccine that poses some well-known risks."[160]

Civilian vaccinations did not materialize as planned. By mid-July 2003, about thirty-eight thousand civilian healthcare workers had been vaccinated.[161] The IOM and the CDC's Advisory Committee on Immunization Practices urged the CDC to limit the civilian rollout, but the agency did not relent and moved forward with its plan to vaccinate ten million civilians.[162] Few civilians complied, and by June 2004, the total number of vaccinated civilians was less than forty thousand.[163] The rate of cardiac adverse events among civilians was four times higher than that of military personnel, likely due to the fact that service members were a younger and healthier cohort.[164]

Healthcare workers questioned the merits of the smallpox vaccination program, contending that political and corporate interests usurped public health concerns.[165] Although the entire vaccination program was delayed until Congress enacted legal shields for manufacturers and vaccine clinics, the program commenced without any protections for vaccine recipients or individuals impacted by secondary infections.[166] Many civilians refused the vaccine because of the lack of adequate safeguards for vaccine-related injuries, and union leaders encouraged their members to refuse vaccination until the liability concerns were addressed.[167] Meanwhile, articles and editorials in prominent newspapers—including *The New York Times*, the *Boston Herald*, and *The Washington Post*—chastised vaccine-hesitant physicians and first responders, characterizing them as "vaccine dodgers" who were "unpatriotic" and "deplorable."[168]

The insults dissipated as commentators trended toward viewing the immunization program as a hasty political move to support the Bush administration's desire to invade Iraq, rather than a well-reasoned public health initiative. Although vaccine effectiveness could not be evaluated because there was no smallpox attack, the implementation costs and reputational harms were substantial, including time and money to acquire and administer the vaccine and treat adverse events, the personal and professional losses due to vaccine-related deaths and adverse events, and loss of trust in military and government leaders. The public also lost faith in public health agencies—including CDC and HHS—given concerns that leaders of these departments did not have complete independence to make evidence-based decisions.[169] As one member of the CDC's advisory committee on vaccines publicly stated, "The sense was that the course was already set and we wouldn't make any difference."[170]

In 2011, HHS Director Kathleen Sebelius wrote an editorial in *The New York Times* wherein she argued that a smallpox threat continues to exist because the virus could be reproduced in the laboratory and "it is quite possible that undisclosed or forgotten stocks exist."[171] Three years later, American authorities found undocumented smallpox vials in what they oddly characterized as "an unused portion of a storage room" at the FDA.[172] The discovery highlighted the risks of allowing continued maintenance of smallpox stockpiles, and bolstered the argument for maintaining preparedness to counter an intentional or inadvertent smallpox outbreak.

Enhancing Biosecurity

Apart from creating the anthrax and smallpox immunization programs, military and civilian leaders considered how to restructure laws and regulations to facilitate the development and administration of CBRN countermeasures. As detailed in previous chapters, most of the twentieth-century development of CBRN countermeasures involved the use of humans in myriad experiments and field tests. Conducting such studies became increasingly difficult following codification of federal guidelines for research with human subjects. Although investigations still were needed to analyze countermeasure safety and efficacy, exposing humans to lethal substances was deemed unethical in light of the new rules. To help overcome the hurdles, Congress and the FDA established special regulatory mechanisms that limited the need for testing with human subjects. The EUA pathway—discussed earlier in this chapter—was one. Another was the Animal Rule, which grants the FDA the authority to approve a medical product without human trials in instances where it would be unethical to perform them.[173] Along with new regulatory pathways to bring medical countermeasures to market, the government increased biosecurity funding and created a new department, the Biomedical Advanced Research and Development Authority (BARDA), with an initial budget of more than $1 billion dedicated to countermeasure research and development.[174]

The FDA finalized the Animal Rule a few months after 9/11 and the anthrax letters attacks, but the rule had been contemplated since the countermeasure controversy from the Persian Gulf War. In 1992, the FDA created an accelerated approval pathway whereby drugs could come to market without the need to demonstrate a clinical endpoint, such as longer survival, improved quality of life, or better health outcomes. Rather, pharmaceutical companies could rely entirely on a surrogate endpoint, which is a biomarker that is used as a substitute for a clinical endpoint on the presumption that the surrogate endpoint has a reasonable likelihood of correlating with a clinical benefit to patients.[175] An example is the use of lower blood pressure measurements as a surrogate for decreasing one's risk of a stroke. Surrogate endpoint trials typically are cheaper and less rigorous than clinical endpoint trials, and clinical researchers have raised concerns about the reliability of surrogate endpoints as accurate markers for actual health benefits for patients.[176]

In 1996, the military asked the FDA to apply the new pathway and reevaluate PB as prophylaxis against Soman by using efficacy data from animal studies as a surrogate endpoint.[177] The FDA denied the application, indicating that no "reasonable person" could predict clinical benefit in humans based on animal studies alone.[178] In the years that followed, however, the FDA changed its position.

In a series of internal memoranda written in 1998, FDA officials considered whether animal studies might ever be sufficient for countermeasure approvals,[179] and in October 1999 the agency issued a request for public comment on the Animal Rule.[180] The rule became effective in July 2002.[181] According to Theresa Allio, a scientist with the National Institute of Allergy and Infectious Diseases, "the history surrounding the Animal Rule and its implementation involves the convergence of science, politics, and public health," and the rule might not have been enacted if the DoD did not face such challenges with its application to approve PB as a Soman countermeasure.[182] The Animal Rule permits the approval of new products if animal studies demonstrate that it is "reasonably likely" that humans will receive a clinical benefit.[183] This represents a stark departure from standard FDA approval requirements, which require "substantial evidence" of efficacy based on human clinical trials.[184]

The first product approved via the Animal Rule was PB.[185] The 2003 approval was based on studies involving guinea pigs and monkeys.[186] Once PB became an FDA-approved prophylaxis for Soman, informed consent was no longer necessary for military personnel. As one FDA scientist explained, "The legal situation for use of PB has changed, but the safety profile of PB has of course not changed at all."[187] The second product approved under the Animal Rule was cyanokit, which is used to treat cyanide poisoning, and the primary evidence for the 2006 approval came from one placebo-controlled study in dogs.[188]

Thereafter, seven products were approved between 2012 and 2016. Three were antibiotics to treat the plague, and two drugs were approved to treat radiation exposure. Raxibacumab was approved as a post-exposure treatment for inhalation anthrax, and the first vaccine approved via the Animal Rule was the anthrax vaccine at the center of the AVIP controversy.[189] While, in 2005, the FDA approved the vaccine as pre-exposure prophylaxis for inhalation anthrax, in 2012, the FDA utilized the Animal Rule to approve the vaccine for post-exposure prophylaxis when administered with antibiotics.[190] The 2012 vaccine approval was based on studies with rabbits and nonhuman primates.[191]

The Animal Rule remains controversial. Chief among the concerns are ongoing reliability issues in using animal studies as markers of safety and effectiveness for humans.[192] There are also practical challenges of using animal studies to establish appropriate dosing for humans. Given these uncertainties, products approved via the Animal Rule may put human lives in danger rather than protect them. These risks are borne disproportionally by military personnel, since informed consent is not required and prophylaxis can be mandated as a requirement of service.

Compounding the health risks are laws that limit legal recourse in the event of countermeasure-induced harms. The *Feres* doctrine prohibits lawsuits against the military or government for service-related injuries, including adverse events from medical countermeasures. Additionally, in 2005, Congress created an expansive legal shield for manufacturers and others involved in the development and administration of medical products authorized for use via the EUA pathway. Specifically, the Public Readiness and Emergency Preparedness Act of 2005 (PREP Act) precludes "all claims for loss caused by, arising out of, relating to, or resulting from the administration to or the use by an individual of a covered countermeasure."[193]

The window for lawsuits is so small as to be practically nonexistent. Companies face no liability if they are grossly negligent or engage in careless disregard for the safety and effectiveness of their product.[194] The sole exception to immunity is in cases of "willful misconduct," which encompasses instances where a person or entity acts "(i) intentionally to achieve a wrongful purpose; (ii) knowingly without legal or factual justification; and (iii) in disregard of a known or obvious risk that is so great as to make it highly probable that the harm will outweigh the benefit."[195] In lieu of lawsuits, the law indicates that Congress shall establish a compensation fund to address injuries.[196] However, the HHS Secretary has the sole authority to identify which injuries are eligible for compensation, the amount of compensation, and the procedures governing compensation requests.[197] Moreover, the PREP Act does not mandate that Congress appropriate money to fund the compensation program.

The PREP Act was enacted as part of the DoD Appropriations Act for 2006. According to Senator Chris Dodd, Republican lawmakers inserted broad limited liability protections into the bill "at the last minute, in the middle of the night, without the opportunity for discussion and debate, and without the knowledge or consent" of many lawmakers.[198] Some lawmakers had signed off on the legislation before the liability shield was included.[199]

Lawmakers, first responders, veterans groups, public health officials, and others protested the perverse incentives in the PREP Act. These groups—along with the American Public Health Association, American Nurses Association, and the American Federation of State, County, and Municipal Employees—further argued that individuals would refuse to be administered emergency countermeasures because the law grants extensive legal immunities but fails to guarantee fair and adequate compensation for countermeasure-induced harms.[200] Indeed, PREP Act legal shields mirrored those from the smallpox immunization program, where healthcare providers and first responders refused immunization in large part due to inadequate protections should they encounter vaccine-induced injuries.[201]

Shortly after the PREP Act was enacted, twenty Democratic lawmakers sent a letter to the Republican leadership in Congress to protest what they characterized as a "dead of night" inclusion of overly broad legal immunities that were likely to have a negative impact on public health and emergency preparedness.[202] The lawmakers endorsed a repeal of the law and proposed a revised version that would provide additional protections for individuals. The proposed bill also mandated that once the HHS Secretary issued a legal immunity declaration, the Secretary also must contract with the IOM to produce a report that outlines which injuries should be eligible for compensation.[203] The bill did not gain sufficient support and was abandoned.

In 2013, Congress expanded several provisions of the 2004 BioShield Act.[204] The amendments provided the FDA with new mechanisms in the event of a public health, domestic, or military emergency involving CBRN agents, such as extending the expiration date of countermeasure stockpiles and waiving manufacturing practice requirements to accommodate emergency needs. The 2013 amendments also allowed the FDA to waive post-market risk evaluation requirements. Congressional funding for stockpiling countermeasures grew enormously, ensuring a steady stream of revenue for biopharmaceutical companies that incentivizes countermeasure development and production.[205]

For the DoD, however, the FDA was not moving fast enough to authorize medical products of special import to the military. In 2017, the military floated a proposal, which had some congressional support, whereby it would establish a military health panel that could approve drugs and medical devices independent of the FDA.[206] The proposed military health panel—which mirrored a recommendation that emerged following the countermeasure and informed consent controversy during the Persian Gulf War—sought to bypass altogether the FDA approval process.[207]

The FDA protested the DoD's maneuver, and as a compromise, Congress enacted a law that expanded the realm of medical products for which an EUA could be used to encompass health concerns that pose a "specific risk to U.S. military forces."[208] The law also granted the DoD a separate means of requesting expedited review and approval for medical products that address military exigencies. Shortly thereafter, the DoD and FDA signed a memorandum of understanding that detailed a work plan for increased FDA-DoD collaborations and expedited review of medical products for military health.[209]

For nearly fifteen years, the EUA mechanism remained an obscure and rarely utilized pathway for bringing medical products to market. During the coronavirus pandemic that began in 2020, however, EUAs were an integral component of America's medical and public health response. The FDA issued more than four hundred EUAs for various products, including vaccines, therapies, diagnostic tests, and medical devices. Although historically EUAs were authorized for use in limited populations, during the pandemic EUAs were widely utilized and administered. More than two hundred million American adults and children received an EUA vaccine, and millions more were administered EUA therapies. Several EUA therapies came to market with little data on safety and effectiveness, and some EUAs were rescinded when post-market evaluation revealed serious adverse events and insufficient evidence of effectiveness.[210]

The coronavirus pandemic underscored the long-held understanding that disease can pose a national security threat. Several reports indicated that the virus might have leaked, perhaps unintentionally, from a Chinese biosecurity lab in Wuhan that was studying novel coronaviruses and had received grants from the National Institutes of Health to do so.[211] Notwithstanding the source of the virus, the pandemic vividly illustrated the substantial health, operational, and economic challenges that may stem from a novel biological agent. There have been more than 775 million documented cases of COVID-19, resulting in over 7 million deaths.[212] Although only nine nations maintain nuclear weapons, there are thousands of biological laboratories across the world. Synthetic biology, which can be utilized to engineer novel agents not found in nature, expands the landscape of potential risks.[213] The coronavirus pandemic also highlighted how an existential threat can result in a breakdown of the international order, whereby a nation takes all steps it deems necessary to protect the homeland.

Summary

A potpourri of post-9/11 laws was enacted to facilitate the development and administration of medical countermeasures for public health emergencies and military contingencies. As a practical matter, the relaxation of premarket approval standards via the Animal Rule and EUA pathway has decreased the confidence level in the safety and effectiveness of marketed products. These risks are tolerated on the premise that an unproven but promising countermeasure is better than no countermeasure at all. However, these risks are borne disproportionally by military personnel since the countermeasures may be mandatory components of health readiness protocols.

As detailed in chapter 18, the laws could be strengthened without significantly curtailing military preparedness. Legal shields could be recalibrated to allow lawsuits in instances where a manufacturer fails to exercise reasonable care in the development and production of countermeasures. The safety net could be expanded to better alleviate the burdens that fall upon those who suffer countermeasure-induced injuries. More nuance could be structured into the Animal Rule and EUA pathway: for example, a lower bar to market may be defensible for treatments to combat life-threatening conditions, but less so for prophylaxis against CBRN agents where there is a small likelihood of attack. In each area, striking the right balance requires a robust analysis of health and operational risks for administering—and choosing not to administer—a countermeasure. Such decisions should account for the mistrust that has developed over the history of military medicine.

Notes

1. US Department of Homeland Security, *The World Trade Center Bombing: Report and Analysis* (1993), 1–5.
2. US Congress Office of Technology Assessment, *Proliferation of Weapons of Mass Destruction* (1993), 53–55.
3. *Global Proliferation of Weapons of Mass Destruction: Hearings Before the Senate Committee on Governmental Affairs*, 104th Cong. 2–3 (statement of Senator William V. Roth, Jr.), 6 (statement of Senator Sam Nunn), 69–81 (staff statement) (October 31, 1995).
4. *Global Proliferation*, 104th Cong. 111 (statement of Kyle B. Olson, Senior Staff, Arms Control and Proliferation Analysis Center, TASC, Inc.).
5. Michael Janofsky, "Looking for Motives in Plague Case," *The New York Times*, May 28, 1995.
6. "Poison Probe," *Newsweek*, May 28, 1995.
7. Barnaby, *The Plague Makers*, 43.
8. Manuela Oliveira et al., "Biowarfare, Bioterrorism and Biocrime: A Historical Overview on Microbial Harmful Applications," *Forensic Science International* 314 (2020): 6.
9. *Global Proliferation*, 104th Cong. 2–3 (statement of Senator William V. Roth, Jr.).

10. "Bin Laden's Fatwa," *PBS NewsHour*, August 23, 1996; Usama bin Laden, "Jihad Against Jews and Crusaders," *Federation of American Scientists*, February 23, 1998.
11. William J. Broad and Judith Miller, "Thwarting Terror: Germ Defense Plan in Peril as Its Flaws Are Revealed," *The New York Times*, August 7, 1998.
12. James C. McKinley, "Bombs Rip Apart 2 U.S. Embassies in Africa," *The New York Times*, August 8, 1998; "1998 US Embassies in Africa Bombings: Fast Facts," *CNN*, October 12, 2023.
13. Rolf Mowatt-Larssen, "Al Qaeda's Pursuit of Weapons of Mass Destruction," *Foreign Policy*, January 25, 2010.
14. See a series of twenty-one articles published in *Emerging Infectious Diseases* 5, no. 4 (July–August 1999): 491–565.
15. US Department of Justice, *Amerithrax Investigative Summary* (February 19, 2010), 2–3.
16. *Germs, Toxins and Terror: The New Threat to America: Hearings Before the Senate Subcommittee of Technology, Terrorism, and Government Information of the Committee on the Judiciary*, 107th Cong. 61–67 (November 6, 2001).
17. *Doe v. Rumsfeld*, 297 F. Supp. 2d 119, 124 (D.D.C. 2003) (quoting letter from Anna Johnson-Winegar to Robert Myers, attached to complaint as exhibit G).
18. Food and Drug Administration: Biological Products; Bacterial Vaccines and Toxoids; Implementation of Efficacy Review, 50 Fed. Reg. 51,002, 51,058–059 (December 13, 1985).
19. Michael J. Gilbreath, "Is the Current Anthrax Vaccination Regimen Necessary?," *Department of Defense Information Paper* (November 10, 1995): 1–2, quoted in Fourth Report by the Committee on Government Reform, *The Department of Defense Anthrax Vaccine Immunization Program* (2000), 52.
20. *Doe v. Rumsfeld*, 297 F. Supp. 2d at 124–25.
21. Miller et al., *Germs*, 202.
22. *Doe v. Rumsfeld*, 297 F. Supp. 2d at 125 (quoting letter from Stephen Joseph to Mark Friedman).
23. Dr. Michael A. Friedman, Letter to Dr. Stephen C. Joseph (March 13, 1997), quoted in Fourth Report, *The Department of Defense*, 53.
24. *Doe v. Rumsfeld*, 297 F. Supp. 2d at 132 (citing 21 C.F.R. § 10.85(k)).
25. Fourth Report, *The Department of Defense*, 53.
26. Fourth Report, *The Department of Defense*, 7.
27. Kathryn C. Zoon, Director of FDA Center for Biologics Evaluation and Research, Letter to Robert Myers, Michigan Biologic Products Institute (March 11, 1997).
28. Zoon, Letter to Myers, 6–8.
29. Memorandum from Michael Gilbreath, Joint Program Office for Biological Defense, to Gen. John Doesburg, Walter Busbee, Lt. Col. David Danley, and Lt. Col. James Estep (March 25, 1997), quoted in Miller et al., *Germs*, 204.
30. Fourth Report, *The Department of Defense*, 12.
31. Bradley Graham, "Military Chiefs Back Anthrax Inoculations," *The Washington Post*, October 2, 1996.
32. Graham, "Military Chiefs."
33. A. P. Pomerantsev et al., "Expression of Cereolysine AB Genes in *Bacillus Anthracis* Vaccine Strain Ensures Protection Against Experimental Hemolytic Anthrax Infection," *Vaccine* 15, no. 17–18 (December 1997): 1846–50.
34. Broad and Miller, "Thwarting Terror."
35. Graham, "Military Chiefs."
36. Department of Defense Press Release, "Defense Department to Start Immunizing Troops Against Anthrax" (December 15, 1997).

37. Institute of Medicine, *An Assessment of the CDC Anthrax Vaccine Safety and Efficacy Research Program* (2003), 1.
38. *The Anthrax Immunization Program: Hearing before the House Subcommittee on National Security, Veterans Affairs, and International Relations of the Committee on Government Reform,* 106th Cong. 28–29, 42 (March 24, 1999); Fourth Report, *The Department of Defense,* 12.
39. Fourth Report, *The Department of Defense,* 7.
40. Meryl Ness, "The Anthrax Vaccine Program: An Analysis of the CDC's Recommendations for Vaccine Use," *American Journal of Public Health* 92, no. 5 (May 2002): 716.
41. *Department of Defense's Sole-Source Anthrax Vaccine Procurement: Hearing before the Subcommittee on National Security, Veterans Affairs, and International Relations of the House Committee on Government Reform,* 106th Cong. 6–7 (June 30, 1999) (statement of Louis J. Rodrigues).
42. *Department of Defense's Sole-Source Anthrax Vaccine Procurement,* 106th Cong. 38–51 (statement of Fuad El-Hibri, President and CEO, BioPort Corporation); Martin Meyer Weiss, Peter D. Weiss, and Joseph B. Weiss, "Anthrax Vaccine and Public Health Policy," *American Journal of Public Health* 97, no. 11 (November 2007): 1946–47.
43. Weiss et al., "Anthrax Vaccine," 1946–47.
44. Fourth Report, *The Department of Defense,* 8.
45. *Department of Defense Anthrax Vaccine Immunization Program: Hearings Before the Senate Committee on Armed Services,* 106th Cong. 174 (April 13 and July 12, 2000) (statement of Robert J. Lieberman, Assistant Inspector General for Auditing, Department of Defense).
46. *Department of Defense's Sole-Source,* 106th Cong. 1–2 (introductory remarks by Subcommittee Chairman, Christopher Shays).
47. *Department of Defense's Sole-Source,* 106th Cong. 1–2 (introductory remarks by Subcommittee Chairman, Christopher Shays).
48. *Department of Defense Anthrax Vaccine,* 106th Cong. 117.
49. *Department of Defense Anthrax Vaccine,* 106th Cong. 115–16, 119; Weiss et al., "Anthrax Vaccine," 1946.
50. *Department of Defense Anthrax Vaccine,* 106th Cong. 174 (statement of Robert J. Lieberman, Assistant Inspector General for Auditing, Department of Defense).
51. *Department of Defense's Sole-Source,* 106th Cong. 1–2 (introductory remarks by Subcommittee Chairman, Christopher Shays).
52. Bob Evans, "How a Company Cashed In on Anthrax," *Newport News Daily Press,* December 7, 2005.
53. Evans, "How a Company Cashed In on Anthrax."
54. Fourth Report, *The Department of Defense,* 8.
55. *Department of Defense Anthrax Vaccine,* 106th Cong. 174 (statement of Robert J. Lieberman, Assistant Inspector General for Auditing, Department of Defense).
56. Fourth Report, *The Department of Defense,* 8–9.
57. Fourth Report, *The Department of Defense,* 27.
58. *Department of Defense Anthrax Vaccine,* 106th Cong. 66–71 (statement of Carol R. Schuster).
59. Fourth Report, *The Department of Defense,* 12.
60. Fourth Report, *The Department of Defense,* 7.
61. Congressional Research Service Report for Congress, "Department of Defense Anthrax Vaccination," (November 7, 2000): 1–6.
62. Congressional Research Service, "Department of Defense Anthrax Vaccination," 1–6; Miller, Engelberg, and Broad, *Germs,* 269.

63. US General Accounting Office, "Anthrax Vaccine: GAO's Survey of Guard and Reserve Pilots and Aircrew" (September 2002).
64. Fourth Report, *The Department of Defense*, 2.
65. Fourth Report, *The Department of Defense*, 1–3, 20.
66. Fourth Report, *The Department of Defense*, 2.
67. Fourth Report, *The Department of Defense*, 38.
68. Fourth Report, *The Department of Defense*, 38.
69. Fourth Report, *The Department of Defense*, 41.
70. Fourth Report, *The Department of Defense*, 3–4.
71. Institute of Medicine, *An Assessment of the Safety of the Anthrax Vaccine: A Letter Report* (Washington, DC: National Academy Press, 2000).
72. American Military Health Protection Act, H.R. 2543, 106th Cong. (1999).
73. Department of Defense Anthrax Vaccination Moratorium Act, H.R. 2548, 106th Cong. (1999).
74. *Department of Defense's Sole-Source*, 106th Cong. 43 (statement of Fuad El-Hibri).
75. US General Accounting Office, "Anthrax Vaccine: Changes to the Manufacturing Process" (October 23, 2001), 5 (statement of Nancy Kingsbury, Managing Director, Applied Research and Methods, U.S. General Accounting Office, before the Subcommittee on National Security, Veterans Affairs, and International Relations, House Committee on Government Reform).
76. US General Accounting Office, "Anthrax Vaccine," 2–3.
77. US General Accounting Office, "Anthrax Vaccine," 2–3.
78. US General Accounting Office, "Anthrax Vaccine," 2–3.
79. US General Accounting Office, "Anthrax Vaccine," 2–5.
80. US General Accounting Office, "Anthrax Vaccine," 6.
81. US General Accounting Office, "Anthrax Vaccine," 8.
82. "FDA Licenses Anthrax Vaccine," *UPI*, January 31, 2002.
83. CDC Advisory Committee on Immunization Practices, "Use of Anthrax Vaccine in the United States," *Morbidity and Mortality Weekly Report* 49, no. RR15 (December 15, 2000): 1–20.
84. "FDA Licenses Anthrax Vaccine."
85. Ness, "The Anthrax Vaccine Program," 715–21.
86. *Doe v. Rumsfeld*, 297 F. Supp. 2d at 125.
87. Emergent BioSolutions Website (accessed April 24, 2024).
88. Sharon LaFraniere and Noah Weiland, "Factory Mix-Up Ruins Up to 15 Million Vaccine Doses from Johnson & Johnson," *The New York Times*, March 31, 2021; Chris Hamby, Sharon LaFraniere, and Sheryl Gay Stolberg, "U.S. Bet Big on Covid Vaccine Manufacturer Even as Problems Mounted," *The New York Times*, April 6, 2021.
89. Staff Report, *The Coronavirus Vaccine Manufacturing Failures of Emergent BioSolutions* (May 2022); Sheryl Gay Stolberg et al., "Emergent Hid Evidence of Covid Vaccine Problems at Plant, Report Says," *The New York Times*, May 10, 2022.
90. Staff Report, *The Coronavirus Vaccine*.
91. US DOJ, *Amerithrax Investigative Summary*, 1–11.
92. National Research Council, *Review of the Scientific Approaches Used During the FBI's Investigation of the 2001 Anthrax Letters* (2011).
93. US Government Accountability Office, "Anthrax: Agency Approaches to Validation and Statistical Analysis Could Be Improved" (December 2014).
94. Congressional Research Service, "Department of Defense Anthrax Vaccination," 1–2; *Department of Defense Anthrax Vaccine*, 106th Cong. 151.
95. Mark Thompson, "The Buck Stops (the Anthrax Shots) Here," *Time*, January 12, 2001.

96. "Buck Fined, Avoids Jail, Can Stay in Service," *Biloxi Sun Herald*, May 23, 2001, quoted in Ness, "The Anthrax Vaccine Program," 718.
97. Ness, "The Anthrax Vaccine Program," 718.
98. *See, e.g., United States v. Johnson*, 2004 WL 720153 (N-M. Ct. Crim. App. 2004); *United States v. Washington*, 57 M.J. 394 (C.A.A.F. 2002); *Perry v. Wesely*, 2000 WL 1775249 (N-M. Ct. Crim. App. 2000); *Ponder v. Stone*, 54 M.J. 613 (N-M. Ct. Crim. App. 2000).
99. *Ponder*, 54 M.J. at 616–17; *see also United States v. Schwartz*, 61 M.J. 567, 571 (N-M. Ct. Crim. App. 2005); *Perry*, 2000 WL 1775249, at *3. The cases reference various directives and instructions, as set forth by the DoD and as implemented by the various branches of the military. For example, Department of Navy Instruction 6230.4 (dated April 29, 1998) implements Department of Defense Directive 6205.3 (DoD Immunization Program for Biological Warfare Defense) and the Secretary of Defense's December 15, 1997 order regarding mandatory anthrax immunizations.
100. *Doe v. Rumsfeld*, 341 F. Supp. 2d 1, 3 (D.D.C. 2004).
101. *Doe v. Rumsfeld*, 297 F. Supp. 2d at 128–31.
102. *Doe v. Rumsfeld*, 297 F. Supp. 2d at 125.
103. *Doe v. Rumsfeld*, 341 F. Supp. 2d at 4–5.
104. *Doe v. Rumsfeld*, 297 F. Supp. 2d at 132.
105. *Doe v. Rumsfeld*, 297 F. Supp. 2d at 126.
106. *Doe v. Rumsfeld*, 297 F. Supp. 2d at 127.
107. *Doe v. Rumsfeld*, 297 F. Supp. 2d at 135–36.
108. *Doe v. Rumsfeld*, 297 F. Supp. 2d at 134.
109. *Doe v. Rumsfeld*, 297 F. Supp. 2d at 134.
110. *Food and Drug Administration: Biological Products; Bacterial Vaccines and Toxoids; Implementation of Efficacy Review*, 69 Fed. Reg. 255, 260 (January 5, 2004).
111. *Doe v. Rumsfeld*, 341 F. Supp. 2d at 7 (quoting FDA Press Release).
112. *Doe v. Rumsfeld*, 341 F. Supp. 2d at 9.
113. Project BioShield Act of 2004, Pub. L. No. 108-276, 118 Stat. 835 (2004).
114. Food and Drug Administration, *Emergency Use Authorization of Medical Products and Related Authorities: Guidance for Industry and Other Stakeholders* (January 2017), 7–8; New Drugs, 21 U.S.C. § 355(d).
115. *Doe v. Rumsfeld*, 341 F. Supp. 2d at 16.
116. *Doe v. Rumsfeld*, 341 F. Supp. 2d at 19.
117. *Doe v. Rumsfeld*, 341 F. Supp. 2d at 19.
118. Food and Drug Administration, *Authorization of Emergency Use of Anthrax Vaccine Adsorbed for Prevention of Inhalation Anthrax by Individuals at Heightened Risk of Exposure Due to Attack with Anthrax; Availability*, 70 Fed. Reg. 5452, 5453 (February 2, 2005).
119. FDA, *Authorization of Emergency Use*, 70 Fed. Reg. 5453.
120. FDA, *Authorization of Emergency Use*, 70 Fed. Reg. 5455.
121. Authorization for Medical Products for Use in Emergencies, 21 U.S.C. § 360bbb-3 (2004).
122. Stuart L. Nightingale, Joanna M. Prasher, and Stewart Simonson, "Emergency Use Authorization (EUA) to Enable Use of Needed Products in Civilian and Military Emergencies, United States," *Emerging Infectious Diseases* 13, no. 7 (July 2007): 1050.
123. Food and Drug Administration, *Biological Products; Bacterial Vaccines and Toxoids; Implementation of Efficacy Review; Anthrax Vaccine Adsorbed; Final Order*, 70 Fed. Reg. 75,180 (December 19, 2005).
124. *Doe v. Rumsfeld*, 172 Fed. Appx. 327, 328 (D.C. Cir. 2006).
125. Christopher Lee, "Mandatory Anthrax Shots to Return," *The Washington Post*, October 17, 2006.

126. Quoted in Lee, "Mandatory Anthrax Shots."
127. Lee, "Mandatory Anthrax Shots."
128. *Rempfer v. Von Eschenback*, 535 F. Supp. 2d 99, 101–2 (D.D.C. 2008).
129. *Rempfer v. Von Eschenbach*, 535 F. Supp. 2d at 107.
130. *Rempfer v. Sharfstein*, 583 F.3d 860, 861–62 (D.C. Cir. 2009).
131. Jennifer Gordon Wright et al., "Use of Anthrax Vaccine in the United States: Recommendations of the Advisory Committee on Immunization Practices (ACIP), 2009," *Morbidity and Mortality Weekly Report* 59, no. RR-6 (July 23, 2010): 11.
132. Wright et al., "Use of Anthrax Vaccine," 11.
133. Denise Pica-Branco and Ronald P. Hudak, "U.S. Military Service Members' Perceptions of the Anthrax Vaccine Immunization Program," *Military Medicine* 173, no. 5 (May 2008): 429–33.
134. Pica-Branco and Hudak, "U.S. Military," 430.
135. Pica-Branco and Hudak, "U.S. Military," 430.
136. Smith, *American Biodefense*, 121.
137. *The Smallpox Vaccination Plan: Challenges and Next Steps: Hearing of the Senate Committee on Health, Education, Labor, and Pensions*, 108th Cong. 2 (statement of Senator Judd Gregg), 61 (statement of William J. Bicknell) (January 30, 2003).
138. Gregory A. Poland, John D. Grabenstein, and John M. Neff, "The US Smallpox Vaccination Program: A Review of a Large Modern Era Smallpox Vaccination Implementation Program," *Vaccine* 23, no. 17–18 (March 2005): 2078–81.
139. Leonard A. Cole, *The Anthrax Letters* (2003), 135–37.
140. Alibek and Handelman, *Biohazard*, 111.
141. Anita Manning and Steve Sternberg, "Officials Ponder Timing, Sequence of Immunizations," *USA Today*, October 7, 2002, quoted in Institute of Medicine, *The Smallpox Vaccination Program* (2005), 31.
142. IOM, *The Smallpox Vaccination Program*, 29.
143. Pamela Sankar, Cynthia Schairer, and Susan Coffin, "Public Mistrust: The Unrecognized Risk of the CDC Smallpox Vaccination Program," *American Journal of Bioethics* 3, no. 4 (Fall 2003): W22–W25.
144. IOM, *The Smallpox Vaccination Program*, 16–17.
145. Thomas Mack, "A Different View of Smallpox and Vaccination," *New England Journal of Medicine* 348, no. 5 (January 30, 2003): 460–63.
146. Mack, "A Different View," 460.
147. IOM, *The Smallpox Vaccination Program*, 12.
148. Mack, "A Different View," 460.
149. Mack, "A Different View"; IOM, *The Smallpox Vaccination Program*, 18.
150. John D. Grabenstein and William Winkenwerder, "US Military Smallpox Vaccination Program Experience," *Journal of the American Medical Association* 289, no. 24 (June 25, 2003): 3278–82.
151. Oliver Burkeman, "No Evidence of Smallpox," *The Guardian*, September 19, 2003; Douglas Jehl and Judith Miller, "The Struggle for Iraq: The Weapons; Draft Report Said to Cite No Success in Iraq Arms Hunt," *The New York Times*, September 25, 2003.
152. Central Intelligence Agency, "Comprehensive Report of the Special Advisor to the Director of Central Intelligence on Iraq's WMD" (2004), quoted in IOM, *The Smallpox Vaccination Program*, 71.
153. Poland et al., "The US Smallpox Vaccination Program," 2079.
154. John Neff et al., "Monitoring the Safety of a Smallpox Vaccination Program in the United States," *Clinical Infectious Diseases* 46, Supp. 3 (2008): S264 (indicating a hospitalization rate of 0.1 percent); Poland et al., "The US Smallpox Vaccination Program," 2079.

155. IOM, *The Smallpox Vaccination Program*, 70.
156. Neff et al., "Monitoring the Safety," S261; Grabenstein and Winkenwerder, "US Military Smallpox Vaccination," 3280.
157. Neff et al., "Monitoring the Safety," S262.
158. Neff et al., "Monitoring the Safety," S265–67; Jeffrey S. Halsell et al., "Myopericarditis Following Smallpox Vaccination among Vaccinia-Naïve US Military Personnel," *Journal of the American Medical Association* 289, no. 24 (June 25, 2003): 3283–89.
159. Neff et al., "Monitoring the Safety," S260–65.
160. IOM, *The Smallpox Vaccination Program*, 1.
161. Sankar et al., "Public Mistrust," W22.
162. Sankar et al., "Public Mistrust," W22.
163. Neff et al., "Monitoring the Safety," S259.
164. Neff et al., "Monitoring the Safety," S262.
165. Sankar et al., "Public Mistrust."
166. IOM, *The Smallpox Vaccination Program*, 43; Sankar et al., "Public Mistrust," W23–24.
167. Sankar et al., "Public Mistrust."
168. Editorial, "Doctors' Orders," *The Washington Post*, December 19, 2002; Editorial, "Ducking Smallpox Vaccinations," *The New York Times*, December 22, 2002; Editorial, "Vaccine Dodgers Grow," *Boston Herald*, January 23, 2003; Ceci Connolly, "Bush Smallpox Inoculation Plan Near Standstill," *The Washington Post*, February 24, 2003.
169. Sankar et al., "Public Mistrust"; IOM, *The Smallpox Vaccination Program*, 30–33.
170. Quoted in IOM, *The Smallpox Vaccination Program*, 32.
171. Kathleen Sebelius, "Why We Still Need Smallpox," *The New York Times*, April 25, 2011.
172. "CDC Media Statement on Newly Discovered Smallpox Specimens" (July 8, 2014).
173. Food and Drug Administration, "Product Development Under the Animal Rule: Guidance for Industry" (October 2015), 1.
174. Melanie C. Trull, Tracey V. du Laney, and Mark D. Dibner, "Turning Biodefense Dollars into Products," *Nature Biotechnology* 25, no. 2 (February 2007): 179–84.
175. Food and Drug Administration, New Drug, Antibiotic, and Biological Drug Product Regulations; Accelerated Approval, 57 Fed. Reg. 58,942 (December 11, 1992).
176. Oriana Ciani et al., "A Framework for the Definition and Interpretation of the Use of Surrogate Endpoints in Interventional Trials," *The Lancet* 65 (November 2023): 1–12; Robert Kemp and Vinay Prasad, "Surrogate Endpoints in Oncology: When are they Acceptable for Regulatory and Clinical Decisions, and are they Currently Overused?," *BMC Medicine* 15, no. 134 (July 21, 2017): 1–7; Thomas R. Fleming, "Surrogate Endpoints and FDA's Accelerated Approval Process," *Health Affairs* 24, no. 1 (January/February 2005): 67–78.
177. Paul Aebersold, "FDA Experience with Medical Countermeasures Under the Animal Rule," *Advances in Preventive Medicine* (2012): 2.
178. Aebersold, "FDA Experience," 2.
179. Aebersold, "FDA Experience," 2–3.
180. Food and Drug Administration, New Drug and Biological Drug Products; Evidence Needed to Demonstrate Efficacy of New Drugs for Use Against Lethal or Permanently Disabling Toxic Substances When Efficacy Studies in Humans Ethically Cannot Be Conducted, 64 Fed. Reg. 53,960 (October 5, 1999).
181. Food and Drug Administration, New Drug and Biological Drug Products; Evidence Needed to Demonstrate Effectiveness of New Drugs When Human Efficacy Studies Are Not Ethical or Feasible, 67 Fed. Reg. 37,988 (May 31, 2002).

182. Theresa Allio, "Product Development Under FDA's Animal Rule: Understanding FDA's Expectations and Potential Implications for Traditional Development Programs," *Therapeutic Innovation & Regulatory Science* 50, no.5 (2016): 668.
183. FDA, "Product Development Under the Animal Rule," 2.
184. New Drugs, 21 U.S.C. § 355(d).
185. National Research Council, *Animal Models for Assessing Countermeasures to Bioterrorism Agents* (2011), 44.
186. Allio, "Product Development," 663.
187. Aebersold, "FDA Experience," 9.
188. Aebersold, "FDA Experience," 5–6.
189. CDER, Drug and Biologic Animal Rule Approvals (accessed December 5, 2023).
190. CDER, Drug and Biologic Animal Rule Approvals; David W. C. Beasley, Trevor L. Brasel, and Jason E. Comer, "First Vaccine Approval Under the FDA Animal Rule," *NPJ Vaccines* 1 (2016): 1–3.
191. Beasley et al., "First Vaccine Approval," 2.
192. NRC, *Animal Models*, 3.
193. 42 U.S.C. § 247d-6d(a)(1) (2005).
194. 152 Cong. Rec. S1360–61 (February 15, 2006) (statement of Senator Chris Dodd).
195. 42 U.S.C. § 247d-6d(c)(1)(A) (2005).
196. 42 U.S.C. § 247d-6e (2005).
197. 42 U.S.C. § 247d-6e (2005).
198. 152 Cong. Rec. S1360 (February 15, 2006) (statement of Senator Chris Dodd).
199. 152 Cong. Rec. S1360–61 (February 15, 2006) (statement of Senator Chris Dodd); Bill Theobald, "Hastert, Frist Said to Rig Bill for Drug Firms," *Gannett News Service* (Feb. 9, 2006).
200. 152 Cong. Rec. S1360–61 (February 15, 2006) (statement of Senator Chris Dodd).
201. 152 Cong. Rec. S1360–61 (February 15, 2006) (statement of Senator Chris Dodd).
202. Tom Harkin, "Harkin Calls on Frist and Hastert to Repeal "Dead of Night" Vaccine Liability Provision and Enact Real Protections," *Press Release* (February 15, 2006).
203. Responsible Public Readiness and Emergency Preparedness Act, S. 2291, 109th Cong. (2006); 152 Cong. Rec. S1361–63 (February 15, 2006).
204. Pandemic and All-Hazards Preparedness Reauthorization Act of 2013, Pub. L. No. 113-5, 127 Stat. 161 (2013).
205. "The Strategic National Stockpile: Overview and Issues for Congress," *Congressional Research Service Report* (January 25, 2023); Shondra M. Neumeister and Joshua P. Gray, "The Strategic National Stockpile: Identification, Support, and Acquisition of Medical Countermeasures for CBRN Incidents," *Toxicology Mechanisms and Methods* 31, no. 4 (May 2021): 308–21.
206. Dan Diamond, "Lawmakers Defend 'Unprecedented' Pentagon Health Panel, Which Could Undermine FDA," *Politico*, November 6, 2017.
207. Diamond, "Lawmakers Defend"; Rettig, *Military Use*, 15.
208. An Act to Amend the Federal Food, Drug, and Cosmetic Act to Authorize Additional Emergency Uses for Medical Products to Reduce Deaths and Severity of Injuries Caused by Agents of War, and for Other Purposes, Pub. L. No. 115-92, 131 Stat. 2023 (2017).
209. Food and Drug Administration, "FDA and DoD Formalize Collaboration to Advance Medical Products in Support of American Military Personnel," *News Release* (December 2, 2018).
210. For an overview, see Efthimios Parasidis, Micah L. Berman, and Patricia J. Zettler, "Assessing COVID-19 Emergency Use Authorizations," *Food and Drug Law Journal* 76, no. 3 (2021): 441–501.

211. Max Matza and Nicholas Yong, "FBI Chief Christopher Wray Says China Lab Leak Most Likely," *BBC News*, March 1, 2023; Michael R. Gordon and Warren P. Strobel, "Lab Leak Most Likely Origin of Covid-19 Pandemic, Energy Department Now Says," *The Wall Street Journal*, February 26, 2023; Max Kozlov, "NIH Reinstates Grant for Controversial Coronavirus Research," *Nature* 617 (May 18, 2023): 449.
212. Kaiser Family Foundation, Global COVID-19 Tracker (April 29, 2024).
213. National Academies of Sciences, Engineering, and Medicine, *Biodefense in the Age of Synthetic Biology* (2018).

15
Twenty-First-Century Conflicts and the Military Biomedical Complex

Introduction

America's military biomedical complex has evolved to become a colossal enterprise. As of 2024, the military and veterans' health systems each serve a patient population of more than nine million beneficiaries. The government allocates billions of dollars annually in research funding for military science, investments that intimately link the military with universities and industry across the country. Although this vast biomedical empire provides world-class healthcare and conducts cutting-edge research, the health concerns from post-9/11 wars in Afghanistan and Iraq have challenged the military and veterans' health systems in ways that differ from past conflicts.

In the two decades following 9/11, more than two million service members were deployed to various conflict zones, over half of whom served multiple deployments. Compared to veterans from previous wars, veterans from post-9/11 wars have experienced longer deployments, higher levels of combat exposure, higher rates of survival from combat injuries, higher incidence of disabilities, and more complex medical treatments due to the nature of their injuries.[1] Of particular concern have been the signature injuries of post-9/11 conflicts—war lung injury, traumatic brain injury (TBI), and affiliated physical and mental health ailments. More than 400,000 military personnel have experienced a TBI, and a significant percentage have suffered continuing post-concussive symptoms. Suicide among veterans and service members has been at an all-time high. More than thirty thousand veterans of post-9/11 wars are estimated to have taken their own lives, four times the number of warfighters killed in combat during the same period.[2] The physical and mental health ravages of war continue to plague military personnel during combat and upon their return to civilian life.

Apart from the unique health impact of post-9/11 wars, the sweeping justifications for the global war on terror influenced legal and ethical deliberations regarding war preparations, intelligence operations, and military missions. To further its national security, the United States altered its policies and practices for rendition, detention, and interrogation, and relied extensively on drone surveillance and drone strikes. Of these programs, the CIA's expansive rendition, detention, and interrogation (RDI) program raised complex questions regarding the moral obligations of health professionals who participated in interrogations. A 2014 congressional report lambasted the secret RDI program and called for changes to CIA policies.

Contemporaneously, scientific and technological advances expanded the domain of warfare into cyberspace. Various components of the military and the intelligence community have created comprehensive surveillance programs that track individuals worldwide, including Americans not suspected or linked to any illicit enterprise. Some cyberprojects leverage biometric data, artificial intelligence, and biomedical

America's Military Biomedical Complex. Efthimios Parasidis, Oxford University Press.
 DOI: 10.1093/9780199351473.003.0016

techniques. Others are structured to utilize cybertools to disrupt critical infrastructure and engage in cyberattacks. These new means and methods of war have raised novel questions regarding the legal and ethical limits of the national security state.

The World Is a Battlefield

America's twenty-first-century conflicts are primarily grounded in an Authorization for Use of Military Force (AUMF), a joint congressional resolution passed on September 14, 2001, by a combined vote of 518–1. The crux of the AUMF is distilled to one sentence, which grants the president authorization "to use all necessary and appropriate force against those nations, organizations, or persons he determines planned, authorized, committed, or aided the terrorist attacks that occurred on September 11, 2001, or harbored such organizations or persons, in order to prevent any future acts of international terrorism against the United States by such nations, organizations or persons."[3] Barbara Lee, the lone dissenting vote, warned that the law would grant the president a blank check for military action.[4] Her concerns were summarily ignored.[5]

On September 17, 2001, the day before President George W. Bush signed the AUMF, he signed a secret Memorandum of Notification that authorized covert action and granted the CIA the authority to capture and detain suspected militants anywhere in the world.[6] The secret authorization supported a variety of CIA programs, including a global network of black sites for detention and interrogation, and a clandestine drone program for surveillance and targeted assassinations. As John Rizzo, a CIA attorney who joined the agency in 1976 and worked there for thirty-four years, explained: "I had never in my experience been part of or even seen a presidential authorization as far-reaching and as aggressive in scope. It was simply extraordinary."[7]

One day before Bush signed the secret order, Vice President Dick Cheney publicly signaled the crux of its pronouncements: "We've got to spend time in the shadows of the intelligence world. A lot of what needs to be done here will have to be done quickly, without any discussion, using sources and methods that are available to our intelligence agencies, if we're going to be successful."[8] Proclaiming that "the world is a battlefield" and that the United States is engaged in a "global war on terror" was more than mere rhetoric: it suggested that the law of armed conflict applies throughout the world and that American forces can kill, capture, or detain enemies anywhere and at any time.

Over the next few months, a potpourri of domestic laws expanded the scope of the national security state. Coupled with the AUMF and the secret presidential authorization issued days after 9/11, several additional laws and presidential orders empowered law enforcement and the intelligence community with broad discretion to adopt policies and practices to further national security.[9] The legality of individual projects has often been reviewed in secret by the Foreign Intelligence Surveillance Court (FISC), whose broad jurisdiction includes reviewing applications made by the US government to approve electronic surveillance, physical search, and certain other forms of investigative actions for foreign intelligence purposes.[10]

A separate presidential order, issued November 13, 2001, created a military commission authorized to try suspected terrorists: individuals brought before the commission are characterized as detainees, illegal enemy combatants, or unlawful combatants—classifications that sought to evade protections afforded by the US Constitution and Geneva Conventions.[11] In 2002, the United States established a detention facility at its naval base in Guantanamo Bay, Cuba;[12] the United States acquired a lease to use the land as a naval base following the 1898 Spanish-American War.[13]

In 2006, the US Supreme Court deemed the military commissions unconstitutional and in violation of military law and the Geneva Conventions. The high court also suggested that tactics employed during detention and interrogation may be reviewed as possible war crimes.[14] Following the decision, Congress created new laws that empowered military tribunals to try detainees and eliminated the ability of US courts to review the decisions of the tribunals.[15]

Every post-9/11 president has broadly interpreted and widely utilized these legal tools. Over the past quarter century, the United States has engaged in military action in Afghanistan, Iraq, Libya, Yemen, Syria, and more than a dozen other countries.[16] The United States also has created a network of military installations throughout the African continent, spanning no less than fifteen African nations. These bases include staging outposts for special operations, military training facilities, and airfields from which to launch drone strikes and other military aircraft.[17] In the decade following 9/11, on any given day, US special forces were deployed to nearly seventy countries, working on their own or training or fighting with allied militaries or proxy armies.[18] Hundreds of thousands of private military contractors have supplemented the ranks of the armed forces—modern mercenaries who work and fight alongside military personnel. The military biomedical complex has evolved in parallel to these engagements to support twenty-first-century conflicts.

Post-9/11 Wars

Within days of 9/11, the CIA assembled a seven-man special operations team to enter Afghanistan, link with Afghan allies within the Northern Alliance, and convince Northern Alliance leaders to cooperate with US forces.[19] The CIA had collaborated with the Northern Alliance and other Afghan groups during the 1980s to counter the Soviet invasion of Afghanistan. In the late 1990s, the Northern Alliance assisted the agency in its search for Osama bin Laden.[20] After 9/11, armed with millions of dollars in cash, the CIA team bought the support of tribal leaders, some of whom were provided multiple payments of hundreds of thousands of dollars. Fueled in large part by control over local opium production and drug trafficking, these men wielded significant power in their respective regions.[21] As the CIA team leader, Gary Schroen, later explained: "You cannot buy an Afghan's loyalty, but you can rent it."[22]

A turf war between the CIA and the DoD followed. DoD special forces joined CIA paramilitary personnel in Afghanistan, and the DoD expanded its intelligence and espionage activities to limit its dependence on the CIA.[23] Meanwhile, the dollars provided to tribal leaders often were funneled to local drug kingpins and warlord groups, some of whom fought against US interests in the region.[24]

Complicating the war efforts was Pakistan, which had ceded control of land along the Pakistan-Afghanistan border to Taliban-linked tribal chiefs.[25] The Taliban are comprised primarily of Afghan Pashtuns, who share religious, cultural, and economic ties with Pakistani Pashtuns, millions of whom live in Pakistan's tribal areas.[26] The United States provided Pakistan with $2 billion a year in military funding and collaborated with the Inter-Services Intelligence (ISI), Pakistan's intelligence agency.[27] Despite the collaborations, some ISI components were sabotaging US efforts. Specifically, an ISI section known as Directorate S was financing and nurturing the Taliban, al-Qaeda, and other Islamist groups.[28] For example, Directorate S agents often warned the Taliban of imminent drone attacks so they could flee to safety.[29]

In part, Pakistani dual loyalties stemmed from the concern that, once America exited Afghanistan, India would fill a governance vacuum and threaten Pakistan's national security. India maintained intelligence activities in Afghanistan and plotted how to exercise control and influence.[30] There is a long history here. Following the Indo-Pakistani War, which divided India and Pakistan, India sided with the Soviet Union and Pakistan with America. During the Soviet invasion of Afghanistan (1979–1989), the United States supported the mujahideen, Islamic resistance fighters, with nearly $1 billion annually in arms, cash, and logistical support.[31] The mujahideen included several subgroups, one of which was a small outfit led by bin Laden and named al-Qaeda.[32] Of the multitude of arms provided to the mujahideen were Stinger missiles—portable, easy-to-use shoulder-fired weapons with a heat-seeking guidance system that could take down a fighter jet. After the war, the CIA orchestrated a buyback program, offering up to $150,000 per missile, in an attempt to stem the flow of these arms into groups hostile to America. Despite the program, hundreds of Stinger missiles remained in rebel hands and were used against the United States during the post-9/11 Afghanistan war.[33]

Meanwhile, after the Soviet withdrawal from Afghanistan in 1989, a civil war erupted within Afghanistan, with the Taliban, the Northern Alliance, and other groups battling for power. Pakistan supported the Taliban and viewed the Northern Alliance as having ties with India and Russia. The United States simply left the region. In 1996, the Taliban won the civil war and ruled Afghanistan as an Islamic Emirate.[34]

Years later, bin Laden and al-Qaeda turned against the United States, contending that America sought to colonize Muslims and exploit natural resources throughout the Middle East.[35] Many mujahideen leaders viewed America's post-9/11 wars as analogous to the Soviet invasion from two decades earlier.[36] It thus is not surprising that some Pakistani elements continued to support the Taliban, and that bin Laden was able to live within Pakistan for years until the United States killed him during a nighttime raid of his secret compound in 2011.[37]

This patchwork of evolving loyalties and alignments complicated military planning and operations. Pakistan was an imperfect US ally that had to balance its relationship with Afghans and Americans while being mindful of India's strategic intentions.[38] Meanwhile, the CIA created, trained, and fought alongside a three-thousand-man covert army in Afghanistan named the Counterterrorism Pursuit Teams, which employed many elite Afghan warfighters.[39] The war in Afghanistan against al-Qaeda and the Taliban—Operation Enduring Freedom—was fought mainly by paramilitary and special forces and relied heavily on relentless air attacks and coalitions with Afghan groups. The operation lasted less than four weeks, during which the United States suffered thirteen deaths.[40]

Thereafter, the US-backed government of Hamid Karzai failed to unify Afghanistan, and the nation plunged into civil war. The CIA and DoD remained in the country for nearly twenty years, but could not gain any meaningful headway into defeating the Taliban and ensuring that a stable government could unite and lead the nation. Military gains were fragile and temporary, the Taliban remained a powerful force, corruption and bribes were rampant throughout Afghan society, and billions of US dollars were wasted or stolen.[41] Similar to the Soviet invasion decades earlier, Afghanis became entrenched in a quest to fight for Islam and to resist foreign occupation. America's military tactics—such as night raids, drone strikes, violent and intrusive searches of homes, and countless other indignities—insulted Afghani honor and encouraged further resistance.[42]

America also struggled to comprehend the military significance of tribal, ethnic, and religious dynamics within Afghanistan's various groups and regions. With bags of cash, the Americans sought to entice local leaders—many of whom were war criminals and narcotics traffickers—in an attempt to have them identify individuals loyal to the Taliban. As some tribal elders explained to American officials, this was a fool's errand, as countless locals sought to enrich themselves and address long-standing local feuds by falsely identifying their enemies as Taliban. Despite the warnings, the CIA and military continued to trade cash for information, much of which was false or misleading. Meanwhile, Afghan money launderers carried suitcases full of dollars to Dubai and other offshore venues to invest their gains.[43]

According to a comprehensive report from the Special Inspector General for Afghanistan Reconstruction (SIGAR)—which the US government unsuccessfully attempted to keep secret by suing to block public access—for over a decade, several senior American officials privately expressed the view that the war was an unmitigated disaster despite repeated public statements that misled the public to believe that the war was on ongoing success. The release of the SIGAR report was akin to the release of the Pentagon Papers, the military's top-secret reflections on the Vietnam War, though with Afghanistan, the public backlash was far more limited. In large part, this was because, unlike with Vietnam, the war in Afghanistan was not a significant feature of American life—there was little political opposition, no draft, and few war protests. Although the wars in Afghanistan and Iraq were defining hallmarks for a generation of US warfighters, less than 1 percent of Americans served in the wars; 2,443 were killed in battle and 20,666 were injured.[44]

Meanwhile, the Afghan economy tumbled, leading locals to rely increasingly on opium sales. By 2006 Afghanistan was supplying nearly 90 percent of the world's opium, and several reports linked the widespread availability and low cost of Afghan opium to the opioid drug crisis in the United States.[45] Between 2002 and 2017, the acreage devoted to growing opium poppies quadrupled.[46] President Bush sought to douse the opium fields with glyphosate—a chemical herbicide deployed extensively during the Vietnam War to damage Vietnamese crops—but Afghan President Karzai refused to consent to the request. Bush considered whether to proceed without Karzai's support—in part, because US officials believed that he was collaborating with Afghan drug kingpins—but the program was reportedly called off after internal memos warned that the spraying would violate international prohibitions on chemical warfare.[47] Officials were also mindful of the US experience with the deployment of herbicides during the Vietnam War, and were concerned about potential health risks to American military personnel.[48]

Throughout the war, the Taliban relied extensively on improvised explosive devices (IEDs) detonated by pressure or remote control, such as roadside bombs, footpath bombs, bicycle bombs, suicide bombs, and donkey bombs.[49] IEDs caused thousands of casualties among US military personnel and were universally feared.[50] The United States maintained healthcare facilities at base camps and used a network of medevacs that endeavored to provide assistance within one hour of an injury. Since many injuries occurred in remote areas, the "bleed-out" time—the time a wounded individual would have to wait for medical assistance—often was closer to two hours.[51] Such delays hindered the military's ability to provide timely emergency care.

The DoD also created a Joint Trauma System to account for the wartime health challenges. The system included a trauma registry that collected copious details of a warfighter's injury, treatment, and outcomes. The evaluation of registry data revealed suboptimal aspects of care, and factors associated with positive or poor health outcomes. This information was then applied to reorient trauma care and to create new techniques to address the health concerns. The military's robust analysis of trauma registry data resulted in a continuous improvement cycle that was heralded throughout the medical profession and that served as a model to improve civilian trauma care.[52]

The trauma system also helped facilitate treatment of injuries from IEDs, including severely injured extremities and TBIs.[53] Between 2001 and 2018, thousands of service members suffered significant limb injuries, and more than fifteen hundred required an amputation.[54] Upon their return to the United States, injured service members and veterans were assigned to special units called Wounded Warrior Battalions. Although these support units provided much-needed care and helped wounded warfighters assimilate back into civilian life, some reports revealed patterns of overmedication with psychotropic drugs.[55] TBIs also were correlated with increased posttraumatic stress disorder (PTSD) and dementia among service members and veterans.[56]

Coupled with the war in Afghanistan, by mid-2002, the Bush administration determined that it would go to war against Iraq, ostensibly because Iraq maintained weapons of mass destruction and was violating a 1991 UN resolution that prohibited Iraq from developing or maintaining chemical or biological weapons. The Bush administration pressured the CIA to produce intelligence reports corroborating these positions. The agency complied by relying on unverified and untrustworthy sources.[57] Mainstream media outlets—including *The New York Times* and *The Washington Post*—supported the war efforts despite evidence that suggested the bases for war were specious.[58] In October 2002, a joint congressional resolution granted authorization for a war against Iraq.[59] The war commenced in March 2003, without UN authorization or approval, with an invasion force of approximately 150,000 American service members, more than ten times the number deployed to Afghanistan.[60] An additional 190,000 US military personnel were deployed to the broader Persian Gulf region.[61]

Saddam Hussein was overthrown within weeks of the invasion. Although initially the majority of Iraqis welcomed his removal, as America continued the war and sought to purge Baath Party members from government positions and demobilize the Iraqi Army, the Sunni elite in Iraq (a minority of the population, but one that had held power for decades under Hussein) rebelled, plummeting the nation into a devastating civil war and leading to the rise of the Islamic State of Iraq and Syria (ISIS) and other rebel groups.[62] Meanwhile, inspectors did not find any evidence of weapons of mass destruction, and years later, it was confirmed that American leaders manufactured evidence to support their decision to go to war.[63]

As with Afghanistan, insurgency tactics were a hallmark of the Iraq war. Exiled Baath Party members and the Iraqi military (many of whom were part of the Sunni elite in Iraq) supported ISIS, which also received support from several Sunni-majority Arab nations, including Turkey, Saudi Arabia, and Qatar. ISIS and its collaborators battled Kurdish Iraqis (who comprise 20 percent of the Iraqi population), Shiites, and the US military. The fight against ISIS brought an unlikely alliance between America, Russia, and Iran, with Israel supporting Sunni warfighters to blunt Iran's regional power.[64] America was not expert in building governance in impoverished landscapes with sectarian violence, or winning asymmetrical conflicts with guerilla fighters,[65] and the war continued for nearly two decades. In the fall of 2022, the CIA met with Taliban leaders to discuss a joint counterterrorism engagement to counter ISIS, a group both had been battling for years.[66]

In Iraq, Afghanistan, and other nations where the United States pursued suspected terrorists—such as Pakistan and Yemen—America increasingly relied on drones for surveillance and air strikes. America's drone guidelines limit strikes to individuals who pose a "continuing, imminent threat to U.S. persons," although the classification has been liberally interpreted.[67] The CIA and the DoD operate separate drone programs, and oftentimes individuals are tracked and attacked via electronic surveillance of metadata, SIM cards, and mobile phones.[68] According to Eliot Ackerman, a former CIA paramilitary officer, many of his colleagues expressed discomfort at using drones in targeting killings "because it felt like we were doing something, on a large scale, that we'd sworn not to. Most of us felt as though we were violating Executive Order 12333," which prohibits the US government from involvement in assassinations.[69] Notwithstanding the "discomfort" expressed by some CIA and DoD officials, in 2002, President Bush's National Security Advisor Condoleezza Rice stated, "I can assure you that no constitutional questions are raised" by the president authorizing drone assassinations during wartime.[70] Reports estimate that drone strikes have killed thousands of individuals.[71]

Although American leaders have widely heralded drones as precise instruments that minimize civilian casualties, opposition leaders characterize drones as "cruel and cowardly" and contend that the United States is "not fighting with honor."[72] Legal and human rights questions have clouded drone operations, and drone attacks on American citizens suspected of terror links have been particularly controversial. In 2011, the United States assassinated Anwar al-Aulaqi, an imam from Denver who moved to Yemen, advocated for jihad against America, and was suspected of being a regional commander for al Qaeda. The American Civil Liberties Union (ACLU) filed a lawsuit that questioned the military's authority to assassinate al-Aulaqi, arguing that the action violated the constitutional rights of an American citizen such as the right to due process of law. A federal judge dismissed the case, ruling that a claim involving overseas killing of an American citizen deemed by US authorities to be an active enemy involves a question of national security that civilian courts are precluded from second-guessing since it would "impermissibly draw the Court into the heart of executive and military planning and deliberation."[73] Within two weeks of al-Aulaqi's assassination, his sixteen-year-old American-born son was killed in a drone strike in Yemen, and US forces later killed al-Aulaqi's eight-year-old American-born daughter during a raid in Yemen.[74]

There likewise are significant health risks to drone operators, who often are witness to gruesome acts such as torture, mass killings, beheadings, and other horrors, and then watch as families and medical personnel come to the aid of the wounded or dead. It is a high-stress environment with long work hours, low morale, and a high attrition rate. Studies have found that the psychological impact on drone operators is equivalent to combatants fighting in person.[75] Notwithstanding the health risks, the use of drones is accelerating. There are more than five hundred types of drones and other unmanned systems: some insect-sized drones can travel ten meters per second, hover in place for an extended time, climb up walls or pipes, or carry tiny sensors, cameras, and weapons.[76]

Another hallmark of twenty-first-century conflicts is an increasing reliance on private contractors hired to build bases, cook meals, provide security, engage in combat and intelligence operations, and more. Companies such as Halliburton, KBR, Triple Canopy, DynCorp, and Blackwater (renamed Xe, then Academi) often poach individuals with specific expertise or security clearances to be able to staff a diverse range of military contracts. Of private contractors that engage in combat, many are former Navy SEALS or other special forces. It is common to find former high-ranking military officials on the boards of the companies. In addition to assisting in procuring government contracts, these individuals signal that the company knows how the military thinks and acts.[77]

In 2010, the United States deployed 175,000 service members and 207,000 contractors to war zones, and contractor casualties surpassed military casualties. Privatization of the armed forces has continued to increase, notwithstanding reports that fraud and abuse are estimated at billions of dollars annually.[78] It is an opaque industry not subject to Freedom of Information Act (FOIA) requests or direct legislative oversight. Between 2001 and 2022—with significant assistance from private contractors—the United States built more than 500 bases in Iraq and 450 in Afghanistan, ranging from combat outposts to camps the size of small American towns with fast-food restaurants and power generation plants.[79] Lax standards and minimal oversight sometimes have resulted in shoddy work product with devastating consequences for service members. For example, hundreds of military personnel have suffered electric shock due to faulty wiring on American bases, and no less than eighteen have died from electrocution while taking showers.[80]

In Afghanistan and Iraq, KBR's contract work included waste disposal, and the company created hundreds of burn pits that used jet fuel to burn a wide variety of products including plastic, paint, medical waste, oil, batteries, appliances, and computers (see Figure 15.1). The largest burn pit was a massive ten-acre facility that burned up to 250 tons of waste each day. The military approved the use of burn pits in lieu of incinerators or other safe methods of waste disposal.[81] It later was discovered that

Figure 15.1 Smoldering rubbish from a large burn pit constructed near a US forward operating base in Afghanistan.
Photograph from the Special Inspector General for Afghanistan Reconstruction.

personnel who worked at the burn pits suffered from unusually high rates of asthma, bronchitis, cancer, and other health conditions. Studies conducted by the Centers for Disease Control and Prevention (CDC) and the Department of Veterans Affairs (VA) revealed that many burn pit workers had high levels of titanium and unusual biomasses in their lungs. The disease would later be called "war lung injury" and would become one of the most devastating health ailments from America's post-9/11 wars.[82]

As cases of war lung injury were mounting, the CDC and VA refused to admit that a causal link existed between burn pits and the health concerns. At the same time, DoD officials asked CDC researchers to stall their efforts to uncover the causes of war lung injury.[83] According to congressional testimony provided by Steven Coughlin, an epidemiologist who worked at the CDC and the VA, his supervisors instructed him "not to look at data regarding hospitalizations and doctors' visits."[84] Coughlin further testified that the VA downplayed any connection between the burn pits and war lung injury, and noted that "on the rare occasions when embarrassing study results are released, data are manipulated to make them unintelligible."[85] He analogized the cover-up to Gulf War Illness and adverse health consequences from the use of Agent Orange in Vietnam. Coughlin also emphasized the relevance of the VA's conflict of interest—whereby the department would have to pay more benefits should war lung injury be deemed combat-related.[86]

In January 2013, after increasing pressure from veterans groups, President Barack Obama signed legislation that ordered the VA to create a burn pit registry.[87] Five years later, a federal court dismissed a class action lawsuit against KBR—brought by hundreds of veterans who were suffering from war lung injury—ruling that the political question doctrine precluded court review of KBR's practices because the military exercised direction and control over KBR's work.[88] In 2022—four years after the lawsuit was dismissed—the administration of President Joe Biden announced additional federal programs to assist veterans suffering from war lung injury.[89] Biden's son, Beau, an Iraqi war veteran, died at the age of forty-six from brain cancer that is alleged to have been caused by his exposure to burn pits.[90]

The VA's efforts to suppress the nature and extent of war lung injury paralleled the department's suppression of data on suicide rates throughout post-9/11 wars. Coughlin and others were deeply concerned about the increased rates, yet the VA stifled efforts to track mental health concerns over time. At one point, Coughlin received a formal written reprimand for insubordination due to his continuing efforts to research veterans' health concerns that his supervisors had no interest in exploring.[91] Further troubling were revelations about squalid conditions at Walter Reed Army Medical Center, where staff neglected veterans with disabilities and housed them in cockroach-infested rooms with black mold on the walls.[92] A 2012 study found that, of service members and veterans who served in Iraq or Afghanistan and screened positive for PTSD, nearly half did not receive any treatment. For those who did receive care, the study found that efficacy data for many treatments was lacking.[93] These reports underscored a pattern, spanning hundreds of years, whereby the US government often has been deficient in affording veterans with appropriate healthcare to address their service-related injuries.

By the early 2020s, the United States had spent trillions of dollars in its war efforts in Iraq and Afghanistan, casualties were mounting, and veterans were suffering from health ailments and high rates of suicide. The American public was growing weary of the wars, and public opinion shifted largely to support full withdrawal of US forces. After a series of surges and pullbacks, American forces withdrew from Afghanistan in August 2021, thus ending the longest war in US history. Days before the exit, the Taliban quickly reclaimed leadership and control of Afghanistan, while to the west Iraq remained in a violent civil war. The number of terror groups throughout the Middle East has escalated, and questions remain whether the United States can properly claim to have won the war on terror when it failed to win post-9/11 wars in Afghanistan and Iraq.[94] As of March 2024, the 2001 AUMF remains in effect and has been invoked no less than forty-one times to authorize military action in no less than nineteen countries.[95] One of the most controversial aspects of America's global war on terror has been the use of rendition, detention, and interrogation.

Rendition, Detention, and Interrogation

Following a series of attacks from al Qaeda in the early 1990s, President Bill Clinton authorized the CIA to apprehend terror suspects anywhere in the world and detain and interrogate them in secret black sites.[96] The frequency of rendition, detention, and interrogation increased significantly following 9/11, and largely was conducted without regard to rights outlined in international conventions and bilateral agreements. Although rendition typically occurs pursuant to an extradition treaty whereby an individual is sent to another country for criminal prosecution, extraordinary rendition occurs outside this framework, and an individual is secretly transferred to a country to be interrogated.[97] With America's extraordinary rendition program, suspects were interrogated by the CIA—sometimes in collaboration with the Federal Bureau of Investigation (FBI), DoD, or foreign agencies—in countries that reportedly included Afghanistan, Algeria, Egypt, Jordan, Iraq, Lithuania, Morocco, Pakistan, Poland, Romania, Saudi Arabia, Slovenia, and Thailand.[98] To support the extraordinary rendition program, the CIA provided millions of dollars in cash payments to encourage the participation of foreign officials, and invested more than $300 million in non-personnel costs to build and maintain secret detention facilities around the world.[99]

Interrogation techniques included waterboarding, shackling, prolonged stress positions, sleep deprivation, cramped confinement, slamming a person against a wall, face slaps, sexual humiliation, and threats to harm a detainee's family.[100] Some individuals within the CIA and the DoD questioned the legality of the interrogation methods, and some military lawyers within the Judge Advocate General's Corps argued for adhering to the Geneva Conventions and Army Field Manual (which prohibited certain types of interrogation methods). These positions were rejected by the DoD, and in a series of classified reports, Department of Justice (DOJ) attorneys concluded that enhanced interrogation techniques (EITs) were permissible and did not violate any laws or international treaties.[101] These reports became known as the Torture Memos.

The Torture Memos concluded that various interrogation techniques could be used, and that the infliction of pain during interrogation amounted to torture only if the pain was "equivalent in intensity to the pain accompanying serious physical injury, such as organ failure, impairment of a bodily function, or even death."[102] This conclusion contradicted the Army Field Manual and previous positions taken by the United States. For example, in 1983, the United States specifically identified waterboarding as torture.[103] In addition, a 1984 UN convention defined torture as "any act by which severe pain or suffering, whether physical or mental, is intentionally inflicted on a person for such purposes as obtaining from him or a third person information or a confession, punishing him for an act he or a third person has committed or is suspected of having committed, or intimidating or coercing him or a third person."[104] Indeed, the understanding that waterboarding equates to torture dates back at least to World War II (WWII). When Japanese military personnel used waterboarding against US prisoners of war, American prosecutors at the Tokyo trials characterized it as torture and prosecuted Japanese interrogators for war crimes.[105]

EITs were grounded in a psychological theory called "learned helplessness." After 9/11, the CIA's Psychological Operations group embraced the theory, which dates back to the 1960s and was honed by University of Pennsylvania researchers who experimented with dogs to examine under what circumstances the animals would make efforts to evade repeated electrical shocks.[106] After consulting with psychologists and other experts on learned helplessness, the CIA concluded that by subjecting a detainee to a series of hostile and uncontrollable interrogation techniques, the detainee would cooperate and provide truthful information in order to avoid the continuous harassment.[107]

Prior to the post-9/11 EIT program, the concept of learned helplessness was integrated into the military's survival, evasion, resistance, and escape (SERE) program. The SERE program was developed during the Cold War to help warfighters develop survival skills, evade capture, resist interrogation techniques, and escape captivity.[108] After 9/11, psychologists James Mitchell and Bruce Jessen received more than $80 million and comprehensive legal indemnity from the CIA to help the agency reverse engineer SERE tactics to create interrogation techniques. The men previously had worked in the Air Force's SERE program but had no experience as interrogators or in counterterrorism, and no specialized knowledge of al-Qaeda.[109]

The CIA also consulted with several other experts, including renowned psychologist Martin Seligman, who had worked on the 1960s dog studies and was a pioneer in the study of learned helplessness.[110] In 2003, Seligman wrote: "A science advisor to the President today needs to help direct natural science and social science toward winning our war against terrorism. First and foremost."[111] Seligman's sentiment was eerily reminiscent of Fritz Haber's—Germany's World War I (WWI) gas warfare mastermind—who unabashedly declared that during wartime, a scientist belongs first to his country. Seligman's group at the University of Pennsylvania later would receive a $31 million no-bid contract from the Army to assist with resilience training.[112]

At the time EITs were developed, experts debated whether such harsh tactics were an effective means of eliciting truthful information. Several studies have found that violent interrogations often lead individuals to say whatever they think their interrogator wants to hear so that the pain does not continue. Other studies have found that EITs create a mental state that impairs memory and promotes inaccurate answers.[113]

The CIA's adoption of unproven psychological theories in its post-9/11 interrogation program is analogous to its reliance on LSD, hypnosis, and other unproven methods during interrogation programs developed during the 1950s–1970s. Yet, the CIA failed to heed the lessons from the Cold War programs. Following the mid-century interrogation studies, in 1983, the CIA's *Human Resource Exploitation Manual* explained: "The routine use of torture lowers the moral caliber of the organization that uses it and corrupts those that rely on it as the quick and easy way out. We strongly disagree with this approach and instead emphasize the use of psychological techniques designed to persuade the subject to want to furnish us with the information we desire."[114] The 1983 CIA manual further stated that the "use of force is a poor technique," a conclusion reiterated by the 1987 *Army Interrogation Field Manual*.[115] In the post-9/11 landscape, as EITs were being utilized to elicit information from detainees, interrogators and their colleagues who questioned the veracity of the information provided by detainees were widely ignored by leaders in the military and the intelligence community.[116]

Embedding psychologists into secret interrogation programs of questionable legality raised novel ethical concerns. In 1985, the American Psychological Association and the American Psychiatric Association issued a joint statement that condemned torture and the use of "psychological knowledge and techniques . . . to design and carry out torture."[117] After 9/11, however, the American Psychological Association

(APA) amended its code of ethics to account for situations where a psychologist faced a conflict between an ethical principle and the law.

The new principle had the practical effect of allowing psychologists to override their ethical responsibilities and participate in EITs as long as a "governing legal authority" had indicated that the techniques were legal.[118] The latter component of this ethical equation encapsulated the Torture Memos, and the new APA principle created a symbiotic relationship between the profession and the CIA: psychologists could participate in EITs and still be within APA's ethical code as long as the government said EITs were legal, while the involvement of psychologists in EITs reinforced the DOJ's conclusion that EITs were legal since they were being conducted in concert with health professionals.

Years later, reports indicated that the APA instituted this ethical shift as part of a larger effort to further the professional relationship between psychologists and the national security establishment. One component of the APA's strategic efforts involved prescribing privileges: unlike psychiatrists, psychologists are not medical doctors and do not generally have prescribing privileges for prescription medications such as psychiatric drugs. The DoD began to afford some psychologists with prescribing privileges if they were treating patients in military hospitals. This was an important professional milestone that the APA sought to protect by cooperating with EITs.[119]

In addition to psychologists, physicians and other medical professionals—including individuals from the CIA's Office of Medical Services—provided healthcare advice and medical treatment during detentions and interrogations.[120] Health professionals also provided information that was used to set the severity of EITs, and at times military pathologists withheld or misstated information on autopsy reports to conceal that EITs played a role in causing a detainee's death. Some detainees claimed that they were drugged during interrogations, a position disputed by the CIA and military.[121] Several military medical personnel remained silent despite witnessing or participating in this system of neglect and abuse.[122]

As bioethicist George Annas observed, the net result was a collaborative relationship whereby "physicians would not engage in torture unless the Justice Department lawyers granted them legal immunity for their actions, and the Justice Department lawyers would not grant them (or, as importantly, their CIA colleagues) immunity unless the physicians agreed to actively participate in the torture and be on hand to stop it if medically necessary."[123] As one CIA official commented, following an internal CIA's Medical Services Office report that questioned the ethics of EITs: "Just hope our myopic view of the interrogation process doesn't come back to haunt us."[124]

Between 2002 and 2004, the International Committee of the Red Cross (ICRC)—an independent agency charged under the Geneva Conventions with investigating the treatment of prisoners—inspected American-run prisons and detention centers in Afghanistan, Iraq, and Guantanamo Bay. In a series of classified meetings with senior American officials, the ICRC voiced concerns with what it viewed as mistreatment of prisoners and detainees. The ICRC also noted that the United States had improperly denied ICRC employees access to certain sites and interrogation facilities.[125] Some of these concerns were leaked in 2004. Thereafter, the American College of Physicians, the American Public Health Association, and the American Nursing Association vocalized concerns regarding medical complicity in EITs and abuse of prisoners and detainees.[126]

The issue of detainee abuses by the American military was cast further into the public limelight following the April 2004 release of photographs of prisoner abuse at the Abu Ghraib detention facility in Iraq. The influential American Medical Association (AMA) did little to protest these practices; instead, it focused its lobby efforts on increasing Medicare reimbursements and legislation to limit malpractice lawsuits and compensation for injured patients. AMA medical journals, including the prestigious *Journal of the American Medical Association*, did not publish articles on human rights abuses and medical complicity in EITs until late 2005, more than a year after other journals, such as the *New England Journal of Medicine* and the *British Medical Journal*.[127] As Steven Miles—a physician and bioethicist who was one of the first to describe the participation and complicity of health professionals in EITs—explained, "A torturing nation uses fear, persuasion, and propaganda to secure the assent to torture from society in general and from members of its legal, academic, journalistic, and medical professions."[128]

One of the first detainees upon whom EITs were utilized was Abu Zubaydah, a Palestinian national who was captured during a March 2002 raid in Pakistan. Zubaydah was deemed a high-value detainee due to his suspected links to al Qaeda; he was held and interrogated in various CIA black sites and then transferred to Guantanamo. Zubaydah reportedly was waterboarded more than eighty times in one month; he also was repeatedly slammed into a wall, placed into a small confinement box for several days, deprived of sleep, and kept on an all-liquid diet. The CIA recorded the detention and interrogation of Zubaydah and at least one other person.[129]

In 2003, an internal DoD report warned Defense Secretary Donald Rumsfeld that "Participation by U.S. military personnel in interrogations which use techniques that are more aggressive than those appropriate for POWs would constitute a significant departure from traditional U.S. military norms and could have an adverse impact on the cultural self-image of U.S. military forces."[130] In a separate internal report, the CIA Inspector General raised several concerns with the agency's interrogation program.[131] Thereafter, CIA Clandestine Services Director Jose Rodriguez secretly ordered the destruction of more than ninety interrogation video recordings, even though congressional leaders and presidential counsel Alberto Gonzalez informed the CIA that the law required that the agency keep a copy of the recordings.[132] The Rodriguez order was similar to the unlawful destruction of the MKULTRA files during the 1970s. As an internal CIA disciplinary review into the tape destruction candidly concluded, however, Rodriguez acted in "the best interests of the CIA and its officers, particularly the latter."[133] Evidence later revealed that Rodriquez had warned his agents not to question the legality of EITs in writing.[134]

In 2007, the ICRC concluded that EITs "amounted to torture and/or cruel, inhuman, or degrading treatment" in violation of international law.[135] The secret memo was shared with the CIA, and excerpts of the report were publicly revealed in 2009.[136] The US government elected not to punish anyone involved in EITs or those involved in destroying evidence.[137] Instead, the person who drafted Rodriguez's destruction order, Gina Haspel, would rise to become director of the CIA.[138]

In 2014, the Senate Select Committee on Intelligence (SSCI)—a committee created after the mid-1970s Church Committee hearings that called for more exacting congressional oversight over the intelligence community—released a heavily redacted 500-page Executive Summary of a classified 6,700-page report on the CIA's RDI program. Over four years, the committee and its staff reviewed more than six million pages of CIA documents, some of which were so sensitive that the CIA only permitted review in a special facility rather than at SSCI's offices on Capitol Hill.[139] At one point, the CIA spied on SSCI staff reviewing the documents, seeking to uncover what the SSCI had found and determine whether the agency should revoke access to specific files. Although, at first, the CIA denied the allegation, the agency later confirmed that it was true and issued an apology.[140]

The SSCI's exhaustive examination concluded that the CIA's justification for using EITs "rested on inaccurate claims of their effectiveness" and that EITs were "not an effective means of acquiring intelligence or gaining cooperation from detainees."[141] According to the committee, the volume of evidence revealed that detainees subjected to EITs "fabricated information on critical intelligence issues, including the terrorist threats which the CIA identified as its highest priorities."[142] The SSCI also found that "detainees provided significant accurate intelligence prior to, or without having been subjected to these techniques."[143] The committee determined that the "CIA repeatedly provided inaccurate information to the Department of Justice," which impeded a proper legal analysis of the detention and interrogation program, "actively avoided or impeded congressional oversight of the program," "impeded effective White House oversight and decision-making," "impeded oversight by the CIA's Office of Inspector General," and impeded the national security missions of the FBI, State Department, and the Office of the Director of National Intelligence (ODNI).[144]

With respect to EITs, the SSCI recounted the horrid treatment of detainees, which the committee characterized as "brutal and far worse than the CIA represented to policymakers and others."[145] According to the committee, in at least three instances, the CIA used EITs "despite warnings from CIA medical personnel that the techniques could exacerbate physical injuries."[146] Coupled with the CIA's widespread deception directed at the highest levels of government, the SSCI found that the agency "coordinated the release of classified information to the media, including inaccurate information concerning the effectiveness of" EITs.[147] In one communication, the deputy director of the CIA's Counterterrorism Center candidly wrote to a colleague, "We either put out our story or we get eaten. There is no middle ground."[148] The CIA's selective leaking of classified information, much of which was inaccurate, included information provided to the director of the blockbuster film *Zero Dark Thirty*, a dramatization of the hunt for Osama bin Laden.[149]

Due to incomplete recordkeeping, the SSCI noted that a "full accounting of CIA detentions and interrogations may be impossible."[150] The committee also found that the CIA's internal review procedures were woefully inadequate, that the agency marginalized individuals who criticized the program, and that the agency rarely reprimanded individuals for serious and significant violations of CIA policy.[151] The CIA contested many of the SSCI's findings, stating that the committee's report "fails in significant and consequential ways to correctly portray and analyze" the agency's RDI program.[152] The CIA was resolute in its conclusion that the RDI program "produced unique intelligence that helped the US disrupt plots, capture terrorists, better understand the enemy, prevent another mass casualty attack, and save lives."[153] Yet, the agency admitted that it should improve its management and assessment of covert actions, improve its recordkeeping, and create mechanisms for reevaluating legal guidance for ongoing covert actions.[154]

Within the RDI program, more than 85 percent of the CIA's interrogators were private contractors.[155] Of the individuals who conducted the interrogations, many later suffered from PTSD and substance abuse. Some interrogators have died by suicide, while others remain shell-shocked and dehumanized by their actions. Many report living with guilt for not doing more to stop EITs, while others relive the pain that they imparted to the detainees. Some have been granted full disability benefits due to the enormity of their ailments. Studies have found that interrogators within the military health system fear speaking with healthcare professionals due to concerns that discussing their mental health issues might hinder their career advancement, or that their medical records might be used against them if they ever are charged with violating military law.[156]

The SSCI concluded that the RDI program "caused immeasurable damage to the United States' public standing" and "resulted in other significant monetary and non-monetary costs."[157] Although the report was exhaustive, it did not adequately discuss the shared responsibility for the RDI program—namely, how Congress could have exercised better oversight over the intelligence community, and the respective roles of the White House and the DOJ in supporting the CIA's efforts. The report also focused more on rank-and-file employees than agency leadership and other Executive branch officials, thus leaving unanswered questions as to who should be held responsible for the improper acts.[158]

Within months of the publication of the SSCI report, the APA repudiated its post-9/11 ethics guidelines that were relied on to permit psychologists to engage in EITs, and a new resolution was adopted that recommended against participation in government interrogation programs.[159] A 2015 report—commissioned by the APA in light of the ethics scandal—concluded that "APA officials colluded with DoD officials" and collaborated with the CIA to adopt and maintain APA ethics guidelines favorable to the RDI program, in large part due to the APA's desire "to foster the growth of the profession of psychology by supporting military and operational psychologists, rather than restricting their work in any way."[160]

More than two decades after 9/11, the United States still engages in rendition, detention, and interrogation, though some EITs are no longer practiced, and US policy is to adhere to the interrogation guidelines outlined in the *Army Field Manual*. As of mid-2024, the detention facility at Guantanamo Bay remains open and still houses Zubaydah and other individuals deemed high-value detainees; nearly eight hundred detainees have been housed at Guantanamo over the past two decades, including several children between the ages of thirteen and eighteen.[161]

In black sites that have been part of the RDI program, some individuals who were detained but later released were secretly marked with a bioreactive taggant so they could be remotely and continuously monitored and tracked. A person is secretly swabbed with a chemical that emits a signal, which is then used to assist in intelligence gathering and identification of drone targets.[162] The clandestine biotagging device dovetails with America's rapidly advancing cyberoperations.

Cyberoperations

In the late 1960s, the DoD's Advanced Research Projects Agency launched the ARPANET, the world's first network for accessing information across geographically separate computers. One of the key ambitions underlying the development of the ARPANET was the ability to share confidential information across the country, particularly for military and intelligence purposes.[163] From the outset, however, developers recognized that nefarious parties could infiltrate and manipulate such systems.[164] In the following decades, networked computer communications evolved dramatically, as did the ability to exploit and disrupt the systems and information contained therein.

Much like chemical and biological warfare, the research and development of cyberoperations includes offensive and defensive measures, and the line between the two is amorphous. There are no clear or binding guidelines to distinguish between cyberdefense, espionage, and cyberwarfare, and each nation maintains discretion to institute cyberoperations and respond to cyberattacks. It is incredibly difficult to identify cyberattackers or attribute cyberattacks, factors that allow cyberoperations to flourish in the shadows. America reportedly spends more than $27 billion annually on cyberoperations, surpassing the entire defense budget of Canada, Israel, and dozens of other nations.[165] One leaked National Security Agency (NSA) document described the agency's goal as achieving "global network dominance."[166]

In 2009—months after analysts discovered that hackers had breached the classified computer network of the DoD's Central Command, which was spearheading the wars in Iraq and Afghanistan—the DoD created a dedicated Cyber Command, and cyberspace was deemed a domain of warfare just like air, land, and sea. Previously, cyberdefense and cyberoperations were handled by various departments and divisions spread throughout the military and the intelligence community. Although Cyber Command was initially created as a subdivision of Strategic Command—the division that oversees America's nuclear weapons—in 2018, it was elevated to full combatant command status, a move that freed Cyber Command from the constraints of working through a separate military entity.[167]

Cyber Command is headquartered at Fort Meade, Maryland, in the same sprawling complex that houses the NSA. Established in 1952 to spearhead interception of foreign communications, the NSA has evolved to become America's leading agency for cryptology, signals intelligence, cybersecurity, and cyberoperations. With a staff of more than thirty thousand, the NSA surpasses the CIA in terms of personnel and budget. As the NSA details, its mission is to use cryptology tools "to gain a decisive advantage for the nation and our allies."[168] Cyber Command and the NSA are preeminent cyberoperators. One motto attributed to the NSA is: "In God we trust, all others we monitor."[169]

Cyberoperations are intricately woven into war plans and the national security landscape. For example, in 2002, the Defense Advanced Research Projects Agency

(DARPA) publicly disclosed a project, *Total Information Awareness*, which sought to utilize data mining and artificial intelligence to amass all information on individuals across the world. Once publicized, pressure from civil rights groups and Congress reportedly led to the program's defunding, though program elements were integrated into other projects.[170]

The ubiquity of digital communications and Internet-connected devices has created a fertile landscape for cyberoperations. Cyberspace has evolved to encompass critical infrastructure, such as healthcare, water, power, communications systems, banking, and transportation. Meanwhile, state and non-state actors are engaged in a near-constant barrage of cyberattacks against government, corporate, military, and university systems. No less than 97 percent of Fortune 500 companies have been hacked.[171]

By the early 2010s, the DoD feared a "cyber-Pearl Harbor," a digital attack that would impact computer systems and cause widespread death and destruction. In response to the fears, the United States established policies that set guidelines for the offensive use of cyberattacks.[172] The NSA greatly expanded its cyberoperations, and created dragnet surveillance programs to monitor, store, and mine global digital information. The NSA also developed breakthrough eavesdropping capabilities that could quickly acquire, store, and disseminate intercepted communications from mobile phones, emails, and other devices, without the device owner knowing they were being surveilled.[173] Often such programs gained legal approval after a secret review by the FISC, a court established in 1978 to provide classified judicial oversight over intelligence activities such as surveillance, physical searches, and compelled production.[174] Few government requests have been denied: according to a report prepared by the Electronic Privacy Information Center, a nonprofit research and advocacy institute, between 2001 and 2014, the FISC approved 23,711 surveillance orders and denied 12.[175]

In structuring its programs, the NSA has collaborated with more than three hundred industry partners, including leading companies such as Google, Apple, Amazon, Microsoft, AT&T, and Verizon.[176] NSA projects leverage America's global leadership in technology and telecommunications. As of 2007, nearly 80 percent of the world's digital communications passed through the United States.[177] Communication surveillance collaborations between the NSA and industry have a long history. From the 1940s through the 1970s, leading companies such as RCA and ITT stored telegrams and other communications on magnetic tapes, which they shared with the NSA, who used the tapes to monitor American citizens and foreigners.[178]

The NSA's contemporary programs benefit from laws that permit a wide range of activities. For example, in 2008, Congress amended the Foreign Intelligence Surveillance Act (FISA) to permit blanket authorizations for electronic surveillance of individuals worldwide. The new law also eliminated the requirement that FISA judges scrutinize the factual basis for a FISA warrant. Instead, judges could presume that the government's rationale was factually accurate, and could approve a warrant without an independent factual investigation. The law also provided broad legal immunities for telecommunications companies that participated in cyberactivities—including retroactive immunity dating back to September 11, 2001—regardless of whether the activities were legal.[179]

In 2013, Edward Snowden, an NSA contractor working with Booz Allen Hamilton, helped bring several secret NSA projects into the public limelight after his attempts to raise ethical concerns within agency channels were ignored. One program, code-named PRISM, collected metadata from all phone calls and text messages sent through the servers of US telecommunications companies, which included large swaths of personal information regarding who called who, locations from where calls were made, length of calls, and other details. The vast cyber collection was not limited to individuals suspected of wrongdoing or terrorist activities. The NSA also tracked website history and metadata from emails, spying on leaders worldwide.[180]

Officials from the Bush and Obama administrations stated that Congress was repeatedly briefed on the programs and did not raise any objections.[181] Within government circles, the programs were widely deemed a necessary trade-off between privacy and national security. As former NSA Director Keith Alexander explained: "You need the haystack to find the needle."[182] Following the Snowden disclosures, nations around the globe—including China and several European countries—labeled America as the world's greatest cybersecurity threat.[183] The disclosures of the NSA's far-reaching dragnet programs motivated public debate on the limits of government surveillance and the national security state, debates that are ongoing due to the exponential growth of digital information and network-connected devices.

Snowden shared classified information with four journalists, who then published some of the information in *The Guardian*, *The Washington Post*, and other news outlets. Although the journalists would earn the Pulitzer Prize for their reporting, the US government deems Snowden a felon and has sought to prosecute him for the disclosures. Snowden fled to Hong Kong and then to Russia, where he now lives outside the reach of US extradition. He has dedicated his career to advocating for limits on the national security state. Some view Snowden as a criminal who has harmed US national security, while others herald him as a brave whistleblower who helped unmask illegal government surveillance. The dueling characterizations of Snowden are not unlike the portrayal of revelations that emerged from the 1975 Church Committee investigations. Many applauded the 1975 hearings as a means of revealing governmental overreach and wrongdoing, whereas others viewed the hearings as undermining national security and jeopardizing the safety of the nation.[184] According to James Clapper, former Director of National Intelligence and an intelligence advisor to the Obama administration, following the Snowden disclosures, "most disturbingly, we'd lost the trust of the American people, which questioned what we were doing globally on their behalf and what we were doing domestically to them."[185]

Shortly after the Snowden disclosures, hackers accessed and published a detailed catalog of the NSA's elite cybertools and techniques, placing in the public domain a variety of methods that could be adopted by actors across the globe. The number of cyberattacks increased by more than 400 percent from 2009 to 2018, and some of the most destructive cyberattacks in the world have been based on techniques created by the NSA.[186]

To date, Stuxnet remains one of the most controversial cyberattacks. A malicious code first deployed in Iran's nuclear facilities, Stuxnet destroyed 3,000 Iranian centrifuges and caused substantial damage to Iran's nuclear program. The malware then spread beyond Iran's nuclear facilities and infiltrated more than 100,000 computer systems throughout the world. Data security experts were able to decode the program and trace its origins. Given the sophistication of the code, and in light of information leaked to the public, it is widely suspected that the United States and Israel jointly developed and deployed Stuxnet, though neither country has confirmed its involvement.[187]

Cyberoperations from China, Russia, and North Korea have increased substantially over the past decade. In one operation, China stole millions of documents that provided details of the F-35 Joint Strike Fighter and was able to build a knock-off of the highly sophisticated jet.[188] China also hacked the US Office of Personnel Management, stealing personnel files of more than four million Americans, many of whom had undergone detailed security clearances.[189] Still more, in early 2021, the former director of the US National Counterintelligence and Security Center indicated that China has stolen personal information from more than 80 percent of Americans by infiltrating the computer systems of healthcare institutions, social media sites, Internet-connected devices, and other platforms, and that China has triangulated the data with information scraped from publicly available websites.[190] Other reports have underscored the risks of Chinese spies embedded in American universities, some of whom have received NIH funding for their work in the United States and have created shadow labs in China that expropriate research and intellectual property.[191] The existence of sophisticated cyberoperators in nearly every corner of the world gives new meaning to the Cold War concept of mutually assured destruction. According to Michael Hayden, a former director of the NSA and the CIA: "Adult nations steal information from one another, and steal my secrets, shame on me, not shame on you."[192]

The vast troves of information collected by cyberoperations fuel machine learning and artificial intelligence, which sometimes are synthesized with advancements in cognitive and biological sciences.[193] Advanced facial recognition programs can identify and track individuals from myriad vantage points, such as public streets, buildings, and drones.[194] A separate program seeks to track individuals based on a person's unique "thermal fingerprint" by measuring the heat that a person emits. In addition, remote sensors can detect a person's heart rate from one hundred meters away.[195] Artificial intelligence programs are integrated into devices that bridge neuroscience and nanotechnology, and offer avenues for designing devices that augment human abilities.[196] As detailed in chapter 16, human enhancements are one of the military biomedical complex's core areas for research and development.

Summary

America's post-9/11 global war on terror launched the military into an extraordinary state of proliferation. The total war-related costs are estimated to exceed $6 trillion, a figure that does not include ongoing costs to provide healthcare and benefits to wounded warriors and their families. More than 40 percent of veterans from post-9/11 wars have been approved for lifetime disability benefits, a rate nearly double from WWII and the wars in Vietnam and Korea. The healthcare and benefit costs are estimated to be over $2 trillion through 2050.[197]

Post-9/11 national security concerns also restructured America's geopolitical priorities and rekindled a Cold War mindset that prioritized the pursuit of national security at any price. As the global war on terror intensified, civil rights were curtailed, government surveillance expanded, and ethical principles among healthcare professionals were reformulated and weakened to accommodate military aims. In the wake of the worst attack on US soil since WWII, physicians and researchers were intimately intertwined with the military and intelligence community to support wartime health needs, but also to create a controversial RDI program that was cloaked in secrecy and utilized questionable interrogation methods in an attempt to gain a military advantage.

While constructed with the intent to promote national security during what was viewed as an existential threat, the expansive RDI program garnered intelligence of questionable value, served to diminish America's reputation, created animus against the United States, and curtailed America's ability to contend that the nation protects fundamental human rights and abides by the rule of law. Drone warfare, cyberoperations, and other new means and methods of war have also raised ethical and legal concerns regarding the scope and limits of the national security state. These issues, as discussed in the following chapter, are particularly salient in the realm of military human enhancements.

Notes

1. Linda J. Bilmes, "The Long-Term Costs of United States Care for Veterans of the Afghanistan and Iraq Wars," *Brown University: Costs of War Project* (August 18, 2021).
2. Thomas H. Suitt, "High Suicide Rates Among United States Service Members and Veterans of the Post-9/11 Wars," *Brown University: Costs of War Project* (June 21, 2021).
3. Authorization for Use of Military Force, Pub. L. No. 107-40, 115 Stat. 224 (2001).
4. Philip Shenon, "In One Vote, a Call for Restraint," *The New York Times*, September 16, 2001.
5. Catie Edmondson, "Barbara Lee's Long Quest to Curb Presidential War Powers Faces a New Test," *The New York Times*, September 14, 2021.
6. *Report of the Senate Select Committee on Intelligence: Committee Study of the Central Intelligence Agency's Detention and Interrogation Program* (Declassified Revisions, December 3, 2014), Executive Summary, 11–17.
7. Quoted in Sarah Moughty, "John Rizzo: The Lawyer Who Approved CIA's Most Controversial Programs," *PBS Frontline Interview, Edited Transcript*, September 6, 2011.
8. Dick Cheney, "Interview with Tim Russert," *Meet the Press*, September 16, 2001.
9. For a summary, see Kim Lane Scheppele, "Law in a Time of Emergency: States of Exception and the Temptations of 9/11," *University of Pennsylvania Journal of Constitutional Law* 6, no. 5 (May 2004): 1001–83.
10. U.S. Foreign Intelligence Surveillance Court, About the Foreign Intelligence Surveillance Court (accessed April 4, 2024).
11. Military Order of November 13, 2001, Detention, Treatment, and Trial of Certain Non-Citizens in the War Against Terrorism, 66 Fed. Reg. 57,833 (November 16, 2001).
12. U.S. Government Accountability Office, "Guantánamo Bay Detainees" (November 2012).
13. "Naval Station Guantanamo Bay: History and Legal Issues Regarding Its Lease Agreements," *Congressional Research Service Report* (August 1, 2022).

14. *Hamdan v. Rumsfeld*, 548 U.S. 577 (2006).
15. Detainee Treatment Act of 2005, Pub. L. No. 109-148 (2005); Military Commissions Act of 2006, Pub. L. No. 109-366, 120 Stat. 2600 (2006).
16. *The Authorizations for the Use of Military Force: Hearing before the Senate Committee on Foreign Relations*, 115th Cong. 2 (October 30, 2017).
17. U.S. Africa Command, *United States Africa Command: The First Ten Years* (2018); Nick Turse, "Pentagon's Own Map of U.S. Bases in Africa Contradicts Its Claim of 'Light' Footprint," *The Intercept*, February 27, 2020.
18. Nick Turse, *The Changing Face of Empire* (2012), 1–16.
19. Gary Schroen, *First In* (2005), 15–19, 321.
20. Rizzo, *Company Man*, 162–66.
21. Schroen, *First In*, 28, 88, 321, 354–56.
22. Schroen, *First In*, 359.
23. James Risen, *State of War* (2006), 69–71.
24. Risen, *State of War*, 156–59.
25. Bob Woodward, *Obama's Wars* (2010), 3–12.
26. Craig Whitlock, *The Afghanistan Papers* (2021), 83.
27. Woodward, *Obama's Wars*, 3–12.
28. Steve Coll, *Directorate S* (2018), 46–50.
29. Woodward, *Obama's Wars*, 3–12.
30. Woodward, *Obama's Wars*, 3–12, 43–47.
31. Robert L. Grenier, *88 Days to Kandahar* (2015), 37–45.
32. Carter Malkasian, *The American War in Afghanistan* (2021), 46–50.
33. Steve Coll, *Ghost Wars* (2004), 11–12, 337.
34. Nasreen Ghufran, "The Taliban and the Civil War Entanglement in Afghanistan," *Asian Survey* 41, no. 3 (May/June 2001): 462–87; Zalmay Khalilzad, "Afghanistan in 1995: Civil War and a Mini-Great Game," *Asian Survey* 36, no. 2 (February 1996): 190–95.
35. Usama bin Laden, "Jihad Against Jews and Crusaders," *Federation of American Scientists*, February 23, 1998.
36. Bruce Riedel, "Comparing the U.S. and Soviet Experiences in Afghanistan," *West Point Combating Terrorism Center Sentinel* 2, no. 5 (May 2009): 1–3.
37. Peter Baker, Helene Cooper, and Mark Mazzetti, "Bin Laden Is Dead, Obama Says," *The New York Times*, May 1, 2011.
38. Coll, *Directorate S*, 90–92.
39. Woodward, *Obama's Wars*, 8.
40. Coll, *Directorate S*, 110–17, 143–46.
41. Special Inspector General for Afghanistan Reconstruction (SIGAR), *What We Need to Learn: Lessons from Twenty Years of Afghanistan Reconstruction* (August 2021).
42. Malkasian, *The American War*, 4–8.
43. Whitlock, *The Afghanistan Papers*, 19–23, 183–88.
44. SIGAR, *Lessons*, 1.
45. Coll, *Directorate S*, 266–79.
46. Whitlock, *The Afghanistan Papers*, 254.
47. Coll, *Directorate S*, 266–79.
48. Whitlock, *The Afghanistan Papers*, 139.
49. Coll, *Directorate S*, 329.
50. Andrew Cockburn, "Search and Destroy: The Pentagon's Losing Battle Against IEDs," *Harper's Magazine*, November 2011, 71–77.
51. Coll, *Directorate S*, 332.

52. Todd E. Rasmussen and Arthur L. Kellermann, "Wartime Lessons—Shaping a National Trauma Action Plan," *New England Journal of Medicine* 375, no. 17 (October 27, 2016): 1612–15.
53. Rasmussen and Kellermann, "Wartime Lessons."
54. Lynn G. Stansbury et al., "Amputations in U.S. Military Personnel in the Current Conflicts in Afghanistan and Iraq," *Journal of Orthopaedic Trauma* 22, no. 1 (January 2008): 43–46; Whitney Delbridge Nichels, "Soldier Amputees Have More Options Than Ever for Redeployment," *U.S. Army News*, July 27, 2018.
55. Tom Bowman, "Wounded Warriors Face New Enemy: Overmedication," *NPR*, April 26, 2012.
56. Zara Raza, Syeda F. Hussain, Suzanne Ftouni, Gershon Spitz, et al., "Dementia in Military and Veteran Populations: A Review of Risk Factors," *Military Medical Research* 8, no. 1 (October 2021): 1–13.
57. Draper, *To Start a War*, 138–56, 201–19, 271–300.
58. "The Case for Action," *The Washington Post*, February 4, 2003; "Disarming Iraq," *The New York Times*, February 15, 2003; "'Drumbeat' on Iraq?: A Response to Readers," *The Washington Post*, February 26, 2003; "The Times and Iraq," *The New York Times*, May 26, 2004; Ari Paul and Julie Hollar, "20 Years Later, NYT Still Can't Face Its Iraq War Shame," *Fairness & Accuracy in Reporting*, March 22, 2023; Carroll Doherty and Jocelyn Kiley, "A Look Back at How Fear and False Beliefs Bolstered U.S. Public Support for War in Iraq," *Pew Research Center*, March 14, 2023.
59. Joint Resolution to Authorize the Use of United States Armed Forces Against Iraq, Pub. L. No. 107-243, 116 Stat. 1498 (October 16, 2002).
60. David E. Sanger and John F. Burns, "Bush Orders Start of War on Iraq," *The New York Times*, March 20, 2003.
61. Timothy Andrews Sayle, "US War in Iraq Since 2003," *Oxford Research Encyclopedia* (March 26, 2019).
62. Risen, *State of War*, 135–36.
63. Max Fisher, "20 Years On, a Question Lingers About Iraq: Why Did the U.S. Invade?," *The New York Times*, March 18, 2023.
64. Vali Nasr, "All Against All," *Foreign Affairs* 101, no. 1 (January/February 2022): 128–38.
65. Coll, *Directorate S*, 666–67.
66. David Cole, "Who Are the Taliban Now?," *The New York Review of Books*, June 22, 2023, 24–28.
67. Presidential Policy Guidance, "Procedures for Approving Direct Action Against Terrorist Targets Located Outside the United States and Areas of Active Hostilities" (May 22, 2013): 11.
68. Jeremy Scahill, "The Drone Legacy," in Jeremy Scahill, *The Assassination Complex* (2016), 1–12; Jeremy Scahill, "Find, Fix, Finish," in Scahill, *The Assassination Complex*, 41–52; Jeremy Scahill and Glenn Greenwald, "Death by Metadata," in Scahill, *The Assassination Complex*, 95–106.
69. Jane Harman, "Disrupting the Intelligence Community: America's Spy Agencies Need an Upgrade," *Foreign Affairs* 94, no. 2 (March/April 2015): 99–107.
70. Sue Chan, "CIA's License to Kill," *CBS News*, December 3, 2002.
71. Micah Zenko, "Obama's Final Drone Strike Data," *Council on Foreign Relations* (January 20, 2017); Pauline Muchina and Michael Merryman-Lotze, "The U.S. has Killed Thousands of People with Lethal Drones—It's Time to Put a Stop to It," *American Friends Service Committee* (May 16, 2019); Owen Fiss, *A War Like No Other* (2015), 260.
72. M. Shane Riza, *Killing Without Heart* (2013), 119.
73. *Al-Aulaqi v. Panetta*, 35 F. Supp. 3d 56, 79 (D.D.C. 2014) (internal quotations omitted).

74. "Awlaki Family Protests U.S. Killing of Anwar Awlaki's Teen Son," *ABC News*, October 18, 2011; Spencer Ackerman, Jason Burke, and Julian Borger, "Eight-year-old American Girl Killed in Yemen Raid Approved by Trump," *The Guardian*, February 1, 2017.
75. Denise Chow, "Drone Wars: Pilots Reveal Debilitating Stress Beyond Virtual Battlefield," *Scientific American*, November 7, 2013.
76. P. W. Singer, *Wired for War* (2009), 116–20, 241.
77. Sean McFate, *The Modern Mercenary* (2014).
78. McFate, *The Modern Mercenary*, 19–22.
79. Tom Engelhardt, *Shadow Government* (2014), 69–70.
80. Jeremy Scahill, "Another Electrocution Death in Iraq," *NPR*, September 10, 2009.
81. *In re: KBR Inc.*, 893 F.3d 241, 254–56 (4th Cir. 2018).
82. Timothy Olsen et al., "Iraq/Afghanistan War Lung Injury Reflects Burn Pits Exposure," *Scientific Reports* 12, no. 14671 (December 22, 2022): 1–14; Anthony Szema et al., "Proposed Iraq/Afghanistan War-Lung Injury (IAW-LI) Clinical Practice Recommendations: National Academy of Sciences' Institute of Medicine Burn Pits Workshop," *American Journal of Men's Health* 11, no. 6 (2017): 1153–63; Megan K. Stack, "The Soldiers Came Home Sick. The Government Denied It Was Responsible," *The New York Times*, January 11, 2022.
83. *Gulf War: What Kind of Care are Veterans Receiving 20 Years Later?: Hearing Before the Subcommittee on Oversight and Investigations of the House Committee on Veterans' Affairs*, 113th Cong. 6–9 (statement of Steven S. Coughlin) (March 13, 2013).
84. *Gulf War*, 113th Cong. 7 (statement of Steven S. Coughlin).
85. *Gulf War*, 113th Cong. 7 (statement of Steven S. Coughlin).
86. *Gulf War*, 113th Cong. 6–9 (statement of Steven S. Coughlin).
87. Dignified Burial and Other Veterans' Benefits Improvement Act of 2012, Pub. L. No. 112-260, § 201, 126 Stat. 2422 (2013).
88. *In re: KBR Inc.*, 893 F.3d at 259–62.
89. Michael D. Shear, "Biden Signs Bill to Help Veterans Who Were Exposed to Toxic Burn Pits," *The New York Times*, August 10, 2022.
90. Peter Baker, "'It's Personal': Biden Highlights Law Helping Veterans Exposed to Burn Pits," *The New York Times*, December 16, 2022.
91. James Risen, *Pay Any Price* (2014), 145–51.
92. Dana Priest and Anne Hull, "Soldiers Face Neglect, Frustration, at Army's Top Medical Facility," *The Washington Post*, February 18, 2007.
93. Institute of Medicine, *Treatment for Posttraumatic Stress Disorder in Military and Veteran Populations* (2012), 12, 232–74.
94. Nelly Lahoud, "Bin Laden's Catastrophic Success," *Foreign Affairs* 100, no. 5 (September/October 2021): 10–21; Thomas Hegghammer, "Resistance is Futile," *Foreign Affairs* 100, no. 5 (September/October 2021): 44–52; Elliot Ackerman, "Winning Ugly," *Foreign Affairs* 100, no. 5 (September/October 2021): 66–74.
95. Matthew Weed, "Presidential References to the 2001 Authorization for Use of Military Force in Publicly Available Executive Actions and Reports to Congress," *Congressional Research Service Memorandum* (February 16, 2018); Jennifer Shutt, "The U.S. Senate Repealed Iraq War Authorizations a Year Ago. In the House, They're Frozen," *Missouri Independent*, March 22, 2024.
96. Scahill, *Dirty Wars*, 23–30; Coll, *Directorate S*, 160–71.
97. Fiss, *A War*, 179–80.
98. Mark Mazzetti, *The Way of the Knife* (2013), 117–19; Risen, *State of War*, 21, 30, 145; Woodward, *Obama's Wars*, 53–56; Scahill, *Dirty Wars*, 23–30; Mohamedou Ould Slahi,

Guantanamo Diary (2015), xxxvi, 20–21, 132–44, 154–55, 267–68; Adam Goldman, "The Hidden History of the CIA's Prison in Poland," *The Washington Post*, January 23, 2014.

99. *Report of the SSCI*, Findings and Conclusions, 16–17.
100. *Report of the SSCI*, Executive Summary.
101. George J. Annas, *Worst Case Bioethics* (2010), 47–49; David Cole, "The Sacrificial Yoo: Accounting for Torture in the OPR Report," *Journal of National Security Law & Policy* 4, no. 2 (2010): 455–64.
102. Jay S. Bybee, Memorandum to Alberto Gonzalez, Counsel to the President (August 1, 2002): 1.
103. Mark Danner, *Spiral* (2016), 18–21.
104. U.N. Convention Against Torture and Other Cruel, Inhuman, or Degrading Treatment or Punishment, Part 1, Art. 1, § 1 (December 10, 1984).
105. Johnson, *Spy Watching*, 198.
106. Martin E. P. Seligman and Steven F. Maier, "Failure to Escape Traumatic Shock," *Journal of Experimental Psychology* 74, no. 1 (May 1967).
107. *Report of the SSCI*, Findings and Conclusions, 11, 19 (note 32).
108. Risen, *Pay Any Price*, 176–201; Maria Konnikova, "Trying to Cure Depression, But Inspiring Torture," *The New Yorker*, January 14, 2005.
109. *Report of the SSCI*, Findings and Conclusions, 11–12; Risen, *Pay Any Price*, 176–201.
110. Martin Seligman, "The Hoffman Report, the Central Intelligence Agency, and the Defense of the Nation: A Personal View," *Health Psychology Open* 5, no. 2 (July–December 2018); Risen, *Pay Any Price*, 176–201.
111. Martin Seligman, "Reply to Inquiry: 'What Are the Pressing Scientific Issues for the Nation and World, and What is Your Advice on How I Can Begin to Deal with Them? – GWB,'" *The Edge* (2003).
112. Mark Benjamin, "'War on Terror' Psychologist Gets Giant No-Bid Contract," *Salon*, October 14, 2010.
113. Risen, *Pay Any Price*, 176–201; Steven H. Miles, *Oath Betrayed* (2006), 14–18.
114. Central Intelligence Agency, *Human Resource Exploitation Training Manual* (1983), A-2.
115. CIA, *Human Resource Exploitation*, iv; Army Field Manual, *Intelligence Interrogation* (May 1987), 1-1.
116. Miles, *Oath Betrayed*, 14–17.
117. *Against Torture: Joint Resolution of the American Psychiatric Association and the American Psychological Association* (December 1985).
118. American Psychological Association "Ethical Principles of Psychologists and Code of Conduct," *American Psychologist* 57, no. 12 (December 2002): 1063.
119. Risen, *Pay Any Price*, 194–201.
120. *Report of the SSCI*, Executive Summary, 111–15.
121. Jonathan H. Marks, "Neuroskepticism: Rethinking the Ethics of Neuroscience and National Security," in James Giordano, editor, *Neurotechnology in National Security and Defense* (2015), 183–84.
122. Miles, *Oath Betrayed*, 119–39.
123. Annas, *Worst Case Bioethics*, 56.
124. Quoted in John Prados, *The Ghosts of Langley* (2017), 19.
125. *ICRC Report on the Treatment of Fourteen "High Value Detainees" in CIA Custody* (February 2007).
126. Miles, *Oath Betrayed*, 126–39.
127. Miles, *Oath Betrayed*, 126–39.
128. Miles, *Oath Betrayed*, xii.
129. *Report of the SSCI*, Findings and Conclusions, 3–11, Executive Summary, 17–49.

130. DoD Working Group Report, *Detainee Interrogations in the Global War on Terrorism: Assessment of Legal, Historical, Policy, and Operational Considerations* (April 4, 2003), 69.
131. *Report of the SSCI*, Executive Summary, 121–28.
132. Prados, *The Ghosts of Langley*, 14–20.
133. Michael J. Morell, Memorandum to Central Intelligence Agency Director, "Disciplinary Review Related to Destruction of Interrogation Tapes" (December 20, 2011), 4.
134. Prados, *The Ghosts of Langley*, 14.
135. *ICRC Report*, 27.
136. Mark Danner, "The Red Cross Torture Report: What It Means," *The New York Review of Books*, April 30, 2009.
137. Mark Mazzetti and Charlie Savage, "No Criminal Charges Sought Over C.I.A. Tapes," *The New York Times*, November 9, 2010.
138. Julian E. Barnes and Scott Shane, "Cables Detail C.I.A. Waterboarding at Secret Prison Run by Gina Haspel," *The New York Times*, August 10, 2018; Brian Tashman, "The Government Has Information on Gina Haspel's Torture Record: The Senate Can't See It," *American Civil Liberties Union*, May 15, 2018.
139. *Report of the SSCI*, Forward by Chairman Diane Feinstein, 1–6. Approximately 9,400 CIA documents were withheld from the committee by the White House, which claimed that Executive Privilege precluded the SSCI's review of the information. *Report of the SSCI*, Executive Summary, 9 (note 2).
140. CIA Inspector General, *Summary of Report Prepared at the Request of the Congressional Intelligence Committees* (July 31, 2014); Mark Mazzetti and Carl Hulse, "Inquiry by C.I.A. Affirms It Spied on Senate Panel," *The New York Times*, July 31, 2014.
141. *Report of the SSCI*, Findings and Conclusions, 2–3.
142. *Report of the SSCI*, Findings and Conclusions, 2.
143. *Report of the SSCI*, Findings and Conclusions, 2.
144. *Report of the SSCI*, Findings and Conclusions, 4–8.
145. *Report of the SSCI*, Findings and Conclusions, 3–4.
146. *Report of the SSCI*, Findings and Conclusions, 3–4.
147. *Report of the SSCI*, Findings and Conclusions, 8–9.
148. *Report of the SSCI*, Findings and Conclusions, 8–9.
149. *Report of the SSCI*, Minority Views of Vice Chairman Chambliss, Senators Burr, Risch, Coats, Rubio, and Coburn, 3 (note 9); Charlie Savage, *Power Wars* (2015), 338–39.
150. *Report of the SSCI*, Findings and Conclusions, 12–13.
151. *Report of the SSCI*, Findings and Conclusions, 12–15.
152. *CIA Comments on the Senate Select Committee on Intelligence Report on the Rendition, Detention, and Interrogation Program* (Declassified and Approved for Release, December 8, 2014), 2.
153. *CIA Comments on the SSCI Report*, 12–15. In an appendix to its comments, the CIA provided twenty examples of "Value of Intelligence Acquired From Detainees."
154. *CIA Comments on the SSCI Report*, 16–18.
155. *Report of the SSCI*, Findings and Conclusions, 10–12.
156. Eric Fair, *Consequence* (2016), 118–25, 135, 204–5; Miles, *Oath Betrayed*, 18; Risen, *Pay Any Price*, 164–76.
157. *Report of the SSCI*, Findings and Conclusions, 16–17.
158. David Cole, "Did the Torture Report Give the CIA a Bum Rap?," *The New York Times*, February 20, 2015; Brian Greer, "Examining the Shortcomings of the Senate Intelligence Committee's 'Torture Report,'" *Lawfare*, January 17, 2020.

159. American Psychological Association, "New APA Policy Bans Psychologist Participation in National Security Interrogations" (September 2015).
160. Sidney Austin LLP, *Report to the Special Committee of the Board of Directors of the American Psychological Association: Independent Review Relating to APA Ethics Guidelines, National Security Interrogations, and Torture* (Revised Version, September 4, 2015), 10–11.
161. "The Guantánamo Docket," *The New York Times*, February 12, 2024; "Guantánamo by the Numbers," *American Civil Liberties Union*, May 2018.
162. Scahill, *Dirty Wars*, 172–74.
163. Weinberger, *The Imagineers of War*, 109, 122–24.
164. Willis H. Ware, "Security and Privacy in Computer Systems," Paper Presentation, Spring Joint Computer Conference, Atlantic City (April 1967).
165. David Hollingworth, "US Government Sets Aside US$27.5bn for Cyber Security Spending," *Cyber Daily* (March 12, 2024); "Countries with the Highest Military Spending Worldwide in 2022," *Statista* (accessed April 5, 2024).
166. Quoted in Alexander Klimburg, *The Darkening Web* (2017), 138.
167. Sue Gordon and Eric Rosenbach, "America's Cyber-Reckoning," *Foreign Affairs* 101, no. 1 (January/February 2022): 10–20.
168. National Security Agency, Mission, www.nsa.gov (accessed December 12, 2023).
169. Quoted in James Bamford, *Body of Secrets* (2001), xiii.
170. Weinberger, *The Imagineers of War*, 302–15; Singer, *Wired for War*, 275–76.
171. P. W. Singer and Allan Friedman, *Cybersecurity and Cyberwar* (2014), 2–15.
172. Gordon and Rosenbach, "America's Cyber-Reckoning."
173. Woodward, *Obama's Wars*, 7–12.
174. Foreign Intelligence Surveillance Act of 1978, Pub. L. No. 95-511, 92 Stat. 1783 (October 25, 1978).
175. Electronic Privacy Information Center, *Foreign Intelligence Surveillance Act Court Orders 1979-2022: FISA Application and Order Statistics, Traditional FISA Surveillance Orders* (accessed December 12, 2023).
176. National Security Agency, *Cybersecurity Year in Review* (2022), 9–15; Glenn Greenwald and Ewen MacAskill, "NSA Prism Program Taps in to User Data of Apple, Google and Others," *The Guardian*, June 7, 2013; Julia Angwin, Charlie Savage et al., "AT&T Helped U.S. Spy on Internet on a Vast Scale," *The New York Times*, August 15, 2015; Farhad Manjoo, "Et Tu, Silicon Valley?: How PRISM Could Ruin Apple, Google, and Every Other Big Tech Company," *Slate*, June 7, 2013; Declan McCullagh, "Surveillance 'Partnership' Between NSA and Telcos Points to AT&T, Verizon," *CNET*, June 27, 2013; Chris Mills Rodrigo, "Amazon Awarded Secret $10B NSA Cloud Computing Contract," *The Hill*, August 10, 2021.
177. Fred Kaplan, *Dark Territory* (2016), 191.
178. Johnson, *A Season of Inquiry*, 110–13.
179. FISA Amendments Act of 2008, Pub. L. No. 110-261, 122 Stat. 2436 (July 10, 2008); Edward C. Liu, "Retroactive Immunity Provided by the FISA Amendments Act of 2008," *Congressional Research Service* (July 25, 2008).
180. Greenwald and MacAskill, "NSA Prism Program"; Glenn Greenwald, Ewen MacAskill, and Laura Poitras, "Edward Snowden: The Whistleblower Behind the NSA Surveillance Revelations," *The Guardian*, June. 11, 2013; Ewen MacAskill and Gabriel Dance, "NSA Files: Decoded," *The Guardian*, November 1, 2013.
181. Nina Totenberg, "Why the FISA Court Is Not What It Used To Be," *NPR*, June 18, 2013.
182. Quoted in J. D. Tuccille, "Why Spy on Everybody?: Because 'You Need the Haystack to Find the Needle,' Says NSA Chief," *Reason Magazine*, July 19, 2013.

183. Gordon and Rosenbach, "America's Cyber-Reckoning."
184. Johnson, *A Season of Inquiry*, 171.
185. Clapper, *Facts and Fears*, 250.
186. Gordon and Rosenbach, "America's Cyber-Reckoning."
187. Klimburg, *The Darkening Web*, 180–84.
188. Kaplan, *Dark Territory*, 224–25.
189. Gordon and Rosenbach, "America's Cyber-Reckoning."
190. Oona A. Hathaway, "Keeping the Wrong Secrets," *Foreign Affairs* 101, no. 1 (January/February 2022): 85–98.
191. Gideon Lewis-Kraus, "Dangerous Minds," *The New Yorker*, March 21, 2022, 43–55.
192. Quoted in Klimburg, *The Darkening Web*, 140–41.
193. Thom Dixon, "The Grey Zone of Cyber-Biological Security," *International Affairs* 97, no. 3 (May 2021): 685–702; Sherali Zeadally, Erwin Adi, Zubair Baig, and Imran A. Khan, "Harnessing Artificial Intelligence Capabilities to Improve Cybersecurity," *IEEE Access* 8 (January 20, 2020): 23, 817–37.
194. Denise Almeida, Konstantin Shmarko, and Elizabeth Lomas, "The Ethics of Facial Recognition Technologies, Surveillance, and Accountability in an Age of Artificial Intelligence," *AI and Ethics* 2 (2022): 377–87; Steven Feldstein, "The Global Expansion of AI Surveillance," *Carnegie Endowment for International Peace* (September 2019).
195. Scahill, *Dirty Wars*, 172–74; Patrick Tucker, "Tomorrow Soldier: How the Military is Altering the Limits of Human Performance," *Defense One*, July 12, 2017.
196. Gabriel A. Silva, "A New Frontier: The Convergence of Nanotechnology, Brain Machine Interfaces, and Artificial Intelligence," *Frontiers in Neuroscience* 12, no. 843 (November 15, 2018): 1–8.
197. Bilmes, "The Long-Term Costs," 1–3.

16
Biomedical Enhancements and the Modern Warfighter

Introduction

The demand for super soldiers has never been greater. As the Department of Defense (DoD) candidly states, one of its primary goals is to exploit technological advancements to create warfighters with exceptional physical, physiological, and cognitive abilities. The military and the intelligence community sponsor or conduct cutting-edge research, often through projects led by the Defense Advanced Research Projects Agency (DARPA), the Central Intelligence Agency (CIA) Directorate of Science and Technology, or specialized departments within the Army, Navy, and Air Force. Examples include drugs that increase endurance, neurostimulation devices that enhance focus, and brain-computer interfaces that allow warfighters to control weapons via thought alone. The work builds on contemporary research in biomedical sciences, computer science, nanotechnology, and artificial intelligence. Biomedical enhancements aim to provide service members with distinctive advantages, but they also raise concerns that strike at the heart of several key questions explored throughout this book regarding *jus in praeparatione bellum*: how to measure and assess benefits and risks of new military technologies, and what safeguards are appropriate during research and operational integration.

At the outset, it is important to recognize that there is no scholarly consensus on how to define the term "enhancement." Some ethicists adopt a broad characterization whereby enhancements encompass all interventions that augment a person's natural abilities, such as vaccines, antibiotics, or eyeglasses. Others use the term to capture "nontraditional" interventions such as gene editing and neurostimulation devices. As to the former, although a broad characterization may have some philosophical appeal, it does little to help frame a distinct field of intellectual inquiry. Regarding the latter, while the approach may have some intuitive appeal, the distinction between traditional and nontraditional is ambiguous and varies depending on context, culture, and time.[1]

Another approach involves distinguishing therapeutic from nontherapeutic enhancements—drawing a line between interventions that provide health benefits from those that amplify physical or cognitive abilities beyond a human baseline—and structuring distinct protocols for each category. While this categorical approach may be useful in cases where a clear distinction can be made regarding the purpose of an intervention, some interventions will have more than one primary purpose. For example, a helmet that provides neurostimulation that suppresses the feeling of a traumatic event as it occurs may serve two goals: allowing a warfighter to continue battle despite witnessing gruesome casualties, and providing long-term health benefits such as a decreased likelihood of developing post-traumatic stress. In such instances, it

America's Military Biomedical Complex. Efthimios Parasidis, Oxford University Press.
 DOI: 10.1093/9780199351473.003.0017

may be difficult to determine the extent to which therapeutic or nontherapeutic protocols should apply.

One must be aware of the definitional ambiguity surrounding the term "enhancement" but should not be paralyzed by it. Rather, each intervention should be assessed on its merits throughout research, field testing, and operational use. This chapter examines contemporary and emerging advancements in pharmacological enhancements, precision medicine, and military applications of neuroscience. I explore how military human enhancement projects raise unique concerns in medico-legal concepts such as autonomy, privacy, and human dignity. I also examine whether biomedical enhancements might alter one's ability to control or be responsible for their actions.

Pharmacological Enhancements

In a landmark 2009 report, the National Academies recommended that the military establish close relationships with the pharmaceutical industry, the National Institutes of Health (NIH), and academic institutions to "keep abreast of advances in neuropharmacology, cellular and molecular neurobiology, and neural development and to identify new drugs that have the potential to sustain or enhance performance in military-unique circumstances."[2] The report highlighted that drugs could be utilized to alter mood, motivation, memory, and cognition, and could "restore function, mitigate pain or other responses to trauma, or facilitate recovery from injury or trauma."[3]

The 2009 report was not the first to herald the use of pharmacological enhancements for military missions. Nor did the findings represent a drastic shift in military norms and practices. For centuries, drugs have been used to enhance human performance during war. For example, Vikings ate hallucinogenic mushrooms, Inca warriors chewed coca leaves, and Nazi soldiers regularly consumed methamphetamine tablets and cocaine-infused chewing gum. During World War II (WWII), military personnel from Australia, Britain, Finland, Italy, Japan, and the United States commonly ingested amphetamine pills.[4] British soldiers dubbed the tablets "wakey-wakey pills."[5] Adolf Hitler's personal doctor, Theodor Morell, frequently injected the Führer with concoctions that included hormones, steroids, methamphetamine, cocaine, opioids, barbiturates, and other substances.[6] Winston Churchill and John F. Kennedy reportedly were administered drug cocktails that included amphetamines to help boost their performance during key political events.[7]

During the wars in Korea and Vietnam, the US military distributed hundreds of millions of dextroamphetamine tablets to enhance warfighter performance even though the drug has a high potential for dependency, is FDA-approved solely to treat narcolepsy and attention-deficit/hyperactivity disorder (ADHD), and can lead to heart failure, seizures, hallucinations, and psychosis.[8] Service members colloquially called the tablets "pep" pills.[9] As one Vietnam War veteran recalled, the pills "gave you a sense of bravado as well as keeping you awake. Every sight and sound was heightened. You were wired into it all and at times you felt really invulnerable."[10] In addition to promoting stamina, the drugs were administered to alleviate the psychological toll of warfighting. Military physicians also commonly administered antipsychotic medications to help relieve stress and anxiety, and countless warfighters self-medicated with illicit drugs such as heroin, hallucinogens, and opium.[11]

Dextroamphetamine continued to be used as a performance-enhancing drug after the Vietnam War. For example, military pilots have used the drug as a fatigue countermeasure even though civilian pilots generally are prohibited from flying while taking dextroamphetamine since the side effects of the drug increase the risk of in-flight accidents. During the Persian Gulf War, one comprehensive survey found that 65 percent of pilots used dextroamphetamine, and 60 percent believed that the pills played a crucial role in helping to fulfill military missions.[12] The Air Force banned the use of the pills following the war due to concerns about drug abuse and addiction.[13] The Air Force also was concerned that commanders might abuse their discretion and seek to get "superhuman effort out of their subordinates."[14] The ban was repealed in the mid-1990s during the Yugoslav Wars, as the Air Force viewed the drug risks as manageable and dextroamphetamine as a mission-essential countermeasure for fatigue, particularly for long-haul flights.[15]

Fatigue is a significant operational limitation that impacts not only pilots but also special operations forces and other military personnel who endure long duty periods, unpredictable work hours, circadian disruptions (due to time zone transitions), and insufficient sleep. The deleterious effects of fatigue are well documented: people think more slowly, are less alert, have memory and learning difficulties, and are more prone to error and miscalculations. Studies have found that fatigue-related performance degradation has been responsible for a significant percentage of aviation accidents, while a sleep deprivation study among Navy SEALs and Army Rangers found that lack of sleep impacted physical and cognitive ability more than if a warfighter was intoxicated or had ingested sedatives.[16]

By 2021, the Army, Navy, and Air Force permitted the use of dextroamphetamine as a "go" pill for fatigue management in certain military operations. The Air Force also permits the use of modafinil, which is FDA-approved to treat narcolepsy and other sleep disorders, has fewer and less serious side effects than dextroamphetamine, and is less likely to lead to addiction. Studies have found that dosing with modafinil or dextroamphetamine between intervals of four and eight hours can lead to sustained alertness for forty to forty-eight hours.[17] The use of go pills is strictly regulated and governed by various military policies. For example, Air Force pilots must ground-test the pills under physician supervision to assess adverse reactions, and policies created in consultation with military doctors set in-flight dosages.[18] The use of go pills is voluntary, although a publicly released informed consent document indicates that commanders may consider use, or refusal to use, as a factor in determining duty assignments.[19] Go pills often must be counterbalanced with "no-go" sleeping aids such as Ambien or Restoril. Long-haul pilots and warfighters conducting special operations are frequent users of go and no-go pills; upon their return to civilian life, they often find it difficult to break the habit of using no-go pills to fall asleep.[20]

Utilization of go pills was cast into the public limelight in 2002 following a "friendly fire" incident at Tarnak Farm in Afghanistan, where an American pilot fired a laser-guided bomb that killed four Canadian soldiers and injured eight others. During a nighttime flight, Major Harry Schmidt believed that he and his flight lead, Major William Umbach, were being attacked from adversaries on the ground. The pilots reported the ground fire to the Air Force controller, who ordered the pilots to "hold fire" so he could attempt to obtain additional information before granting the pilots permission to respond with force. As the pilots circled the area and waited for a response, the ground fire continued. Schmidt announced he was "rolling in, in self-defense" and launched a bomb.[21] Within seconds of the bomb hitting its target, the controller notified the pilots that the ground fire was from "friendlies"—Canadians who were engaged in a night training exercise.[22]

The incident received considerable media attention—the four men were the first Canadian deaths in a combat zone since the Korean War. The Americans were charged under the Uniform Code of Military Justice (UCMJ) with dereliction of duty, violations of the rules of engagement, and manslaughter; news reports also highlighted the lack of coordination between the American and Canadian militaries. As part of his defense, Schmidt argued that he was pressured into taking dextroamphetamine and that the drug was a contributing factor to his fog-of-war decision to respond to a perceived attack. Schmidt was flying back to base after a ten-hour mission during which he ingested dextroamphetamine in accordance with Air Force policy.[23]

Whether the drug played a role in Schmidt's actions remains unclear, but the Tarnak Farm incident raised broader questions of whether warfighters should utilize pharmacological enhancements and whether the drugs impact the ability of warfighters to control or be personally responsible for their actions. As the debate continued, pilots and other military personnel supported continued use of go pills, contending that the pills are sensible fatigue countermeasures that save lives and support military missions. Advocates referenced thousands of instances where go pills were ingested with no adverse operational consequences. At the same time, critics pointed to anecdotal evidence of improper conduct while military personnel were under the influence of the pills. The charges against the pilots eventually were dropped, though each received a reprimand for neglecting combat duties.[24] Go pills remain available for use across the US military.

One go pill—Captagon—has been particularly controversial. Captagon is an amphetamine-based pharmaceutical that was initially marketed in the 1960s as a treatment for narcolepsy and ADHD. It was removed from the US and European markets due to severe adverse health concerns and high rates of addiction. More recently, athletes have used Captagon to enhance their performance, and over the past two decades illegally produced knock-offs have been used regularly by warfighters in Syria, Iraq, Afghanistan, and elsewhere. Several countries produce illicit drugs—including Turkey, Lebanon, Syria, and Afghanistan—often with varying ingredients such as amphetamines, caffeine, ephedrine, and other substances.[25]

Tens of millions of illicit pills have been confiscated by law enforcement agencies throughout Europe, yet the pills remain widely used by the Taliban, ISIS, and others.[26] According to a Lebanese distributor who sells illicit Captagon: "No matter how tired you are, it makes you wake up. Your senses become very sharp. Sometimes you don't sleep for 24 or 48 hours, depending on how many pills you take. If you shoot someone on Captagon, they don't feel it. And if someone takes many pills, like 30 or so, they become violent and crazy, paranoid, unafraid of anything."[27]

In several countries, Captagon knock-offs or similar drugs are regularly provided to child soldiers. The superiors tell the children that the pills will give them strength, magic powers, and divine protection. Children become addicted to the substances and will do nearly anything their superiors ask of them in exchange for more pills.[28] These tactics have been used by the Lord's Resistance Army—a nonstate actor responsible for violence throughout Central Africa since the 1980s—which has forcibly constricted tens of thousands of child soldiers.[29]

Apart from pharmaceuticals that counter fatigue, increase aggressiveness, or reduce pain, one drug—propranolol—may be able to reduce posttraumatic stress. Propranolol is a beta-blocker that is FDA-approved to treat anxiety and heart conditions. If administered during or immediately after a traumatic event, some studies have found that propranolol blocks the biochemical processes that lead to the consolidation of memory.[30] This helps a person detach a memory of an event from the emotional response to the event. Although a person will still remember the traumatic episode, they may not have a strong emotional response upon recall. This desensitization may help decrease the effects of posttraumatic stress, though it also may increase a person's willingness to cause harm due to the desensitization effects. For example, if WWII German military personnel working in concentration camps were provided propranolol, might their conduct have been even more gruesome? Additional research is necessary to unmask propranolol's safety and effectiveness, including the potential for increasing untoward conduct.

A host of other pharmacological enhancements also are being researched for military use. Via its Metabolically Dominant Soldier program, DARPA examined the development of a super-nutritional pill that could provide "continuous peak performance and cognitive function for 3 to 5 days, 24 hours per day, without the need for calories."[31] Medications licensed as ADHD treatments—such as Adderall, Ritalin, and Provigil—are being used and evaluated for nonmedical purposes to enhance focus and concentration.[32] Oxytocin is being evaluated as a means of increasing trust. It is a naturally occurring neurochemical that helps a person feel close to and trust those around them; mothers with newborn children are found to have high levels of oxytocin while breastfeeding. One theory is that oxytocin can aid interrogations if administered to a person being questioned.[33]

The use of pharmacological enhancements is not limited to the military. Some musicians and performing artists utilize propranolol to counter stage fright. Liquid Trust, a private company, sells aerosolized oxytocin, marketed as a perfume additive that increases the likelihood that others will find you attractive. Drugs such as Ritalin, amphetamines, and other stimulants are used widely by students and adults to improve focus or for a recreational high, despite serious side effects that include aggression, seizures, psychosis, and heart attacks.[34] Contemporary use of stimulants is not unlike what occurred in WWII Germany, where methampthetamine was commonly ingested by warfighters and civilians. Even German chocolates were laced with methamphetamine and advertised as a perfect pick-me-up to help a person speed through their work.[35]

Precision Medicine

In 2015, President Barack Obama launched the Precision Medicine Initiative, a program designed to leverage emerging biomedical advancements to promote individualized healthcare tailored to each person's unique characteristics. Precision medicine incorporates advancements from omics technologies, such as genomics (study of the genome and genetic variants), epigenomics (study of epigenetic changes, including the role of societal and environmental factors that impact gene expression), transcriptomics (study of RNA and how it relates to pathological or physiological conditions), proteomics (study of proteins and how they relate to physiological functions), metabolomics (study of metabolite levels and metabolic functioning), and microbiomics (study of the microbiome and human health). Through multiomics research—which utilizes sophisticated statistical analysis of large data sets to synthesize multiple omics technologies—scientists endeavor to better understand the dynamics of health, disease, and cognitive and physical abilities. Machine learning and artificial intelligence bolster these analyses.[36]

In addition to using multiomics research to personalize medical care for warfighters, the military is researching the use of precision medicine biomarkers to inform operational decisions and military assignments.[37] Within the biomedical research community, the US military is at the forefront of multiomics research and evaluation of expression circuits, which are biomarkers correlated with cognitive, behavioral, or physical traits. This new line of inquiry represents an innovative leap beyond genetics research—which seeks to identify genetic variants that are correlated with disease or functionality—to a systems biology assessment that takes a multifaceted approach that has the potential to better understand disease and human abilities.[38]

One DARPA program, Measuring Biological Aptitude, is working to decipher expression circuits that could be measured over time to assess warfighter performance and promote operational readiness. Parallel research is exploring whether implantable devices could be utilized to provide instantaneous biometric assessments of expression circuits and enhance warfighter performance. As DARPA explains, expression circuits have the potential to alter how the military recruits, evaluates, and assigns individual warfighters to specific missions.[39]

Analogous work is being conducted throughout the military. For example, the Army Research Lab's Human Variability Project is developing a system to outfit warfighters with interactive sensors that continuously measure biometric data and environmental elements. The Naval Health Research Center's Warfighter Performance Department is studying the measurement, maintenance, restoration, enhancement, and modeling of human performance in military environments. This includes examining how to maximize operational performance, enhance physiological and psychological resilience, and improve warfighter assessment, diagnosis, and rehabilitation strategies to prepare for and maintain operational readiness The Air Force's Total Exposure Health program is analyzing molecular and other health information to help predict and maintain peak warfighter performance. The Air Force also has developed specialized helmets that track blood-oxygen levels, heart rate, eye movement, and other biometric information.[40]

These projects aim to improve individual performance by obtaining microphysiological details about individuals during military training and operations. Other precision medicine programs utilize a multiomics research model to study the prevention, diagnosis, and treatment of cancer, diabetes, cardiovascular disease, and mental illness. For example, the Air Force's Total Exposure Health initiative and the Defense Health Agency's Defense Occupational & Environmental Health Readiness System take a multiomics research approach to examining environmental and occupational health hazards experienced by warfighters. A separate program, the Army Study to Assess Risk and Resilience in Servicemembers, investigates possible genetic or biometric predictors of suicide or stress resilience.[41]

The DoD and the Department of Veterans Affairs (VA) are key partners in the Precision Medicine Initiative, but the program's reach extends beyond the armed forces. The NIH's All of Us program has collected more than one million genomes from volunteers nationwide, along with their associated phenotypes and other social and environmental information.[42] Since genes are often expressed only if triggered by physiological, pharmacological, or environmental factors, a database like All of Us, which links genomic and phenotypic information, holds great potential for precision medicine research.[43] In addition, the DoD and VA are collaborating on a parallel project, the Million Veteran Program, which examines potential links between genomics and health outcomes—data from the voluntary program can be triangulated with health data from the military and veterans' health systems to better understand disease etiology.[44]

Although precision medicine and omics technologies hold great promise for promoting warfighter health and performance, adversaries may also use the data to tailor precision pathologies based on vulnerabilities or sensitivities of certain groups. Particularly since commercial entities own, sell, or exploit a significant amount of health-related data, there are substantial biosecurity risks. Adversaries—who may be able to acquire the data legally or illegally—can utilize the information to disrupt human health, military capabilities, and social and economic structures. For example, policy experts have warned about the potential harms to US interests from China's expansive Military-Civil Fusion initiative, which leverages China's extraordinary talents in science and technology to further the nation's military, economic, and national security aims.[45]

Precision medicine also dovetails with gene editing techniques, which can provide precise genetic alterations. Gene editing has been heralded as a breakthrough technology that can prevent or treat a wide range of monogenetic diseases, such as hemophilia, sickle cell anemia, and cystic fibrosis, as well as other conditions with a genetic component.[46] In 2018, researchers in China announced that they created human embryos that were genetically edited to resist HIV infection. The embryos were implanted into two women, one of whom had one child and the other who had twins.[47] During the research, the scientists acknowledged the ethical controversy of human gene editing but focused on what they described as the medical and societal benefits of the procedure.[48] Once published, the work was widely criticized by academic leaders within the research and bioethics communities, and three of the researchers were prosecuted by Chinese authorities.[49]

Calls for a moratorium on human gene editing have received considerable—though not universal—support.[50] Of particular concern is germline gene editing, whereby the edited trait is heritable. Heritable genetic edits raise concerns about eugenic practices and preferential treatment for modified individuals. Gene editing may also be used to enhance physical or cognitive abilities to create elite warfighters or intelligence agents. Scientists have created a genetically engineered mouse that is much stronger than other mice,[51] and it is likely that human enhancement via gene editing is being explored. For example, some publications have outlined how to use genetic engineering to increase alertness, augment strength, and reduce pain sensation.[52] As precision medicine and omics research lead to a more detailed understanding of human traits, gene editing may be viewed as a natural step in the quest for scientific and military superiority. According to a 2017 National Academies report: "Heritable genome-editing trials must be approached with caution, but caution does not mean they must be prohibited."[53]

Military Applications of Neuroscience

Coupled with pharmaceutical enhancements and precision medicine initiatives, the military is investing considerable resources in military applications of neuroscience. DARPA has funded neuroscience projects since the 1970s. In one of DARPA's first projects in this area—which began in 1974 and was titled Close-Coupled Man/Machine Systems—researchers worked on developing systems that permitted direct communications between humans and machines, and programs that could monitor neural states to assess human vigilance, fatigue, emotions, decision-making, perception, and cognitive ability. Over the past half-century, some DARPA projects have sought to create neurotechnologies that can restore neural or behavioral functions, while others have focused on improving human cognition, training, and performance.[54]

In a 2019 report, *Cyborg Soldier 2050: Human/Machine Fusion and the Implications for the Future of the DoD*, a DoD council outlined four areas of military human enhancement that were technically feasible by 2050: (1) ocular enhancements to imaging, sight, and situational awareness; (2) restoration and programmed muscular control through an optogenetic bodysuit sensor web; (3) auditory enhancement for communication and protection; and (4) direct neural enhancement of the human brain for two-way data transfer. Regarding the fourth area, the council described neuroenhancements as "a revolutionary advancement" that "would allow warfighters direct communication with unmanned and autonomous systems, as well as with other humans, to optimize command and control systems and operations."[55] This would be accomplished through neurodevices that would increase a warfighter's situational awareness, strength, speed, sight, communication, physiology, attention, memory, learning, and sense of smell, among other capabilities.[56]

Echoing sentiments set forth by proponents of the chemical and biological warfare research programs from the first half of the twentieth century, the 2019 biotechnology council called for a "whole-of-nation approach to human/machine enhancement technologies" that integrates efforts with universities and industry in order to outpace rivals, such as China, and "sustain U.S. dominance in cyborg technologies."[57] These findings were underscored in a 2022 NATO report, *Neuroenhancement in Military Personnel*, which advocated for the identification of "new training and technological approaches to enable sustained, optimized, and/or enhanced performance of military personnel."[58]

Warfighters are routinely tasked with performing complex skills under tremendous stress in a dynamic and unpredictable environment. These demands require high-quality cognition, physical strength, and an ability to adeptly utilize highly sophisticated military equipment and weapons. "Decision superiority" is essential to mission success—an ability to make decisions faster and better than one's adversaries.[59] Since the operational demand for decision speed and data volume processing is increasingly becoming mismatched with native human abilities, neuroenhancements are viewed as a means of ensuring operational readiness and overcoming the challenges of integrating modern technology into warfighting.[60]

The military and the intelligence community are researching neurotechnologies that aim to allow individuals to learn faster, train more efficiently, process complex and multisourced informational inputs, counteract performance degradation, and improve decision-making and threat assessment (see Figure 16.1). Achieving these goals requires the development of neurotechnologies that can accurately monitor brain activity and make reliable assessments of how brain activity relates to cognition or physiology. It further entails creating neurostimulation devices that can augment human performance.

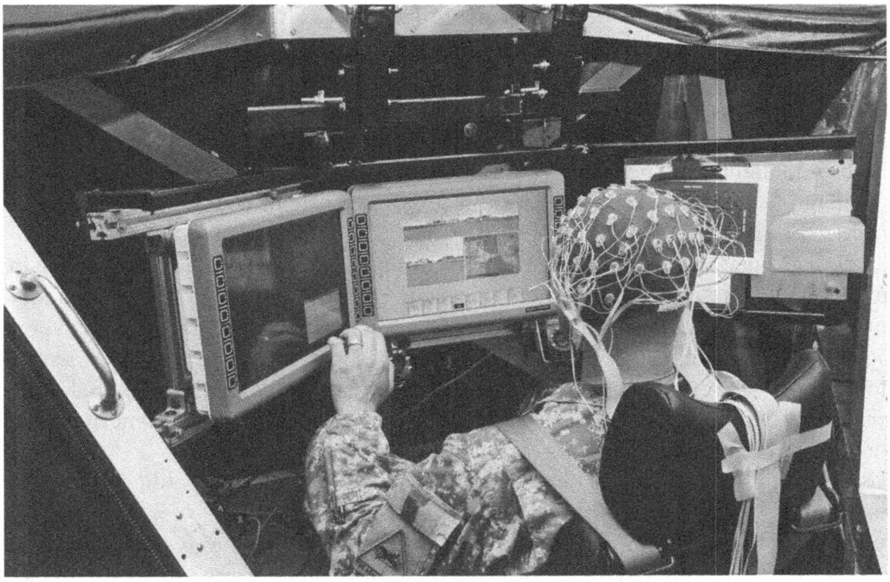

Figure 16.1 Brain-computer interface research conducted by the US Army.
U.S. Army photograph.

Neurotechnologies rely on a variety of instruments—such as functional magnetic resonance imaging (fMRI), electroencephalogram (EEG), and brain-to-computer interfaces—and often are integrated with artificial intelligence, machine learning, and nanotechnologies.[61] Neurostimulation devices include noninvasive caps worn on the head or invasive devices embedded in the brain. The external caps contain EEG sensors for monitoring brain activity and electrodes for stimulating it. The Army Research Lab is designing individualized EEG helmets for biometric and performance monitoring, 3-D printed to fit perfectly for each warfighter.[62] Some neurotechnologies integrate additional biometric monitoring such as eye movements, breathing, and cardiovascular functioning. Assessing the relationship between brain activity and cognition or physiology is incredibly complex due to scientific unknowns regarding how the human brain functions. The work requires years of study, and promising early findings often are found to be inaccurate or unreliable.[63]

Noninvasive stimulation techniques include transcranial direct current stimulation (tDCS), transcranial magnetic stimulation, and transcranial pulsed current stimulation, among others. Of these, tDCS is currently viewed as the most promising.[64] tDCS modulates brain activity via electrical currents that emanate from electrodes placed on a person's head. Electrodes on some devices are as thin as a strand of human hair.[65] The effectiveness of tDCS depends on a variety of factors, such as electrode placement, magnitude of the current, stimulation duration, direction of the current, and whether the stimulation was applied before, during, or after a task. Studies have found that tDCS enhances learning, creativity, attention span, motor skills, decision-making, vision, auditory perception, tactile perception, navigation, working memory, long-term memory, planning ability, probabilistic assessment, and problem-solving.[66]

While tDCS may provide cognitive benefits, it also comes with risks. Some studies have found that the effects of tDCS can be highly individualized, while others have noted that neurostimulation can have a detrimental effect on individuals who are already high performers. Side effects from tDCS include itching and a burning sensation at the point of contact with the electrode, as well as headache, fatigue, nausea, and dizziness. Other participants in tDCS research reported unintended changes in mood, personality, and behavior. Additional research is necessary to examine if there are latent effects, cumulative harms from prolonged use, or the potential to be addicted to brain stimulation.[67]

It is also unclear whether targeting certain brain activity may impact activity in nontargeted areas, or if enhancement of certain neural activity will lead to degradation elsewhere. This may pose significant risks, since the unpredictability of adverse cognitive functioning may negatively impact a warfighter and the military mission. More serious side effects have been observed in animal studies with tDCS, including permanent tissue damage. Adverse effects also may come from equipment malfunction or if stimulation levels are not calibrated accurately.[68]

Beyond the military, tDCS devices are widely marketed to the public to enhance learning, memory, relaxation, and focus, and many are available for under $150. Considering the FDA's limited jurisdiction in this area—which only covers medical devices that are intended to cure, mitigate, treat, or prevent disease—companies can sell tDCS devices without FDA approval as long as the company limits their marketing claims to improvement in focus and productivity and does not state that the device can cure, mitigate, treat, or prevent disease.[69] This regulatory gap extends to the military in that the intended use for tDCS devices impacts whether FDA medical device guidelines apply to research, field testing, and operational use.

Besides neurostimulation devices that use noninvasive technology, electrodes can be surgically embedded into a person's brain. Deep brain stimulation devices can measure brain activity and release currents that seek to increase or decrease certain brain activity to achieve a desired cognitive or physical state, a process known as intracortical microstimulation. Known risks of deep brain stimulation include hemorrhage, infection, mania, and psychosis.[70] Via its Reliable Neural-Interface Technology program, DARPA funded research to develop a stentrode, a less-intrusive means of placing electrodes inside a person's brain. Modeled from heart stent procedures, the stentrode weaves through a blood vessel in the neck and into the brain.[71] A separate area, optogenetics, uses light to stimulate neurons that have been modified to express light-sensitive genes.[72] In some studies, researchers could selectively inhibit fear responses by shining light on certain brain regions.[73]

Researchers are also examining brain stimulation devices that seek to sustain and improve warfighter performance by countering a range of metabolic, physiological, or psychological stressors faced before, during, or after a military operation. Scientists are exploring real-time neuroimaging assessments that can monitor an individual's stressors and trigger notifications that provide the warfighter and their commander with stressor information. These devices can be linked with a neurostimulation device that provides targeted brain stimulation as a stressor countermeasure.[74]

Brain stimulation devices are also being examined for therapeutic means of regulating emotions or treating posttraumatic stress. In one study, individuals diagnosed with depression and posttraumatic stress who underwent deep brain stimulation reported a significant decrease in the severity of nightmares.[75] Through its Systems-Based Neurotechnology for Emerging Therapies program, DARPA collaborated with hospitals and university labs to examine whether deep brain stimulation can ameliorate mental health conditions.[76] Other studies are utilizing implanted electrodes that monitor and stimulate brain activity to treat anxiety, PTSD, and other mental illnesses.[77]

Scientists are likewise studying neurotechnologies to predict decision-making capacity, optimize decision-making, and match decision makers with appropriate roles. Humans often err in estimating probabilities and risks, and differences in assessments are highly idiosyncratic and linked to an individual's personality, physiology, and neurology. Particularly when a person is under pressure, decision-making is rife with biases that cloud judgment and lead to suboptimal risk assessments and decisions. For example, some people may be more impulsive, while others may be more risk-averse. From an institutional perspective, one tendency does not necessarily lead to better or worse outcomes, though the bias may make certain people a better fit for certain tasks.

Accurately measuring individual decision-making variability can be used to optimize group performance.[78] And, decision-making could be enhanced via the utilization of neural stimulation. Neurotechnologies seek to measure what data elements a person is reading, seeing, reviewing, or analyzing—and what elements they are missing or setting aside—to assess if they are examining a situation objectively. Such technologies could also be used to determine a person's confidence level or level of confusion. Thereafter, a person's decision-making can be mediated by a neurodevice that relays to that person additional data elements, either consciously or unconsciously. One developmental project, Auto-Diagnostic Adaptive Precision Training, utilizes eye-tracking software, EEG imaging, and various means of enhancing analysis and decision-making.[79]

Several additional projects aim to improve warfighter cognition. DARPA's Cognitive Technology Threat Warning System is a signal-processing system coupled with a helmet-mounted EEG device that monitors and stimulates brain activity to augment a warfighter's ability to detect threats in a wide field of vision.[80] A separate DARPA program, Neuroscience for Intelligence Analysts, uses EEG to detect brain activity that reflects conscious or nonconscious attention that might be indicative of a threat. Part of the theory underlying this program is that threat perception may be nonconscious and that neurotechnologies may be utilized to capture the nonconscious observation and rapidly provide stimulation to the warfighter to act.[81] In 2018, the Army publicly announced that one of its key priorities is to enhance a warfighter's ability to make accurate "kill/do not kill" decisions and follow through on those decisions with precision.[82]

Brain-computer interfaces also seek to synthesize human reasoning and sensory perception with access to encyclopedic information. The military and the intelligence community have access to troves of information that can be integrated into neurotechnologies to help warfighters make smarter decisions during military operations.[83] Researchers are also examining the use of augmented reality technologies whereby a warfighter wears glasses or contact lenses that allow them to see the natural world augmented with information-rich virtual elements.[84] Other projects include overlaying ocular tissue or replacing a human eyeball with an ocular enhancement system that transmits data streams directly into a person's brain based on what they are seeing.[85]

Some brain-computer interfaces allow warfighters to control weapons systems via thought alone.[86] This work falls under the field of neuroergonomics, whereby brain activity is translated into a signal that can be transduced into a control input of an external system.[87] Some electrodes are as small as a grain of rice and utilize wireless technology.[88] Via DARPA-funded work at several universities, researchers have created brain-computer interfaces whereby individuals who are paralyzed can use their thoughts to move a robotic hand, send email, move a cursor on a computer screen, draw, or play video games.[89] Outside healthcare and rehabilitation settings, such technologies could be integrated to facilitate a warfighter-system symbiosis that, as the National Research Council explained in a 2009 report, "measurably outperforms conventional human-system interfaces."[90] For example, a brain-computer interface could be linked to orthotic exoskeletons that substantially enhance a person's strength and endurance.[91] In one DARPA-funded study, a quadriplegic woman with a brain-computer interface and ninety-six microelectrodes implanted in her brain used only her thoughts to control a fighter jet in a flight simulator.[92]

As of 2022, more than six hundred military projects were examining how to incorporate artificial intelligence into warfighting. Via DARPA's Air Combat Evolution program, scientists are exploring the use of artificial intelligence to improve performance during aerial combat.[93] DARPA's Mind's Eye Program is building a smart camera with machine-based visual intelligence capable of analyzing human interactions in real-world environments.[94] Other projects aim to address "temporal dissonance"—the period between a decision and action—to minimize the time to milliseconds. Just like automated systems are relied upon to make decisions on when to deploy an airbag in a car, researchers seek to use artificial intelligence to guide flight maneuvers and other operational decisions, including when to shoot. Artificial intelligence has defeated experienced fighter pilots in several aerial combat simulations, and additional work is underway to translate simulation success into battlefield success.[95]

Artificial intelligence alters the nature of war because it introduces nonhuman decision-making into military operations. New technologies are beginning to blur the line between semiautonomous and autonomous weapons, and concerns have been raised about allowing a machine to determine when to kill a human. Israel has developed a drone that independently searches for a target and then dives in to destroy it.[96] During the Israeli war in the Gaza Strip, which began in 2023 and has continued as of the writing of this book, the Israeli Defense Forces utilized a program dubbed Habsora (which translates to "The Gospel" in English), an artificial intelligence algorithm that rapidly synthesizes vast amounts of data from a variety of sources, including drone footage, intercepted communications, and surveillance data. The program produces automated recommendations for attacking targets, including individuals and private homes.[97]

A report in *The Guardian* quoted an anonymous source from within the Israeli military who expressed some concerns, stating that "it really is like a factory. We work quickly and there is no time to delve deep into the target. The view is that we are judged according to how many targets we manage to generate."[98] Several security experts expressed alarm over automation bias, high error rates, and the large increase in targets pursued by Israel during the war as compared to previous attacks within the Gaza Strip.[99] Commenting on Israel's actions, one former US official candidly remarked that "other states are going to be watching and learning."[100]

Although some groups have called for a ban on lethal autonomous weapons, America and other major powers, including China and Russia, are not signatories to any such treaties.[101] In 2017, the US military created Project Maven, an "algorithmic warfare cross-functional team" that works closely with leading technology firms and serves to consolidate "all initiatives that develop, employ, or field artificial intelligence, automation, machine learning, deep learning, and computer vision algorithms."[102] Researchers are also examining whether artificial intelligence can be applied to national security policy and strategic planning of military operations. The goal is to augment human decision-making by obtaining machine consultations.[103]

The DoD recently created a Chief Digital and Artificial Intelligence Office to analyze the benefits and risks posed by artificial intelligence in warfighting. The office is also charged with establishing a set of ethical principles to govern how artificial intelligence should be incorporated into DoD projects. Yet, concerns have been raised by inadequate funding for the group and the lack of DoD staff with the technical expertise needed to test, evaluate, procure, and manage artificial intelligence systems.[104]

Although governments around the world have publicly called for the responsible use of artificial intelligence in warfighting and societal applications, the long history of military science endeavors suggests that nations might be quick to disregard the guardrails should they face a real or perceived existential threat to their economic or geopolitical interests. Of particular concern is the utilization of artificial intelligence to create more devastating chemical or biological weapons. In May 2023, the directors of dozens of artificial intelligence labs signed a letter that warned about technological innovations that posed an existential threat to humanity.[105] Some commentators have called on the DoD to create a Digital Corps—modeled on the Army Medical Corps—which would leverage American strengths in technology companies and universities to create a division that would spearhead training, research and development, procurement, and organization of digital technologies into military missions.[106]

In addition to serving as cognitive or physical enhancements, some neurotechnologies may be used to surveil thoughts. The military and the intelligence community are examining neurophysiological technologies that observe neural processes to detect deception and psychological states. These scans might be able to elicit when a person is lying or predict a person's propensity to engage in harmful conduct. One prototype under development is akin to an airport body scanner. Rather than scanning for metal objects, the device scans brain activity to identify individuals contemplating criminal conduct or terrorist acts. Researchers are also examining whether cognitive states and intentions can be altered or controlled—for example, disrupting an adversary's motivation to fight, or making a person under interrogation trust the interrogator and reveal truthful information. As to the latter, oxytocin has been explored as a means of increasing trust.[107]

Some technologies—such as those that disrupt or impair cognition—may be properly characterized as neuroweapons.[108] This includes bioregulators, which are chemicals that can cause temporary incapacitation. In 2002, when Chechen rebels held nine hundred people captive in a Moscow theater for several days, Russian special forces stormed the theater after pumping a noxious gas into the room to incapacitate the terrorists. President George W. Bush praised Russian President Vladimir Putin's use of the gas—which was a derivative of fentanyl—to help end the siege.[109] More than a decade later, experts characterized neurotoxins and neurobiological agents as "a new class of weapons that can damage the nervous system, alter mood, trigger psychological changes and kill."[110] The hostage crisis, and Bush's response, suggests that nations maintain stockpiles of bioregulators that can be deployed quickly.

There likewise have been reports of odd cognitive ailments among American diplomats, military personnel, and intelligence agents stationed in Austria, China, Cuba, India, Russia, and Taiwan. First observed in Cuba in 2016, the mysterious condition is commonly referred to as Havana Syndrome. Individuals with Havana Syndrome inexplicably suffer from vertigo, severe headaches, nausea, head pressure, and piercing directional noises. As of mid-2024, the cause remains unknown, though some experts suspect that the injured are victims of a directed energy attack, perhaps caused by pulsed electromagnetic energy.[111] Bioregulators and directed energy attacks illustrate that neuroweapons have become components of national security arsenals.

Summary

Biomedical enhancements may provide warfighters with special competitive advantages, but they also come with special risks. Making an accurate risk-benefit analysis is complicated by the fact that known risks are difficult to quantify, and there may be unknown risks. There likewise are challenging moral considerations, such as whether biomedical enhancements will negatively impact individual autonomy or one's sense of worth or personal identity. At the same time, in light of the changing face of warfare and the technological sophistication of contemporary military systems, some policy experts have argued that there is a moral obligation to utilize an enhancement if it provides a health or operational advantage for military personnel.[112]

As an Army training manual explains, military operations "place a premium on the Soldier's strength, stamina, agility, resiliency and coordination. Victory—and even the Soldier's life—so often depend upon these factors."[113] Because biomedical enhancements raise complex ethical, legal, and operational concerns, it is imperative that each enhancement be assessed throughout the lifecycle of the intervention, and that appropriate mechanisms be implemented to account for health, familial, and societal concerns that may materialize over time. The following chapters explore this recommendation.

Notes

1. Julian Savulescu and Nick Bostrom, eds., *Human Enhancement* (2011), 1–4.
2. National Research Council, *Opportunities in Neuroscience for Future Army Applications* (2009), 5.
3. NRC, *Opportunities in Neuroscience*, 5.
4. Kamienski, *Shooting Up*, 145–51.
5. Heloise Goodley, "Pharmacological Performance Enhancement and the Military," *Chatham House International Security Programme* (November 2020): 6–8.
6. Ohler, *Blitzed*, 107–16, 157–63, 180–83.
7. Kamienski, *Shooting Up*, 145–51.
8. Mayo Clinic, Dextroamphetamine (Oral Route) (accessed April 24, 2024).
9. Nicolas Rasmussen, "America's First Amphetamine Epidemic: 1929-1971," *American Journal of Public Health* 98, no. 6 (June 2008): 974–85; Andrew Golub, Alex S. Bennett, and Luther Elliott, "Beyond America's War on Drugs: Developing Public Policy to Navigate the Prevailing Pharmacological Revolution," *AIMS Public Health* 2, no. 1 (2015): 142–60.
10. Quoted in Chris H. Gray, *Postmodern War: The New Politics of Conflict* (New York: Guilford Press, 1997), 209.
11. Kamienski, *Shooting Up*, 191–208.
12. David L. Emonson and Rodger D. Vanderbeek, "The Use of Amphetamines in U.S. Air Force Tactical Operations During Desert Shield and Storm," *Aviation, Space, and Environmental Medicine* 66, no. 3 (March 1995): 260–63.
13. Rhonda Cornum, John Caldwell, and Kory Cornum, "Stimulant Use in Extended Flight Operations," *Airpower Journal* (Spring 1997): 53–58.
14. Cornum et al., "Stimulant Use," 56.
15. United States Air Force Scientific Advisory Board, "Report on United States Air Force Expeditionary Forces," vol. 3, app. I (February 1998): 178–80; David R. Jones and Royden W. Marsh, "Flight Surgeon Support to United States Air Force Fliers in Combat," *United States Air Force School of Aerospace Medicine* (May 2003): 140–55.
16. John A. Caldwell et al., "Fatigue Countermeasures in Aviation," *Aviation, Space, and Environmental Medicine* 80, no. 1 (January 2009): 29–59; Elizabeth Stanley,

"Neuroplasticity, Mind Fitness, and Military Effectiveness," in Robert E. Armstrong et al., eds., *Bio-Inspired Innovation and National Security* (2010), 260–61; Rajee Olaganathan et al., "Fatigue and Its Management in the Aviation Industry, with Special Reference to Pilots," *Journal of Aviation Technology and Engineering* 10, no. 1 (2021): 45–57.

17. Caldwell et al., "Fatigue Countermeasures"; Olaganathan et al., "Fatigue."
18. Commander, Moody Air Force Base, "Go Pills, No Go Pills, Ciprofloxacin, and Doxycycline for Flying Operations," *Moody Air Force Base Instruction* 44–101 (July 14, 2010).
19. "Informed Consent for Operational Use of Dexedrine," in United States Navy, *Performance Maintenance During Continuous Flight Operations* (January 1, 2000): 21.
20. Mark Owen, *No Easy Day* (2012), 194–99, 278–86.
21. Quoted in Michael Friscolanti, *Friendly Fire* (2005), 437.
22. Quoted in Friscolanti, *Friendly Fire*, 205.
23. Friscolanti, *Friendly Fire*, 430–53.
24. Friscolanti, *Friendly Fire*, 430–53.
25. "Captagon: Understanding Today's Illicit Market," *European Monitoring Centre for Drugs and Drug Addiction* (2018).
26. "Captagon: Understanding Today's Illicit Market"; Sulome Anderson, "These are the People Making Captagon, the Drug ISIS Fighters Take to Feel 'Invincible,'" *New York Magazine*, December 9, 2015; Jack Guy et al., "Italian Police Seize Over $1 Billion of 'ISIS-Made' Captagon Amphetamines," *CNN*, July 1, 2020.
27. Quoted in Anderson, "These are the People Making Captagon."
28. Kamienski, *Shooting Up*, 243–56.
29. Phuong N. Pham, Patrick Vinck, and Eric Stover, "The Lord's Resistance Army and Forced Conscription in Northern Uganda," *Human Rights Quarterly* 30, no. 2 (May 2008): 404–11; Rosa Ehrenreich, "The Stories We Must Tell: Ugandan Children and the Atrocities of the Lord's Resistance Army," *Africa Today* 45, no. 1 (1998): 79–102; Opiyo Oloya, *Child to Soldier: Stories from Joseph Kony's Lord's Resistance Army* (Toronto: University of Toronto Press, 2013).
30. Alain Brunet et al., "Reduction of PTSD Symptoms with Pre-Reactivation Propranolol," *American Journal of Psychiatry* 175, no. 5 (May 1, 2018): 427–33; Pascal Roullet et al., "Traumatic Memory Reactivation with or Without Propranolol for PTSD and Comorbid MD Symptoms," *Neuropsychopharmacology* 46 (February 2021): 1643–49.
31. Quoted in Maxwell J. Mehlman, *The Price of Perfection* (2009), 20.
32. James Giordano and Rachel Wurzman, "Neurotechnologies as Weapons in National Intelligence and Defense: An Overview," *Synesis: A Journal of Science, Technology, Ethics, and Policy* 2, no. 1 (2011): 55–71.
33. Laura Sanders, "Brains May Be War's Battlegrounds," *Science News* 180, no. 13 (December 17, 2011): 14.
34. Substance Abuse and Mental Health Services Administration, *Prescription Stimulant Misuse and Prevention Among Youth and Young Adults* (2021).
35. Ohler, *Blitzed*, 34–35.
36. Yehudit Hasin, Marcus Seldin, and Aldons Lusis, "Multi-omics Approaches to Disease," *Genome Biology* 18, no. 83 (May 5, 2017): 1–15; Uniformed Services University Center for the Study for the Study of Traumatic Stress, *Precision Medicine in Human Performance* (May 8–9, 2019).
37. Patrick Tucker, "Tomorrow Soldier: How the Military Is Altering the Limits of Human Performance," *Defense One*, July 12, 2017.
38. Hasin et al., "Multi-omics Approaches to Disease"; DARPA, *Measuring Biological Aptitude* (accessed December 14, 2023); Diane DiEuliis and James Giordano, "Balancing

Act: Precision Medicine and National Security," *Military Medicine* 187, Supp. 1 (January–February 2022): 32–35.
39. DARPA, *Measuring Biological Aptitude.*
40. Tucker, "Tomorrow Soldier."
41. Lucas Poon et al., "A Review of Genome-Based Precision Medicine Efforts Within the Department of Defense," *Military Medicine* 187, Supp. 1 (January–February 2022): 25–31.
42. The All of Us Research Program Investigators, "The 'All of Us' Research Program," *New England Journal of Medicine* 381, no. 7 (August 15, 2019): 668–76.
43. DiEuliis and Giordano, "Balancing Act: Precision Medicine and National Security."
44. Bruce Doll et al., "Precision Medicine—A Demand Signal for Genomics Education," *Military Medicine* 187, Supp. 1 (January–February 2022): 40–46.
45. U.S. Department of Defense Joint Chiefs of Staff, *Chinese Strategic Intentions: A Deep Dive into China's Worldwide Activities,* edited by Nicole Peterson (December 2019); DiEuliis and Giordano, "Balancing Act: Precision Medicine and National Security."
46. R. Alta Charo, "Rogues and Regulation of Germline Editing," *New England Journal of Medicine* 380, no. 10 (March 7, 2019): 976–80; Lisa Rosenbaum, "The Future of Gene Editing—Toward Scientific and Social Consensus," *New England Journal of Medicine* 380, no. 10 (March 7, 2019): 971–75; George Q. Daley, Robin Lovell-Badge, and Julie Steffann, "After the Storm—A Responsible Path for Genome Editing," *New England Journal of Medicine* 380, no. 10 (March 7, 2019): 897–99; FDA News Release, *FDA Approves First Gene Therapies to Treat Patients with Sickle Cell Disease* (December 8, 2023); Gina Kolata, "FDA Approves Sickle Cell Treatments, Including One That Uses CRISPR," *The New York Times,* December 8, 2023.
47. Dennis Normile, "Chinese Scientist Who Produced Genetically Altered Babies Sentenced to 3 Years in Jail," *Science,* December 30, 2019.
48. Dana Goodyear, "Dangerous Designs," *The New Yorker,* September 11, 2023, 30–43.
49. Normile, "Chinese Scientist."
50. Amy Gutmann and Jonathan D. Moreno, "Keep CRISPR Safe," *Foreign Affairs* 97, no. 3 (May/June 2018): 171–76; Charo, "Rogues and Regulation of Germline Editing."
51. European Parliament Directorate General for Internal Policies, *Human Enhancement Study* (May 2009), 11.
52. Goodyear, "Dangerous Designs."
53. National Academies of Sciences, Engineering, and Medicine, *Human Genome Editing* (2017), 134.
54. Robbin A. Miranda, William D. Casebeer, Amy M. Hein, et al., "DARPA-Funded Efforts in the Development of Novel Brain-Computer Interface Technologies," *Journal of Neuroscience Methods* 244 (April 2015): 52–67.
55. Peter Emanuel et al., "Cyborg Soldier 2050: Human/Machine Fusion and the Implications for the Future of the DoD," *U.S. Army Combat Capabilities Development Command, Chemical Biological Center* (October 2019), v.
56. Emanuel et al., *Cyborg Soldier 2050,* 2.
57. Emanuel et al., *Cyborg Soldier 2050,* vii.
58. Tad T. Brunye et al., "Neuroenhancement in Military Personnel: Conceptual and Methodological Promises and Challenges," *NATO* (February 4, 2022), at 1-1.
59. Kevin S. Oie and Kaleb McDowell, "Neurocognitive Engineering for Systems' Development," in Giordano, ed., *Neurotechnology,* 35.
60. Steven E. Davis and Glen A. Smith, "Transcranial Direct Current Stimulation Use in Warfighting: Benefits, Risks, and Future Prospects," *Frontiers in Human Neuroscience* 13, no. 114 (April 2019): 1–18.

61. National Research Council, *Emerging Cognitive Neuroscience and Related Technologies* (2008), 1–12.
62. Tucker, "Tomorrow Soldier."
63. Kathryn A. Feltman et al., "Viability of tDCS in Military Environments for Performance Enhancement: A Systematic Review," *Military Medicine* 185, no. 1–2 (January–February 2020): e53–e60.
64. Feltman et al., "Viability of tDCS."
65. Davis and Smith, "Transcranial Direct Current Stimulation"; Singer, *Wired for War*, 70–74.
66. Gary Sheftick, "Army Researchers Looking to Neurostimulation to Enhance, Accelerate Soldiers' Abilities," *U.S. Army News Service*, May 31, 2018; Feltman et al., "Viability of tDCS"; Davis and Smith, "Transcranial Direct Current Stimulation."
67. Feltman et al., "Viability of tDCS"; Davis and Smith, "Transcranial Direct Current Stimulation"; Sheftick, "Army Researchers Looking to Neurostimulation."
68. Bernhard Sehm and Patrick Ragert, "Why Non-Invasive Brain Stimulation Should Not Be Used in Military and Security Services," *Frontiers in Human Neuroscience* 7, art. 553 (September 2013): 1–3; Feltman et al., "Viability of tDCS"; Davis and Smith, "Transcranial Direct Current Stimulation"; Sheftick, "Army Researchers Looking to Neurostimulation."
69. Federal Food, Drug, and Cosmetic Act, § 201(h) (2023); F. Fregni et al., "Regulatory Considerations for the Clinical and Research Use of Transcranial Direct Current Stimulation (tDCS): Review and Recommendations from an Expert Panel," *Clinical Research and Regulatory Affairs* 32, no. 1 (March 2015): 22–35.
70. European Parliament, *Human Enhancement Study*, 89.
71. DARPA, "Minimally Invasive 'Stentrode' Shows Potential as Neural Interface for Brain" (February 8, 2016).
72. Michael N. Tennison and Jonathan D. Moreno, "Neuroscience, Ethics and National Security: The State of the Art," *PLoS Biology* 10, no. 3 (March 2012): 1–4.
73. Melanie Segado, "Military Applications of Invasive Brain Stimulation," *Technology and Society*, June 29, 2017.
74. Davis and Smith, "Transcranial Direct Current Stimulation"; Karl E. Friedl, "Military Applications of Soldier Physiological Monitoring," *Journal of Science and Medicine in Sport* 21, no. 11 (November 2018): 1147–53.
75. Jean-Philippe Langevin et al., "Deep Brain Stimulation of the Basolateral Amydala: Targeting Technique and Electrodiagnostic Findings," *Brain Sciences* 6 (September 2016): 1–9.
76. DARPA, Systems-Based Neurotechnology for Emerging Therapies (accessed April 8, 2024).
77. Jackob N. Keynan et al., "Limbic Activity Modulation Guided by Functional Magnetic Resonance Imaging-Inspired Electroencephalography Improves Implicit Emotion Regulation," *Biological Psychiatry* 80, no. 6 (September 2016): 490–96; Mark S. George, "Is Functional Magnetic Resonance Imaging-Inspired Electroencephalogram Feedback the Next New Treatment in Psychiatry?," *Biological Psychiatry* 80, no. 6 (September 2016): 422–23; Feltman et al., "Viability of tDCS"; Segado, "Military Applications."
78. NRC, *Opportunities in Neuroscience*, 3–4, 36–43.
79. Kay M. Stanney et al., "Neural Systems in Intelligence and Training Applications," in Giordano, ed., *Neurotechnology*, 23–31.
80. Bruce Sterling, "Augmented Reality: DARPA Cognitive Technology Threat Warning System," *Wired*, September 19, 2012.
81. NRC, *Opportunities in Neuroscience*, 68–72; Miranda et al., "DARPA-Funded Efforts," 60–64.
82. Feltman et al., "Viability of tDCS," e54.

83. NRC, *Emerging Cognitive Neuroscience*, 1–12, 141.
84. NRC, *Opportunities in Neuroscience*, 76–79.
85. Emanuel et al., *Cyborg Soldier 2050*, 3–5.
86. NRC, *Emerging Cognitive Neuroscience*, 141.
87. NRC, *Opportunities in Neuroscience*, 6.
88. Singer, *Wired for War*, 70–74.
89. Miranda et al., "DARPA-Funded Efforts," 52–67.
90. NRC, *Opportunities in Neuroscience*, 6.
91. Thomas X. Hammes, "Biotech Impact on the Warfighter," in Armstrong et al., eds., *Bio-Inspired Innovation and National Security*, 1–7; Tennison and Moreno, "Neuroscience, Ethics and National Security."
92. Nick Stockton, "Woman Controls a Fighter Jet Sim Using Only Her Mind," *Wired*, March 5, 2015.
93. Sue Halpern, "Flying Aces," *The New Yorker*, January 24, 2022, 18–24.
94. Rachel Wurzman and James Giordano, "'NEURINT' and Neuroweapons: Neurotechnologies in National Intelligence and Defense," in Giordano, ed., *Neurotechnology*, 79–109.
95. Kenneth Payne, *I, Warbot* (2021), 92–93.
96. Payne, *I, Warbot*, 12, 83.
97. Harry Davies, Bethan McKernan, and Dan Sabbagh, "'The Gospel': How Israel Uses AI to Select Bombing Targets in Gaza," *The Guardian*, December 1, 2023; Geoff Brumfiel, "Israel Is Using an AI System to Find Targets in Gaza: Experts Say It's Just the Start," *NPR*, December 14, 2023.
98. Quoted in Davies et al., "'The Gospel.'"
99. Davies et al., "'The Gospel'"; Brumfiel, "Israel is Using an AI System."
100. Quoted in Davies et al., "'The Gospel.'"
101. "Defense Primer: U.S. Policy on Lethal Autonomous Weapon Systems," *Congressional Research Service* (February 1, 2024); Mary Ellen O'Connell, "Banning Autonomous Weapons: A Legal and Ethical Mandate," *Ethics and International Affairs* 37, no. 3 (Fall 2023): 287–98; Brian Stauffer, "Stopping Killer Robots: Country Positions on Banning Fully Autonomous Weapons and Retaining Human Control," *Human Rights Watch* (August 10, 2020).
102. Memorandum from Deputy Secretary of Defense, "Establishment of an Algorithmic Warfare Cross-Functional Team (Project Maven)" (April 26, 2017).
103. Payne, *I, Warbot*, 1–3, 137–46.
104. Michele A. Flournoy, "AI is Already at War," *Foreign Affairs* 102, no. 6 (November/December 2023): 56–69.
105. Statement on AI Risk, Center for AI Safety (May 30, 2023); Kevin Roose, "A.I. Poses 'Risk of Extinction,' Industry Leaders Warn," *The New York Times*, May 30, 2023.
106. Flournoy, "AI is Already at War."
107. NRC, *Emerging Cognitive Neuroscience*, 1–17.
108. Giordano and Wurzman, "Neurotechnologies as Weapons."
109. Mike Allen, "Bush Defends Putin in Handling of Siege," *The Washington Post*, November 19, 2002.
110. S. Bokan, "The Toxicology of Bioregulators as Potential Agents of Bioterrorism," *Arh Hig Rada Toksiko* 56 (2005): 205, quoted in Kelle, Nixdorff, and Dando, *Preventing a Biochemical*, 23–24; Wurzman and Giordano, "'NEURINT' and Neuroweapons," in Giordano, ed., *Neurotechnology*, 79–109.

111. Kattie Bo Lillis, "CIA Doctor hit by Havana Syndrome says he was in 'Disbelief' as he Suffered what he was Investigating," *CNN*, September 25, 2022; Adam Entous, "Are U.S. Officials Under Silent Attack?" *The New Yorker*, May 31, 2021.
112. John Harris, "Enhancements Are a Moral Obligation," in Savulescu and Bostrom, *Human Enhancement*, 131–54; Ioana Maria Puscas, "Military Human Enhancement," in William H. Boothby, ed., *New Technologies and the Law in War in Peace* (2019), 182–229.
113. Department of the Army, *Army Physical Readiness Training* (March 2010), 1–1.

PART V
LOOKING AHEAD

Jus in Praeparatione Bellum

17

Jus in Praeparatione Bellum

A Normative Framework

Introduction

Just war theory is a centuries-old tradition of military ethics that assesses what actions are proper or improper in the context of war. The two foundational elements of just war theory are *jus ad bellum*, which considers justice in the reasons for entering war, and *jus in bello*, which examines justified conduct during war.[1] Subsequent developments in just war theory have added *jus ad vim*, which investigates moral principles governing the use of limited force, and *jus post bellum*, which explores fair treatment after a war has ended.[2] Legal frameworks that codify these principles are found in customary and conventional international law, most prominently the Charter of the United Nations, the Hague and Geneva Conventions, and the International Criminal Court. Key concepts include, for example, maintaining the political sovereignty and territorial integrity of nations, ensuring that nations that engage in war do so for legitimate reasons, and setting standards of warfighting that limit the type of aggressive conduct and against whom such actions can be taken.[3]

Principles of just war theory have evolved alongside societal mores, technological and scientific innovations, and civilian-military relationships. By strengthening global institutions and incentivizing compliance with international norms related to legitimate uses of force, just war theory aims to uphold an international order, reduce the incidence of war, and ensure greater protections for human rights before, during, and after war. There is no single principle that links all just war theorists. Rather, the link is a method of thinking about just and unjust actions in military affairs.[4]

Just war theory adopts a common law tradition that utilizes general rules but analyzes each situation on its own merits.[5] The doctrine relies on moral judgment but does not proscribe debate about questions of morality. As an analogy, one may support the view that there are fundamental human rights but acknowledge that reasonable minds can differ on what rights qualify as fundamental and who gets to decide.

Within the just war tradition, in this book, I introduce the concept of *jus in praeparatione bellum*—justice in war preparations. The concept of *jus in praeparatione bellum* represents an expansion of just war theory. Whereas other elements of just war theory focus primarily on justice in how one nation treats another before, during, or after war, *jus in praeparatione bellum* is inward looking for each nation and evaluates justice principles in how a nation treats its own service members and citizens in the context of war preparations.

In this chapter, I outline the contours of *jus in praeparatione bellum*. I discuss how the concept is grounded in just war theory but also draws on key elements of liberal internationalism and realism, two doctrines of international relations that have substantially influenced US foreign policy and military affairs for over a century. I then

America's Military Biomedical Complex. Efthimios Parasidis, Oxford University Press.
 DOI: 10.1093/9780199351473.003.0018

explain how *jus in praeparatione bellum* can be applied to military science, and how the theory can help guide laws and ethical codes applicable to the military biomedical complex.

Contours of the Theory

Just war theory is a doctrine of restraint, and *jus in praeparatione bellum* is a theory of self-restraint in a nation's war preparations. Each command within the US military maintains an operational doctrine—an institutional framework for planning and conducting military operations—and the concept of *jus in praeparatione bellum* can help develop and evolve doctrine for war preparations.

Nations are constantly preparing for war, through military training, weapons procurement, military science pursuits, or other endeavors. With *jus in praeparatione bellum*, the key inquiry is whether, during war preparations, a nation balances its military aims with fair treatment of its own citizens. This includes an examination of whether a nation bypasses or curbs certain fundamental human rights to further military affairs, and an assessment of whether war preparations cause unjustifiable harms. *Jus in praeparatione bellum* also considers how war impacts the health and human dignity of a nation's own service members—both during and after service—and how the government can assist its service members in reconciling their military responsibilities with their life outside the armed forces.

Although *jus in praeparatione bellum* appraises the relationship between a nation and its own citizens in war preparations—as opposed to other aspects of just war theory, which examine relationships between nations—traditional just war principles can be adopted from nations to individuals. Specifically, a nation ought to respect each person's autonomy and bodily integrity, and appropriate standards must be set that limit interference with those rights. As with other areas within the just war tradition, a key focus of *jus in praeparatione bellum* is intention: whether a nation has intended to act honorably and to balance military aims and the collective good with the rights and welfare of individuals. This includes consideration of whether an activity has a reasonable hope of success, whether harms caused by an activity are proportional to the intended benefits, and whether an activity is likely to cause unnecessary harm. To be sure, war preparations go far beyond military science and include a wide range of activities, such as espionage, weapons testing, and tactical and strategic operations. Although the concept of *jus in praeparatione bellum* can be applied to these areas, in this book I examine *jus in praeparatione bellum* in the context of military science.

For the military biomedical complex, *jus in praeparatione bellum* is a synthesis of normative theories from bioethics and international relations. Bridging these two disciplines is essential because each offers important insights into the legal, ethical, and operational foundations of military science. Bioethics is a discipline that examines ethical, legal, and societal issues in medicine, public health, and biotechnology. Bioethical issues include, for example, the parameters of the doctor-patient relationship, what protections should be afforded to individuals enrolled in research studies, and the ethical and societal implications of new biotechnologies. Many bioethical issues require assessing how to align risks and benefits, and balance trade-offs between individual rights and the common good.

Although bioethics is a rich and diverse discipline, it is not a field tailored to tackling complex issues in war, foreign policy, and national security. Indeed, as detailed throughout this book, physicians and researchers often have deferred to military and government leaders to identify instances where legal and ethical frameworks for military science should diverge from civilian standards. Unpacking this divergence requires an intimate look at theories of international relations—specifically, liberal internationalism and realism, two doctrines that have influenced national security policy in the United States and worldwide.

Liberal Internationalism and Realism

The concept of *jus in praeparatione bellum* is grounded in just war theory, but it also incorporates principles of liberal internationalism and realism. This is because US foreign policy has relied extensively on tenets of liberal internationalism and realism to guide international relations and military decision-making. Liberal internationalists emphasize the importance of global institutions as bodies that set and enforce rules-based norms that promote liberal democracy, free markets, cooperative security, and global trade. Examples include the United Nations, European Union, International Monetary Fund, Organization for Economic Cooperation and Development, and the North Atlantic Treaty Organization (NATO). Liberal internationalism emphasizes collaborative decision-making and shared obligations. For liberal internationalists, war can be justified to promote a liberal world order regardless of whether it entails armed intervention for humanitarian purposes, regime change, or nation-building.[6]

Realists question the power of global institutions to guide international relations, contending that such institutions are not autonomous but rather reflect the political will of their most powerful members.[7] For realists, the supremacy of nation-states is paramount, and there is no "government over governments."[8] Realists contend that nations are in a constant state of negotiation—a continuing struggle for advantage and survival—where the powerful have unequal bargaining power and the international system is marked by anarchy. For realists, anarchy does not refer to chaos, but instead refers to the notion that no sovereign political authority maintains a monopoly on determining legitimate uses of force. Realists maintain that conflicts between nations are inevitable and that nations engage in acts deemed essential to preserve and expand their national interests.[9]

Although the doctrine of realism maintains that the international system is anarchic in that nations have no binding central authority above them and that powerful nations have unequal bargaining power in the global world order, realists consider moral, ethical, and legal concerns in war and international relations. Realists also believe that cooperation and mutual goals can exist, and that nations engage in security balancing based on their respective interests. Realists recognize the co-binding of nations based on national security priorities, but they deem self-interest as the driving force of decision-making in matters of national security.[10]

In light of this worldview, realists deem power balances as essential to preserving peace. For instance, mutually assured destruction via nuclear weapons is a powerful deterrent to all nuclear powers. Moreover, realists contend that, to preserve its homeland and economic or political interests, a nation may find it in its best interest to adopt moral and legal limits to war and national security policy.[11] Realists recognize that there are costs to adopting unjust geopolitical policies. These may include, for example, national security threats to retaliate for the unjust conduct, isolation from international trade or treaties, or implementation of sanctions to punish and deter certain practices.[12]

Attempts to bring foundational elements of liberal internationalism onto the world stage accelerated during the early twentieth century, with an increased focus on establishing multinational treaties and dispute resolution mechanisms for conflicts between nations. As detailed in chapters 4 and 6, the League of Nations was formed during this period and several treaties were enacted, including the Hague Conventions of 1899 and 1907, the 1925 Geneva Protocol, and the 1928 Kellogg-Briand Pact. While these efforts helped progress a liberal internationalist agenda—as German and Japanese aggression during the 1930s revealed—they failed to implement binding agreements that were backed by international institutions that could adequately address violations.

As political scientist John Ikenberry explains, "The liberal internationalism envisaged by [Woodrow] Wilson was a historical failure," since it "did not involve the construction of deeply transformative legally-binding political institutions" and did not maintain "underlying conditions needed for a collective security system to function."[13] Coupled with the institutional shortcomings, while American leaders saw benefits to maintaining military and economic ties with other nations, the American public favored isolationism and focused its political attention on domestic affairs.[14] Moreover, after World War I (WWI), a narrative arose in the popular press that blamed bankers and arms manufacturers for nudging the United States into war solely to pad corporate profits.[15] These opinions carried forward to the beginning of World War II (WWII), and the US public was hesitant to enter the conflict.[16] This sentiment changed after the Japanese bombing of Pearl Harbor.[17]

The worldwide devastation caused by WWII motivated a shift in public opinion: a desire to avoid another world war underscored the importance of functional international institutions that could mediate conflicts, decrease the likelihood of war, and limit methods of warfare that were repugnant. At the same time, the Allied victory in WWII underscored the advantages of a technologically superior military, and how a nation that is quick to capitalize on advancements in military science gains a strategic advantage over its adversaries. A corollary lesson was that failing to keep pace with scientific and technological innovations creates substantial geopolitical and national security risks.[18]

After WWII and throughout the Cold War, great power competition between the United States and the Soviet Union reflected a blend of liberal internationalism and realism. The United Nations was established to mediate international conflicts and promote universal norms for political, economic, and environmental matters, while NATO and the Warsaw Pact were created as cooperative security agreements of the United States and the USSR, respectively.[19] Each nation dedicated substantial resources to military science and amassed a fearsome weapons arsenal. America and the Soviet Union dominated global politics and international affairs, setting rules and policies within their spheres of influence that sought primarily to increase their respective power.

One key component of America's efforts centered on the liberal economic principles outlined during the 1944 Bretton Woods conference, a three-week gathering at a secluded New Hampshire resort that set new rules for a postwar international monetary system and laid the groundwork for the creation of the International Monetary Fund and the World Bank.[20] Buttressing these economic institutions was the Marshall Plan, whereby the United States provided nations around the world with financial assistance for postwar reconstruction. The Marshall Plan also sought to promote liberal democracies and fight against the spread of communism.[21] These efforts were widely successful, and the US dollar and American market system evolved to dominate global governance.[22]

Throughout the Cold War, however, American policy often prioritized free market development over democracy and civil rights. American leaders promoted democracy as a superior form of government, but for much of the Cold War the United States was a segregated nation that did not maintain equal rights for all of its citizens. Also during this period—while America championed democratic principles and chastised authoritarian regimes—to further liberal economic policies, the United States frequently supported autocratic leaders. It helped orchestrate coups of democratically elected officials who campaigned for anticolonial measures, nationalization of natural resources, and a social safety net for citizens. Many characterized America's illiberal policies and state-sanctioned discrimination as evidence of US hypocrisy.[23]

Although the United States and the Soviet Union were, in large part, decoupled economically, each maintained an expansive military and a broad sphere of influence. The two great powers competed relentlessly and engaged in a series of worldwide proxy battles, but they also entered into several nonproliferation agreements and put significant pressure on other nations to join them. These treaties were enacted to establish mutually beneficial arms control policies between the world's two superpowers, but they also sought to limit the rise of other great powers.[24]

For many policy experts, the demise of the Soviet Union and the end of the Cold War validated America's neoliberal endeavors.[25] During the 1990s and early 2000s—with no great power competition—the United States expanded its economic model of free trade and globalization, most notably by increasing its reliance on imports from China, Mexico, and other nations with low labor and manufacturing costs. In large part this shift was facilitated by trade agreements, such as the North American Free Trade Agreement (NAFTA), and the creation of the Geneva-based World Trade Organization (WTO), a powerful international entity that grew out of the General Agreement on Tariffs and Trade (GATT), which was established in 1948.[26]

Also, after the end of the Cold War, the United States led efforts to enlarge NATO's military reach by including in the alliance former Soviet republics and Eastern European nations that had been members of the Warsaw Pact.[27] Shortly after the dissolution of the Soviet Union, Russia sought to join NATO, but the United States rebuffed the proposal.[28] Thereafter, Russia saw NATO expansion as a significant threat to Russian security, a factor that influenced its decision to invade Ukraine in 2022.[29]

Meanwhile, although the United States heralded the importance of international institutions and a liberal world order, it also exploited its superpower status to reject treaties, use military force at will, and respect allies and international law as it wished. As political scientist Beate Jahn explains, America's success "fed American hubris, arrogance and exceptionalism, which in turn found expression in insensitive and high-handed policies that ultimately undermined the community of liberal states and contributed to a shift of power relations away from the United States and its European allies." The spread of "fundamentalist capitalism," Jahn adds, created "the very inequalities that now feed resistance against globalization, free trade, banks and common markets."[30] In the United States, two decades of war in Iraq and Afghanistan rekindled criticism of America's foreign policy apparatus and renewed calls for US isolationism.

By the early 2020s, the American-led model was in crisis. By privileging international institutions over domestic considerations, hyperglobalization increased corporate wealth and global trade and investment. But within the United States—and especially outside major metropolitan areas—millions of workers had lost manufacturing jobs, wages were declining when adjusted for inflation, income inequality grew enormously, and countless American communities were devastated.[31] These factors contributed to a growing public inclination to retreat from global institutions and focus America's political efforts on building back the homeland.[32] Analogous sentiments blossomed throughout the world, leading to nationalist movements in Britain and throughout the European Union.[33] America's nationalist movement expanded during the coronavirus pandemic, particularly as the nation experienced a shortage of essential medical supplies and staple household goods due to the country's supply chain vulnerabilities and overreliance on imports.[34]

America's geopolitical strategy was not based on ad hoc decisions or an expression of an illiberal foreign policy. Rather, it tracked realist concepts of a global order where the powerful set the rules and choose which rules to abide by. In practice, however, rather than expand America's global dominance, America's post-Cold War domestic and foreign policy choices facilitated an end to the nation's unipolar moment. US grand strategy also contributed to an increase in trade wars, the rise of Russia and China as great powers, and a shift in global alliances in economic and military matters.[35] This confluence of events has unfolded at a time when military science pursuits—such as bioengineered pathogens, military artificial intelligence technologies, and human enhancements—have created new and substantial threats to humanity that urgently call for sensible doctrines of restraint.

Bridging Military Science with Just War Theory, Liberal Internationalism, and Realism

Just war theory, liberal internationalism, and realism are epistemological lenses through which one can assess the moral, economic, political, and strategic reasons for foreign policy and armed interventions. Each theory has its critics, and each can be invoked in a self-righteous manner to support conduct that one nation finds necessary and another finds excessive. The doctrines have been interpreted to justify war in a variety of circumstances, including for purposes of colonization and exploitation, and with a myriad of means, including atomic and chemical weapons. Of the theories, none is a panacea for addressing the moral and legal limits of war or war preparations.

The impact of liberal internationalism is compromised when moral or legal principles are not universally adopted or enforced, or when global institutions are captured by their most powerful members. Liberal internationalism—particularly when it entails interventions that aim to transform a nation into a liberal democracy—can be interpreted as a form of social and political engineering that is prone to lead to domestic backlash in both the nation that is seeking to engineer and that which is the subject of the engineering: America's post-9/11 wars in Iraq and Afghanistan illustrate the perils.

Realism presumes that nations accurately perceive and respond to international situations, a presumption that belies human nature and experience, particularly in times of crisis or in the face of a perceived existential threat.[36] And, just war theory is intimately tied to Western notions of morality and law, as well as to the Anglo- and Euro-centric notion of sovereign nations as the sole legitimate sources of war. In large part, the modern-day doctrine of just war theory originates from medieval Christian theorists, most notably Saint Augustine and Saint Thomas Aquinas, whose concepts were utilized as moral justifications for the crusades and other wars, some of which exploited vulnerable groups that posed no imminent threat.[37] In short, it is not clear what standards should be implemented in assessing just and unjust wars, who should decide, and what penalties should or can be imposed on violators.

Notwithstanding their respective limitations, just war theory, liberal internationalism, and realism capture salient elements of international relations and national security policy. They are doctrines that have dominated the principles underlying geopolitical decision-making in matters of national security and foreign policy, and have guided the legal and ethical frameworks governing military science. Although often characterized as competing theoretical models, the doctrines maintain several shared assumptions and can be complementary.[38] One example is the concept of structural liberalism, which builds on the strengths of realism and liberal internationalism and seeks to go beyond the limitations of each.[39] The goal of this book is not to debate the intricacies of these theories but rather to draw on them, in combination with principles in bioethics, in thinking about how to elucidate conduct that is appropriate or inappropriate in the context of war preparations and military science.

This begins with a recognition that nations are deeply interconnected in military matters and national security concerns. The global system contains extensive consensual and reciprocal relationships. By setting norms regarding the development of military science, international institutions set rules that identify how states cooperate and compete with each other, thereby incentivizing certain state behavior. On a global scale, this includes adherence to legacy agreements such as the 1925 Geneva Protocol, the 1972 Biological Weapons Convention, and the 1993 Chemical Weapons Convention. It likewise includes faithful compliance with Additional Protocol I to the Geneva Conventions, which will be discussed in chapter 19, a 1977 amendment that aims to set limits to means and methods of warfare, including limitations on the integration of biomedical technologies into military missions. The United States is a signatory to the 1977 protocol, but as of mid-2024, has yet to ratify it.[40]

International institutions matter, but as experience has revealed they are likely to be captured by their most powerful members. In addition, nations might be quick to disregard global norms when they are faced with a real or perceived existential threat. These aspects of geopolitics need not translate to policies that unilaterally favor the most powerful. Nor must the exception be the paramount driving force behind the rule. International treaties that limit the type of military science pursuits that can be pursued or utilized serve an important function by establishing universal norms, though nations must remain mindful that norms can be breached. An important goal is to convince nations that cooperation and restraint can increase security more than competition and aggression.

In the realm of military science, one key challenge is maintaining appropriate international pressure on nations to restrain from breaching international doctrines and pursuing inappropriate or unethical military science endeavors. Another challenge is devising and implementing adequate punitive measures that can be enforced fairly when one nation breaks the rules. Equally as important are rules that a nation applies to its own war preparations. The spirit of international treaties can and should be integrated into a nation's domestic laws, policies, and ethical frameworks.

To be sure, the history of military science reveals that nations will pursue scientific endeavors that are calculated to provide a competitive advantage, and that nations will go to great efforts to conceal their efforts. History also has repeatedly shown that nations may cooperate and have marriages of convenience, but alliances shift, allies become enemies, and each nation can never be certain about the intentions of another nation. This relentless strategizing and hedging impacts decision-making and often leads nations down a path where scientific pursuits are pursued to maintain a balance of power, even if a nation does not intend to use the technology or would do so under very limited circumstances. Political scientist Robert Jervis termed this predicament the "offense-defense balance," which contends that, when a nation can build offensive measures more readily than it can defend against a threat, it will seek to bolster its offensive capabilities, and competition between nations will intensify.[41] Research and development of nuclear, biological, and chemical weapons are prominent examples. More recently, as discussed in chapter 16, this calculus has influenced decisions regarding the extent to which nations are pursuing and integrating human enhancements into military research, training, and missions.

Fundamentally, as the scores of examples detailed throughout this book reveal, each nation will set its own standards under which military science will evolve. Within this decision-making calculus, a doctrine of *jus in praeparatione bellum* can help guide assessments of what military science projects to pursue, and the proper parameters for research, field testing, and operational use. In the context of medical practices, preventive health measures, and war-related biomedical innovations, *jus in praeparatione bellum* is a theory of restraint that can help leaders assess moral, legal, and operational elements.

While the quest for military superiority often drives military science, a nation's moral values must be scrutinized in context to balance human rights and societal welfare with military goals and national security concerns. History provides ample examples where real or perceived existential threats have driven nations to engage in untoward military science endeavors. It is within these challenging moments where a nation's moral values are tested, and a doctrine of *jus in praeparatione bellum* can help guide decision-making.

Summary

The concept of *jus in praeparatione bellum* draws on foundational principles within the fields of bioethics and foreign relations, but the doctrine is more than a mere theoretical exercise. Rather, it provides a framework that can be applied to assess and guide the development of laws and ethical codes governing military science pursuits. Applying *jus in praeparatione bellum* to specific examples within the military biomedical complex is the focus of the next three chapters. In chapter 18, I recommend amendments to the regulatory regime regarding approval and administration of medical countermeasures, advocate for strengthening the medical autonomy of military personnel, and provide a rationale for eliminating informed consent waivers. In chapter 19 I offer a framework for standardizing and expanding military science ethics review, and in chapter 20, I propose a reformulation of the legal shields within the state secrets privilege, the political question doctrine, and sovereign immunity. Each of the recommendations is grounded on normative principles of *jus in praeparatione bellum.*

Notes

1. Michael Walzer, *Just and Unjust Wars* (1977), 21.
2. Jai Galliott, ed., *Force Short of War in Modern Conflict* (2019); Gary J. Bass, "Jus Post Bellum," *Philosophy and Public Affairs* 32, no. 4 (2004): 384–412.
3. Michael Walzer, "The Triumph of Just War Theory (and the Dangers of Success)," *Social Research* 69, no. 4 (Winter 2002): 925–44.
4. Gregory M. Reichberg, "Jus ad Bellum," in Larry May, ed., *War* (2008), 11–29.
5. Walzer, *Just and Unjust Wars*, xii–xvi
6. Daniel Deudney and G. John Ikenberry, "Getting Restraint Right: Liberal Internationalism and American Foreign Policy," *Survival: Global Politics and Strategy* 63, no. 6 (December 2021–January 2022): 63–100.
7. Kenneth N. Waltz, "Structural Realism After the Cold War," *International Security* 25, no. 1 (Summer 2000): 5–41. Robert Jervis, "Realism, Neoliberalism, and Cooperation: Understanding the Debate," *International Security* 24, no. 1 (Summer 1999): 42–63.
8. John J. Mearsheimer, "The False Promise of International Institutions," *International Security* 19, no. 3 (Winter 1994/95): 10.
9. Posen, *Restraint*, 21.
10. Daniel Deudney and G. John Ikenberry, "The Nature and Sources of Liberal International Order," *Review of International Studies* 25, no. 2 (April 1999): 179–96.
11. Waltz, "Structural Realism."
12. Posen, *Restraint*, 5–23.

13. G. John Ikenberry, "Liberal Internationalism 3.0: America and the Dilemmas of Liberal World Order," *Perspectives in Politics* 7, no. 1 (March 2009): 75.
14. E. H. Carr, *The 20 Years' Crisis, 1919-1939* (1948), 224–39; Toft and Kushi, *Dying by the Sword*, 110–13.
15. H. C. Engelbrecht and F. C. Hanighen, *Merchants of Death* (Garden City, NY: Garden City Publishing, 1934); "American Isolationism in the 1930s," U.S. Department of State, Office of the Historian (accessed March 5, 2024).
16. Waldo Heinrichs, *Threshold of War: Franklin D. Roosevelt and American Entry into World War II* (New York: Oxford University Press, 1988), 6–12.
17. Craig Nelson, *Pearl Harbor: From Infamy to Greatness* (New York: Scribner's Sons, 2016).
18. Andrew F. Krepinevich, *The Origins of Victory* (2023), 3–5.
19. Sara Hellmüller and Martin Wählisch, "Reflecting about the Past, Present, and Future of UN Mediation," *International Negotiation* 27 (2022): 1–9; Betty Goetz Lall, "A NATO-Warsaw Détente?," *Bulletin of the Atomic Scientists* 20, no. 9 (November 1964): 37–39.
20. Michael D. Bordo and Barry Eichengreen, editors, *A Retrospective on the Bretton Woods System: Lessons for International Monetary Reform* (Chicago: University of Chicago Press, 1993).
21. Diane B. Kunz, "The Marshall Plan Reconsidered: A Complex of Motives," *Foreign Affairs* 76, no. 3 (May-June 1997): 162–70.
22. Ikenberry, "Liberal Internationalism," 76.
23. Francis A. Boyle, "The Hypocrisy and Racism Behind the Formulation of U.S. Human Rights Foreign Policy," *Social Justice* 16, no. 1 (Spring 1989): 71–93; Mary L. Dudziak, *Cold War Civil Rights* (2000).
24. John J. Mearsheimer, "Bound to Fail: The Rise and Fall of the Liberal International Order," *International Security* 43, no. 4 (Spring 2019): 10–13.
25. Francis Fukuyama, "The End of History?," *The National Interest* 16 (Summer 1989): 3–18; William C. Wohlforth, "Realism and the End of the Cold War," *International Security* 19, no. 3 (Winter 1994/95): 91–129; John Lewis Gaddis, "International Relations Theory and the End of the Cold War," *International Security* 17, no. 3 (Winter 1992/93): 5–58.
26. Malcolm Fairbrother, "Economists, Capitalists, and the Making of Globalization: North American Free Trade in Comparative-Historical Perspective," *American Journal of Sociology* 119, no. 5 (March 2014): 1324–79; James Thuo Gathii, "The Neoliberal Turn in Regional Trade Agreements," *Washington Law Review* 86, no. 3 (2011): 421–74; Thomas Piketty, *Capital and Ideology* (2020), 23–39.
27. Mearsheimer, "Bound to Fail," 21–27.
28. Thomas L. Friedman, "Yeltsin Says Russia Seeks to Join NATO," *The New York Times*, December 21, 1991; "NATO Membership for Russia Doubted," *The New York Times*, September 10, 1994.
29. Gabrielle Tétrault-Farber and Tom Balmforth, "Russia Demands NATO Roll Back from East Europe and Stay out of Ukraine," *Reuters*, December 17, 2021; Becky Sullivan, "How NATO's Expansion Helped Drive Putin to Invade Ukraine," *NPR*, February 24, 2022.
30. Beate Jahn, "Liberal Internationalism: Historical Trajectory and Current Prospects," *International Affairs* 94, no. 1 (2018): 58.
31. Thomas Piketty, *Capital in the Twenty-First Century* (2014), 20–33.
32. Mearsheimer, "Bound to Fail," 8.
33. Miatta Fahnbulleh, The Neoliberal Collapse," *Foreign Affairs* 99, no. 1 (January/February 2020): 38–43; Rana Foroohar, "After Neoliberalism: All Economics Is Local," *Foreign Affairs* 101, no. 6 (November/December 2022): 134–45; Mitchell A. Orenstein and Bojan Bugarič, "Work, Family, Fatherland: The Political Economy of Populism in Central and Eastern Europe," *Journal of European Public Policy* 29, no. 2 (2002): 176–95.

34. The White House, *Issue Brief: Supply Chain Resilience* (November 30, 2023).
35. Mearsheimer, "Bound to Fail," 7–9.
36. Waltz, "Structural Realism."
37. Walzer, "The Triumph of Just War Theory"; Debra B. Bergoffen, "The Just War Tradition: Translating the Ethics of Human Dignity into Political Practices," *Hypatia* 23, no. 2 (April–June 2008): 72–94.
38. Valerie Morkevičius, "Power and Order: The Shared Logics of Realism and Just War Theory," *International Studies Quarterly* 59, no. 1 (March 2015): 11–22.
39. Deudney and Ikenberry, "The Nature and Sources of Liberal International Order," 180.
40. International Committee of the Red Cross, International Humanitarian Law Databases: Protocol Additional to the Geneva Convention of 12 August 1949, and Relating to the Protection of Victims of International Armed Conflicts (Protocol I), of 8 June 1977 (accessed April 11, 2024).
41. Robert Jervis, "Cooperation Under the Security Dilemma," *World Politics* 30, no. 2 (January 1978): 167–214.

18
Recalibrating Regulatory Review and Strengthening the Medical Autonomy of Service Members

Introduction

Applying the doctrine of *jus in praeparatione bellum*, this chapter proposes several amendments to the regulatory regime surrounding the approval and administration of medical countermeasures. The recommendations center on two expedited Food and Drug Administration (FDA) review protocols established after 9/11 and the 2001 anthrax letter attacks: Emergency Use Authorizations (EUAs) and the Animal Rule. Although these rules have allowed medical countermeasures to come to market faster and with less data on safety and efficacy than otherwise would be required under standard FDA guidelines, individuals who are administered countermeasures that are marketed via these expedited pathways may be exposed to undue risks, and the countermeasures may not work as anticipated.

Considering these concerns, I advocate bolstering the scientific data requirements for authorization, approval, and continued use of countermeasures that come to market utilizing the EUA pathway or Animal Rule. I also recommend recalibrating the legal shields available to pharmaceutical companies to allow lawsuits against companies in instances where a company fails to exercise reasonable care during countermeasure research and development. Furthermore, I recommend that medical products approved via the expedited mechanisms be characterized as "conditionally approved" medical products until the FDA's standard protocols for safety and efficacy are satisfied. For medical products that are conditionally approved, service members should maintain the medical autonomy to refuse administration, and this decision should be deemed private health information and should not impact a service member's career or duty assignments. In addition, Congress should revoke the informed consent waiver rule, which allows the president to waive informed consent when the Department of Defense (DoD) seeks to force military personnel to submit to medical treatments not FDA-approved for the use intended by the DoD.

Recalibrating the Countermeasure Approval Process

As detailed in chapters 13 and 14, in preparation for the Persian Gulf War and to combat rising chemical, biological, radiological, and nuclear (CBRN) threats during the 1990s and early 2000s, the US government created new rules regarding the evaluation and administration of medical products intended for use in military missions. These new rules included the EUA pathway, which allows for expedited review of

America's Military Biomedical Complex. Efthimios Parasidis, Oxford University Press.
 DOI: 10.1093/9780199351473.003.0019

medical products by applying lower standards for safety and efficacy, and the Animal Rule, which permits approval of medical products without clinical trials with human participants. Although both rules afford the government and military flexibility to address national security and public health concerns, the rules also increase the risk that individuals will be administered unsafe or ineffective medical products. The rules should be amended to require ongoing evaluation and an increasingly stringent evaluation of safety and effectiveness, and a new rule should be created that explicitly grants individuals the medical autonomy to determine whether to consent to the administration of medical products approved via these specialized mechanisms.

Compared to traditional standards for FDA approval, the EUA pathway has a lower bar to market. Standard FDA drug approval guidelines require a demonstration of safety through "adequate tests by all methods reasonably applicable" and "substantial evidence" of effectiveness,[1] but under the EUA rule a product may come to market if "it is reasonable to believe" that "the product may be effective" and that "the known and potential benefits of the product . . . outweigh known and potential risks."[2] This is a substantial distinction regarding data required to approve a product. Although the lower bar for EUAs provides the FDA flexibility during an emergency, it also permits the marketing of medical products where safety and efficacy data are less robust than with full approval.

One way the FDA attempts to detail this distinction to the public is to state that EUA products are authorized for emergency use rather than FDA-approved for use. However, as experienced throughout the coronavirus pandemic—during which the FDA authorized more than four hundred medical products via the EUA pathway—this minor linguistic distinction was largely lost on the public and mainstream media, and both often conflated authorization with approval.[3]

Medical products authorized for emergency use via the EUA pathway have had varying degrees of success. While some were essential elements of the health and public health responses to the coronavirus pandemic, others were pulled from the market months after being authorized because they ultimately proved unsafe or ineffective.[4] Consider the ramifications if an EUA-authorized chemical weapon countermeasure fails to provide its purported benefit or has significant side effects that were not revealed during expedited review. This could jeopardize a military mission and expose service members to serious short- and long-term adverse health consequences.

The issuance of EUAs also negatively impacts clinical trials, slowing or preventing research needed to understand the safety and effectiveness of EUA products and potential competitor products. This has a compounding effect. Drugs that produce promising results from *in vitro* studies often fail during human clinical trials with a control group. Furthermore, once a medical product is made available to the public, it is incredibly difficult to structure randomized control trials and recruit research volunteers. The absence of data from randomized control trials—the gold standard for obtaining data on safety and efficacy—hinders the ability to conclude whether a treatment is safe and beneficial for patients.[5]

The coronavirus pandemic provides a powerful reminder of these hazards. During the pandemic, the FDA authorized several treatments for emergency use without requiring companies to produce safety or efficacy data from a controlled clinical trial. As an article in the *Journal of the American Medical Association* explained:

> The administration of any unproven drug as a "last resort" wrongly assumes that benefit will be more likely than harm. However, when a drug with unknown clinical effects is given to patients who have severe illness from a new disease (like COVID-19), there is no way to know whether the patients had benefited or were harmed if they were not compared to a concurrent control group. A common interpretation of off-label use and compassionate use of drugs is that if the patient died, they died from the disease, but if the patient survived, they survived because of the given drug. This is not true.[6]

Analogous concerns are raised by countermeasures authorized for use via the EUA pathway where no randomized control trials are conducted: conclusions regarding efficacy cannot be made, even if a patient benefits when administered an EUA-authorized countermeasure. The risks of EUA countermeasures become more evident when one considers the failure rate for drugs undergoing FDA review: only 12 percent of drugs that begin clinical trials are ultimately approved.[7] For those that enter Phase 3 trials (the last stage before market approval), the success rate is slightly less than 50 percent.[8] A lower bar to market increases uncertainty and the chance for adverse health outcomes.

In light of these concerns—and particularly for countermeasures that will be utilized by the military, where an adverse event not only affects the patient but also can negatively impact a military mission—the FDA should require more exacting clinical trials prior to issuing an EUA, with clinical endpoints that more precisely capture salient health goals. Although the EUA guidelines permit the FDA to apply a low standard when considering an EUA application, they do not require the FDA to do so. Rather, the FDA maintains the discretion to demand whatever safety and efficacy data that agency believes is necessary to meet the authorization threshold.

Experience during the coronavirus pandemic supports raising the evidence bar higher than the minimum requirements set forth in the EUA guidelines. Arguably, the most beneficial EUAs issued during the coronavirus pandemic were for vaccines, for which the FDA applied a more stringent standard than that minimally required by the law.[9] Of the least beneficial were EUAs for COVID-19 treatments—such as chloroquine phosphate, hydroxychloroquine sulfate, and COVID-19 convalescent plasma—each of which initially was judged by the FDA to satisfy the minimum EUA threshold despite an absence of data on safety and efficacy. A massive surge in the administration of these treatments followed the issuance of the EUAs, but each treatment was later found ineffective. In the interim, tens of thousands of patients were administered the treatments, and in some cases, the treatments caused significant adverse side effects and may have hastened death.[10]

There likewise have been concerns that one EUA treatment, molnupiravir—designed to kill the virus by inducing mutations in the viral genome—actually had a perverse effect and contributed to creating new virus variants that were more pathogenic and transmissible.[11] Also problematic were several EUA-authorized tests to diagnose if a person was infected with COVID-19, some of which had false-negative rates of more than 20 percent. Despite test inaccuracies, many of which were linked to several outbreaks (including an event at the White House), the FDA nevertheless allowed the tests to remain on the market.[12]

Although Congress granted the FDA a flexible legal standard that affords the agency substantial discretion in authorizing EUAs, the FDA should apply that discretion judiciously to ensure that all marketed medical products are safe and effective, even if they are authorized for use via expedited review. In addition to bolstering the required premarket data regarding safety and efficacy, the FDA also should condition each EUA on active post-market surveillance, comprehensive post-market observational studies, and mandatory data reevaluation periods in short intervals (such as weekly or biweekly). The numerous post-market concerns raised by several EUA products during the coronavirus pandemic underscore the importance of robust post-authorization assessments.

Under its existing authority, the FDA can incorporate the aforementioned pre- and post-market measures into its review and approval of EUA countermeasures, but Congress should cement these requirements into law by amending the EUA rules within the Federal Food, Drug, and Cosmetic Act (FDCA). Alternatively, the military can, on its own accord, issue guidelines for military physicians, requiring that all EUA countermeasures administered to service members comply with heightened data and post-market surveillance requirements, regardless of whether the FDA has determined that the products have met the minimum EUA thresholds.

While the proposed measures may lead to a longer wait before products can be marketed and higher research costs after the issuance of an EUA, the products that come to market and stay on the market are more likely to be safe and beneficial. Keep in mind that these additional requirements still would represent a lower bar than the standard FDA approval process; they would not eliminate expedited review under the EUA pathway but rather would recalibrate the expedited process to increase the likelihood that EUA products are safe and effective.

Analogous pre- and post-market protocols also should be incorporated into the Animal Rule. The FDA can invoke the Animal Rule to approve medical products without consideration of safety and efficacy data from clinical trials with humans. For years, the FDA took the position that animal efficacy data could never be an adequate gauge of human efficacy.[13] Notwithstanding that long-standing position, the Animal Rule was established shortly after 9/11 and the anthrax letter attacks to facilitate the approval of CBRN countermeasures.[14] Over the past two decades, the FDA has used the rule to approve countermeasures for anthrax, nerve agents, radiation exposure, and other threats.[15] Because the products have been approved without human trials—let alone randomized control trials with human participants—there is substantial uncertainty as to whether the products are safe and effective when utilized by humans. More robust pre- and post-market requirements can help ensure that medical products approved via the Animal Rule are reliable countermeasures.

Another lever that can be reformulated to help ensure that marketed medical products are reasonably safe and effective is PREP Act immunity, discussed in chapter 14, which precludes "all claims for loss caused by, arising out of, relating to, or resulting from the administration to or the use by an individual of a covered countermeasure" except in cases of "willful misconduct."[16] The PREP Act was enacted as a "dead of night" amendment to the 2006 DoD Appropriations Act following substantial backroom lobbying by pharmaceutical companies, seventeen months after Congress created the EUA mechanism and had initially decided not to afford companies with such broad immunities for EUA countermeasures.[17] Several groups—including first responders, veterans groups, and public health officials—protested the immunity provisions, but to no avail.[18]

Under the PREP Act, companies face no liability even if they are grossly negligent or engage in careless disregard for the safety and effectiveness of their product.[19] Instead of lawsuits, the law calls for creating a compensation fund. However, it does not mandate that Congress appropriate money to fund the program, leaving substantial discretion to the Department of Health and Human Services (HHS) to identify what injuries are compensable and at what amounts.[20] Several studies have highlighted the inadequacies of the compensation program, and during the coronavirus pandemic, the program was lambasted as a grossly inadequate alternative to litigation.[21]

Although Congress need not eliminate PREP Act immunity, Congress should amend the immunity rule and specify that companies can only benefit from immunity if they exercise reasonable care in structuring pre-market clinical trials, conducting the trials, interpreting the clinical data, and producing the countermeasure. If a company engages in unreasonable conduct and a patient who is administered an EUA countermeasure suffers harm because of that unreasonable conduct, the patient should have the right to sue the company. By recalibrating the PREP Act legal shield, companies will be incentivized to act reasonably, and if they do not, they will bear the legal consequences of their unreasonable conduct. This will help reset the balance of risks, which under the current system are borne disproportionally by those individuals who are administered countermeasures marketed under expedited pathways.

Enhancing the Medical Autonomy of Service Members for "Conditionally Approved" Medical Products

In addition to adopting new rules that require a more stringent assessment of countermeasure safety and effectiveness and an ongoing data reevaluation, a new policy should be adopted that requires all medical products approved via the EUA pathway or Animal Rule to be labeled as medical products with "conditional approval." The conditional approval category is a means of documenting and disclosing that the medical products were not reviewed pursuant to the FDA's standard approval protocols. It also signals to individuals that there is less data on whether the products are safe and effective when compared to products that earn full FDA approval. Congress can integrate the conditional approval categorization into the FDCA, or the military can adopt the category on its own accord via a DoD directive without any action from Congress or the FDA.

For a medical product to move from conditional approval to full approval, additional studies should be required to evaluate real-world safety and effectiveness. Moreover, conditionally approved medical products should not be subject to mandatory use during the conditional approval period. Nor should a service member's decision to consent to medical products that are conditionally approved play a role in duty assignments or career advancement. Rather, a service member's decision to consent to treatment with a product that has been conditionally approved should be made through a confidential discussion between the service member and a military physician, and should not be subject to the military command exception to the federal health privacy rule, which allows commanders access to private health information of subordinates.

The military command exception allows commanders to bypass standard federal health privacy protections to permit access to the health information of a subordinate in instances where the health information may help determine the subordinate's "fitness for a duty," "fitness to perform any particular mission, assignment, order, or duty," or ability "to carry out any other activity necessary to the proper execution of the mission of the Armed Forces."[22] This includes health information shared during a private doctor-patient engagement, and encompasses information regarding the service member's physical or mental health. Concerning the latter, disclosure of mental health information is required under a long list of circumstances, including, but not limited to, evidence where a service member may commit harm to self or others, or if their mental health condition may potentially harm a military mission.[23] These broad categories capture nearly any aspect of a service member's military career, and thus may apply to almost any aspect of their private healthcare decisions.

The military command exception to standard federal health privacy laws was created to balance individual rights with military goals. As a Marine Corps memorandum details, "Too often we learn of a Marine or Sailor struggling with a medical issue or with the unanticipated effects of prescribed medications only after a tragic event. This order is intended to strike an appropriate balance between the individual's right to privacy and the obligation of leaders to train, equip and prepare forces for the execution of their assigned missions."[24] It is representative of a broader ethical concept in military medicine, referred to as "dual loyalty" or "mixed agency," whereby military physicians may be presented with conflicting obligations since they have legal and ethical duties to serve the best interests of their patients and the military mission.[25]

A service member's decision whether to accept the administration of a conditionally approved countermeasure should not be subject to the military command exception. As detailed, medical products authorized for use via the EUA pathway or approved via the Animal Rule have not been subjected to randomized control trials as would be required under standard FDA approval protocols. Since the FDA cannot make a conclusive determination that a conditionally approved medical product is safe or effective for its intended use, there exists reasonable doubt as to whether the administration of a conditionally approved medical product will, in fact, bring health benefits to warfighters in a military mission. Given the data gap, administration—or refusal—of a conditionally approved medical product arguably does not help determine an individual's fitness for duty or fitness for a military assignment.

Experience during the anthrax vaccine immunization program (AVIP) helps provide context. As outlined in chapter 14, a federal court halted the program because the FDA did not adhere to its vaccine approval guidelines when it expanded the on-label indication to include prophylaxis against weaponized aerosol anthrax.

Congress created the EUA mechanism to assist the DoD in continuing with the AVIP, and the anthrax vaccine was the first medical product to receive an EUA. When the anthrax vaccine was authorized for emergency use via the EUA pathway—and not FDA-approved as prophylaxis for weaponized aerosol anthrax—DoD policy was to grant service members the option to submit or not to the administration of the vaccine, and the service member's choice did not impact their duty assignment or career. Moreover, the legislative history of EUAs reveals that Congress's intent in creating the expedited pathway was to permit medical products to come to market and allow individuals (including military personnel) the option whether to submit to administration.[26]

There likewise is a recent precedent for building exceptions into the military command exception to the federal health privacy rule. During the post-9/11 wars in Afghanistan and Iraq, where traumatic brain injuries (TBIs) and posttraumatic stress were signature injuries, countless service members opted not to seek care for their mental health concerns due to stigma and fear that doing so would negatively impact their duty assignments and military career. A mental health epidemic followed, including thousands of service members who died by suicide.[27] In response, the DoD instituted policies to encourage service members to seek mental health care, including guidelines that limited the scope of private mental health information that could be shared with commanders via the military command exception to instances where there is an "exigent circumstance."[28]

An analogous policy should be implemented in the case of conditionally approved countermeasures. The proposed autonomy and confidentiality components regarding conditionally approved medical products seek to ensure that a service member's decision does not negatively impact duty assignments or their military career. The components also seek to account for the lower standard of review and data gap on safety and effectiveness that comes with products that utilize the EUA pathway or Animal Rule.

Insofar as the DoD believes that a conditionally approved medical product may provide benefits to service members engaged in a particular military mission, the DoD can provide service members with evidence to support its recommendation and allow them the autonomy to decide whether to accept the recommendation. At the same time, military physicians should track the health and operational impact on those who accept or reject a conditionally approved product to unearth data on whether the product provides benefits or comes with risks. These studies can be a component of the research needed to shift a product from conditional approval to full approval.

Although service in the armed forces requires that individuals cede a degree of medical autonomy to further unit health priorities and military goals, this relinquishment must have reasonable limits. These limits should include affording service members the medical autonomy to decide whether to submit to medical products that have undergone expedited review under the EUA pathway or Animal Rule. As discussed in the following section, it likewise should include expanding this medical autonomy to any medical product that has not been FDA-approved for the use intended by the military.

Eliminating the Informed Consent Waiver Rule for Medical Products That Are Not FDA-Approved for Their Intended Use by the Military

The informed consent waiver rule, 10 U.S.C. § 1107(f), grants the US president the authority to mandate that service members comply with the forced administration of a medical product under military exigencies. This rule encompasses medical products that have not been FDA-approved for any use, off-label uses of FDA-approved products, and products that have been authorized for emergency use under the EUA pathway. As detailed in chapter 13, the law was created in 1997 following the controversy surrounding the FDA's issuance of informed consent waivers for PB and the BT vaccine during the Persian Gulf War. Since the law's inception, no president has publicly indicated that the law has been utilized to issue an informed consent waiver for military personnel. It is possible that a president has invoked the rule but has kept invocation of the waiver classified, as is permitted under the law.[29] But, given the negative political implications that may flow from such a decision, the chance that a classified invocation has occurred is highly unlikely.

There are only two publicly known instances where invocation of the informed consent waiver rule was under consideration. During the early 2000s, President George W. Bush contemplated the issuance of an informed consent waiver for the anthrax vaccine after the FDA issued an EUA for the vaccine, but he refused to invoke the rule in light of the ongoing legal and political controversy surrounding the DoD's anthrax vaccine mandate.[30] In 2021, President Joe Biden considered whether to issue an informed consent waiver for the COVID-19 vaccine while it was authorized for emergency use under an EUA.[31] He refused to do so, but once a COVID-19 vaccine was FDA-approved, an informed consent waiver was no longer required, and the vaccine was mandated via a military policy pursuant to standard military medical protocols governing medical preparedness.

Elimination of the informed consent waiver rule will not impact the military's ability to issue a mandate for medical products that are FDA-approved. It will only affect the president's ability to authorize a mandate for a medical product that has not earned FDA approval for the military's intended use. FDA approval is not merely a rubber stamp; it entails a comprehensive evaluation to ensure that a medical product is safe and effective for a specific health concern. Eliminating the informed consent waiver rule will underscore the value of FDA review and approval, and will nudge pharmaceutical companies to conduct appropriate pre-market studies to support on-label indications for their products. Eliminating the informed consent waiver rule will also signal to service members and the public that it is not morally appropriate to have special rules that permit the forced administration of medical products on military personnel when such mandates would be illegal for civilians because the products have not earned FDA approval for the intended use.

Over the past century, there have been several controversies surrounding mandates for medical products that have not earned FDA approval for their intended use, and in many instances, the products caused substantial adverse events, jeopardized the health of the fighting force, or negatively impacted military recruitment or retention. As outlined throughout this book, these controversies include the World War II (WWII) yellow fever vaccine mandate, the mandates for PB and the BT vaccine during the Persian Gulf War, and the anthrax vaccine mandate that began in the late 1990s and continued through post-9/11 wars.

Eliminating the informed consent waiver rule is a sensible measure. The law has never been invoked by a president, and revocation of the law will not impact the DoD's authority to mandate medical products that are FDA-approved for the military's intended use. Elimination of the informed consent waiver rule will underscore the value of FDA approval, and will incentivize pharmaceutical companies to conduct additional safety and efficacy studies to support additional uses of their medical products. As with the proposed category of conditional approval that was introduced in the previous section, insofar as the DoD believes that an EUA medical product or off-label use of an FDA-approved medical product may provide benefits to warfighters engaged in a particular military mission, the DoD can provide service members with evidence to support its recommendation and allow service members the autonomy to decide whether to submit to the treatment.

Summary

Current guidelines under the expedited review protocols for EUAs and the Animal Rule allow medical products to come to market in instances where the safety and effectiveness of the products are uncertain. This not only creates undue risks to service members because of the potential for unknown short- or long-term adverse health effects, but it also creates risks to military missions since service members may not be protected from a CBRN threat in the way commanders think they are. Prudence mandates more exacting regulatory review prior to the administration of CBRN countermeasures, and more robust post-market surveillance and analysis of real-world uses of the countermeasures. It also entails holding companies legally accountable when they engage in unreasonable conduct and a patient suffers harm because of the company's unreasonable conduct.

Due to uncertainties regarding the safety and efficacy of countermeasures marketed under the EUA pathway or Animal Rule—and since the products have not undergone standard FDA approval protocols—the countermeasures should be labeled as conditionally approved medical products. In addition, the choice to submit to the administration of a conditionally approved countermeasure should be left to each service member without fear that their decision will negatively impact mission assignments or their career. Congress should also eliminate the informed consent waiver rule, a law that has never been invoked and is based on unconvincing moral and ethical principles regarding the compelled use of medical products that have not been FDA-approved for their intended use by the military. These proposed measures aim to recalibrate the balance between individual rights and military goals, and promote the principles of *jus in praeparatione bellum.*

Notes

1. 21 U.S.C. § 355(d) (2024).
2. Food and Drug Administration, *Emergency Use Authorization of Medical Products and Related Authorities: Guidance for Industry and Other Stakeholders* (January 2017), 7–8.
3. Zachary Brennan, "Why FDA's Issuance of EUAs are Not 'Approvals' and Why that Matters," *Regulatory News*, April 3, 2020.
4. Efthimios Parasidis, Micah L. Berman, and Patricia J. Zettler, "Assessing COVID-19 Emergency Use Authorizations," *Food and Drug Law Journal* 76, no. 3 (2021): 441–501.
5. Andre C. Kalil, "Treating COVID-19—Off-Label Drug Use, Compassionate Use, and Randomized Clinical Trials During Pandemics," *Journal of the American Medical Association* 323, no. 19 (March 24, 2020): 1897–99.
6. Kalil, "Treating COVID-19," 1897.
7. Congressional Budget Office, *Research and Development in the Pharmaceutical Industry* (April 2021), 2.
8. Derek Lowe, "The Latest on Drug Failure and Approval Rates," *Science*, May 9, 2019.
9. Joshua M. Sharfstein, Jesse L. Goodman, and Luciana Borio, "The US Regulatory System and COVID-19 Vaccines: The Importance of a Strong and Capable FDA," *Journal of the American Medical Association* 325, no. 12 (February 15, 2021): 1153–54.
10. Parasidis et al., "Assessing COVID-19 Emergency Use Authorizations," 462–73.
11. Robert F. Service, "Could a Popular Covid-19 Antiviral Supercharge the Pandemic?" *Science* 379, no. 6632 (February 10, 2023): 526.
12. Rachana Pradhan, "As Problems Grow with Abbott's Fast Covid Test, FDA Standards are Under Fire," *Kaiser Health News*, June 22, 2020; Katherine J. Wu, "The White House Bet on Abbott's Rapid Tests: It Didn't Work Out," *The New York Times*, October 6, 2020.
13. Paul Aebersold, "FDA Experience with Medical Countermeasures Under the Animal Rule," *Advances in Preventive Medicine* (2012): 2.
14. Food and Drug Administration, New Drug and Biological Drug Products; Evidence Needed to Demonstrate Effectiveness of New Drugs When Human Efficacy Studies Are Not Ethical or Feasible, 67 Fed. Reg. 37,988 (May 31, 2002).
15. CDER, Drug and Biologic Animal Rule Approvals (accessed December 5, 2023).
16. 42 U.S.C. § 247d-6d (2005).
17. Tom Harkin, "Harkin Calls on Frist and Hastert to Repeal 'Dead of Night' Vaccine Liability Provision and Enact Real Protections," *Press Release* (February 15, 2006).
18. 152 Cong. Rec. S1360-61 (February 15, 2006) (statement of Senator Chris Dodd).
19. 152 Cong. Rec. S1360-61 (February 15, 2006) (statement of Senator Chris Dodd).
20. 42 U.S.C. § 247d-6e (2005).
21. Katharine Van Tassel, Carmel Shachar, and Sharona Hoffman, "Covid-19 Vaccine Injuries—Preventing Inequities in Compensation," *New England Journal of Medicine* 384, no. 10 (March 11, 2021): e34(1–3); Tom Hals, "COVID-19 Era Highlights U.S. 'Black Hole' Compensation Fund for Pandemic Vaccine Injuries," *Reuters*, August 21, 2020.
22. Department of Defense, *DoD Health Information Privacy Program*, 68 Fed. Reg. 17, 357–58 (April 9, 2003). See also 45 C.F.R. § 164.512(k)(1)(i) (2024).
23. Defense Health Agency, "The Military Command Exception and Disclosing PHI of Armed Forces Personnel" (accessed January 30, 2024); DoD Instruction 6490.08, *Command Notification Requirements to Dispel Stigmas in Providing Mental Health Care to Service Members* (September 6, 2023); DoD Manual 6025.18, *Implementation of the Health Insurance Portability and Accountability Act (HIPAA) Privacy Rule in DoD Health Care Programs* (March 13, 2019).

24. United States Marine Corps, Command Element, "Release of Protected Health Information to Commanding Officers and their Designated Representatives," (June 21, 2011): 2.
25. Edmund G. Howe, "Mixed Agency in Military Medicine: Ethical Roles in Conflict," in Thomas E. Beam and Linette R. Sparacino, eds., *Military Medical Ethics*, vol. 1 (2003), 331–60
26. Project BioShield Act of 2004, Cong. Rec. H5721–H5741 (July 14, 2004); Efthimios Parasidis and Aaron S. Kesselheim, "Assessing the Legality of Mandates for Vaccines Authorized via an Emergency Use Authorization," *Health Affairs Forefront*, February 16, 2021.
27. Defense Health Board, *Achievable Mission: Report of the Department of Defense Task Force on Mental Health* (June 2007); Marie-Louise Sharp et al., "Stigma as a Barrier to Seeking Health Care Among Military Personnel with Mental Health Problems," *Epidemiological Reviews* 37 (2015): 144–62.
28. DoD Instruction 6490.08, *Command Notification Requirements to Dispel Stigmas in Providing Mental Health Care to Service Members* (September 6, 2023); Deborah A. Gibbs and Kristine L. Rae Olmsted, "Preliminary Examination of the Confidential Alcohol Treatment and Education Program," *Military Psychology* 23, no. 1 (2011): 97–111.
29. 21 C.F.R. § 50.23(d)(1)(xvii) (2024).
30. Robert Pear, "Judge Halts Military's Required Anthrax Shots," *The New York Times*, December 23, 2003; Dee Ann Divis, "BioWar: Mandatory Anthrax Shots Loom Again," *UPI*, December 23, 2004.
31. Ellie Kaufman, "Lawmakers Ask Biden to Issue Waiver to Make Covid-19 Vaccination Mandatory for Members of Military," *CNN*, March 24, 2021.

19

Standardizing and Expanding Military Science Ethics Reviews

Introduction

The military emphasizes moral character and ethical analysis as indispensable elements of military culture.[1] To further these goals, ethics education and ethics consultations are integrated throughout the military, and military policies outline moral principles to guide decision-making. For example, the Department of Defense (DoD) Joint Ethics Regulation contains numerous guidelines regarding professional obligations, such as adherence to rules governing financial disclosures, outside employment, and conflicts of interest. These policies are compliance-based protocols concerning the legal obligations of military professionals.[2]

The military also conducts values-based ethics reviews that provide ethical assessments of military projects and military research. These include, for example, institutional review board (IRB) review of military research involving human participants, which under federal law must include an evaluation of a study's potential benefits and risks.[3] Values-based ethics reviews are also conducted during deliberations concerning the acquisition, testing, evaluation, and implementation of military technologies. Unlike compliance-based ethics reviews—which are bound by a core set of rules outlined in the Joint Ethics Regulation and centralized within the DoD Standards of Conduct Office—the military does not maintain a centralized department for values-based ethics reviews and has not fully implemented recommendations from a series of reports over the past two decades that have called for more robust values-based ethics programs for military science.[4]

The DoD should fill this gap by creating a department that spearheads ethics reviews for military science. Similar to the Standards of Conduct Office for compliance-based ethics reviews, a central DoD department for values-based military science ethics reviews would create and implement policies that apply throughout the military. Such a department would represent an expansion of existing programs, such as the DoD Medical Ethics Program and the DoD Medical Ethics Integrated Product Team. Established in 2017, these ventures increased opportunities and requirements for ethics training and expanded access to ethics consultations for healthcare professionals.[5] While they remain important aspects of patient care, the scope of the ethics work should be expanded beyond clinical care to encompass centralized ethics reviews for military science. Indeed, a 2019 DoD biotechnology council underscored the lack of adequate ethical frameworks within the military, and the importance of creating "dynamic legal, security, and ethical frameworks" and "forward-leaning policies (internal and external) that protect individual privacy, sustain security, and manage personal and organizational risk, while maximizing defined benefits."[6]

America's Military Biomedical Complex. Efthimios Parasidis, Oxford University Press.
 DOI: 10.1093/9780199351473.003.0020

Figure 19.1 Organizational chart for Office of Military Science Ethics Review.

The moral principles that guide values-based ethics reviews should mirror those that the DoD has identified as principles to guide compliance-based ethics reviews: honesty, integrity, loyalty, accountability, fairness, caring, respect, promise-keeping, responsible citizenship, and pursuit of excellence.[7] From an organizational perspective, as detailed in Figure 19.1, a military science ethics review department should have three core components: research ethics review, field test ethics review, and operational use ethics review. These components need not be divided among the military services, but each service should have representatives in each component to ensure that the various needs of each service are fully considered. Within this framework, specialized consideration should be afforded to operationalizing military human enhancements. Specifically, enhancement ethics reviews should include a holistic framework for evaluating biomedical interventions that analyzes ethical, legal, and societal concerns. This chapter outlines the contours of these proposals.

A Life-cycle Framework for Military Science Ethics Reviews

Values-based ethics reviews should be conducted throughout the life cycle of military science projects. A life-cycle approach to ethics reviews includes considering whether a military science project should be pursued, the parameters of research and field testing, protocols for operational integration, and post-operational assessment of risks and benefits. This work should be centralized in a military science ethics review department. In creating a single department for military science ethics reviews, the DoD can centralize resources and standardize procedural mechanisms and substantive considerations. Centralization will promote the efficient use of DoD resources by eliminating duplicative roles across services and bringing together experts across multiple departments. This framework will facilitate the ethical development and use of new technologies, and will help promote the progress of military science.

The first stage of ethics review—research ethics review—should encompass an ethical assessment of military science projects regardless of whether ethics review is legally required pursuant to the Common Rule or other applicable law. The current ethics gap includes, for example, exemptions from IRB review for certain research using biospecimens and clinical data science research. As the DoD Office of General Counsel has explained, health surveillance and medical quality assurance also fall outside the Common Rule. Thus, federal research protections do not apply to instances where a commander orders a subordinate to participate in these activities. Also exempt are operational test and evaluation projects, where the test aims to assess weapons, equipment, or munitions.[8]

The ethics gap extends beyond the several types of research exempt from federal IRB review requirements and includes limitations on the scope of IRB review. When a research study must undergo IRB review, federal law prohibits the IRB from considering certain types of risks, including societal and occupational risks. Specifically, the Common Rule states that IRBs "should not consider possible long-range effects of applying knowledge gained in the research."[9] IRBs are not legally required to consider risks to communities in which the research subjects reside, or occupational risks to researchers engaged in a research project.[10] These exclusions from the risk-benefit analysis are particularly problematic with some types of military research—such as bioengineered pathogens and human enhancements—where there may be substantial long-term health, occupational, or societal harms that stem from the work.

One influential federal advisory committee recommended that IRBs should sometimes consider the potential impact of research on communities and on individuals who might be exposed to research hazards but are not explicitly identified as research subjects. However, the committee acknowledged that this analysis is "extra-regulatory and will lead to IRBs assuming a role that is not specifically written into the U.S. regulatory lexicon."[11] The committee did not recommend an amendment to the Common Rule to allow IRB consideration of societal, community, or occupational risks, but rather suggested that IRBs should exercise discretion in deciding when to consider them.[12] Taken together, there is nothing in federal law that requires IRBs to consider societal, community, or occupational risks, and studies have found that many IRBs do not go beyond the mandatory federal requirements to assess these additional considerations.[13] Moreover, several scholars have advocated for IRBs to refrain from considering community or societal harms, since such consideration might be speculative and policy-focused.[14]

Rather than allowing categorical and topical exemptions from research ethics reviews, all military science research projects should require an ethics review that examines the anticipated benefits and risks of a project, participant recruitment methods, informed consent protocols, project implementation, and requirements for evaluation of research risks as the project is ongoing and after it is completed.

Mandatory ethics review for all military science research is a sensible approach to addressing vulnerabilities related to military studies. The DoD acknowledges that service members are a vulnerable population in the context of research, and sets forth additional protections beyond those legally required under the federal guidelines outlined in the Common Rule.[15] These additional protections include, for example, a requirement that commanding officers not be present during the solicitation of research participants and the appointment of medical monitors for research involving greater than minimal risk to research participants.[16] For service members enrolled in a study that involves "greater than minimal risk" and when recruitment occurs in a group setting, DoD guidelines require that the IRB appoint an ombudsperson to monitor "that the recruitment and informed consent explain that participation is voluntary and that the information provided about the research is consistent with the IRB-approved script and materials."[17]

DoD research protections are robust, but they do not apply to field testing. Under DoD Instruction 3216.02, activities related to an operational test and evaluation (OT&E) are excluded from the definition of "research involving human subjects."[18] An OT&E activity is defined as a "field test, under realistic combat conditions, of any item of (or key component of) weapons, equipment, or munitions for the purpose of determining the effectiveness and suitability of the weapons, equipment, or munitions for use in combat by typical military users."[19] As the 1995 Human Radiation Experiments committee noted, field testing "may often be hazardous, may involve the use of volunteers, but may not be considered human research."[20] Rather, field testing aims to validate and verify the benefits and risks of a new military technology.

It is unclear whether the DoD definition of OT&E activities includes medical products, though nothing in the policy affirmatively states that medical products are excluded. Furthermore, the terms "weapons" and "equipment" arguably would include combat-related medical devices, such as brain-to-computer interfaces or other neurotechnologies. The DoD broadly defines "military equipment" as "all weapons systems, weapon platforms, vehicles, and munitions of the Department of Defense, and the components of such items." "Weapon systems" are defined as "items that can be used directly by the Armed Forces to carry out combat missions."[21] These definitions are broad enough to capture brain-to-computer interfaces and other neurotechnologies, since a key goal of the work is to create equipment that can assist in military operations.

Not all research projects are field tested, since research sometimes concludes that a particular medical product or technology is not likely to be safe or effective when used in the field. When the military believes that a medical product or technology shows promise, it should conduct a field test ethics review prior to field testing. This adds a second layer of protection for service members, though it encompasses an ethics review that differs from an ethics review for research.

With research ethics review, the risk-benefit calculus examines the risks to the human participants in the study and the benefits that may flow to the military should the research provide promising results. Research participants themselves may not gain benefits from participation in a research study; rather, research participants are dedicating their bodies to help produce generalizable knowledge that furthers military science. With field test ethics review, the general benefits to the military still are the focus, but the benefit side of the ethics equation looks more closely at whether the individual participants who are field testing a technology receive benefits that help them to fulfill a military mission. Field test ethics reviews should assess the distribution of benefits and risks within the field test protocol, to examine whether specific subpopulations of service members are subject to undue risks that may impact their careers and lives. Field test ethics reviews should also consider what metrics will be utilized to assess the risks and benefits of the underlying technology, what re-evaluation markers will be applied, and under what conditions the field test will be halted. Also relevant is consideration of short- and long-term health monitoring for field test participants.

A separate ethics review should be conducted for technologies that move from field testing to operational use. Several substantive principles should guide operational use ethics reviews. These include ensuring that a new technology is utilized to achieve a legitimate military purpose and does not improperly impede the autonomy or dignity of warfighters. Operational use ethics review should also include an evaluation of what metrics will be utilized to measure the impact of the new technology, and what assessments will be conducted to measure and address short- or long-term harms.

There is overlap among the principles that should be considered during research, field test, and operational use ethics reviews. However, the scope of operational use ethics reviews differs from research and field test ethics reviews in at least three key areas. The first is scale: implementation involves operationalizing a new technology on a much larger scale than field testing. Coupled with administrative challenges that are inherent to scaling new technologies, project managers and commanders must be cognizant of the chance for rare adverse events that might materialize once a technology is implemented across a large population that may not have presented during research or field testing.

The second distinction involves scope: operational use of new technologies often evolves with shifting military needs, and the uses for which a technology was researched or field tested may differ from those encountered during military missions. Operational use ethics reviews must account for the evolving uses by maintaining assessment protocols that co-evolve with operational uses.

The third distinction concerns the latent or cumulative effects of new technologies. Implementation and operational use of a new product may generate harms that materialize slowly over time, or only if the new product is combined with other products or in certain circumstances. Operational use ethics review protocols should account for these possibilities by structuring long-term observational studies to evaluate technology-related harms. Also important is a mitigation plan to account for long-term harms that may arise. With operational use ethics reviews, the ultimate goal is to validate and verify the benefits and risks of the technology while maintaining an appropriate safety net for affiliated harms. Table 19.1 provides a summary of key considerations for research, field test, and operational use ethics reviews.

As an Army leadership treatise succinctly states: "Ethical reasoning must occur in everything leaders do—in planning, preparing, executing, and assessing operations."[22] Ethical analysis must be taught and honed over time, just like other aspects of military training. Most people believe they have a strong moral compass and can make ethical decisions, but few individuals are trained in ethics theory and applied ethics. This adds to the ethics imbalance because many leaders instinctively believe ethics review is unnecessary. To be sure, ethicists often are focused more on theory than practice, and ethics consultations often do not lead to definitive conclusions or actionable advice. Ethicists must be mindful that their job is not simply to espouse principles, but to apply principles to concrete questions and offer advice that speaks to the benefits and risks of various courses of action.

A centralized department for military science ethics reviews furthers the goals of *jus in praeparatione bellum*. Structured ethics reviews represent an acknowledgment that legal assessments are a necessary but insufficient means of addressing whether a new technology should be researched, field tested, or operationalized. Ethics reviews

Table 19.1 Key Considerations in Military Science Ethics Reviews

Military Science Ethics Reviews Key Considerations		
Research Ethics Review	**Field Test Ethics Review**	**Operational Use Ethics Review**
o Are anticipated benefits of research proportional to risks faced by research participants? o Are research participant recruitment methods and informed consent protocols appropriate? o Do research implementation procedures appropriately balance research goals and the safety of research participants, researchers, and the community? o Are there appropriate risk-benefit reevaluation guidelines and markers throughout the course of the research, and under what conditions will the research be halted?	o Are anticipated benefits to the military and potential users of the technology proportional to the risks faced by field test users? o Is there fair selection of field test users, and are certain military subpopulations subject to undue risks from field testing? o What short- and long-term health monitoring is appropriate? o What metrics will be utilized to assess the risks and benefits of the technology, what field test reevaluation markers will be applied, and under what conditions will the field test be halted?	o Are anticipated benefits to the military and operational users of the technology proportional to the risks faced by the operational users? o Has the scope of operational use altered the risk-benefit calculus? o What short- and long-term health monitoring is appropriate, keeping in mind the potential for latent or cumulative effects to operational users and harms that may materialize to nonusers? o What metrics will be utilized to assess the risks and benefits of operational use of the technology, and under what conditions will operational use be halted?

also serve to increase the accountability of military science decision makers. As a 2019 DoD biotechnology council concluded, ethics reviews should track a technology's life cycle and be appropriately tailored to each life-cycle stage.[23]

As this book has detailed, it is essential to conduct robust ethical inquiries during national security exigencies, as historically these are the moments when decision makers often have sidestepped ethical concerns to address military needs. This will be a challenging norm to shift, but it can be accomplished by instituting a legal requirement to conduct ethics reviews, where failure to discuss ethical concerns adequately can have legal consequences for decision makers. This is not to say that decision makers will face legal consequences for using their judgment in operational decisions—rather, consequences will stem from failing to conduct an appropriate ethics review. The distinction is akin to that performed by federal courts in reviewing actions of an administrative agency: courts can sanction an agency if the agency fails to follow the appropriate procedures, selects a course of action that is arbitrary or capricious, or exceeds its delegated authority. But, when Congress grants an agency discretion to make factual determinations within the agency's area of expertise, courts will respect the agency's decision so long as it is reasonable.[24] In structuring mandatory ethics reviews for military science, an analogous standard should apply to the military and national security establishments.

Precedent for legally mandated ethics reviews is found in international law. Article 36 of Additional Protocol I of the Geneva Conventions states that nations must conduct a legal review "in the study, development, acquisition or adoption of a new weapon, means or method of warfare" to determine if the new weapon, means or method of warfare is compliant with the law of armed conflict.[25] Also codified in Additional Protocol I is the Martens Clause, which states: "In cases not covered by this Protocol or by other international agreements, civilians and combatants remain under the protection and authority of the principles of international law derived from established custom, from the principles of humanity and from the dictates of public conscience."[26]

The Martens Clause, which was formulated in 1899, endeavors to create moral limits to war, and has been characterized by the International Court of Justice as particularly relevant to "addressing the rapid evolution of military technology."[27] The United States was a signatory to Additional Protocol I at its inception in 1977. As of mid-2024, 174 nations have ratified Additional Protocol I; the United States, India, Iran, Israel, Pakistan, and Turkey have refused to do so.[28] Nevertheless, the United States has indicated that it voluntarily abides by several aspects of Additional Protocol I, many of which it acknowledges have become aspects of customary international law, including Article 36.[29]

Article 36 does not specify how a legal review should be conducted, nor does it detail substantive considerations that should be considered. The International Committee of the Red Cross has published recommended guidelines, but the Article 36 review procedure and substantive determinations are left to each nation.[30] Just as Article 36 serves as a legal requirement for weapons reviews but leaves open specifics related to procedural and substantive determinations during the review, the proposed military science ethics review requirements would mandate ethics reviews but leave individual decisions at the discretion of the reviewers. Within this nation-specific review process, the doctrine of *jus in praeparatione bellum* can help guide decision-making.

Since values may conflict in a pluralistic society, an ethics forum can help guide assessments when more than one course of action is ethically justifiable. The goal is not to proscribe a set ethical dogma, but to embed ethics reviews as routine components of research, field testing, and operational use of military technologies throughout their life cycle. Within this realm, military human enhancements raise particular concerns; thus, additional procedures and protections should exist.

Operationalizing Human Enhancements

It is one thing to develop and test human enhancements in a laboratory or controlled environment, and another to integrate human enhancements into military missions. This concern is not unique to twenty-first-century biomedical enhancements, but rather reflects a long-standing understanding that operationalizing new technologies requires a comprehensive integration framework. In the realm of human enhancements, helpful guidance is found in a 1988 National Academy of Sciences report that evaluated then-current psychological techniques for enhancing military human performance: the research included neurolinguistic programming, sleep-learning, telepathy, bioenergy transfer, psychic healing, and psychic control. Army officials sought to use these methods to create enhanced soldiers and "warrior monks," and the report discussed the state of the research and whether the techniques should be operationalized.[31]

The 1988 report recognized that the military might accept risks that would be unacceptable in civilian endeavors. It recommended that the military create a two-part standardized process to evaluate enhancements: (1) scientific evaluation of the safety and effectiveness of the intervention and (2) field testing to further analyze risks and benefits and determine whether operational implementation is feasible.[32] Both remain integral steps to operationalizing new technologies. A third essential element—not discussed in the 1988 report—is conducting a comprehensive assessment of the technology once it has been operationalized, including evaluating ethical, legal, societal, and national security concerns.

Each stage—research, field testing, and operational assessment—entails a different calculus. A model akin to pharmaceutical and medical device development is appropriate for research of human enhancements. Namely, prior to experimentation in humans, adequate studies should be conducted *in vitro* (laboratory research involving microorganisms or human or animal cell cultures), *in vivo* with animals (laboratory studies involving live animals), and *in silico* (computational models that analyze biological and other medical data to facilitate development and predict risks and benefits). These research methods may be pursued contemporaneously to generate data that can be used to determine whether an intervention should proceed to human clinical trials. *In vitro*, animal *in vivo*, and *in silico* studies are particularly relevant for military studies where conducting research with human participants is challenging, if not unethical.

Field testing would occur for enhancements where research generates data that show promising operational benefits with acceptable risks. One key goal of field testing is to observe whether the risks and benefits observed in research settings manifest in the field, and whether new risks or benefits materialize. Field testing is a common and essential element in the development of new military technologies. In cases where field testing confirms that an enhancement provides operational benefits with acceptable risks, broader operational integration of a new enhancement may be justifiable.

The military and the intelligence community maintain acquisition programs to bring new technologies into operational use. Each service has its own acquisition program and priorities, and acquisition agencies conduct due diligence to evaluate product effectiveness, suitability, cost, quality, reliability, producibility, and supportability.[33] This is a continuous and iterative process. As such, during operational use ethics reviews, there is an ongoing obligation to monitor and assess risks and benefits and to provide service members appropriate care should an enhancement be linked to short- or long-term harms. To evaluate the long-term impact of enhancements, it is imperative to conduct robust health monitoring to uncover latent or rare adverse events.

To be sure, instituting a procedural mechanism for evaluating whether a biomedical enhancement should be utilized does not ensure a robust substantive evaluation. And, even if a substantive evaluation is comprehensive, the ultimate decision on whether to utilize the enhancement may come down to a judgment of whether the potential benefits outweigh the risks.

In assessing military human enhancements, it is essential to conduct a holistic risk-benefit calculus that considers health and operational concerns. This includes evaluating whether an enhancement is reasonably necessary to achieve a legitimate military purpose, if it would negatively impact the dignity of warfighters, and how to appropriately hold superiors accountable for the decision to utilize an enhancement. Additional considerations include the societal impact of the enhancement should it be utilized in nonmilitary settings, whether having enhanced and unenhanced warfighters in the same unit might negatively affect unit cohesion, and what short- or long-term accommodations are appropriate if an enhancement negatively impacts a person's life outside the military or after service.[34] Table 19.2 summarizes these special considerations for military human enhancements.

In his book, *In Search of the Warrior Spirit*, psychologist and military consultant Richard Strozzi-Heckler contends that an overemphasis on biometric measurements of human performance will detract from traditional military virtues of courage, heroism, endurance, and bravery. As Strozzi-Heckler details, this will result in the spiritual poverty of warfighters, and may stunt how warfighters develop the strength to assess their conduct, face their fears, and cope with traumatic experiences.[35]

Emotional responses to experiences are an important aspect of learning at the individual, group, and institutional levels.[36] In some cultures, returning warriors undergo a purification ritual that serves as a ceremonial cleanse for the warfighter, to help the individual rid themselves from wartime stress or guilt, and for the community, which welcomes the warfighter and shows gratitude for their efforts.[37] According to David Grossman, a psychology professor at West Point and former Army Ranger, "Commanders, families, and society need to understand the soldier's desperate need

Table 19.2 Special Considerations for Military Human Enhancements

- Is the enhancement reasonably necessary to achieve a legitimate military purpose?
- Can enhancement research be structured to appropriately balance research goals with risks to research participants?
- Would the enhancement negatively impact the dignity of a warfighter?
- Would having enhanced and unenhanced warfighters in the same unit negatively impact unit cohesion or the military mission?
- How will the health impact of the enhancement be monitored in the short- and long-term?
- What short- and long-term accommodations are appropriate for enhanced service members?
- What is the anticipated societal impact of the enhancement should it be utilized in nonmilitary settings?

for recognition and acceptance, his vulnerability, and his desperate need to be constantly reassured that what he did was right and necessary."[38]

As Grossman further explains, service members pay a psychological price, as they "are recruited at a psychologically malleable age. They are distanced from their enemy psychologically, taught to hate and dehumanize. They are given the threat of authority, the absolution and pressure of groups."[39] Historically, there may have been cultural mechanisms to help service members process these emotions. However, contemporary society must be mindful of the risk of overrelying on biomedical interventions to serve as modern fixes to long-standing challenges endured by service members and veterans.

A holistic approach to analyzing the risks and benefits of enhancements also helps account for the unique medico-legal aspects of military service. Military personnel cannot refuse medical care or preventive health measures that make them fit for a mission.[40] A warfighter's medical autonomy is less than a civilian's because the collective goals of the unit can usurp individual preferences. Along with these characteristics that are unique to the military, warfighters are faced with dangers and risks not comparable to most components of civilian life: the mission's success is paramount, with the understanding that a service member may be injured or killed in the line of duty.

At the same time, commanders at all levels are responsible for the health and safety of their command, and ensuring that those under their command are not subjected to undue risks not commensurate with anticipated operational benefits.[41] In conducting these analyses, commanders can draw on principles of public health ethics, which emphasize balancing individual autonomy and the common good. This includes integrating elements of procedural and distributive justice into the decision-making process, being transparent about known risks and benefits, highlighting data and information gaps, and conducting an ongoing assessment of risks and benefits by actively monitoring implementation of a biomedical intervention. Building and maintaining trust is essential.[42]

The decision to pursue research, field testing, or operational use of a biomedical enhancement should include a continuing assessment of the reasons why the enhancement is desired and a comprehensive risk-benefit profile for the enhancement. Each intervention must be analyzed on its own merits, whether it is an EEG cap that enhances focus, an acoustic device that amplifies sound, or a special diet for special forces. Robust structural mechanisms and substantive frameworks that account for the historical gaps in warfighter protections are necessary steps, but equally important is the use of sound judgment when determining the extent to which biomedical enhancements should be operationalized.

Summary

A series of DoD advisory committees has underscored the urgent need for more robust military science ethics reviews. To promote the efficient use of resources, the DoD should standardize military science ethics reviews in a centralized department that draws on the resources and strengths of the various branches of the armed forces.

Ethics review protocols should be established across the life cycle of military technologies, and should be tailored to the individual ethical concerns raised by research, field testing, and operational use. For military human enhancements, particular care should be undertaken to ensure a holistic ethical analysis that contemplates potential military and civilian uses of the enhancement technologies, as well as associated ethical, legal, and societal concerns. Ethics reviews should be legally mandated to comport with international treaties and normative principles embedded in customary international law, and faithful compliance with the legal requirement to conduct ethics reviews should be a factor that courts consider when presented with a case that alleges harm caused by a military science pursuit.

Notes

1. Department of the Army, *ADP 6-22: Army Leadership and the Profession* (July 2019).
2. Department of Defense Directive 5500.07, *Standards of Conduct* (November 29, 2007), incorporating Secretary of Defense, DoD 5500.7-R, *Joint Ethics Regulation* (August 1993).
3. 32 C.F.R. § 219 (2024); Department of Defense Instruction 3216.02, *Protection of Human Subjects and Adherence to Ethical Standards in DoD-Conducted and -Supported Research* (April 15, 2020).
4. As detailed by the GAO, this includes, for example, reports from 2005, 2008, 2009, 2010, 2012, 2013, and 2015. Government Accountability Office, *Military Personnel: Additional Steps are Needed to Strengthen DOD's Oversight of Ethics and Professionalism Issues* (September 2015): 10–17.
5. Department of Defense Instruction 6025.27, *Medical Ethics in the Military Health System* (November 8, 2017).
6. Peter Emanuel et al., "Cyborg Soldier 2050: Human/Machine Fusion and the Implications for the Future of the DoD," *U.S. Army Combat Capabilities Development Command, Chemical Biological Center* (October 2019), vi.
7. GAO, *Military Personnel: Additional Steps Are Needed*, 5–6.
8. John A. Casciotti, Associate Deputy General Counsel (Health Affairs), DoD Office of General Counsel, Memorandum for Director of Biological Systems, *Applicability of Human Research Subject Protections to Certain Activities* (October 22, 2004).
9. 21 C.F.R. § 56.111(a)(2) (2024).
10. Secretary's Advisory Committee on Human Research Protections (SACHRP), "The Protection of Non-Subjects from Research Harm" (March 2022).
11. SACHRP, "The Protection of Non-Subjects."
12. SACHRP, "The Protection of Non-Subjects."
13. Robert L. Klitzman, "How IRBs View and Make Decisions About Social Risks," *Journal of Empirical Research on Human Research Ethics* 8, no. 3 (July 2013): 58–65; Daniel M. Hausman, "Third-Party Risks in Research: Should IRBs Address Them?," *IRB: Ethics and Human Research* 29, no. 3 (May-June 2007): 1–5.
14. Alan Fleischman et al., "Dealing with the Long-Term Social Implications of Research," *American Journal of Bioethics* 11, no. 5 (May 2011): 5–9.
15. DoD Instruction 3216.02, § 3.9.
16. DoD Instruction 3216.02, § 3.9(f).
17. DoD Instruction 3216.02, § 3.9(f).
18. DoD Instruction 3216.02, 29.

19. 10 U.S.C. § 139(a)(2)(A) (2022).
20. Final Report, *Human Radiation Experiments*, 304.
21. Nancy L. Spruill, Director of Acquisition Resources & Analysis, Memorandum for Assistant Secretary of the Army, et al., *Military Equipment Definition* (January 24, 2007).
22. Department of the Army, *Army Leadership and the Profession*, 2–7.
23. Emanuel et al., *Cyborg Soldier 2050*, 21–22.
24. *Skidmore v. Swift & Co.*, 323 U.S. 134 (1944); *Loper Bright Enterprises v. Raimondo*, 144 S. Ct. 2244 (2024). As the US Supreme Court indicated in *Loper Bright*, "when a particular statute delegates authority to an agency consistent with constitutional limits, courts must respect the delegation, while ensuring that the agency acts within it." *Loper Bright*, 144 S. Ct. at 2273.
25. Protocol Additional to the Geneva Convention of 12 August 1949, and Relating to the Protection of Victims of International Armed Conflicts (Protocol I), of 8 June 1977, art. 36 (June 8, 1977).
26. Additional Protocol I, art. 1, ¶ 2.
27. Legality of the Threat or Use of Nuclear Weapons, Advisory Opinion, 1996 I.C.J. Reports 257 (July 8, 1996).
28. International Committee of the Red Cross, International Humanitarian Law Databases: Protocol Additional to the Geneva Convention of 12 August 1949, and Relating to the Protection of Victims of International Armed Conflicts (Protocol I), of 8 June 1977 (accessed April 11, 2024).
29. Department of Defense Office of General Counsel, *Law of War Manual* (July 2023); International Committee of the Red Cross, *A Guide to the Legal Review of New Weapons, Means and Methods of Warfare: Measures to Implement Article 36 of Additional Protocol I of 1977* (January 2006), 4–5.
30. ICRC, *A Guide to the Legal Review of New Weapons*, 1–5.
31. National Research Council, *Enhancing Human Performance* (1988).
32. NRC, *Enhancing Human Performance*, 24–35.
33. Steve Murray and Matthew A. Yanagi, "Transitioning Brain Research: From Bench to Battlefield," in Giordano, ed., *Neurotechnology*, 11–21.
34. Patrick Lin, Maxwell J. Mehlman, and Keith Abney, *Enhanced Warfighters: Risk, Ethics, and Policy* (January 1, 2013).
35. Richard Strozzi-Heckler, *In Search of the Warrior Spirit* (2007), 250–51.
36. Goodley, "Pharmacological Performance Enhancement," 14–15.
37. Grossman, *On Killing*, 273–75.
38. Grossman, *On Killing*, 296.
39. Grossman, *On Killing*, 327.
40. Army Command Policy, Army Regulation 600-20, "Section 5-4: Command Aspects of Medical Readiness and Medical Care" (July 24, 2020).
41. Army Command Policy, Army Regulation 600-20, "Section 5-4."
42. James F. Childress et al., "Public Health Ethics: Mapping the Terrain," *Journal of Law, Medicine & Ethics* 30, no. 2 (Summer 2002): 170–78; Efthimios Parasidis and Amy L. Fairchild, "Closing the Public Health Ethics Gap," *New England Journal of Medicine* 387, no. 11 (September 15, 2022): 961–63.

20
Reformulating Governmental Immunities

Introduction

In the context of military science pursuits, the government and the military have frequently invoked three legal doctrines to shield untoward conduct from court review: the state secrets privilege, the political question doctrine, and sovereign immunity. The doctrines are a means of protecting military secrets and national security interests, and are a facet of America's tripartite system of government that separates powers between the Executive, Legislative, and Judiciary. Although the doctrines serve important goals, they also reduce governmental accountability and impede the ability of injured citizens to seek redress through the courts.

Congress should enact legislation—a National Security Accountability Act—that reformulates the doctrines to better harmonize national security goals with *jus in praeparatione bellum*. The legislation should have three key provisions: providing courts with more authority to assess whether policy decisions made by the Executive or Legislative infringe upon an individual's constitutional rights, precluding government use of the doctrines when government conduct infringes constitutional rights, and establishing a mediation program that provides claimants with an alternative mechanism for redress when a governmental immunity applies and court review is precluded. This chapter discusses the history of the state secrets privilege, the political question doctrine, and sovereign immunity. It also sets forth recommendations for reformulating the doctrines to promote justice in war preparations.

The State Secrets Privilege

The state secrets privilege affords the government a means of withholding evidence from a lawsuit if the government provides a written affirmation that the evidence, if disclosed, might reveal sensitive information that might impact national security. The privilege extends to lawsuits brought against the government and lawsuits involving private parties where a government secret might be revealed. The government can intervene in any lawsuit to assert the privilege and prevent evidence from being disclosed. Moreover, a judge can rely entirely on a government affirmation, and need not conduct an independent review of the evidence to assess whether national security would be negatively impacted by disclosure of the evidence.[1] Although the state secrets privilege technically does not involve a bar on lawsuits, since the withheld information typically is central to a plaintiff's ability to prove their case, when a judge excludes state secrets from a case, it often has the practical effect of dismissing the case altogether.

The state secrets privilege is a common law doctrine dating back over 150 years. One landmark case from the Civil War era, *Totten v. United States*, involved the estate

America's Military Biomedical Complex. Efthimios Parasidis, Oxford University Press.
 DOI: 10.1093/9780199351473.003.0021

of William Lloyd, a Union spy who claimed that, in July 1861, he contracted with President Abraham Lincoln to conduct secret wartime activities for which he would be paid $200 a month. For years, Lloyd infiltrated various Southern positions and relayed information back to the Union Army, but at the end of the war, he was reimbursed only for his expenses and did not receive the monthly stipend.[2] The Supreme Court dismissed the case, finding that "public policy forbids the maintenance of any suit in a court of justice, the trial of which would inevitably lead to the disclosure of matters which the law itself regards as confidential." The court characterized the agreement as "a secret service," noted that "such services are sometimes indispensable to the government," and indicated that, in some instances, disclosure of secret contracts "might compromise or embarrass our government in its public duties."[3]

Totten set a precedent that judges could not compel the government to disclose military secrets.[4] Over the years, courts have consistently reaffirmed this position, albeit with substantial controversy. In an influential 1953 Supreme Court case, *United States v. Reynolds*, three widows sued the government for money damages after their husbands were killed when a B-29 bomber crashed during an Air Force training exercise in Georgia after a fire broke out mid-flight. Invoking the state secrets privilege, the Air Force refused to produce the official accident report, claiming that the report contained sensitive information regarding secret electronic equipment being evaluated during the flight. The lower court ordered that the Air Force produce the report for an *in camera* court review so the judge could privately evaluate the military's asserted privilege. The Air Force refused the court order, and the judge ruled that the withheld information would support the plaintiffs' position that the military was negligent in maintaining the aircraft.[5]

In a 6–3 decision, the Supreme Court overruled the lower court decision. The high court acknowledged the delicate balance between secrecy and accountability, and sought to establish a "formula of compromise" between "judicial inquiry into the claim of privilege" that "would force disclosure of the thing the privilege was meant to protect" and "a complete abandonment of judicial control" that "would lead to intolerable abuses." The majority noted that "judicial control over the evidence in a case cannot be abdicated to the caprice of executive officers" but also explained that it "will not go so far as to say that the court may automatically require a complete disclosure to the judge before the claim of privilege will be accepted in any case." The court explained that the state secrets privilege should not be "lightly invoked" but also highlighted the Cold War mentality that had gripped the nation, observing that "this is a time of vigorous preparation for national defense" and "air power is one of the most potent weapons." As the court further detailed, "newly developing electronic devices have greatly enhanced the effective use of air power. It is equally apparent that these electronic devices must be kept secret if their full military advantage is to be exploited in the national interests."[6]

More than half a century after the crash, the Air Force declassified the accident report. It contained evidence that the airplane was unsafe for flight and described reasons why the airplane crashed during the training exercise—specifically, the airplane did not maintain standard safeguards to prevent the engine from overheating. Notably, the report contained no sensitive national security information.[7] According to the American Civil Liberties Union (ACLU), the Air Force simply used the state secrets privilege "to cover up its own negligence."[8]

Although public records cannot provide an accurate count, since the *Reynolds* case, scholars have estimated that the government has asserted the state secrets privilege hundreds of times.[9] After 9/11, the privilege has been increasingly utilized by the government in a wide range of cases, including extraordinary rendition, enhanced interrogation, targeted killing, and warrantless electronic surveillance.[10] It likewise has been invoked by the government in cases brought against private military contractors where American service members were injured or killed due to helicopter malfunctions, lax toxic waste disposal practices, radiation exposure, faulty wiring, and defective weapons and military equipment.[11]

As the government's use of the privilege has expanded, lawsuits challenging the scope of the privilege have increased. One case, decided by the Supreme Court in 2022, involved Abu Zubaydah (discussed in chapter 15), one of the first enemy combatants to be subjected to extraordinary rendition and enhanced interrogation, who has been detained at Guantanamo Bay for more than twenty years. Zubaydah sought information regarding his treatment while he was detained at a CIA black site in Poland; he also sought to depose two civilian contractors who allegedly interrogated him at the behest of the CIA. Some of the information sought by Zubaydah was in the public domain, though the US government did not confirm or deny the truthfulness of the information. Rather, during the case the government invoked the state secrets privilege to prevent Zubaydah from deposing the CIA contractors and asking questions about his detention.[12]

The Supreme Court ruled in favor of the government. The court indicated that, even if secret information has been publicly revealed, this does not eliminate the government's ability to assert the state secrets privilege when the government itself has not confirmed or denied the truthfulness of the information. The court also reaffirmed the long-standing rule that a judge need not conduct an independent evaluation of the evidence, but rather can rely solely on the government's affirmation that the evidence is confidential and disclosure would negatively impact national security.[13]

For decades, scholars and civil liberties advocates have called on Congress to set procedural and substantive guidelines to govern the state secrets privilege.[14] In 2007, following the disclosure of several post-9/11 national security programs that the government sought to keep confidential by utilizing the state secrets privilege, the American Bar Association (ABA) issued a resolution that called on Congress to enact legislation that requires judicial review of all evidence in cases where the state secrets privilege is invoked, and a bar on use of the privilege to cover up illegal government actions. As the ABA explained, "It is critically important that courts act as an independent check on the government when it asserts the state secrets privilege, and that courts conduct a meaningful review of the evidence that the government claims must remain secret."[15] In the same year as the ABA report, the State Secrets Protection Act was proposed and received bipartisan support. The bill sought to codify many of the protections outlined in the ABA resolution, but it was not enacted.

Following the ABA report and proposed bill, the Department of Justice (DOJ) issued a policy that adopted some of the recommendations. For example, the DOJ policy states that the agency will not defend the government's use the privilege to "conceal violations of the law, inefficiency, or administrative error" or "prevent embarrassment to a person, organization, or agency of the United States government." However, the DOJ noted that its policy does not create an enforceable legal right.[16]

Given the legal gap, Congress should enact legislation that instills sensible and enforceable legal limits on when the state secrets privilege can be used, and mandates independent judicial review of all state secret assertions. In addition to support from the ABA and civil rights advocates, there is judicial support for such legislation. For example, in the Supreme Court's 2022 *Zubaydah* case, Justices Neil Gorsuch and Sonia Sotomayor joined in a dissenting opinion that advocated for judicial review of all government claims of state secrets. In their opinion, the Justices highlighted a lesson that should have been learned from the 1953 *Reynolds* Air Force crash case—unverified affirmations allow courts to be deceived and the government to evade responsibility for wrongful conduct.[17] By not conducting an *in camera* review each time the state secrets privilege is invoked, courts allow themselves the opportunity to be misled.

The exponential growth of classified information over the past two decades underscores the urgency of the need for congressional action. Executive branch officials estimate that "between 50% to 90% of classified material does not merit that treatment."[18] In part, mass overclassification is a product of risk-averse government officials. There is no penalty for overclassifying documents, but underclassifying them might have severe consequences and might threaten national security. There is a ripple effect, as documents that incorporate classified information must also be classified.[19] If classification standards are broadly written and loosely enforced, it leaves ample room for the system to conceal mistakes or untoward conduct under the guise of national security. Overclassification hinders public discourse, erodes trust, and undermines democratic accountability. Moreover, as the 9/11 Commission concluded, too much classification can have the perverse effect of jeopardizing national security.[20]

To be sure, there will be instances where the government has a legitimate claim to prevent disclosure of state secrets from litigants, and courts should uphold that secrecy. Indeed, military science is one crucial area where secrecy is essential. But, invocation of the state secrets privilege need not automatically end judicial inquiry into governmental actions that should be kept secret. In these circumstances, judges should have the discretion to rule that information can be withheld from the litigants due to the state secrets privilege, but that the legality of the governmental action will be independently evaluated by a federal court that has clearance to review the matter. This balances the government's need to protect state secrets with governmental accountability and the societal goal of ensuring that the government does not engage in conduct that violates constitutional rights.

Such a balancing mechanism could be accomplished in several ways. One is to grant federal judges the authority to assess the merits of claims that involve state secrets without revealing the secrets to the parties in the lawsuit. Congress has already enacted legislation that allows federal courts to hold *in camera* hearings to review challenges to the legality of electronic surveillance.[21] Other laws permit federal courts to review classified information in criminal cases,[22] while DoD regulations state that federal judges do not need security clearances and "may be granted access to DoD classified information to the extent necessary to adjudicate cases being heard before these individual courts."[23] More than four million Americans hold security clearances; 1.3 million hold top-secret clearances.[24] From the perspective of access to classified information, there is little risk in granting federal judges access to information alleged by the government to constitute state secrets, particularly since federal judges routinely evaluate sensitive information, are bound by strict confidentiality rules, and are subject to stiff penalties for improper disclosure.

If Congress would prefer to centralize state secrets cases in one docket, it could expand the type of cases reviewed by Foreign Intelligence Surveillance Act (FISA) courts to encompass litigation involving state secrets. Even if specific factual evidence cannot be shared with parties in the litigation or their attorneys, courts can evaluate the claims independently or call on DOJ attorneys or independent advisers with appropriate security clearances to serve as counsel. In short, federal judges regularly review classified information in various cases, and expanding this realm to include assessments of projects involving state secrets would not represent a drastic expansion of judicial authority.[25] Rather, allowing courts to evaluate the legality of projects involving state secrets will promote justice in war preparations. Society affords the government the ability to maintain secrets from the general public to promote national security, but the government must institute appropriate safeguards to help ensure that this trust is not abused or utilized for improper purposes.

The Political Question Doctrine

A separate doctrine that has impeded litigants' ability to seek legal redress is the political question doctrine, which limits the authority of courts to evaluate cases involving policy decisions made by the Executive or Legislative branches of government. Separation of powers is the key rationale underlying the political question doctrine. Although courts can address whether the Executive or Legislative branches have exceeded their constitutional authority, courts may not dictate policy choices or issue value judgments where resolution is more appropriate in the political branches of government that are accountable to the electorate.[26]

Courts and scholars have recognized the amorphous boundary of this distinction, and no clear rule elucidates what qualifies as a political question.[27] Rather, each question is evaluated on a case-by-case basis utilizing several factors, such as a constitutional commitment of an issue to the Legislature or Executive, the impossibility of deciding a case without impeding the authority of the Legislature or Executive, and whether there is a lack of judicial standards to resolve an issue or an inability of a court to decide without making a policy determination that falls outside judicial discretion.[28]

In practice, the political question doctrine frequently has been invoked on matters related to foreign policy, war powers, and national security. This includes considering when hostilities begin and end, decisions to enter into or rescind treaties, military strategy, and military training.[29] As the Supreme Court has indicated, decisions regarding "composition, training, equipping, and control of a military force are essentially professional military judgments, subject always to civilian control of the Legislative and Executive Branches."[30]

The political question doctrine also protects private companies contracting with the government if the government exercises direct control over the contractor or if national defense interests are "closely intertwined with the military's decisions" regarding the contractor's work.[31] For example, in 2018, a federal appellate court ruled that the political question doctrine barred court review of a class action lawsuit brought against KBR and Halliburton regarding injuries suffered by service members exposed to waste management burn pits in Iraq and Afghanistan. The court held that, since the military exercised direct and actual control of the work regarding the burn pits, such as where the pits would be located and what could be burned, the political question doctrine shielded the military and the private contractors.[32]

To be sure, not all military and national security questions fall under the political question doctrine, and not all political cases present political questions.[33] Judges can evaluate the legal authority of Legislature and Executive actions, including federal statutes, executive agreements, and interpretation of treaties.[34] As detailed in chapters 13 and 14, federal judges rejected the government's attempts to use the political question doctrine to preclude court review of the Persian Gulf War countermeasure mandates and the anthrax vaccine mandate.[35] For both, the courts held that judicial review was proper because the cases did not involve a value judgment of whether the mandates should have been adopted, but instead evaluated whether the military and the Food and Drug Administration (FDA) abided by governing statutes and regulations. As the Supreme Court has explained, although Congress and the Executive maintain a "premier role" in foreign affairs, "under the Constitution, one of the Judiciary's characteristic roles is to interpret statutes, and we cannot shirk this responsibility merely because our decision may have significant political overtones."[36]

Although courts and scholars typically characterize the political question doctrine as a means of judicial restraint because it limits a court's ability to pass judgment on policy decisions made by the Executive and Legislature, some scholars view the doctrine as an exercise of judicial power since courts are making the first-order decision of which subjects are within their purview.[37] Furthermore, although courts are not permitted to make political decisions and judges are expected to be apolitical, court opinions often have political implications.

To further the principles of *jus in praeparatione bellum*, Congress should enact legislation that clarifies that the political question doctrine may not be invoked to bar court review of due process claims where the allegation is that the government has unconstitutionally deprived an individual of life, liberty, or property. It is well-settled law that courts are authorized to evaluate whether the Legislature or Executive branches have exceeded their constitutional authority, and this should include whether Legislative or Executive policy decisions have infringed upon the constitutional rights of citizens, "even if they implicate foreign policy decisions."[38] Just like courts can rule that the DoD or the FDA has exceeded its legal authority in the approval or administration of medical countermeasures, courts should be able to decide cases when an individual has been harmed because the government or military has unlawfully deprived a person of their constitutional right to life, liberty, or property.

The new legislation also should afford plaintiffs a nonjudicial remedy when a court rules that the political question doctrine bars judicial review of a case. This could take several forms, but one method that may be particularly beneficial is to create a mediation program that assesses a claimant's allegations that they should be compensated for harms suffered as a result of a policy decision of the Executive or Legislative branch. To be successful, the mediation program should be apolitical and staffed by distinguished attorneys, law professors, or retired judges. A panel of three mediators could hear claims in an informal setting with flexible procedures that would allow claimants to represent themselves and thereby forego the need to spend resources to hire legal counsel. The mediation format should not be confrontational, though the government should be permitted to respond to the claims in writing or via a hearing set before the mediation panel.

A key component of the responsibilities of the mediators would be to issue a recommendation regarding how the political branches of government can implement measures to remedy harms caused by a political decision. As a practical matter, Congress is unlikely to create a binding mediation program since doing so would equate to delegating policymaking authority to the mediators. But even a nonbinding mediation program would be a positive step forward for litigants involved in cases encompassed by the political question doctrine. The mediation program protocols could be structured so that the recommendation of the mediators must be put to a congressional vote within thirty days of the decision.

The proposed mediation program aims to balance continued use of the political question doctrine as a component of separation of powers with the public policy of providing citizens an avenue for redress if they are injured by the government's decision to engage in certain conduct. To promote transparency and accountability, the mediation panel's hearing and decision should be public. Such a mediation program would provide injured citizens with a means of having their claims heard by an independent forum, and will help inform the electorate on the practical implications of policy decisions. Such a forum would also promote democratic accountability and justice in war preparations.

Sovereign Immunity

In addition to the political question doctrine, sovereign immunity precludes court review of military and governmental actions. Sovereign immunity in the United States stems from an English common law principle, *rex non potest peccare*, a Latin maxim that translates to "the king can do no wrong." Even though the Founders rejected royal prerogatives and the principle of sovereign immunity is not found in the US Constitution, courts have consistently held that the government maintains sovereign immunity and cannot be sued unless it explicitly waives its immunity or consents to suit. In 1946, Congress enacted the Federal Tort Claims Act (FTCA), which waives sovereign immunity for certain tort claims against the government.[39] Although the FTCA was landmark legislation that intended to hold the government accountable for negligent conduct, it was enacted with several exceptions, two of which are particularly relevant to military matters: the discretionary function exception and the combatant activities exception, the latter of which has been expanded by courts under the *Feres* doctrine. These exceptions, which are not mutually exclusive, have substantially limited court review of tort claims brought by service members and their families.[40]

The discretionary function exception applies when the government has engaged in conduct deemed discretionary and susceptible to policy analysis. The rule prevents courts from "judicial second guessing of legislative and administrative decisions grounded in social, economic, and political policy through the medium of an action in tort."[41] The discretionary function exception is the broadest and most litigated exception to the FTCA's waiver of sovereign immunity. The government has successfully invoked the exception to maintain sovereign immunity in several cases, including injuries stemming from biological warfare field tests over San Francisco, radiation exposure during atomic test explosions, exposure to Agent Orange during the Vietnam War, and claims brought on behalf of children of service members who were born with congenital disabilities after the service members were administered medical countermeasures and exposed to toxic substances during the Persian Gulf War.[42]

The combatant activities exception preserves sovereign immunity for "any claim arising out of the combatant activities of the military or naval forces, or the Coast Guard, during time of war."[43] In the 1950 case of *Feres v. United States*, the Supreme Court expanded this preservation of sovereign immunity to encompass all injuries that "arise out of or are in the course of activity incident to service."[44] As the Court reasoned: "We know of no American law which ever has permitted a soldier to recover for negligence, against either his superior officers or the Government he is serving."[45] Courts have justified the maintenance of sovereign immunity for military matters because there is a unique relationship between the Executive and members of the armed forces, and court review would "disrupt the unique hierarchical and disciplinary structure of the military" and embroil judges "in sensitive military affairs at the expense of military discipline and effectiveness."[46] Courts seek to avoid legal review of intra-military affairs, and point to the availability of health and other benefits as alternative means of remedying harms incident to service.

The *Feres* doctrine has been invoked to bar a slew of lawsuits, including injury claims related to the military's World War II (WWII) mustard gas experiments,[47] radiation exposure during atomic weapons testing,[48] military training accidents,[49] sexual harassment,[50] racial discrimination,[51] medical malpractice in military hospitals,[52] and more.[53] In a few cases, courts have applied sovereign immunity to service-related injuries but have allowed litigation to proceed if the government committed an independent tort after a service member was discharged from the armed forces. For example, in a 1979 decision involving a surreptitious LSD experiment conducted during the interrogation of an Army private stationed in France who was suspected of stealing classified documents, the court held that sovereign immunity precluded claims related to the experiment but did not automatically preclude claims that the military deliberately failed to provide follow-up medical treatment after the private was discharged.[54] Similarly, in some cases from the early 1980s involving injuries caused by participation in mid-century atomic weapon test blasts, courts held that sovereign immunity might not apply if the government failed to warn or monitor veterans if the military learned of radiation dangers after the service members left the military.[55]

Notwithstanding these limited instances, courts typically have held that sovereign immunity precludes all legal recourse against the government on the grounds that the post-discharge tort was a continuing harm that stemmed from the initial service-related harm.[56] In a landmark 1987 case, *Stanley v. United States*, which was discussed in chapter 11, the Supreme Court ruled that sovereign immunity precluded all claims for harm caused by the military's clandestine LSD experiments, and that immunity extended not only to the military and government but also to individuals working on behalf of the government if an injury arises out of or is in the course of activity incident to service, regardless of whether the individuals are government employees or civilians, such as university scientists.[57]

In *Stanley*, the Supreme Court adopted an expansive definition of the term "service-related," finding that the LSD experiments were incident to service because they were conducted on a military base and were for the benefit of the military.[58] The high court further held that Stanley's claim that the military and researchers failed to warn him of the potential risks of the experiment was barred because the initial conduct that led to the harm—the experiments—took place while Stanley was in the Army; as such, the failure-to-warn claim was not separate and distinct from the alleged harm caused by the experiments.[59]

Courts also have held that sovereign immunity extends to cover private military contractors.[60] In a case where a service member sued a defense contractor for injuries sustained due to an alleged faulty aircraft ejection system, the Supreme Court reasoned that, since litigation against the contractor would implicate fault on the part of the military, and thereby second-guess military decision-making, sovereign immunity applies.[61] The government contractor defense was also used successfully by manufacturers of Agent Orange, who were found to be immune from lawsuits brought by Vietnam War veterans who suffered injuries due to their exposure to the toxic chemical.[62]

A separate doctrine within sovereign immunity bars tort claims against a government contractor if imposing liability on the contractor would conflict with national interests or undesirably interfere with military objectives.[63] In addition, the FTCA does not waive sovereign immunity if a person suffers an injury in a foreign country, and courts have routinely held that conduct occurring on American military bases in foreign countries falls within this exception.[64] Even if a service member's claim can clear all the legal roadblocks of sovereign immunity, the political question doctrine and state secrets privilege may be independent limitations on court review.

Taken together, sovereign immunity is an extraordinarily broad doctrine repeatedly invoked by the military to bar court review of harms suffered by service members and their families. Judges, scholars, and civil rights advocates have widely criticized sovereign immunity, particularly when it is applied to forego court review of alleged constitutional violations.[65] In the 1980s—following unsuccessful lawsuits involving injuries sustained from the military's mustard gas, LSD, and atomic weapons experiments—legislation was proposed that would eliminate sovereign immunity for claims that federal officials violated the constitutional rights of service members, such as the right not to be deprived of life, liberty, or property without due process of law.[66]

The bill was not enacted, but in the *Stanley* case four dissenting Justices contended that sovereign immunity should not extend to claims that individual researchers deprived a service member of their constitutional due process rights. As Justice Sandra Day O'Connor wrote, the clandestine LSD experiment went "so far beyond the bounds of human decency that as a matter of law it simply cannot be considered a part of the military mission."[67] Justices William Brennan, Thurgood Marshall, and John Paul Stevens agreed. Drawing on historical parallels between the case and the Nuremberg Doctors' Trial, the Justices observed that the "practical result of this decision is absolute immunity from liability for money damages for all federal officials who intentionally violate the constitutional rights of those serving in the military."[68] They added: "Having invoked national security to conceal its actions, the Government now argues that the preservation of military discipline requires that Government officials remain free to violate the constitutional rights of soldiers without fear of money damages. What this case and others like demonstrate, however, is that Government officials (military or civilian) must not be left with such freedom."[69]

Following *Stanley*, Justice Antonin Scalia—who wrote the majority opinion in the case during his first term on the Supreme Court—became a staunch critic of the *Feres* doctrine. Just five weeks after the *Stanley* opinion was published, Justice Scalia dissented in a separate *Feres* doctrine case where the Court extended immunity to encompass harms to service members caused by civilians.[70] He wrote: "*Feres* was wrongly decided and heartily deserves the widespread, almost universal criticism it has received."[71] Justice Clarence Thomas also has long been a vocal critic of *Feres*; since 2019, he has authored three dissenting opinions wherein he has lambasted *Feres* as "demonstrably wrong" and underscored how courts have improperly interpreted the combatant activities exception beyond harms suffered during war to include harms "that are even remotely related to the individual's status as a member of the military."[72]

The historical record supports the critiques set forth by Justices O'Connor, Brennan, Marshall, Stevens, Scalia, and Thomas. *Feres* was a consolidated case that brought together three separate negligence claims against the military: a barracks fire in 1947 in a military facility in New York allegedly caused by a defective heating plant, a medical malpractice claim against an army physician stemming from a surgery conducted in a government hospital in Virginia during WWII, and a medical malpractice claim against an army physician for a surgery that occurred in 1947 at an Army hospital located on an Illinois base.[73] Two of the three claims involved injuries that occurred after the conclusion of WWII. The appellate court decisions in those cases were split on whether the combatant activities exception applies: the Tenth Circuit held that the medical malpractice case could proceed,[74] while the Second Circuit ruled that the barracks fire case could not.[75]

Feres represented a broad judicial interpretation of the FTCA's combatant activities exception because it included in that exception negligent conduct that occurred outside wartime combatant activities. The Supreme Court issued the *Feres* ruling in 1950, on the heels of WWII and shortly after Congress took the extraordinary step to waive sovereign immunity for certain tort claims, and the decision may have been driven by practical considerations and a policy preference to limit service-related claims against the government.

But had the Supreme Court in 1950 been presented with Stanley's case—a clandestine LSD experiment that caused severe harm to an unwitting research subject—it would have caused an international uproar for the top American court to rule that sovereign immunity protected the military and medical researchers, some of whom were physicians at the University of Maryland. Justice Robert Jackson penned the

Feres opinion in 1950, shortly after he returned to the Supreme Court after taking a leave of absence to serve as one of America's chief prosecutors during the Nuremberg trials. Recall that the Nuremberg tribunal agreed with American prosecutors that Germany's nonconsensual concentration camp studies violated well-settled legal and ethical principles regarding informed consent and medical experimentation on humans.

Yet, in the decades following WWII and the *Feres* decision, US courts ruled that sovereign immunity precluded claims for relief stemming from the military's mustard gas, biological warfare, and LSD experiments, each of which was conducted without informed consent. From a medical ethics perspective, this may not be surprising since, as detailed in chapter 8, the American medical establishment largely dismissed the Nuremberg Code as a code for barbarians and an unnecessary hindrance to the American research agenda. American lawmakers and courts simply reinforced this medical ethics mindset as the Cold War intensified and military research proliferated throughout the second half of the twentieth century.

Today, Congress can take three concrete steps to balance sovereign immunity with principles of justice in war preparations. First, Congress should clarify the scope of the FTCA's combatant activities exception. If Congress decides that the *Feres* doctrine's interpretation of sovereign immunity is acceptable, it should codify that position in the statute. If not, Congress should clearly articulate that the *Feres* doctrine improperly expands the scope of the combatant activities exception. Second, Congress should waive sovereign immunity for constitutional violations, regardless of whether the violation occurs to further national security or a military mission. Courts should be permitted to assess a claim that the government, or an individual working at the behest of the government, has infringed a claimant's constitutional rights, such as the right not to be deprived of life, liberty, or property without due process of law.

To be sure, such claims already are permitted in certain circumstances—they are termed *Bivens* claims, named after a Supreme Court opinion from 1971 wherein the Court held that, even if the government maintains sovereign immunity, federal officials may be individually responsible for constitutional violations that they commit during their work.[76] *Bivens* claims are very limited, and in *Stanley*, the Supreme Court refused to extend *Bivens* claims to harms that arise incident to military service.[77] Congress can ameliorate the judicial reluctance to extend *Bivens* remedies to military matters by enacting legislation that explicitly waives sovereign immunity in cases where a service member is deprived of their constitutional rights.

A third step Congress can take is similar to the one I discuss in the previous section regarding mediation of political questions: in instances where sovereign immunity applies and court review is precluded, Congress should establish a mediation program that affords claimants an alternative means of obtaining a remedy. A mediation program will help promote justice in war preparations by adding a layer of protection for service members and their families. The program would assess whether to award benefits beyond those ordinarily provided by the military, such as healthcare, disability, and survivor benefits.

Such a program would expand upon congressional action from 2020, which mandated that the DoD establish an administrative program to evaluate and compensate service members for medical malpractice by military healthcare providers. The 2020 law did not permit such claims to go to court, but it provided an alternative method of dispute resolution that is authorized to award compensation that supplements existing military and veterans benefits.[78]

To create the malpractice mediation program, the DoD promulgated a rule that outlined an administrative, nonadversarial procedure conducted within the DoD to determine fair compensation for medical malpractice claims. The DoD estimated that annual payments under the program would be $11.2 million,[79] a minuscule amount in relation to the annual defense appropriation, which was $816.7 billion in 2023.[80]

The medical malpractice administrative program can serve as a model for a broader mediation program for constitutional violations. Such a program is akin to the long-standing practice that allows injured citizens and service members to petition Congress for private legislation when court review of a claim is precluded. Private legislation—which involves Congress passing a bill that provides individualized relief to someone who has suffered harm at the hands of the government—is a practice that dates back over a century. Although private legislation has been utilized to award compensation in some instances related to military science pursuits, such as the military's LSD experiments, the system is inefficient (each person must file a petition for relief), creates a substantial burden on Congress (each petition is reviewed individually), and is plagued by political favoritism.[81]

A mediation program will help promote justice in war preparations by adding a layer of protection for service members and citizens in the context of military or government actions where courts are precluded from assessing whether a Legislative or Executive policy decision infringes constitutional rights. Mediating such claims serves as a means of acknowledging that harms caused by constitutional violations are distinct from injuries sustained during military combat or training. In such instances, it is proper to provide benefits and remedies beyond those traditionally afforded to all service members for service-related injuries.

Summary

The state secrets privilege, the political question doctrine, and sovereign immunity are legal doctrines that represent important elements of the American government's system of separation of powers among the Executive, Legislative, and Judiciary, but the doctrines also have been frequently utilized to shield untoward government conduct and limit the ability of injured citizens to seek legal redress. Each should be reformulated to strike a more just balance between national security and individual rights. Via a National Security Accountability Act, Congress can establish a fair and judicious balance by providing courts with more authority to assess whether policy decisions made by the Executive or Legislative infringe upon constitutional rights. The proposed legislation should preclude the government from using the doctrines when government conduct infringes constitutional rights. It should also establish an impartial mediation program that is invoked when a governmental immunity applies and court review is precluded. These measures will promote justice in war preparations and help set a respectable moral boundary between governmental conduct and an individual's constitutional rights. As Justice Brennan wrote in his dissenting opinion in *Stanley*: "Soldiers ought not be asked to defend a Constitution indifferent to their essential human dignity."[82]

Notes

1. *United States v. Zubaydah*, 595 U.S. 195, 212–13 (2022).
2. *Totten v. United States*, 92 U.S. 105, 105–6 (1875).
3. *Totten*, 92 U.S. at 106–7.
4. *United States v. Reynolds*, 345 U.S. 1, 7 n.11 (1953) (citing cases and legal treatises).
5. *Reynolds*, 345 U.S. at 2–5.
6. *Reynolds*, 345 U.S. at 7–10.
7. State Secrets Protection Act, Report with Minority Views to Accompany S. 2533, 110th Cong. 5 (August 1, 2008).
8. American Civil Liberties Union, "Background on the State Secrets Privilege" (January 31, 2007).
9. Laura K. Donohue, "The Shadow of State Secrets," *University of Pennsylvania Law Review* 159 (2010): 78–87.
10. Congressional Research Service, "The State Secrets Privilege: National Security Information in Civil Litigation" (April 28, 2022).
11. Several cases are detailed in Donohue, "The Shadow of State Secrets," 105–15.
12. *Zubaydah*, 595 U.S. at 198–204.
13. *Zubaydah*, 595 U.S. at 212–13.
14. Frederick A. O. Schwarz, *Democracy in the Dark* (2015), 111–13.
15. American Bar Association, Resolution on the State Secrets Privilege: Revised Report 116A (August 2007), 1.
16. Memorandum from the Attorney General for Heads of Executive Departments and Agencies, Policies and Procedures Governing Invocation of the State Secrets Privilege (September 23, 2009).
17. *Zubaydah*, 595 U.S. at 251–53 (Gorsuch, J., dissenting).
18. *Zubaydah*, 595 U.S. at 252 (Gorsuch, J., dissenting).
19. Oona A. Hathaway, "Keeping the Wrong Secrets," *Foreign Affairs* 101, no. 1 (January/February 2022): 85–98.
20. The 9/11 Commission Report, *Final Report of the National Commission on Terrorist Attacks Upon the United States* (2004), 416–19.
21. 50 U.S.C. § 1806(f) (2023).
22. Classified Information Procedures Act, Pub. L. No. 96-456, § 3, 94 Stat. 2025 (October 15, 1980), as amended, 18 U.S.C. App. 3, § 3.
23. 32 C.F.R. 154.16(d)(5) (2023).
24. Office of the Director of National Intelligence, National Counterintelligence and Security Center, "Fiscal Year 2017 Annual Report on Security Clearance Determinations," 4–5.
25. State Secrets Protection Act, Report with Minority Views to Accompany S. 2533, 110th Cong. 9.
26. *Carmichael v. Kellogg, Brown & Root Services*, 572 F.3d 1271, 1281 (11th Cir. 2009).
27. Jed S. Rakoff, "Don't Count on the Courts," *The New York Review of Books*, April 5, 2018, 46–47.
28. The Supreme Court outlined these and other factors in *Baker v. Carr*, 369 U.S. 186, 217 (1962), and courts have relied upon the factors to varying degrees on a case-by-case basis.

29. *Carmichael*, 572 F.3d at 1293; Jared P. Cole, "The Political Question Doctrine: Justiciability and the Separation of Powers," *Congressional Research Service* (December 23, 2014).
30. *Gilligan v. Morgan*, 413 U.S. 1, 10 (1973) (emphasis in original omitted).
31. *In re: KBR, Inc.*, 893 F.3d 241, 259–60 (4th Cir. 2018).
32. *In re: KBR, Inc.*, 893 F.3d at 259–64.
33. *Japan Whaling Association v. American Cetacean Society*, 478 U.S. 221, 230 (1986); *Bancoult v. McNamara*, 445 F.3d 427, 435 (D.C. Cir. 2006).
34. *Japan Whaling*, 478 U.S. at 230.
35. *Doe v. Sullivan*, 939 F.2d 1370, 1379–81 (D.C. Cir. 1991); *Doe v. Rumsfeld*, 297 F. Supp. 2d 119, 126–29 (D.D.C. 2003).
36. *Japan Whaling*, 478 U.S. at 230.
37. Tara Leigh Grove, "The Lost History of the Political Question Doctrine," *New York University Law Review* 90 (2015): 1913–15.
38. *Bancoult*, 445 F.3d at 435.
39. Federal Tort Claims Act, Pub. L. No. 79-601, § 401, 60 Stat. 842 (August 2, 1946).
40. Michael D. Contino and Andreas Kuersten, "The Federal Tort Claims Act (FTCA): A Legal Overview," *Congressional Research Service* (April 17, 2023).
41. *Berkovitz ex rel. Berkovitz v. United States*, 486 U.S. 531, 536–37 (1988) (internal quotations omitted).
42. *Nevin v. United States*, 696 F.2d 1229 (9th Cir. 1983); *Minns v. United States*, 155 F.3d 445 (4th Cir. 1998); Contino and Kuersten, "The Federal Tort Claims Act," *Congressional Research Service*, 18–24.
43. 28 U.S.C. § 2680(j) (2023).
44. *Feres v. United States*, 340 U.S. 135, 146 (1950).
45. *Feres*, 340 U.S. at 141.
46. *Ortiz v. United States*, 786 F.3d 817, 821 (10th Cir. 2015); *United States v. Johnson*, 481 U.S. 681, 690 (1987).
47. *Schnurman v. United States*, 490 F. Supp. 429 (E.D. Va. 1980).
48. *Jaffee v. United States*, 663 F.2d 1226 (3d Cir. 1981).
49. *Charland v. United States*, 615 F.2d 508 (5th Cir. 1980).
50. *Stubbs v. United States*, 744 F.2d 58 (8th Cir. 1984).
51. *Chappell v. Wallace*, 462 U.S. 296 (1983).
52. *Stanley v. Central Intelligence Agency*, 639 F.2d 1146, 1150 (5th Cir. 1981) (citing several cases involving medical malpractice claims barred by the *Feres* doctrine).
53. Contino and Kuersten, "The Federal Tort Claims Act."
54. *Thornwell v. United States*, 471 F. Supp. 344 (D.D.C. 1979).
55. *Broudy v. United States*, 661 F.2d 125 (9th Cir. 1981); *Cole v. United States*, 755 F.2d 873 (11th Cir. 1985).
56. *Stanley*, 639 F.2d at 1153–56; *In re Agent Orange Product Liability Litigation*, 506 F. Supp. 762, 778–79 (E.D.N.Y. 1980); *Schnurman*, 490 F. Supp. at 436–38.
57. *United States v. Stanley*, 483 U.S. 669, 683–84 (1987).
58. *Stanley*, 639 F.2d at 1152.
59. *Stanley*, 483 U.S. at 672–73.
60. *Carroll v. United States*, 661 F.3d 87 (1st Cir. 2011).
61. *Stencel Aero Engineering Corp. v. United States*, 431 U.S. 666, 673–74 (1977).
62. *In re Agent Orange Product Liability Litigation*, 373 F. Supp. 2d 7, 23–27 (E.D.N.Y. 2005).
63. *Boyle v. United Technologies Corp.*, 487 U.S. 500 (1988); *Saleh v. Titan Corp.*, 580 F.3d 1 (D.C. Cir. 2009); *Koohi v. United States*, 976 F.2d 1328 (9th Cir. 1992).
64. 28 U.S.C. § 2680(k) (2023); *Sosa v. Alvarez-Machain*, 542 U.S. 692, 712 (2004); *Doe v. Meron*, 929 F.3d 153, 167 (4th Cir. 2019).

65. Contino and Kuersten, "The Federal Tort Claims Act," 30–32.
66. *Gaspard v. United States*, 713 F.2d 1097, 1103 n.13 (5th Cir. 1983) (discussing the proposed legislation).
67. *Stanley*, 483 U.S. at 686 (O'Connor, J., concurring in part and dissenting in part).
68. *Stanley*, 483 U.S. at 691 (Brennan, J., concurring in part and dissenting in part).
69. *Stanley*, 483 U.S. at 689 (Brennan, J., concurring in part and dissenting in part).
70. *United States v. Johnson*, 481 U.S. 681, 692–703 (1987) (Scalia, J., dissenting).
71. *Johnson*, 481 U.S. at 700–701 (Scalia, J., dissenting) (internal quotations omitted).
72. *Clendening v. United States*, 143 S. Ct. 11, 11–14 (2022) (Thomas, J., dissenting from denial of certiorari); *Doe v. United States*, 141 S. Ct. 1498, 1498–1500 (2021) (Thomas, J., dissenting from denial of certiorari); *Daniel v. United States*, 139 S. Ct. 1713, 1713–14 (2019) (Thomas, J., dissenting from denial of certiorari).
73. *Feres*, 340 U.S. at 136–37; *Griggs v. United States*, 178 F.2d 1, 2 (10th Cir. 1949); *Jefferson v. United States*, 77 F. Supp. 706, 708–9 (D. Md. 1948).
74. *Griggs*, 178 F.2d at 3.
75. *Feres v. United States*, 177 F.2d 535, 538 (2d Cir. 1949).
76. *Bivens v. Six Unknown Named Agents of Federal Bureau of Narcotics*, 403 U.S. 388 (1971).
77. *Stanley*, 483 U.S. at 684.
78. Department of Defense, Medical Malpractice Claims by Members of the Uniformed Services: Final Rule, 87 Fed. Reg. 52,446 (August 26, 2022).
79. DoD, Medical Malpractice Claims, 87 Fed. Reg. 52,461.
80. Jim Garamone, "Biden Signs National Defense Authorization Act Into Law," *DoD News*, December 23, 2022.
81. Contino and Kuersten, "The Federal Tort Claims Act," 4–5.
82. *Stanley*, 483 U.S. at 707 (Brennan, J., concurring in part and dissenting in part).

Conclusion

In 1887, Friedrich Engels warned that a relentless pursuit for economic superiority between nations would unleash a continuing competition for acquiring the newest and most technologically advanced weapons of war. He envisioned this pursuit leading to a "universal lapse into barbarism" and a "moment when the systematic development of mutual oneupmanship in armaments reaches its climax"—a massive war with widespread devastation that would facilitate the rise of the working class.[1] This prediction manifests in several conflicts, most notably World War I (WWI) and World War II (WWII). However, a vital facet missing from Engels's vision was that the drive for military and scientific superiority would also cause substantial harm to a nation's own citizens. While military science pursuits often have resulted in extraordinary achievements, as illustrated throughout this book, there have also been scores of examples where untoward conduct was rationalized as necessary to promote national security and achieve military goals.

For America, the toll has been extraordinary. Segments of society have been exploited to promote medical research of particular importance to the military. Service members and civilians have been exposed to chemical and biological agents during research studies, often without their knowledge or consent. At the same time, cities and towns across the country have been used as test sites for chemical and biological warfare. Atomic explosions to prepare service members for nuclear war have caused widespread environmental destruction and myriad health ailments in generations of Americans. Health professionals have been intimately involved in enhanced interrogation programs, while warfighters face an emerging scenario where biomedical enhancements may become essential elements of military missions.

Laws and ethical codes have been rewritten to facilitate these endeavors and shield conduct from public view, while a host of policies have hindered the medical autonomy of military personnel. Millions of service members have been administered medical products that have been approved via specialized Food and Drug Administration (FDA) protocols that maintain lower standards for determining safety and efficacy. Due to secrecy mandates, broad governmental immunities, and lackluster health surveillance and access to care, many of those harmed by these actions have been left without recourse to legal remedies or adequate means to address their injuries.

Military science pursuits likewise have caused a shift in the moral compass of society. Government, industry, and academics have coalesced to support a massive military-industrial complex that frequently has invoked a utilitarian and consequentialist frame, where the ends justify the means and a minority of the population can be subjected to risky experiments to further national security. Vast sums of money have been invested in military projects and diverted from other societal priorities such as education, healthcare, economic development, and environmental protection. Covert actions and clandestine operations have violated democratic principles,

America's Military Biomedical Complex. Efthimios Parasidis, Oxford University Press.
 DOI: 10.1093/9780199351473.003.0022

exposed weaknesses in democratic controls, and fractured the trust of the populous in government. To a large extent, universities and the media have increasingly become, as political theorist Hans Morgenthau described nearly a half-century ago, "handmaidens of government" that seek to maintain a "pretense to independence."[2] Taken together, these actions have had a ripple effect within the military that has negatively impacted recruitment, retention, and the willingness of individuals to sacrifice themselves for missions.[3]

A nation that sends its citizens to war has a moral obligation to protect them. This includes a continuing commitment to support service members and veterans who suffer service-related injuries, whether from combat, occupational hazards, or adverse health effects from military research, medical interventions, or biomedical enhancements. It further entails an ongoing duty to provide access to healthcare and opportunities that support a veteran's post-service education and career. Also important is the availability of resources for military families, to help them address their unique hardships. Throughout America's history, funding and resources for this safety net have often been insufficient, and military leaders have frequently underestimated health risks or downplayed the health impact of war and war preparations.

A primary aim of this book has been to provide a historical reflection on how laws and ethical codes have co-evolved with military science, by detailing the moral calculus conducted by decision makers at the time the decisions were made. Reviewing this history exposes the hard truth that military aims often have been used to justify the improper treatment of service members and civilians, and that decision makers recognized that their actions raised significant legal and ethical concerns. Examining this history also helps us understand how science and society have evolved. This opens a window through which we can empathize with the challenges faced by prior generations, and allows us to identify analyses that produced beneficial results and those that did not. Multiple generations have experienced analogous dilemmas and struggles in grappling with national security concerns, and investigating this history provides context for evaluating contemporary military science pursuits. It likewise facilitates a critical assessment of whether modern practices appropriately balance national security goals with individual freedoms and human rights.

A second aim of this book has been to recommend policies that reorient legal and ethical frameworks to better harmonize national security priorities with principles of *jus in praeparatione bellum*. The recommendations include strengthening the countermeasure approval process and enhancing the medical autonomy of warfighters, eliminating the informed consent waiver rule for military personnel, standardizing and expanding military science ethics reviews, and reformulating governmental legal shields within the state secrets privilege, the political question doctrine, and sovereign immunity. Taken together, the proposals create new *ex ante* protections that decrease the likelihood that warfighters will be exposed to undue risks from medical products, provide additional *ex post* remedies in the event of service-related harms, and increase accountability of military and government decision-makers.

For centuries, humanity has debated the morality of war. This book adds to that rich history by introducing the concept of *jus in praeparatione bellum* and using military science as a frame to examine principles of justice in war preparations. Medical professionals and scientists have contributed significantly to maintaining a healthy fighting force and alleviating the health impact of war, but they also have dedicated their efforts to advancing military technologies and creating weapons of mass destruction. While legal doctrines have attempted to guide the research and development of these weapons and instill a global norm against their use, history has repeatedly shown that nations are willing to disregard the guardrails in the face of real or perceived existential threats.

The doctrine of *jus in praeparatione bellum* is forward-looking, but the process of thinking about justice in war preparations also includes an assessment of past actions and reflection on whether decision-makers appropriately balanced the risks and benefits of the various courses of action they were presented with. If structured properly, this analysis can serve as a mechanism for truth and reconciliation within the military establishment, where past wrongs can be acknowledged and harms can be mitigated by instituting appropriate remedies. In an essay he published during the Vietnam War, Hans Morgenthau eloquently captured the danger of not undertaking a comprehensive review: "To lay bare what is wrong is not an idle exercise in ex post facto fault-finding. Rather, it is an act of public purification and rectification. If it is not performed and accepted by government and people alike, faults, undiscovered and uncorrected, are bound to call for new disasters."[4]

There is a social function of military science that is shaped by a nation's political, economic, and cultural goals, and one that contributes to the shaping of a nation's image. Just as science and technology have been essential elements of military preparedness, military goals and national security priorities have largely shaped the scientific establishment. At the same time, as philosopher Hans Jonas cautioned, society "would indeed be threatened by the erosion of those moral values whose loss, possibly caused by too ruthless a pursuit of scientific progress, would make its most dazzling triumphs not worth having."[5] For technologies to be properly integrated into society, technological momentum must appropriately co-evolve with legal and ethical momentum.

Military science pursuits can help a nation achieve an operational advantage, but they also place significant risks and burdens on a nation's own citizens. The costs of war and war preparations extend beyond lives and treasure, and include infringements on individual liberty, abridgments of ethical principles, and societal and environmental harms. As detailed in this book, more nuanced regulation and restraint are necessary to balance national security with principles of justice, fairness, and human dignity.

This can be challenging in the political economy of the United States, where defense and pharmaceutical industries have long been among the top lobby firms, flooding lawmakers with hundreds of millions of dollars annually to support their industries.[6] As experienced with the chemical industry following WWI, the pharmaceutical sector following WWII, and biotechnology firms in recent decades, a large part of these lobby efforts has been directed to military science and has contributed to the exponential growth of the military biomedical complex.

Notwithstanding the substantial influence of money in American politics, a moral imperative remains to continuously strive for justice in war preparations. It is incumbent on the most powerful nations to uphold and model high standards that comply with customary and conventional international law. This entails structuring domestic laws that balance military objectives and war preparations with fundamental human rights. In this realm, a doctrine of *jus in praeparatione bellum* can be beneficial.

Writing in the mid-1800s, Francis Lieber, a law professor and author of the Union Army's guidelines that outlined principles and limits of warfare, wrote that war motivates a "peculiar attribute of greatness of intellect" and creates "the spark of moral electricity."[7] This statement still rings true today. Laws reflect a nation's vision, codify its moral principles, and delineate proper and improper uses of power. Although we may not always live up to our ideals, it is important to identify what they are, set guidelines for achieving them, and clarify what mitigation measures are triggered when we fall short.

The front lines of science will continue to evolve. Nations can promote their national security by acting sensibly and structuring appropriate mechanisms for guiding and assessing military research and the operational integration of new technologies. The drive for geopolitical superiority need not compromise American democracy and constitutional protections, and military goals should not drive the pursuit of science at any price.

Notes

1. Friedrich Engels, "Introduction to Sigismund Borkheim's Pamphlet, *In Memory of the German Blood-and-Thunder Patriots, 1806-1807*" (December 15, 1887), in *Marks and Engels: Collected Works*, vol. 26 (London: Lawrence & Wishart, 2010), 451.
2. Hans J. Morgenthau, *Truth and Power* (1970), 51–55.
3. Phillip N. Ash, "The Bureaucratic Fix to the Military Recruitment Crisis," *Council on Foreign Relations*, December 21, 2013; Richard Sisk, "The Military Recruitment Outlook is Grim Indeed: Loss of Public Confidence, Political Attacks and the Economy Are All Taking a Toll," *Military News*, January 22, 2024; Manuela Lopez Restrepo, "The U.S. Army Is Falling Short of Its Recruitment Goals," *NPR*, October 5, 2023; Ellen Mitchell, "Why Falling Confidence in America's Military Is Creating 'A Real Crisis,'" *The Hill*, August 4, 2023; Juan Quiroz, "The U.S. Military's Personnel Crisis," *Foreign Affairs*, January 5, 2024; Beth J. Asch, *Navigating Current and Emerging Army Recruiting Challenges* (Santa Monica, CA: RAND Corporation, 2019).
4. Morgenthau, *Truth and Power*, 416.
5. Hans Jonas, "Philosophical Reflections on Experimenting with Humans," in Hans Jonas, *Philosophical Essays* (1974), 131.
6. A detailed accounting of money inflows from lobbyists to lawmakers can be found among the many reports issued by OpenSecrets, a nonpartisan and nonprofit organization that tracks money in politics.
7. Francis Lieber, *Manual of Political Ethics*, vol. 2 (Boston: Charles C. Little and James Brown, 1839), 632.

Select Bibliography*

The 9/11 Commission Report, *Final Report of the National Commission on Terrorist Attacks Upon the United States* (New York: W.W. Norton & Company, 2004).
Abella, Alex, *Soldiers of Reason: The RAND Corporation and the Rise of the American Empire* (Orlando, FL: Harcourt, 2008).
Abrams, Jeanne E., *Revolutionary Medicine: The Founding Fathers and Mothers in Sickness and in Health* (New York: New York University Press, 2013).
Alibek, Ken, and Stephen Handelman, *Biohazard* (New York: Delta, 1999).
Annas, George J., *Worst Case Bioethics: Death, Disaster, and Public Health* (New York: Oxford University Press, 2010).
Annas, George J., and Michael A. Grodin, eds., *The Nazi Doctors and the Nuremberg Code* (New York: Oxford University Press, 1992).
Armstrong, Robert E., Mark D. Drapeau, Cheryl A. Loeb, and James J. Valdes, eds., *Bio-Inspired Innovation and National Security* (Washington, DC: National Defense University Press, 2010).
Ashburn, P.M., *A History of the Medical Department of the United States Army* (Cambridge, MA: Riverside Press, 1929).
Bakeless, John, *Turncoats, Traitors, and Heroes* (Philadelphia: Lippincott, 1959).
Bamford, James, *Body of Secrets: Anatomy of the Ultra-Secret National Security Agency from the Cold War through the Dawn of a New Century* (New York: Doubleday, 2001).
Barnaby, Wendy, *The Plague Makers: The Secret World of Biological Warfare* (New York: Continuum: 2002).
Bayne-Jones, Stanhope, *The Evolution of Preventive Medicine in the United States Army, 1607-1939* (Washington, DC: U.S. Government Printing Office, 1968).
Beam, Thomas E., and Linette R. Sparacino, eds., *Military Medical Ethics*, vol. 1 (Falls Church, VA: Office of the Surgeon General, 2003).
Beecher, Henry K., *Research and the Individual* (Boston: Little, Brown and Company, 1970).
Bernard, Claude, *An Introduction to the Study of Experimental Medicine* (1865), trans. Henry C. Greene (New York: Shuman, 1949).
Blum, William, *Killing Hope: U.S. Military and C.I.A. Interventions Since World War II* (Monroe, ME: Common Courage Press, 2004).
Boothby, William H., ed., *New Technologies and the Law in War in Peace* (Cambridge: Cambridge University Press, 2019).
Brands, Hal, *Latin America's Cold War* (Cambridge, MA: Harvard University Press, 2012).
Brophy, Leo P., and George B. Fisher, *The Chemical Warfare Service: Organizing for War* (Washington, DC: Center of Military History, 1958).

* There are more than 1,200 sources identified within the 2,182 endnotes of this book. Rather than relisting all the sources, this Bibliography contains select books I referenced during my research.

Brophy, Leo P., Wyndham D. Miles, and Rexmond C. Cochrane, *The Chemical Warfare Service: From Laboratory to Field* (Washington, DC: Center of Military History, 1959).
Brown, Frederic J., *Chemical Warfare* (Princeton, NJ: Princeton University Press, 1968).
Brown, Harvey E., *The Medical Department of the United States Army from 1775 to 1873* (Washington, DC: Surgeon General's Office, 1873).
Bush, Vannevar, *Pieces of the Action* (New York: William Morrow and Company, 1970).
Bush, Vannevar, *Science: The Endless Frontier* (Washington, DC: National Science Foundation, 1945).
Carr, E. H., *The 20 Years' Crisis, 1919-1939: An Introduction to the Study of International Relations* (London: Macmillan & Company, 1948).
Carroll, Michael C., *Lab 257: The Disturbing Story of the Government's Secret Germ Laboratory* (New York: Harper, 2005).
Chemical Corps Association, *The Chemical Warfare Service in World War II: A Report of Accomplishments* (New York: Reinhold Publishing, 1938).
Cirillo, Vincent J., *Bullets and Bacilli* (New Brunswick, NJ: Rutgers University Press, 2004).
Clapper, James R., *Facts and Fears: Hard Truths from a Life in Intelligence* (New York: Viking 2018).
Coffey, Patrick, *American Arsenal: A Century of Waging War* (New York: Oxford University Press, 2014).
Colby, William, *Honorable Men: My Life in the CIA* (New York: Simon and Schuster, 1978).
Cole, Leonard A., *The Anthrax Letters: A Medical Detective Story* (Washington, DC: Joseph Henry Press, 2003).
Cole, Leonard A., *Clouds of Secrecy: The Army's Germ Warfare Tests over Populated Areas* (Totowa, NJ: Rowman & Littlefield, 1988).
Cole, Leonard A., *The Eleventh Plague: The Politics of Biological and Chemical Warfare* (New York: W.H. Freeman and Company, 1997).
Coll, Steve, *Directorate S: The C.I.A. and America's Secret Wars in Afghanistan and Pakistan* (New York: Penguin Press, 2018).
Coll, Steve, *Ghost Wars: The Secret History of the CIA, Afghanistan, and bin Laden, from the Soviet Invasion to September 10, 2001* (New York: The Penguin Press, 2004).
Conant, James B., *My Several Lives* (New York: Harper & Row, 1970).
Conant, Jennet, *The Great Secret: The Classified World War II Disaster that Launched the War on Cancer* (New York: Norton, 2020).
Corera, Gordon, *The Art of Betrayal: The Secret History of MI6* (New York: Pegasus, 2012).
Craughwell, Thomas J., *The War Scientists* (New York: Metro Books, 2011).
Danner, Mark, *Spiral: Trapped in the Forever War* (New York: Simon & Schuster, 2016).
D'Antonio, Michael, *Atomic Harvest: Hanford and the Lethal Toll of America's Nuclear Arsenal* (New York: Crown Publishers, 1993).
Dean, Eric T., *Shook over Hell: Post-Traumatic Stress, Vietnam, and the Civil War* (Cambridge, MA: Harvard University Press, 1997).
Draper, Robert, *To Start a War: How the Bush Administration Took America into Iraq* (New York: Penguin Press, 2020).
Dudziak, Mary L., *Cold War Civil Rights: Race and the Image of American Democracy* (Princeton, NJ: Princeton University Press, 2000).
Duffy, John, *The Healers: A History of American Medicine* (Urbana: University of Illinois Press, 1979).

Eddington, Patrick G., *Gassed in the Gulf: The Inside Story of the Pentagon-CIA Cover-Up of Gulf War Syndrome* (Washington, DC: Insignia Publishing, 1997).

Emerson, Steven, *Secret Warriors: Inside the Covert Military Operations of the Reagan Era* (New York: G.P. Putnam's Sons, 1988).

Endicott, Stephen, and Edward Hagerman, *The United States and Biological Warfare* (Bloomington: Indiana University Press, 1998).

Engelhardt, Tom, *Shadow Government: Surveillance, Secret Wars, and a Global Security State in a Single-Superpower World* (Chicago: Haymarket Books, 2014).

Engelman, Rose C., ed., *A Decade of Progress: The United States Army Medical Department, 1959-1969* (Washington, DC: U.S. Government Printing Office, 1971).

Fair, Eric, *Consequence: A Memoir* (New York: Henry Holt and Company, 2016).

Faith, Thomas I., *Behind the Gas Mask: The U.S. Chemical Warfare Service in War and Peace* (Urbana: University of Illinois Press, 2014).

Fenn, Elizabeth A., *Pox Americana: The Great Smallpox Epidemic of 1775-82* (New York: Hill & Wang, 2001).

Final Report, *Presidential Advisory Committee on Gulf War Veterans' Illnesses* (Washington, DC: U.S. Government Printing Office, 1996).

Final Report of the President's Advisory Committee, *The Human Radiation Experiments* (New York: Oxford University Press, 1996).

Finkbeiner, Ann, *The JASONS: The Secret History of Science's Postwar Elite* (New York: Viking, 2006).

Finney, J. M. T., *A Surgeon's Life* (New York: G.P. Putnam's Sons, 1940).

Fiss, Owen, *A War Like No Other: The Constitution in a Time of Terror* (New York: The New Press, 2015).

Foster, Stephen, ed., *The Project MKULTRA Compendium: The CIA's Program of Research in Behavioral Modification* (Lexington, KY: Lulu, 2009).

Fourth Report by the Committee on Government Reform, *The Department of Defense Anthrax Vaccine Immunization Program: Unproven Force Protection* (Washington, DC: U.S. Government Printing Office, 2000).

Freund, Paul A., ed. *Experimentation with Human Subjects* (New York: George Braziller, 1970).

Friscolanti, Michael, *Friendly Fire: The Untold Story of the U.S. Bombing that Killed Four Canadian Soldiers in Afghanistan* (Mississauga, Canada: John Wiley & Sons Canada, 2005).

Gaddis, John Lewis, *Strategies of Containment: A Critical Appraisal of American National Security Policy During the Cold War* (Oxford: Oxford University Press, 1982).

Galliott, Jai, ed., *Force Short of War in Modern Conflict: Jus Ad Vim* (Edinburgh: Edinburgh University Press, 2019).

Garrison, Fielding H., *Notes on the History of Military Medicine* (Washington, DC: Association of Military Surgeons, 1922).

Gentile, Gian, Jameson Karns, Michael Shurkin, and Adam Givens, *The Evolution of U.S. Military Policy from the Constitution to the Present*, vol. 1 (Santa Monica, CA: RAND Corporation, 2019).

Gold, Hal, *Unit 731 Testimony* (Tokyo: Tuttle Publishing, 1997).

Gillett, Mary C., *The Army Medical Department, 1775-1818* (Washington, DC: U.S. Government Printing Office, 1981).

Gillett, Mary C., *The Army Medical Department, 1818-1865* (Washington, DC: U.S. Government Printing Office, 1987).
Gillett, Mary C., *The Army Medical Department, 1865-1917* (Washington, DC: U.S. Government Printing Office, 1995).
Gillett, Mary C., *The Army Medical Department, 1917-1941* (Washington, DC: U.S. Government Printing Office, 2009).
Ginn, Richard V. N., *The History of the U.S. Army Medical Service Corps* (Washington, DC: Office of the Surgeon General, 1997).
Giordano, James, ed., *Neurotechnology in National Security and Defense* (Boca Raton, FL: CRC Press, 2015).
Gould, Chandré, and Peter Folb, *Project Coast: Apartheid's Chemical and Biological Warfare Program* (Geneva: United Nations Publication, 2002).
Grandin, Greg, *The Last Colonial Massacre: Latin America in the Cold War* (Chicago: University of Chicago Press, 2011).
Gray, Bradford H., *Human Subjects in Medical Experimentation* (New York: John Wiley & Sons, 1975).
Greenwood, John T., and F. Clifton Berry, *Medics at War: Military Medicine from Colonial Times to the 21st Century* (Annapolis, MD: Naval Institute Press, 2005).
Grenier, Robert L., *88 Days to Kandahar: A CIA Diary* (New York: Simon & Schuster, 2015).
Grossman, Dave, *On Killing: The Psychological Cost of Learning to Kill in War and Society* (New York: Back Bay Books, 2009).
Guillemin, Jeanne, *American Anthrax: Fear, Crime, and the Investigation of the Nation's Deadliest Bioterror Attack* (New York: Henry Holt and Company, 2011).
Guillemin, Jeanne, *Biological Weapons* (New York: Columbia University Press, 2005).
Guillemin, Jeanne, *Hidden Atrocities* (New York: Columbia University Press, 2017).
Haller, John S., *American Medicine in Transition, 1840-1910* (Urbana: University of Illinois Press, 1981).
Harris, Robert, and Jeremy Paxman, *A Higher Form of Killing: The Secret History of Chemical and Biological Warfare* (New York: Random House, 2002).
Harris, Sheldon H., *Factories of Death: Japanese Biological Warfare, 1932-45, and the American Cover-Up* (London: Routledge, 1994).
Hastings, Max, *The Secret War: Spies, Ciphers, and Guerrillas, 1939-1945* (New York: Harper Collins, 2016).
Heller, Charles E., *Chemical Warfare in World War I: The American Experience, 1917-1918* (Washington, DC: U.S. Government Printing Office, 1984).
Helling, Thomas, *The Great War and the Birth of Modern Medicine* (New York: Pegasus Books, 2022).
Henry, Robert S., *The Armed Forces Institute of Pathology: Its First Century, 1862-1962* (Washington, DC: Office of the Surgeon General, 1964).
Hersh, Seymour M., *Chemical and Biological Warfare* (Indianapolis, IN: Bobbs-Merrill Company, 1968).
Hicks, Robert D., ed., *Civil War Medicine* (Bloomington: Indiana University Press, 2019).
Hoffman, David E., *The Dead Hand* (New York: Anchor Books, 2009).
Hopkins, Donald R., *The Greatest Killer: Smallpox in History* (Chicago: University of Chicago Press, 2002).

Hornblum, Allen M., *Acres of Skin: Human Experiments at Holmesburg Prison* (New York: Routledge, 1998).

Hornblum, Allen M., Judith L. Newman, and Gregory J. Dober, *Against Their Will: The Secret History of Medical Experimentation on Children in Cold War America* (New York: Palgrave Macmillan, 2013).

Hunt, Linda, *Secret Agenda: The United States Government, Nazi Scientists, and Project Paperclip, 1945 to 1990* (New York: St. Martin's Press, 1991).

Huntington, Samuel P., *The Soldier and the State* (Cambridge, MA: Harvard University Press, 1957).

Immerwahr, Daniel, *How to Hide an Empire: A Short History of the Greater United States* (New York: Farrar, Straus and Giroux, 2019).

Institute of Medicine, *Adverse Reproductive Outcomes in Families of Atomic Veterans* (Washington, DC: National Academies Press, 1995).

Institute of Medicine, *The Anthrax Vaccine: Is it Safe? Does It Work?* (Washington, DC: National Academies Press, 2002).

Institute of Medicine, *An Assessment of the CDC Anthrax Vaccine Safety and Efficacy Research Program* (Washington, DC: National Academies Press, 2003).

Institute of Medicine, *Assessing Health Outcomes Among Veterans of Project SHAD (Shipboard Hazard and Defense)* (Washington, D.C.: National Academies Press, 2016).

Institute of Medicine, *Long-Term Health Effects of Participation in Project SHAD (Shipboard Hazard and Defense)* (Washington, DC: National Academies Press, 2007).

Institute of Medicine, *The Smallpox Vaccination Program: Public Health in an Age of Terrorism* (Washington, DC: National Academies Press, 2005).

Institute of Medicine, *Treatment for Posttraumatic Stress Disorder in Military and Veteran Populations* (Washington, DC: National Academies Press, 2012).

Institute of Medicine, *Veterans at Risk: The Health Effects of Mustard Gas and Lewisite* (Washington, DC: National Academies Press, 1993).

Interim Report, *Presidential Advisory Committee on Gulf War Veterans' Illnesses* (Washington, DC: U.S. Government Printing Office, 1996).

Jacobsen, Annie, *Operation Paperclip* (New York: Little, Brown and Company, 2014).

Jacobsen, Annie, *Phenomena: The Secret History of the U.S. Government's Investigations into Extrasensory Perception and Psychokinesis* (New York: Little Brown and Company, 2017).

Jacobsen, Annie, *The Pentagon's Brain* (New York: Little, Brown and Company, 2015).

Johnson, Loch K., *A Season of Inquiry Revisited: The Church Committee Confronts America's Spy Agencies* (Lawrence: University Press of Kansas, 2015).

Johnson, Loch K., *Spy Watching: Intelligence Accountability in the United States* (New York: Oxford University Press, 2018).

Jonas, Hans, *Philosophical Essays: From Ancient Creed to Technological Man* (Englewood Cliffs, NJ: Prentice-Hall, 1974).

Jones, James H., *Bad Blood: The Tuskegee Syphilis Experiment* (New York: The Free Press, 1993).

Kaplan, Fred, *The Bomb: Presidents, Generals, and the Secret History of Nuclear War* (New York: Simon and Schuster, 2020).

Kaplan, Fred, *Dark Territory: The Secret History of Cyber War* (New York: Simon and Schuster, 2016).

Kamienski, Lukasz, *Shooting Up: A Short History of Drugs and War* (New York: Oxford University Press, 2016).

Katz, Jay, *Experimentation with Human Beings* (New York: Russell Sage Foundation, 1972).

Kelle, Alexander, Kathryn Nixdorff, and Malcolm Dando, *Preventing a Biochemical Arms Race* (Stanford, CA: Stanford University Press, 2012).

Kinder, John M., *Paying with Their Bodies* (Chicago: University of Chicago Press, 2015).

Klimburg, Alexander, *The Darkening Web: The War for Cyberspace* (New York: Penguin Press, 2017).

Konold, Donald E., *A History of American Medical Ethics, 1847-1912* (Madison: University of Wisconsin Press, 1962).

Krepinevich, Andrew F., *The Origins of Victory: How Disruptive Military Innovation Determines the Fates of Great Powers* (New Haven, CT: Yale University Press, 2023).

Kuhl, Stefan, *The Nazi Connection* (New York: Oxford University Press, 1994).

Ladimer, Irving, and Roger W. Newman, eds., *Clinical Investigation in Medicine: Legal, Ethical, and Moral Aspects* (Boston: Boston University Law-Medicine Research Institute, 1963).

Lederer, Susan E., *Subjected to Science: Human Experimentation in America Before the Second World War* (Baltimore, MD: Johns Hopkins University Press, 1997).

Leffingwell, Albert, *An Ethical Problem* (London: G. Bell and Sons, 1914).

Leffler, Melvyn P., *Safeguarding Democratic Capitalism: U.S. Foreign Policy and National Security, 1920-2015* (Princeton, NJ: Princeton University Press, 2017).

Lepore, Jill, *These Truths: A History of the United States* (New York: Norton, 2018).

Letterman, Jonathan, *Medical Recollections of the Army of the Potomac* (New York: D. Appleton and Company, 1866).

Lovell, Stanley P., *Of Spies and Stratagems* (Englewood Cliffs, NJ: Prentice-Hall, 1963).

Malkasian, Carter, *The American War in Afghanistan: A History* (New York: Oxford University Press, 2021).

Marchetti, Victor, and John D. Marks, *The CIA and the Cult of Intelligence* (New York: Knopf, 1974).

Markel, M. Wade, Alexandra Evans, Miranda Priebe, Adam Givens, Jameson Karns, and Gian Gentile, *The Evolution of U.S. Military Policy from the Constitution to the Present*, vol. 4 (Santa Monica, CA: RAND Corporation, 2019).

Marks, John, *The Search for the "Manchurian Candidate"* (New York: Times Books, 1979).

Masterson, Karen M., *The Malaria Project* (New York: New American Library, 2014).

Materials on the Trial of Former Servicemen of the Japanese Army Charged with Manufacturing and Employing Bacteriological Weapons (Moscow: Foreign Languages Publishing House, 1950).

Maxwell, William Q., *Lincoln's Fifth Wheel: The Political History of the U.S. Sanitary Commission* (New York: Longmans, Green, and Co., 1956).

May, Ernest, *American Cold War Strategy: Interpreting NSC-68* (New York: St. Martin's, 1993).

May, Larry, ed., *War: Essays in Political Philosophy* (New York: Cambridge University Press, 2008).

Mayor, Adrienne, *Greek Fire, Poison Arrows, and Scorpion Bombs: Unconventional Warfare in the Ancient World* (Princeton, NJ: Princeton University Press, 2022).

Mazzetti, Mark, *The Way of the Knife: The CIA, A Secret Army, and a War at the Ends of the Earth* (New York: The Penguin Press, 2013).
McCamley, Nick, *Secret History of Chemical Warfare* (Barnsley, UK: Pen & Sword Military, 2006).
McClintock, Michael, *Instruments of Statecraft: U.S. Guerrilla Warfare, Counterinsurgency, and Counterterrorism, 1940-1990* (New York: Pantheon Books, 1992).
McFate, Sean, *The Modern Mercenary: Private Armies and What They Mean for World Order* (New York: Oxford University Press, 2014).
The Medical and Surgical History of the War of the Rebellion, 2 vols. in 6 parts (Washington, DC: Government Printing Office, 1870–1888).
Mehlman, Maxwell J., *The Price of Perfection: Individualism and Society in the Era of Biomedical Enhancement* (Baltimore, MD: Johns Hopkins University Press, 2009).
Miles, Steven H., *Oath Betrayed: Torture, Medical Complicity, and the War on Terror* (New York: Random House, 2006).
Miller, Judith, Stephen Engelberg, and William Broad, *Germs: Biological Weapons and America's Secret War* (New York: Simon & Schuster, 2002).
Millett, Allan R., Peter Maslowski, and William B. Feis, *For the Common Defense: A Military History of the United States from 1607 to 2012* (New York: Free Press, 2012).
Mitscherlich, Alexander and Fred Mielke, *Doctors of Infamy: The Story of the Nazi Medical Crimes* (New York: Henry Schuman, 1949).
Moreno, Jonathan D., ed., *In the Wake of Terror: Medicine and Morality in a Time of Crisis* (Cambridge, MA: MIT Press, 2003).
Moreno, Jonathan D., *Undue Risk: Secret State Experiments on Humans* (New York: W.H. Freeman & Co., 1999).
Morgenthau, Hans J., *Politics Among Nations* (New York: Alfred A. Knopf, 1967).
Morgenthau, Hans J., *Truth and Power* (New York: Praeger Publishers, 1970).
Morris, R. Crawford, and Alan R. Moritz, *Doctor and Patient and the Law* (Saint Louis, MO: The C.V. Mosby Company, 1971).
Mueller, John, *The Stupidity of War: American Foreign Policy and the Case for Complacency* (Cambridge: Cambridge University Press, 2021).
National Academies of Sciences, Engineering, and Medicine, *Biodefense in the Age of Synthetic Biology* (Washington, DC: National Academies Press, 2018).
National Academies of Sciences, Engineering, and Medicine, *Human Genome Editing: Science, Ethics, and Governance* (Washington, DC: National Academies Press, 2017).
National Research Council, *Animal Models for Assessing Countermeasures to Bioterrorism Agents* (Washington, DC: National Academies Press, 2011).
National Research Council, *Enhancing Human Performance: Issues Theories, and Techniques* (Washington, DC: National Academies Press, 1988).
National Research Council, *Emerging Cognitive Neuroscience and Related Technologies* Washington, DC: National Academies Press, 2008).
National Research Council, *Opportunities in Neuroscience for Future Army Applications* (Washington, DC: National Academies Press, 2009).
National Research Council, *Review of the Scientific Approaches Used During the FBI's Investigation of the 2001 Anthrax Letters* (Washington, DC: National Academies Press, 2011).

National Research Council, *Toxicologic Assessment of the Army's Zinc Cadmium Sulfide Dispersion Tests* (Washington, DC: National Academies Press, 1997).

Nichols, C. F., *Vaccination: A Blunder in Poisons* (Boston: Rockwell and Churchill Press, 1902).

Nie, Jing-Bao, Nanyan Guo, Mark Selden, and Arthur Kleinman, eds., *Japan's Wartime Medical Atrocities* (London: Routledge, 2011).

Nolan, James L., *Atomic Doctors: Conscience and Complicity at the Dawn of the Nuclear Age* (Cambridge, MA: Harvard University Press, 2020).

Nyiszli, Miklos, *Auschwitz: A Doctor's Eyewitness Account*, trans. Tibere Kremer and Richard Seaver (New York: Arcade Publishing, 2011).

Ohler, Norman, *Blitzed: Drugs in the Third Reich*, trans. Shaun Whiteside (New York: Houghton Mifflin Harcourt, 2017).

Owen, Mark, *No Easy Day: The Firsthand Account of the Mission that Killed Osama bin Laden* (New York: Dutton, 2012).

Packard, Francis R., *History of Medicine in the United States*, 2 vols. (New York: Hafner Publishing, 1963).

Pappworth, M. H., *Human Guinea Pigs: Experimentation on Man* (Boston: Beacon Press, 1967).

Payne, Kenneth, *I, Warbot: The Dawn of Artificially Intelligent Combat* (New York: Oxford University Press, 2021).

Pershing, John, *Final Report of General John J. Pershing, Commander-in-Chief American Expeditionary Forces* (Washington, DC: U.S. Government Printing Office, 1920).

Piketty, Thomas, *Capital and Ideology*, trans. Arthur Goldhammer (Cambridge, MA: Harvard University Press, 2020).

Piketty, Thomas, *Capital in the Twenty-First Century*, trans. Arthur Goldhammer (Cambridge, MA: Harvard University Press, 2014).

Pincus, Walter, *Blown to Hell: America's Deadly Betrayal of the Marshall Islanders* (New York: Diversion Books, 2021).

Posen, Barry R., *Restraint: A New Foundation for U.S. Grand Strategy* (Ithaca, NY: Cornell University Press, 2014).

Posner, Gerald L., and John Ware, *Mengele: The Complete Story* (New York: McGraw Hill, 1986).

Powers, Thomas, *The Man Who Kept the Secrets: Richard Helms and the CIA* (New York: Alfred A. Knoff, 1979).

Prados, John, *The Ghosts of Langley: Into the CIA's Heart of Darkness* (New York: The New Press, 2017).

Prados, John, *Presidents' Secret Wars* (Chicago: Elephant Paperbacks, 1996).

Preston, Diana, *A Higher Form of Killing: Six Weeks in World War I that Forever Changed the Nature of Warfare* (New York: Bloomsbury Press, 2015).

Regis, Ed, *The Biology of Doom: The History of America's Secret Germ Warfare Project* (New York: Henry Holt, 1999).

Report of the Commission Appointed by the President to Investigate the Conduct of the War Department in the War with Spain, 8 vols. (Washington, DC: Government Printing Office, 1899).

Report to the President by the Commission on CIA Activities Within the United States (Washington, DC: U.S. Government Printing Office, 1975).

Research Advisory Committee on Gulf War Veterans' Illnesses, *Gulf War Illness and the Health of Gulf War Veterans: Scientific Findings and Recommendation* (Washington, DC: U.S. Government Printing Office, 2008).
Rettig, Richard A., *Military Use of Drugs Not Yet Approved by the FDA for CW/BW Defense* (Santa Monica, CA: RAND Corporation, 1999).
Rhodes, Richard, *The Making of the Atomic Bomb* (New York: Simon & Schuster, 2012).
Richelson, Jeffrey T., *The Wizards of Langley* (Boulder, CO: Westview Press, 2001).
Risen, James, *Pay Any Price: Greed, Power and Endless War* (Boston: Houghton Mifflin Harcourt, 2014).
Risen, James, *State of War: The Secret History of the CIA and the Bush Administration* (New York: Free Press, 2006).
Riza, M. Shane, *Killing Without Heart: Limits on Robotic Warfare in an Age of Persistent Conflict* (Washington, DC: Potomac Books, 2013).
Rizzo, John, *Company Man: Thirty Years of Controversy and Crisis in the CIA* (New York: Scribner, 2014).
Roosevelt, Theodore, *The Rough Riders* (New York: Charles Scribner's Sons, 1899).
Rosenberg, Howard L., *Atomic Soldiers: American Victims of Nuclear Explosions* (Boston: Beacon Press, 1980).
Rostker, Bernard, *Providing for the Casualties of War: The American Experience Through World War II* (Santa Monica, CA: RAND Corporation, 2013).
Rothman, David J., and Sheila M. Rothman, *The Willowbrook Wars* (New York: Harper & Row, 1984).
Rothstein, William G., *American Physicians in the 19th Century* (Baltimore, MD: Johns Hopkins University Press, 1985).
Rubenstein, Leonard, *Perilous Medicine: The Struggle to Protect Health Care from the Violence of War* (New York: Columbia University Press, 2021).
Rush, Benjamin, *Directions for Preserving the Health of Soldiers* (1777).
Saffer, Thomas H., and Orville E. Kelly, *Countdown Zero: GI Victims of U.S. Atomic Testing* (New York: Penguin Books, 1983).
Savage, Charlie, *Power Wars: Inside Obama's Post-9/11 Presidency* (New York: Little, Brown & Company, 2015).
Savulescu, Julian, and Nick Bostrom, eds., *Human Enhancement* (New York: Oxford University Press, 2011).
Scahill, Jeremy, *Dirty Wars: The World Is a Battlefield* (New York: Nation Books, 2013).
Scahill, Jeremy, *The Assassination Complex: Inside the Government's Secret Drone Warfare Program* (New York: Simon & Schuster, 2016).
Schlesinger, Stephen and Stephen Kinzer, *Bitter Fruit: The Story of the American Coup in Guatemala* (New York: Doubleday, 1982).
Schmidt, Elizabeth, *Foreign Intervention in Africa: From the Cold War to the War on Terror* (New York: Cambridge University Press, 2013).
Schmidt, Ulf, *Justice at Nuremberg* (Basingstoke, England: Palgrave Macmillan, 2006).
Schmidt, Ulf, *Secret Science: A Century of Poison Warfare and Human Experiments* (Oxford: Oxford University Press, 2015).
Schroen, Gary, *First In: An Insider's Account of How the CIA Spearheaded the War on Terror in Afghanistan* (New York: Presidio Press, 2005).

Schwarz, Frederick A. O., *Democracy in the Dark: The Seduction of Government Secrecy* (New York: The New Press, 2015).

Shafer, Henry B., *The American Medical Profession, 1783 to 1850* (New York: Columbia University Press, 1936).

Shryock, Richard H., *The Development of Modern Medicine* (Philadelphia: University of Pennsylvania Press, 1936).

Shryock, Richard H., *Medicine and Society in America, 1660-1860* (New York: New York University Press, 1960).

Singer, P. W., *Wired for War: The Robotics Revolution and Conflict in the Twenty-first Century* (New York: The Penguin Press, 2009).

Singer, P. W., and Allan Friedman, *Cybersecurity and Cyberwar: What Everyone Needs to Know* (New York: Oxford University Press, 2014).

SIPRI: Stockholm International Peace Research Institute, *The Problem of Chemical and Biological Warfare*, 6 vols. (Stockholm: Almqvist & Wiksell, 1971–1975).

Slahi, Mohamedou Ould, *Guantanamo Diary*, edited by Larry Siems (New York: Little, Brown & Company, 2015).

Smith, Frank L., *American Biodefense* (Ithaca, NY: Cornell University Press, 2014).

Smith, R. Harris, *OSS: The Secret History of America's First Central Intelligence Agency* (Berkeley: University of California Press, 1972).

Stagg, J. C. A., *The War of 1812: Conflict for a Continent* (Cambridge: Cambridge University Press, 2012).

Stewart, Irvin, *Organizing Scientific Research for War* (Boston: Little, Brown and Company, 1948).

Stille, Charles J., *History of the United States Sanitary Commission, Being the General Report of Its Work During the War of the Rebellion* (Philadelphia: J.B. Lippincott, 1866).

Strozzi-Heckler, Richard, *In Search of the Warrior Spirit: Teaching Awareness Disciplines to the Military* (Berkeley, CA: Blue Snake Books, 2007).

Steuben, Friedrich Wilhelm, *Regulations for the Order and Discipline of the Troops of the United States* (1779).

Surgeon General's Office, *The Medical Department of the United States Army in the World War: Neuropsychiatry*, vol. 10 (Washington, DC: U.S. Government Printing Office, 1929).

Swofford, Anthony, *Jarhead: A Marine's Chronicle of the Gulf War and Other Battles* (New York: Scribner's Sons, 2003).

Tanaka, Yuki, *Hidden Horrors: Japanese War Crimes in World War II* (Boulder, CO: Westview Press, 1996).

Taylor, Telford, *The Anatomy of the Nuremberg Trials* (New York: Alfred A. Knopf, 1992).

Teller, Edward, *Energy from Heaven and Earth* (New York: W.H. Freeman, 1979).

Tencza, Elizabeth, Adam Givens, and Miranda Priebe, *The Evolution of U.S. Military Policy from the Constitution to the Present*, vol. 3 (Santa Monica, CA: RAND Corporation, 2019).

Tenet, George, *At the Center of the Storm: My Years at the CIA* (New York: HarperCollins, 2007).

Tibbets, E.T., *Medical Fashion in the Nineteenth Century; Including a Sketch of Bacteriomania and the Battle of the Bacilli* (London: H.K. Lewis, 1884).

Toft, Monica Duffy and Sidita Kushi, *Dying By the Sword: The Militarization of US Foreign Policy* (New York: Oxford University Press, 2023).
Troy, Thomas F., *Donovan and the CIA* (Frederick, MD: University Publications of America, 1984).
Tucker, Jonathan B., *War of Nerves: Chemical Warfare from World War I to Al-Qaeda* (New York: Pantheon Books, 2006).
Turner, Stansfield, *Burn Before Reading: Presidents, CIA Directors, and Secret Intelligence* (New York: Hyperion, 2005).
Turner, Stansfield, *Secrecy and Democracy: The CIA in Transition* (Boston: Houghton Mifflin, 1985).
Turse, Nick, *The Changing Face of Empire: Special Ops, Drones, Spies, Proxy Fighters, Secret Bases, and Cyberwarfare* (Chicago: Haymarket Books, 2012).
Tutu, Desmond, *No Future Without Forgiveness* (New York: Doubleday, 1999).
U.S. Congress Office of Technology Assessment, *Proliferation of Weapons of Mass Destruction* (Washington, DC: U.S. Government Printing Office, 1993).
Waller, Douglas C., *The Commandos: The Inside Story of America's Secret Soldiers* (New York: Simon and Schuster, 1994).
Waller, Douglas, *Wild Bill Donovan: The Spymaster Who Created the OSS and Modern American Espionage* (New York: Free Press, 2011).
Waltz, Kenneth N., *Man, the State and War* (New York: Columbia University Press, 1959).
Waltz, Kenneth N., *Theory of International Politics* (New York: McGraw-Hill, 1979).
Walzer, Michael, *Just and Unjust Wars* (New York: Basic Books, 1977).
Washington, Harriet A., *Medical Apartheid: The Dark History of Medical Experimentation on Black Americans from Colonial Times to the Present* (New York: Anchor Books, 2006).
Weigley, Russell F., *History of the United States Army* (Bloomington: Indiana University Press, 1984).
Weinberger, Sharon, *The Imagineers of War* (New York: Alfred A. Knopf, 2017).
Weiner, Tim, *Legacy of Ashes: The History of the CIA* (New York: Doubleday, 2007).
Wells, H. G., *The War That Will End War* (New York: Duffield & Company, 1914).
Welsome, Eileen, *The Plutonium Files* (New York: The Dial Press, 1999).
Westad, Odd Arne, *The Cold War: A World History* (New York: Basic Books, 2017).
Westerfield, H. Bradford, ed., *Inside CIA's Private World: Declassified Articles from the Agency's Internal Journal, 1955-1992* (New Haven, CT: Yale University Press, 1995).
Whitlock, Craig, *The Afghanistan Papers: A Secret History of the War* (New York: Simon & Schuster, 2021).
Wilcox, Fred A., *Scorched Earth: Legacies of Chemical Warfare in Vietnam* (New York: Seven Stories Press, 2011).
Willrich, Michael, *Pox: An American History* (New York: Penguin Books, 2011).
Wilson, Sandra, Robert Cribb, Beatrice Trefalt, and Dean Aszkielowicz, *Japanese War Criminals* (New York: Columbia University Press, 2017).
Withington, Charles Francis, *The Relation of Hospitals to Medical Education* (Boston: Cupples, Upham and Company, 1886).
Wolfe, Audra J., *Freedom's Laboratory: The Cold War Struggle for the Soul of Science* (Baltimore, MD: Johns Hopkins University Press, 2018).
Woodward, Bob, *Obama's Wars* (New York: Simon & Schuster, 2010).

Woodward, Bob, *Veil: The Secret Wars of the CIA 1981-1987* (New York: Simon and Schuster, 1987).

Wright, Susan, ed., *Preventing a Biological Arms Race* (Cambridge, MA: MIT Press, 1990).

Zegart, Amy B., *Spies, Lies, and Algorithms: The History and Future of American Intelligence* (Princeton, NJ: Princeton University Press, 2022).

Zeigler, Sean, Alexandra Evans, Gian Gentile, and Badreddine Ahtchi, *The Evolution of U.S. Military Policy from the Constitution to the Present*, vol. 2 (Santa Monica, CA: RAND Corporation, 2019).

Index

For the benefit of digital users, indexed terms that span two pages (e.g., 52–53) may, on occasion, appear on only one of those pages.

Figures are indicated by an italic *f* following the page number.